Annals of Mathematics Studies
Number 215

Moduli Stacks of Étale (φ, Γ)-Modules and the Existence of Crystalline Lifts

Matthew Emerton
Toby Gee

PRINCETON UNIVERSITY PRESS
PRINCETON AND OXFORD
2023

Published by Princeton University Press
41 William Street, Princeton, New Jersey 08540
99 Banbury Road, Oxford OX2 6JX

press.princeton.edu

ISBN 9780691241340
ISBN (pbk.) 9780691241357
ISBN (e-book) 9780691241364

British Library Cataloging-in-Publication Data is available

Editorial: Diana Gillooly, Alena Chekanov, and Kiran Pandey
Production Editorial: Nathan Carr
Production: Danielle Amatucci and Lauren Reese
Copyeditor: Bhisham Bherwani

This book has been composed in LaTeX

10 9 8 7 6 5 4 3 2 1

In memory of Jean-Marc Fontaine.

Contents

Moduli Stacks of Étale (φ, Γ)-Modules and the Existence of Crystalline Lifts

Chapter One

Introduction

In this book we construct moduli stacks of étale (φ, Γ)-modules (projective, of some fixed rank, and with coefficients in p-adically complete rings), and establish some of their basic properties. We also present some first applications of this construction to the theory of Galois representations.

1.1 MOTIVATION

Mazur's theory of deformations of Galois representations [Maz89] is modeled on the geometric study of infinitesimal neighborhoods of points in moduli spaces via formal deformation theory. In the mid-2000s, Kisin suggested that some kind of moduli spaces of local Galois representations should exist; that is, there should be formal algebraic stacks over $\mathbf{Z}_p$ whose closed points correspond to representations $\overline{\rho}\colon G_K \to \mathrm{GL}_d(\overline{\mathbf{F}}_p)$, and whose versal rings at such points should recover appropriate Galois deformation rings. This expectation is borne out by the results of this book. (In fact, Kisin was motivated by calculations of crystalline deformation rings for $\mathrm{GL}_2(\mathbf{Q}_p)$ that had been carried out by Berger–Breuil using the p-adic local Langlands correspondence, and suggested that the versal rings should give crystalline deformation rings. Thus his suggestion is realized by the stacks $\mathcal{X}_d^{\mathrm{crys},\underline{\lambda}}$ of Theorem 1.2.4 below.)

A natural way to construct such a stack would be to consider a literal moduli stack of continuous representations $\rho\colon G_K \to \mathrm{GL}_d(A)$, for K a p-adic field and A a p-adically complete $\mathbf{Z}_p$-algebra; indeed such stacks were constructed by Carl Wang-Erickson [WE18]. However, the stacks constructed in this way are less "global" than one would wish, and in particular the corresponding families of mod p representations $\overline{\rho}\colon G_K \to \mathrm{GL}_d(\overline{\mathbf{F}}_p)$ have constant semisimplification.

In this book, we instead consider moduli stacks of étale (φ, Γ)-modules. These contain Wang-Erickson's stacks as substacks, and coincide with them on the level of $\overline{\mathbf{F}}_p$-points, but their geometry is quite different; in particular, we see much larger families, exhibiting some unexpected features (for example, irreducible representations arising as limits of reducible representations). The relationship between the theory and constructions that we develop here and the usual formal deformation theory of Galois representations, is the same as that between the theory of moduli spaces of algebraic varieties and the formal deformation theory of algebraic varieties: the latter gives valuable local information about the former, but moduli spaces, when they can be constructed, capture global aspects of the situation inaccessible to the purely infinitesimal tools of formal deformation theory.

1.2 OUR MAIN THEOREMS

Our goal in this book is to construct, and establish, the basic properties of moduli stacks of étale (φ, Γ)-modules. More precisely, if we fix a finite extension K of $\mathbf{Q}_p$, and a non-negative integer d (the rank), then we let $\mathcal{X}_d$ denote the category fibred in groupoids over $\operatorname{Spf} \mathbf{Z}_p$ whose groupoid of A-valued points, for any p-adically complete $\mathbf{Z}_p$-algebra A, is equal to the groupoid of rank d projective étale (φ, Γ)-modules with A-coefficients. (See Section 1.3 below for a definition of these.) Our first main theorem is the following. (See Corollary 5.5.18 and Theorem 6.5.1.)

1.2.1 Theorem. *The category fibred in groupoids $\mathcal{X}_d$ is a Noetherian formal algebraic stack. Its underlying reduced substack $\mathcal{X}_{d,\mathrm{red}}$ (which is an algebraic stack) is of finite type over $\mathbf{F}_p$, and is equidimensional of dimension $[K:\mathbf{Q}_p]d(d-1)/2$. The irreducible components of $\mathcal{X}_{d,\mathrm{red}}$ admit a natural labelling by Serre weights.*

We will elaborate on the labelling of components by Serre weights further below. For now, we mention that, under the usual correspondence between étale (φ, Γ)-modules and Galois representations, the groupoid of $\overline{\mathbf{F}}_p$-points of $\mathcal{X}_d$, which coincides with the groupoid of $\overline{\mathbf{F}}_p$-points of the underlying reduced substack $\mathcal{X}_{d,\mathrm{red}}$, is naturally equivalent to the groupoid of continuous representations $\overline{\rho}\colon G_K \to \mathrm{GL}_d(\overline{\mathbf{F}}_p)$. (More generally, if A is any finite $\mathbf{Z}_p$-algebra, then the groupoid $\mathcal{X}_d(A)$ is canonically equivalent to the groupoid of continuous representations $G_K \to \mathrm{GL}_d(A)$.) It is expected that our labelling of the irreducible components can be refined (by adding further labels to some of the components) to give a geometric description of the weight part of Serre's conjecture, so that $\overline{\rho}$ corresponds to a point in a component of $\mathcal{X}_{d,\mathrm{red}}$ which is labeled by the Serre weight $\underline{k}$ if and only if $\overline{\rho}$ admits $\underline{k}$ as a Serre weight; we discuss this expectation, and what is known about it, in Section 1.7 below (and in more detail in Chapter 8).

Again using the correspondence between étale (φ, Γ)-modules and Galois representations, we see that the universal lifting ring of a representation $\overline{\rho}$ as above will provide a versal ring to $\mathcal{X}_d$ at the corresponding $\overline{\mathbf{F}}_p$-valued point. Accordingly we expect that the stacks $\mathcal{X}_d$ will have applications to the study of Galois representations and their deformations. As a first example of this, we prove the following result on the existence of crystalline lifts; although the statement of this theorem involves a fixed $\overline{\rho}$, we do not know how to prove it without using the stacks $\mathcal{X}_d$, over which $\overline{\rho}$ varies.

1.2.2 Theorem (Theorem 6.4.4). *If $\overline{\rho}\colon G_K \to \mathrm{GL}_d(\overline{\mathbf{F}}_p)$ is a continuous representation, then $\overline{\rho}$ has a lift $\rho^\circ\colon G_K \to \mathrm{GL}_d(\overline{\mathbf{Z}}_p)$ for which the associated p-adic representation $\rho\colon G_K \to \mathrm{GL}_d(\overline{\mathbf{Q}}_p)$ is crystalline of regular Hodge–Tate weights.*

We can furthermore ensure that ρ° is potentially diagonalizable.

(The notion of a potentially diagonalizable representation was introduced in [BLGGT14], and is recalled as Definition 6.4.2 below.) In combination with potential automorphy theorems, this has the following application to the globalization of local Galois representations.

1.2.3 Theorem (Corollary 6.4.7). *Suppose that $p \nmid 2d$, and fix $\overline{\rho}\colon G_K \to \mathrm{GL}_d(\overline{\mathbf{F}}_p)$. Then there is an imaginary CM field F and an irreducible conjugate self dual automorphic Galois representation $\overline{r}\colon G_F \to \mathrm{GL}_d(\overline{\mathbf{F}}_p)$ such that for every $v|p$, we have $F_v \cong K$ and either $\overline{r}|_{G_{F_v}} \cong \overline{\rho}$ or $\overline{r}|_{G_{F_{v^c}}} \cong \overline{\rho}$.*

Another key result of the book is the following theorem, describing moduli stacks of étale (φ,Γ)-modules corresponding to *crystalline* and *semistable* Galois representations.

1.2.4 Theorem (Theorem 4.8.12). *If $\underline{\lambda}$ is a collection of labeled Hodge–Tate weights, and if $\mathcal{O}$ denotes the ring of integers in a finite extension E of $\mathbf{Q}_p$ containing the Galois closure of K (which will serve as the ring of coefficients), then there is a closed substack $\mathcal{X}_d^{\mathrm{crys},\underline{\lambda}}$ of $(\mathcal{X}_d)_{\mathcal{O}}$ which is a p-adic formal algebraic stack and is flat over $\mathcal{O}$, and which is characterized as being the unique closed substack of $(\mathcal{X}_d)_{\mathcal{O}}$ which is flat over $\mathcal{O}$ and whose groupoid of A-valued points, for any finite flat $\mathcal{O}$-algebra A, is equivalent (under the equivalence between étale (φ,Γ)-modules and continuous G_K-representations) to the groupoid of continuous representations $G_K \to \mathrm{GL}_d(A)$ which become* crystalline *after extension of scalars to $A \otimes_{\mathcal{O}} E$, and whose labeled Hodge–Tate weights are equal to $\underline{\lambda}$.*

Similarly, there is a closed substack $\mathcal{X}_d^{\mathrm{ss},\underline{\lambda}}$ of $(\mathcal{X}_d)_{\mathcal{O}}$ which is a p-adic formal algebraic stack and is flat over $\mathcal{O}$, and which is characterized as being the unique closed substack of $(\mathcal{X}_d)_{\mathcal{O}}$ which is flat over $\mathcal{O}$ and whose groupoid of A-valued points, for any finite flat $\mathcal{O}$-algebra A, is equivalent to the groupoid of continuous representations $G_K \to \mathrm{GL}_d(A)$ which become semistable *after extension of scalars to $A \otimes_{\mathcal{O}} E$, and whose labeled Hodge–Tate weights are equal to $\underline{\lambda}$.*

1.2.5 Remark. In fact, Theorem 4.8.12 also proves the analogous result for potentially crystalline and potentially semistable representations of arbitrary inertial type, but for simplicity of exposition we restrict ourselves to the crystalline and semistable cases in this introduction.

A crucial distinction between the stacks $\mathcal{X}_d$ and their closed substacks $\mathcal{X}_d^{\mathrm{crys},\underline{\lambda}}$ and $\mathcal{X}_d^{\mathrm{ss},\underline{\lambda}}$ is that while $\mathcal{X}_d$ is a formal algebraic stack lying over $\operatorname{Spf} \mathbf{Z}_p$, it is not actually a p-adic formal algebraic stack (in the sense of Definition A.7);

see Proposition 6.5.2. On the other hand, the stacks $\mathcal{X}_d^{\mathrm{crys},\underline{\lambda}}$ and $\mathcal{X}_d^{\mathrm{ss},\underline{\lambda}}$ *are* p-adic formal algebraic stacks, which implies that their mod p^a reductions are in fact algebraic stacks. This gives in particular a strong interplay between the structure of the mod p fibres of crystalline and semistable lifting rings and the geometry of the underlying reduced substack $\mathcal{X}_{d,\mathrm{red}}$. This plays an important role in determining the structure of this reduced substack, and also in the proof of Theorem 1.2.2. As we explain in more detail in Section 1.7 below, it also allows us to reinterpret the Breuil–Mézard conjecture in terms of the interaction between the structure of the mod p fibres of the stacks $\mathcal{X}_d^{\mathrm{crys},\underline{\lambda}}$ and $\mathcal{X}_d^{\mathrm{ss},\underline{\lambda}}$ and the geometry of $\mathcal{X}_{d,\mathrm{red}}$.

1.3 (φ,Γ)-MODULES WITH COEFFICIENTS

There is quite a lot of evidence, for example from Colmez's work on the p-adic local Langlands correspondence [Col10], and work of Kedlaya–Liu [KL15], that rather than considering families of representations of G_K, it is more natural to consider families of étale (φ,Γ)-modules.

The theory of étale (φ,Γ)-modules for $\mathbf{Z}_p$-representations was introduced by Fontaine in [Fon90]. There are various possible definitions that can be made, with perfect, imperfect, or overconvergent coefficient rings, and different choices of Γ; we discuss the various variants that we use, and the relationships between them, at some length in the body of the book. For the purpose of this introduction we simply let $\mathbf{A}_K = W(k)((T))^\wedge$, where k is a finite extension of $\mathbf{F}_p$ (depending on K), and the hat denotes the p-adic completion. This ring is endowed with a Frobenius φ and an action of a profinite group Γ (an open subgroup of $\mathbf{Z}_p^\times$) that commutes with φ; the formulae for φ and for this action can be rather complicated for general K, although they admit a simple description if $K/\mathbf{Q}_p$ is abelian. (See Definition 2.1.12 and the surrounding material.)

An étale (φ,Γ)-module is then, by definition, a finite $\mathbf{A}_K$-module endowed with commuting semi-linear actions of φ and Γ, with the property that the linearized action of φ is an isomorphism. There is a natural equivalence of categories between the category of étale (φ,Γ)-modules and the category of continuous representations of G_K on finite $\mathbf{Z}_p$-modules.

Let A be a p-adically complete $\mathbf{Z}_p$-algebra. We let $\mathbf{A}_{K,A} := A \widehat{\otimes}_{\mathbf{Z}_p} \mathbf{A}_K$ (where the completed tensor product is taken with respect to the p-adic topology on A and the so-called weak topology on $\mathbf{A}_K$), and define an étale (φ,Γ)-module with A-coefficients just as in the case $A = \mathbf{Z}_p$ described above, but now using the ring $\mathbf{A}_{K,A}$. In the case that A is finite as a $\mathbf{Z}_p$-module there is again an equivalence of categories with the category of continuous representations of G_K on finite A-modules, but for more general A no such equivalence exists. Our moduli stack $\mathcal{X}_d$ is defined to be the stack over $\operatorname{Spf} \mathbf{Z}_p$ with the property that $\mathcal{X}_d(A)$ is the groupoid of projective étale (φ,Γ)-modules of rank d with A-coefficients. (That this is indeed an étale stack, indeed even an *fpqc* stack, follows from results of Drinfeld.) Using the machinery of our paper [EG21] we are able to show that $\mathcal{X}_d$ is an Ind-algebraic stack, but to prove Theorem 1.2.1 we need to go further and make a detailed study of its special fiber and of the underlying reduced substack. This study is guided by ideas coming from Galois deformation theory and the weight part of Serre's conjecture, in a manner that we now describe.

1.4 FAMILIES OF EXTENSIONS

As we have already explained, over a general base A there is no longer an equivalence between (φ, Γ)-modules and representations of G_K. Perhaps surprisingly, from the point of view of applications of our stacks to the study of p-adic Galois representations, this is a feature rather than a bug. For example, an examination of the known results on the reductions modulo p of two-dimensional crystalline representations of $G_{\mathbf{Q}_p}$ (see for example [Ber11, Thm. 5.2.1]) suggests that any moduli space of mod p representations of G_K should have the feature that the representations are generically reducible, but can specialize to irreducible representations. A literal moduli space of representations of a group cannot behave in this way (essentially because Grassmannians are proper), but it turns out that the underlying reduced substack $\mathcal{X}_{d,\mathrm{red}}$ of $\mathcal{X}_d$ does have this property. (See also Section 6.7 and Remark 7.2.19 for further discussions of the relationship between our stacks of (φ, Γ)-modules and stacks of representations of a Galois or Weil–Deligne group.)

More precisely, the results of [Ber11, Thm. 5.2.1], together with the weight part of Serre's conjecture, suggest that each irreducible component of $\mathcal{X}_{d,\mathrm{red}}$ should contain a dense set of $\overline{\mathbf{F}}_p$-points which are successive extensions of characters $G_K \to \overline{\mathbf{F}}_p$, with the extensions being as nonsplit as possible. This turns out to be the case. The restrictions of these characters to the inertia subgroup I_K are constant on the irreducible components, and the discrete data of these characters, together with some further information about peu- and très ramifiée extensions, determines the components. This discrete data can be conveniently and naturally organized in terms of "Serre weights" $\underline{k}$, which are tuples of integers which biject with the isomorphism classes of the irreducible $\overline{\mathbf{F}}_p$-representations of $\mathrm{GL}_d(\mathcal{O}_K)$. The relationship between Serre weights and Galois representations is important in the p-adic Langlands program, and in proving automorphy lifting theorems, and we discuss it further in Section 1.7.

Having guessed that the $\overline{\mathbf{F}}_p$-points of $\mathcal{X}_d$ should be arranged in irreducible components in this way, an inductive strategy to prove this suggests itself. It is easy to see that irreducible representations of G_K are "rigid", in that there are up to twist by unramified characters only finitely many in each dimension; furthermore, it is at least intuitively clear that each such family of unramified twists of a d-dimensional irreducible representation should give rise to a zero-dimensional substack of $\mathcal{X}_d$ (there is a $\mathbf{G}_m$ of twists, but also a $\mathbf{G}_m$ of automorphisms). On the other hand, given characters $\overline{\chi}_1, \dots, \overline{\chi}_d \colon G_K \to \overline{\mathbf{F}}_p^\times$, a Galois cohomology calculation suggests that there should be a substack of $\mathcal{X}_{d,\mathrm{red}}$

of dimension $[K:\mathbf{Q}_p]d(d-1)/2$ given by successive extensions of unramified twists of the $\overline{\chi}_i$. Accordingly, one could hope to construct the stacks corresponding to the Serre weights $\underline{k}$ by inductively constructing families of extensions of representations.

To confirm this expectation, we use the machinery originally developed by Herr [Her98, Her01], who gave an explicit complex which is defined in terms of (φ,Γ)-modules and computes Galois cohomology. This definition goes over unchanged to the case with coefficients, and with some effort we are able to adapt Herr's arguments to our setting, and to prove finiteness and base change properties (following Pottharst [Pot13], we in fact find it helpful to think of the Herr complex of a (φ,Γ)-module with A-coefficients as a perfect complex of A-modules). Using the Herr complex, we can inductively construct irreducible closed substacks $\mathcal{X}^{\underline{k}}_{d,\mathrm{red}}$ of $\mathcal{X}_{d,\mathrm{red}}$ of dimension $[K:\mathbf{Q}_p]d(d-1)/2$ whose generic $\overline{\mathbf{F}}_p$-points correspond to successive extensions of characters as described above (the restrictions of these characters to I_K being determined by $\underline{k}$). Furthermore, by a rather involved induction, we can show that the union of the $\mathcal{X}^{\underline{k}}_{d,\mathrm{red}}$, together possibly with a closed substack of $\mathcal{X}_{d,\mathrm{red}}$ of dimension strictly less than $[K:\mathbf{Q}_p]d(d-1)/2$, exhausts $\mathcal{X}_{d,\mathrm{red}}$. In particular, each $\mathcal{X}^{\underline{k}}_{d,\mathrm{red}}$ is an irreducible component of $\mathcal{X}_{d,\mathrm{red}}$, and any irreducible component that is not one of the $\mathcal{X}^{\underline{k}}_{d,\mathrm{red}}$ is of strictly smaller dimension than these components.

One way to show that the $\mathcal{X}^{\underline{k}}_{d,\mathrm{red}}$ exhaust the irreducible components of $\mathcal{X}_{d,\mathrm{red}}$ would be to show that every representation $G_K\to\mathrm{GL}_d(\overline{\mathbf{F}}_p)$ occurs as an $\overline{\mathbf{F}}_p$-valued point of some $\mathcal{X}^{\underline{k}}_{d,\mathrm{red}}$. We expect this to be difficult to show directly; indeed, already for $d=2$ the paper [CEGS19] shows that the closed points of $\mathcal{X}^{\underline{k}}_{d,\mathrm{red}}$ are governed by the weight part of Serre's conjecture, and the explicit description of this conjecture is complicated (see, e.g., [BDJ10, DDR16]). Furthermore it seems hard to explicitly understand the way in which families of reducible (φ,Γ)-modules degenerate to irreducible ones, or to reducible representations with different restrictions to I_K (phenomena which are implied by the weight part of Serre's conjecture).

Instead, our approach is to show by a consideration of versal rings that $\mathcal{X}_{d,\mathrm{red}}$ is equidimensional of dimension $[K:\mathbf{Q}_p]d(d-1)/2$; this suffices, since our inductive construction showed that any other irreducible component would necessarily have dimension strictly less than $[K:\mathbf{Q}_p]d(d-1)/2$. Our proof of this equidimensionality relies on Theorems 1.2.2 and 1.2.4, as we explain in Remark 1.5.4 below.

1.5 CRYSTALLINE LIFTS

Theorem 1.2.2 solves a problem that has been considered by various authors, in particular [Mul13, GHLS17]. It admits a well-known inductive approach (which is taken in [Mul13, GHLS17]): one writes $\overline{\rho}$ as a successive extension of irreducible representations, lifts each of these irreducible representations to a crystalline representation, and then attempts to lift the various extension classes. The difficulty that arises in this approach (which has proved an obstacle to obtaining general statements along the lines of Theorem 1.2.2 until now) is showing that the mod p extension classes that appear in this description of $\overline{\rho}$ can actually be lifted to crystalline extension classes in characteristic 0. The basic source of the difficulty is that the local Galois H^2 can be nonzero, and nonzero classes in H^2 obstruct the lifting of extension classes (which can be interpreted as classes lying in H^1). In fact, the difficulty is not so much in obtaining crystalline extension classes, as in lifting to any classes in characteristic 0; indeed, it was not previously known that an arbitrary $\overline{\rho}$ had *any* lift to characteristic 0 at all. (Subsequently a different proof of the existence of such a lift has been found by Böckle–Iyengar–Paškūnas [BIP21].)

Our proof of Theorem 1.2.2 relies on the inductive strategy described in the preceding paragraph, but we are able to prove the following key result, which controls the obstructions that can be presented by H^2, and is a consequence of Theorems 5.5.12 and 6.5.1 (see also Remark 1.5.4).

1.5.1 Proposition. *The locus of points $\overline{\rho} \in \mathcal{X}_{d,\mathrm{red}}(\overline{\mathbf{F}}_p)$ at which*

$$\dim H^2(G_K, \overline{\rho}) \geq r$$

is Zariski closed in $\mathcal{X}_{d,\mathrm{red}}(\overline{\mathbf{F}}_p)$, and is of codimension $\geq r$.

Let $R^{\square}_{\overline{\rho}}$ denote the universal lifting ring of $\overline{\rho}$, with universal lifting ρ^{univ}. For each regular tuple of labeled Hodge–Tate weights $\underline{\lambda}$, we let $R^{\mathrm{crys},\underline{\lambda}}_{\overline{\rho}}$ denote the quotient of $R^{\square}_{\overline{\rho}}$ corresponding to crystalline lifts of $\overline{\rho}$ with Hodge–Tate weights $\underline{\lambda}$ (of course, this quotient is zero unless $\overline{\rho}$ admits such a crystalline lift). Then $H^2(G_K, \rho^{\mathrm{univ}})$ is an $R^{\square}_{\overline{\rho}}$-module, and Proposition 1.5.1 implies the following corollary.

1.5.2 Corollary. *For any regular tuple of labeled Hodge–Tate weights $\underline{\lambda}$ the locus of points $x \in \operatorname{Spec} R^{\mathrm{crys},\underline{\lambda}}_{\overline{\rho}}/p$ for which*

$$\dim_{\kappa(x)} H^2(G_K, \rho^{\mathrm{univ}}) \otimes_{R^{\square}_{\overline{\rho}}} \kappa(x) \geq r$$

has codimension $\geq r$.

1.5.3 Remark. Tate local duality, together with the compatibility of H^2 with base change, shows that

$$\begin{aligned}\dim_{\kappa(x)} \big(H^2(G_K, \rho^{\mathrm{univ}}) \otimes_{R^{\square}_{\overline{\rho}}} \kappa(x)\big) &= \dim_{\kappa(x)} H^2\big(G_K, \rho^{\mathrm{univ}} \otimes_{R^{\square}_{\overline{\rho}}} \kappa(x)\big) \\ &= \dim_{\kappa(x)} \operatorname{Hom}_{G_K}\big((\rho^{\mathrm{univ}})^{\vee} \otimes_{R^{\square}_{\overline{\rho}}} \kappa(x), \overline{\epsilon}\big)\end{aligned}$$

(where $\overline{\epsilon}$ denotes the mod p cyclotomic character, thought of as taking values in $\kappa(x)^\times$). Thus the statement of Corollary 1.5.2 is related to the way in which $\operatorname{Spec} R_{\overline{\rho}}^{\mathrm{crys},\underline{\lambda}}/p$ intersects the reducibility locus in $\operatorname{Spec} R_{\overline{\rho}}^{\square}$.

Given Corollary 1.5.2, we prove Theorem 1.2.2 by working purely within the context of formal lifting rings. However we don't know how to prove the corollary while staying within that context. Indeed, as Remark 1.5.3 indicates, this corollary is related to the way in which the special fiber of a potentially crystalline deformation ring intersects another natural locus in $\operatorname{Spec} R_{\overline{\rho}}^{\square}$ (namely, the reducibility locus). Since the special fiber of a potentially crystalline lifting ring is not directly defined in deformation-theoretic terms, such questions are notoriously difficult to study directly. Our proof of the corollary proceeds differently, by replacing a computation on the special fiber of the potentially crystalline deformation ring by a computation on $\mathcal{X}_{d,\mathrm{red}}$; this latter space has a concrete description in terms of families of varying $\overline{\rho}$, whose H^2 we are able to compute, as a result of the inductive construction of families of extensions described in Section 1.4.

In order to deduce Corollary 1.5.2 from Proposition 1.5.1, it is crucial that we know that the natural morphism $\operatorname{Spf} R^{\mathrm{crys},\underline{\lambda}}/p \to \mathcal{X}_d$ is effective, in the sense that it arises from a morphism $\operatorname{Spec} R^{\mathrm{crys},\underline{\lambda}}/p \to \mathcal{X}_d$. More concretely, the universal representation ρ^{univ} gives an étale (φ,Γ)-module over each Artinian quotient of $R_{\overline{\rho}}^{\square}$. By passing to the limit over these quotients, we obtain a "universal formal étale (φ,Γ)-module" over the completion of $(k \otimes_{\mathbf{Z}_p} R_{\overline{\rho}}^{\square}/p)((T))$ with respect to the maximal ideal $\mathfrak{m}$ of $R_{\overline{\rho}}^{\square}$. Since the special fiber of $\mathcal{X}_d$ is formal algebraic but not algebraic (see Section 1.8 below), there is no corresponding (φ,Γ)-module with $R_{\overline{\rho}}^{\square}/p$-coefficients; the φ and Γ actions on the universal formal étale (φ,Γ)-module involve Laurent tails of unbounded degree (with the coefficients of T^{-n} tending to zero $\mathfrak{m}$-adically as $n \to \infty$).

The assertion that $\operatorname{Spf} R^{\mathrm{crys},\underline{\lambda}}/p \to \mathcal{X}_d$ is effective is equivalent to showing that the base change of the universal formal étale (φ,Γ)-module to $R^{\mathrm{crys},\underline{\lambda}}/p$ arises from a genuine (φ,Γ)-module, i.e., from one that involves only Laurent tails of bounded degree. We deduce this from Theorem 1.2.4. Indeed, the ring $R^{\mathrm{crys},\underline{\lambda}}/p$ is a versal ring for the special fiber of the p-adic formal algebraic stack $\mathcal{X}_d^{\mathrm{crys},\underline{\lambda}}$, and (by the very definition of a p-adic formal algebraic stack) this special fiber is an algebraic stack; and the versal rings for algebraic stacks are always effective.

1.5.4 Remark. As our citation of both Theorems 5.5.12 and 6.5.1 for the proof of Proposition 1.5.1 may indicate, our proof of Proposition 1.5.1 is somewhat intricate. Indeed, in Theorem 5.5.12, we show that $\mathcal{X}_{d,\mathrm{red}}$ has dimension at most $[K:\mathbf{Q}_p]d(d-1)/2$, and that the locus considered in Proposition 1.5.1 has dimension at most $[K:\mathbf{Q}_p]d(d-1)/2 - r$. This is in fact enough to deduce Corollary 1.5.2, as $\operatorname{Spec} R^{\mathrm{crys},\underline{\lambda}}/p$ is known to be equidimensional.

Given Corollary 1.5.2, we prove Theorem 1.2.2. In combination with the effective versality of the crystalline deformation rings discussed above we are then able to deduce the equidimensionality of $\mathcal{X}_{d,\mathrm{red}}$, and then also prove Proposition 1.5.1 as stated.

1.6 CRYSTALLINE AND SEMISTABLE MODULI STACKS

We now explain the proof of Theorem 1.2.4; the proof is essentially identical in the crystalline and semistable cases, so we concentrate on the crystalline case. To prove the theorem, it is necessary to have a criterion for a (φ,Γ)-module to come from a crystalline Galois representation. In the case that $K/\mathbf{Q}_p$ is unramified, it is possible to give an explicit criterion in terms of Wach modules [Wac96], but no such direct description is known for general K. Instead, following Kisin's construction of the crystalline deformation rings $R_{\overline{\rho}}^{\mathrm{crys},\lambda}$ in [Kis08], we use the theory of Breuil–Kisin modules. More precisely, Kisin shows that crystalline representations of G_K have finite height over the (non-Galois) Kummer extension K_∞/K obtained by adjoining a compatible system of p-power roots of a uniformizer of K; here being of finite height means that the corresponding étale φ-modules admit certain φ-stable lattices, called Breuil–Kisin modules.

While not every representation of finite height over K_∞ comes from a crystalline representation, we are able to show in Appendix F (jointly written by T. G. and Tong Liu) that a representation $G_K \to \mathrm{GL}_d(\overline{\mathbf{Z}}_p)$ is crystalline if and only if it is of finite height for *every* choice of K_∞, and if the corresponding Breuil–Kisin modules satisfy certain natural compatibilities. (These compatibilities are best expressed in terms of Bhatt–Scholze's prismatic site, as in [BS21], but we do not make use of that perspective in this book. Instead, we write down explicit conditions on the corresponding Breuil–Kisin–Fargues modules; recall that Breuil–Kisin–Fargues modules are a variant of Breuil–Kisin modules introduced by Fargues; see, e.g., [BMS18, §4].)

We use this description of the crystalline representations to prove the existence of the stacks $\mathcal{X}_d^{\mathrm{crys},\underline{\lambda}}$. The proof that $\mathcal{X}_d^{\mathrm{crys},\underline{\lambda}}$ is a p-adic formal algebraic stack relies on an analogue of results of Caruso–Liu [CL11] on extensions of the Galois action on Breuil–Kisin modules, which roughly speaking says that the action of G_{K_∞} determines the action of G_K up to a finite amount of ambiguity. More precisely, given a Breuil–Kisin module over a $\mathbf{Z}/p^a$-algebra for some $a \geq 1$, there is a finite subextension K_s/K of K_∞/K depending only on a, K and the height of the Breuil–Kisin module, such that there is a canonical action of G_{K_s} on the corresponding Breuil–Kisin–Fargues module. This canonical action is constructed by Frobenius amplification, and in the case that the Breuil–Kisin module arises from the reduction modulo p^a of a crystalline representation of G_K, the canonical action coincides with the restriction to G_{K_s} of the G_K-action on the Breuil–Kisin–Fargues module. (In [CL11] a version of this canonical action is used to prove ramification bounds on the reductions modulo p^a of crystalline representations; in Chapter 7, we use analogous arguments in the setting of (φ,Γ)-modules to relate our stacks to stacks of Weil group representations in the rank 1 case.)

There is one significant technical difficulty, which is that we need to define morphisms of stacks that correspond to the restriction of Galois representations from K to K_∞. In order to do this we have to compare (φ,Γ)-modules with A-coefficients (which are defined via the cyclotomic extension $K(\zeta_{p^\infty})/K$) to φ-modules with A-coefficients defined via the extension K_∞/K. We do not know of a direct way to do this; we proceed by proving a correspondence between φ-modules over Laurent series rings with φ-modules over the perfections of these Laurent series rings and proving the following descent result which may be of independent interest; in the statement, $\mathbf{C}$ denotes the completion of an algebraic closure of $\mathbf{Q}_p$.

1.6.1 Theorem (Theorem 2.4.1). *Let A be a finite type $\mathbf{Z}/p^a$-algebra, for some $a \geq 1$. Let F be a closed perfectoid subfield of $\mathbf{C}$, with tilt $F^\flat$, a closed perfectoid subfield of $\mathbf{C}^\flat$. Write $W(F^\flat)_A := W(F^\flat) \otimes_{\mathbf{Z}_p} A$.*

Then the inclusion $W(F^\flat)_A \to W(\mathbf{C}^\flat)_A$ is a faithfully flat morphism of Noetherian rings, and the functor $M \mapsto W(\mathbf{C}^\flat)_A \otimes_{W(F^\flat)_A} M$ induces an equivalence between the category of finitely generated projective $W(F^\flat)_A$-modules and the category of finitely generated projective $W(\mathbf{C}^\flat)_A$-modules endowed with a continuous semi-linear G_F-action.

The existence of the required morphism of stacks follows easily from two applications of Theorem 1.6.1, applied with F equal to respectively the completion of K_∞ and the completion of $K(\zeta_{p^\infty})$. Furthermore, this construction gives an alternative description of our stacks, as moduli spaces of $W(\mathbf{C}^\flat)_A$-modules endowed with commuting semi-linear actions of G_K and φ. It seems plausible that this description will be useful in future work, as it connects naturally to the theory of Breuil–Kisin–Fargues modules (and indeed we use this connection in our construction of the potentially semistable moduli stacks). Note though that the description in terms of (φ,Γ)-modules is important (at least in our approach) for establishing the basic finiteness properties of our stacks.

1.7 THE GEOMETRIC BREUIL–MÉZARD CONJECTURE AND THE WEIGHT PART OF SERRE'S CONJECTURE

We will now briefly explain our results and conjectures relating our stacks to the Breuil–Mézard conjecture and the weight part of Serre's conjecture. Further explanation and motivation can be found throughout Chapter 8. Some of these results were previewed in [GHS18, §6], and the earlier sections of that paper (in particular the introduction) provide an overview of the weight part of Serre's conjecture and its connections to the Breuil–Mézard conjecture that may be helpful to the reader who is not already familiar with them. As in the rest of this introduction, we ignore the possibility of inertial types, and we also restrict to crystalline representations for the purpose of exposition. Everything in this section extends to the more general setting of potentially semistable representations, and indeed as we explain in Section 8.6 when discussing the papers [CEGS19] and [GK14], the additional information provided by non-trivial inertial types is very important.

Let $\overline{\rho}\colon G_K \to \mathrm{GL}_d(\mathbf{F})$ be a continuous representation (for some finite extension $\mathbf{F}$ of $\mathbf{F}_p$), and let $R^{\square}_{\overline{\rho}}$ be the corresponding universal lifting ring. The corresponding formal scheme $\operatorname{Spf} R^{\square}_{\overline{\rho}}$ doesn't carry a lot of evident geometry in and of itself; for example, its underlying reduced subscheme is simply the closed point $\operatorname{Spec} \mathbf{F}$, corresponding to $\overline{\rho}$ itself. On the other hand, $\mathcal{X}_d$ has a quite non-trivial underlying reduced substack $\mathcal{X}_{d,\mathrm{red}}$, which parameterizes all the d-dimensional residual representations of G_K. It is natural to ask whether this underlying reduced substack has any significance in formal deformation theory. More precisely, we could ask for the meaning of the fiber product $\operatorname{Spf} R^{\square}_{\overline{\rho}} \times_{\mathcal{X}_d} \mathcal{X}_{d,\mathrm{red}}$.

This fiber product is a reduced closed formal subscheme of $\operatorname{Spf} R^{\square}_{\overline{\rho}}$ of dimension $d^2 + [K:\mathbf{Q}_p]d(d-1)/2$. It arises (via completion at the closed point) from a closed subscheme of $\operatorname{Spec} R^{\square}_{\overline{\rho}}$ (as does any closed formal subscheme of the Spf of a complete Noetherian local ring), whose irreducible components, when thought of as cycles on $\operatorname{Spec} R^{\square}_{\overline{\rho}}$, are precisely the cycles that (conjecturally) appear in the geometric Breuil–Mézard conjecture of [EG14]. More precisely, we obtain the following qualitative version of the geometric Breuil–Mézard conjecture [EG14, Conj. 4.2.1].

1.7.1 Theorem (Theorem 8.1.4)**.** *If $\overline{\rho}\colon G_K \to \mathrm{GL}_d(\mathbf{F})$ is a continuous representation, then there are finitely many cycles of dimension $d^2 + [K:\mathbf{Q}_p]d(d-1)/2$ in* $\operatorname{Spec} R^{\square}_{\overline{\rho}}/p$*, such that for any regular tuple of labeled Hodge–Tate weights $\underline{\lambda}$, the special fiber* $\operatorname{Spec} R^{\mathrm{crys},\underline{\lambda}}_{\overline{\rho}}/p$ *is set-theoretically supported on the union of some number of these cycles.*

The cycles in the statement of the theorem are precisely those arising from the fiber products $\operatorname{Spf} R^{\square}_{\overline{\rho}} \times_{\mathcal{X}_d} \mathcal{X}^{\underline{k}}_{d,\mathrm{red}}$, where $\underline{k}$ runs over the Serre weights. While Theorem 1.7.1 is a purely local statement, we do not know how to prove it without using the stacks $\mathcal{X}_d$.

The full geometric Breuil–Mézard conjecture of [EG14] makes precise predictions about the multiplicities of the cycles of the special fibres of $\operatorname{Spec} R^{\mathrm{crys},\underline{\lambda}}_{\overline{\rho}}/p$; passing from cycles to Hilbert–Samuel multiplicities then recovers the original Breuil–Mézard conjecture [BM02] (or rather a natural generalization of it to GL_d), which we refer to as the "numerical Breuil–Mézard conjecture". In particular, the multiplicities are expected to be computed in terms of quantities $n^{\mathrm{crys}}_{\underline{k}}(\lambda)$ that are defined as follows: one associates an irreducible algebraic representation $\sigma^{\mathrm{crys}}(\lambda)$ of GL_d/K to $\underline{\lambda}$, defined to have highest weight (a certain shift of) $\underline{\lambda}$. The semisimplification of the reduction mod p of $\sigma^{\mathrm{crys}}(\lambda)$ can be written as a direct sum of irreducible representations of $\mathrm{GL}_d(k)$, and $n^{\mathrm{crys}}_{\underline{k}}(\lambda)$ is defined to be the multiplicity with which the Serre weight $\underline{k}$ occurs.

In Chapter 8 we explain that as we run over all $\overline{\rho}$, the geometric Breuil–Mézard conjecture is equivalent to the following analogous conjecture for the special fibres of our crystalline and semistable stacks. Here by a "cycle" in $\mathcal{X}_{d,\mathrm{red}}$ we mean a formal $\mathbf{Z}$-linear combination of its irreducible components $\mathcal{X}_d^{\underline{k}}$.

1.7.2 Conjecture (Conjecture 8.2.2). *There are cycles $Z_{\underline{k}}$ in $\mathcal{X}_{d,\mathrm{red}}$ with the property that for each regular tuple of labeled Hodge–Tate weights $\underline{\lambda}$, the underlying cycle of the special fiber of $\mathcal{X}_d^{\mathrm{crys},\underline{\lambda}}$ is $\sum_{\underline{k}} n_{\underline{k}}^{\mathrm{crys}}(\lambda) \cdot Z_{\underline{k}}$.*

In fact we expect that the cycles $Z_{\underline{k}}$ are effective, i.e., that they are a linear combination of the irreducible components $\mathcal{X}_d^{\underline{k}'}$ with non-negative integer coefficients. Since there are infinitely many possible $\underline{\lambda}$, the cycles $Z_{\underline{k}}$, if they exist, are hugely overdetermined by Conjecture 1.7.2.

As first explained in [Kis09a], the (numerical) Breuil–Mézard conjecture has important consequences for automorphy lifting theorems; indeed, proving the conjecture is closely related to proving automorphy lifting theorems in situations with arbitrarily high weight or ramification at the places dividing p. Conversely, following [Kis10], one can use automorphy lifting theorems to deduce cases of the Breuil–Mézard conjecture. Automorphy lifting theorems involve a fixed $\overline{\rho}$, and in fact we can deduce Conjecture 1.7.2 from the Breuil–Mézard conjecture for a finite set of suitably generic $\overline{\rho}$.

In particular, we are able to combine results in the literature to show that for GL_2 the cycles $Z_{\underline{k}}$ in Conjecture 1.7.2 must have a particularly simple form: we necessarily have $Z_{\underline{k}} = \mathcal{X}_{d,\mathrm{red}}^{\underline{k}}$ unless $\underline{k}$ is a so-called "Steinberg" weight, in which case $Z_{\underline{k}}$ is the sum of $\mathcal{X}_{d,\mathrm{red}}^{\underline{k}}$ and one other irreducible component. (More precisely, what we show, following [CEGS19, GK14], is that with these cycles $Z_{\underline{k}}$, Conjecture 1.7.2 holds for all "potentially Barsotti–Tate" representations.)

The weight part of Serre's conjecture predicts the weights in which particular Galois representations contribute to the mod p cohomology of locally symmetric spaces. Following [GK14], this conjecture is closely related to the Breuil–Mézard conjecture; indeed, if Conjecture 1.7.2 holds, then the set of Serre weights associated to a representation $\overline{\rho}\colon G_K \to \mathrm{GL}_d(\overline{\mathbf{F}}_p)$ should be precisely the weights $\underline{k}$ for which $Z_{\underline{k}}$ is supported at $\overline{\rho}$. In other words, if we refine our labelling of the irreducible components of $\mathcal{X}_{d,\mathrm{red}}$ by labelling each component by the union of the weights $\underline{k}$ for which that component contributes to $Z_{\underline{k}}$, then we expect the set of Serre weights for $\overline{\rho}$ to be the union of the sets of weights labelling the irreducible components containing $\overline{\rho}$. This expectation holds for GL_2 by the main results of [CEGS19].

1.8 FURTHER QUESTIONS

There are many other questions one could ask about the stacks $\mathcal{X}_d$, which we hope to return to in future papers. For example, we show in Proposition 6.5.2 that $\mathcal{X}_d$ is not a p-adic formal algebraic stack. Indeed, if it were p-adic formal algebraic, then its special fiber would be an algebraic stack, whose dimension would be equal to the dimension of its underlying reduced substack. In turn, this would imply that the versal rings $R_{\overline{\rho}}^{\square}$ would have dimension equal to the dimensions of the crystalline deformation rings $R_{\overline{\rho}}^{\mathrm{crys},\underline{\lambda}}$, and this is known not to be true. In fact, it is a folklore conjecture, recently proved by Böckle–Iyengar–Paškūnas [BIP21], that the lifting rings $R_{\overline{\rho}}^{\square}$ are $\mathbf{Z}_p$-flat local complete intersections of dimension $1+d^2+[K:\mathbf{Q}_p]d^2$, which should imply that the stacks $\mathcal{X}_d$ are $\mathbf{Z}_p$-flat local complete intersections of dimension $1+[K:\mathbf{Q}_p]d^2$ (a notion that we do not attempt to make precise for formal algebraic stacks).

It is natural to ask about the rigid analytic generic fiber of $\mathcal{X}_d$; this should exist as a rigid analytic stack in an appropriate sense. The generic fibres of the substacks $\mathcal{X}_d^{\underline{k}}$ should admit morphisms to the stacks of Hartl and Hellmann [HH20] (although these morphisms won't be isomorphisms, since for any finite extension E of $\mathbf{Q}_p$, the $\mathcal{O}_E$-points of $\mathcal{X}_d^{\underline{k}}$, which would coincide with the E-points of its generic fiber, correspond to lattices in crystalline representations, whereas the stacks of [HH20] parameterize crystalline or semistable representations themselves).

We expect that the $\mathcal{X}_d$ will have a role to play in generalizations of the p-adic local Langlands correspondence. For example, we expect that when $K=\mathbf{Q}_p$ the p-adic local Langlands correspondence for $\mathrm{GL}_2(\mathbf{Q}_p)$ can be extended to give rise to sheaves of $\mathrm{GL}_2(\mathbf{Q}_p)$-representations on $\mathcal{X}_2$. More generally, we expect that there will be a p-adic analogue of the work of Fargues–Scholze on the local Langlands correspondence [FS21] involving the stacks $\mathcal{X}_d$.

1.9 PREVIOUS WORK

The description of local Galois representations in terms of étale (φ,Γ)-modules is due to Fontaine [Fon90]. The importance of "height" as an aspect of the theory was already emphasized in [Fon90], and was further developed by Wach [Wac96], who explored the relationship between the finite height condition and crystallinity of Galois representations in the absolutely unramified context.

The use of what are now called Breuil–Kisin modules as a tool for studying crystalline and semistable representations for general (i.e., not necessarily absolutely unramified) p-adic fields (a study which, apart from its intrinsic importance, is crucial for treating potentially crystalline or semistable representations, even in the absolutely unramified context) was due originally to Breuil [Bre98] and was extensively developed by Kisin [Kis09b, Kis08], who used them to study Galois deformation rings.

The algebro-geometric and moduli-theoretic perspectives that already played key roles in Kisin's work were further developed by Pappas and Rapoport [PR09], who introduced moduli stacks of Breuil–Kisin modules and of étale φ-modules; it is this work of Pappas and Rapoport, which can be very roughly thought of as constructing moduli stacks of representations of the absolute Galois groups of certain perfectoid fields, which is the immediate launching point for our work in this book, as well as for our paper [EG21]. (We should also mention Drinfeld's work [Dri06], which underpins the verification of the stack property for the constructions of [PR09], as well as for those of the present book.) Our use of moduli stacks of Breuil–Kisin–Fargues modules (in the construction of the potentially crystalline and semistable substacks) was in part inspired by the work of Fargues and Bhatt–Morrow–Scholze (see in particular [BMS18, §4]), which taught us not to be afraid of $\mathbf{A}_{\mathrm{inf}}$.

Moduli stacks parameterizing crystalline and semistable representations have already been constructed by Hartl and Hellmann [HH20]; as remarked upon above, these stacks should have a relationship to the stacks $\mathcal{X}_d^{\underline{k}}$ that we construct. See also the related papers of Hellmann [Hel16, Hel13].

As far as we are aware, the first construction of moduli stacks of representations of G_K in which the residual representation $\overline{\rho}$ can vary is the work of Carl Wang-Erickson [WE18] mentioned above, which constructs and studies such stacks in the case that $\overline{\rho}$ has fixed semisimplification. These are literally moduli stacks of representations of G_K; they are isomorphic to certain substacks of our stacks $\mathcal{X}_d$, as we explain in Section 6.7.

1.10 AN OUTLINE OF THE BOOK

We finish this introduction with a brief overview of the contents of this book. The reader may also wish to refer to the introductions to each chapter, as well as to the overview of this book provided by the notes [EG20].

In Chapter 2 we recall several of the coefficient rings used in the theories of (φ, Γ)-modules and Breuil–Kisin modules, and introduce versions of these rings with coefficients in a p-adically complete $\mathbf{Z}_p$-algebra. We also prove almost Galois descent results for projective modules, and deduce Theorem 1.6.1.

In Chapter 3 we recall the results of [EG21] on moduli stacks of φ-modules, and use them to define our stacks $\mathcal{X}_d$ of étale (φ, Γ)-modules. With some effort, we prove that $\mathcal{X}_d$ is an Ind-algebraic stack. Chapter 4 defines various moduli stacks of Breuil–Kisin and Breuil–Kisin–Fargues modules, and uses them to construct our moduli stacks of potentially semistable and potentially crystalline representations, and in particular to prove Theorem 1.2.4.

Chapter 5 develops the theory of the Herr complex, proving in particular that it is a perfect complex and is compatible with base change. We show how to use the Herr complex to construct families of extensions of (φ, Γ)-modules, and we use these families to define the irreducible substack $\mathcal{X}_{d,\mathrm{red}}^{\underline{k}}$ corresponding to a Serre weight $\underline{k}$. By induction on d we prove that $\mathcal{X}_d$ is a Noetherian formal algebraic stack, and establish a version of Proposition 1.5.1 (although as discussed in Remark 1.5.4, we do not prove Proposition 1.5.1 as stated at this point in the argument).

It may help the reader for us to point out that Chapters 4 and 5 are essentially independent of one another, and are of rather different flavor. Chapter 4 involves an interleaving of stack-theoretic arguments with ideas from p-adic Hodge theory and the theory of Breuil–Kisin modules, while in Chapter 5, once we complete our analysis of the Herr complex, our perspective begins to shift: although at a technical level we of course continue to work with (φ, Γ)-modules, we begin to think in terms of Galois representations and Galois cohomology, and the more foundational arguments of the preceding chapters recede somewhat into the background.

In Chapter 6 we combine the results of Chapters 4 and 5 with a geometric argument on the local deformation ring to prove Theorem 1.2.2. Having done this, we are then able to improve on the results on $\mathcal{X}_d$ established in the earlier chapters by proving Theorem 1.2.1. We also deduce Theorem 1.2.3, as well as determining the closed points of $\mathcal{X}_d$, and describing the relationship of our stacks with Wang–Erickson's stacks of Galois representations.

Chapter 7 gives explicit descriptions of various of our moduli stacks in the case $d = 1$, relating them to moduli stacks of Weil group representations. Chapter 8 explains our geometric version of the Breuil–Mézard conjecture, and proves some results towards it, particularly in the case $d = 2$.

Finally the appendices for the most part establish various technical results used in the body of the book. We highlight in particular Appendix A, which summarizes the theory of formal algebraic stacks developed in [Eme], and Appendix F, which combines the theory of Breuil–Kisin–Fargues modules with Tong Liu's theory of $(\varphi, \widehat{G})$-modules to give a new characterization of integral lattices in potentially semistable representations, of which we make crucial use in Chapter 4.

1.11 ACKNOWLEDGMENTS

We would like to thank Robin Bartlett, Laurent Berger, Bhargav Bhatt, Xavier Caruso, Pierre Colmez, Yiwen Ding, Florian Herzig, Ashwin Iyengar, Mark Kisin, Tong Liu, George Pappas, Dat Pham, Léo Poyeton, Michael Rapoport, and Peter Scholze for helpful correspondence and conversations. We would particularly like to thank Tong Liu for his contributions to Appendix F. We would also like to thank the organizers (Johannes Anschütz, Arthur-César Le Bras, and Andreas Mihatsch) of the 2019 Hausdorff School on "the Emerton–Gee stack and related topics", as well as all the participants, both for the encouragement to finish this book and for the many helpful questions and corrections resulting from our lectures.

Our mathematical debt to the late Jean-Marc Fontaine will be obvious to the reader. This book benefited from several conversations with him over the years 2011–2018, and from the interest he showed in our results; in particular, his explanations to us of the relationship between framing Galois representations and Fontaine–Laffaille modules during his visit to Northwestern University in the spring of 2011 provided an important clue as to the correct definitions of our stacks.

We owe special thanks to Colette and Therese for all of their patience and support during the writing and revision of this book.

The first author was supported in part by the NSF grants DMS-1303450, DMS-1601871, and DMS-1902307. The second author was supported in part by a Leverhulme Prize, EPSRC grant EP/L025485/1, Marie Curie Career Integration Grant 303605, ERC Starting Grant 306326, ERC Advanced Grant 884596, and a Royal Society Wolfson Research Merit Award. This project has received funding from the European Research Council (ERC) under the European Union's Horizon 2020 research and innovation programme (grant agreement No. 884596).

1.12 NOTATION AND CONVENTIONS

p-adic Hodge theory

Let $K/\mathbf{Q}_p$ be a finite extension. If ρ is a de Rham representation of G_K on a $\overline{\mathbf{Q}}_p$-vector space W, then we will write $\mathrm{WD}(\rho)$ for the corresponding Weil–Deligne representation of W_K (see, e.g., [CDT99, App. B]), and if $\sigma\colon K \hookrightarrow \overline{\mathbf{Q}}_p$ is a continuous embedding of fields then we will write $\mathrm{HT}_\sigma(\rho)$ for the multiset of Hodge–Tate numbers of ρ with respect to σ, which by definition contains i with multiplicity $\dim_{\overline{\mathbf{Q}}_p}(W \otimes_{\sigma,K} \widehat{\overline{K}}(i))^{G_K}$. Thus, for example, if ϵ denotes the p-adic cyclotomic character, then $\mathrm{HT}_\sigma(\epsilon) = \{-1\}$.

By a *d-tuple of labeled Hodge–Tate weights* $\underline{\lambda}$, we mean a tuple of integers $\{\lambda_{\sigma,i}\}_{\sigma\colon K\hookrightarrow\overline{\mathbf{Q}}_p, 1\le i\le d}$ with $\lambda_{\sigma,i} \ge \lambda_{\sigma,i+1}$ for all σ and all $1 \le i \le d-1$. We will also refer to $\underline{\lambda}$ as a *Hodge type*. By an *inertial type* τ we mean a representation $\tau\colon I_K \to \mathrm{GL}_d(\overline{\mathbf{Q}}_p)$ which extends to a representation of W_K with open kernel (so in particular, τ has finite image).

Then we say that ρ has Hodge type $\underline{\lambda}$ (or labeled Hodge–Tate weights $\underline{\lambda}$) if for each $\sigma\colon K \hookrightarrow \overline{\mathbf{Q}}_p$ we have $\mathrm{HT}_\sigma(\rho) = \{\lambda_{\sigma,i}\}_{1\le i\le d}$, and we say that ρ has inertial type τ if $\mathrm{WD}(\rho)|_{I_K} \cong \tau$.

We often somewhat abusively write that a representation $\rho\colon G_K \to \mathrm{GL}_d(\mathbf{Z}_p)$ is crystalline (or potentially crystalline, or semistable, or ...) if the corresponding representation $\rho\colon G_K \to \mathrm{GL}_d(\mathbf{Q}_p)$ is crystalline (or potentially crystalline, or semistable, or ...).

Serre weights and Hodge–Tate weights

By a *Serre weight* $\underline{k}$ we mean a tuple of integers $\{k_{\overline{\sigma},i}\}_{\overline{\sigma}\colon k\hookrightarrow\overline{\mathbf{F}}_p, 1\le i\le d}$ with the properties that

- $p-1 \ge k_{\overline{\sigma},i} - k_{\overline{\sigma},i+1} \ge 0$ for each $1 \le i \le d-1$, and
- $p-1 \ge k_{\overline{\sigma},d} \ge 0$, and not every $k_{\overline{\sigma},d}$ is equal to $p-1$.

The set of Serre weights is in bijection with the set of irreducible $\overline{\mathbf{F}}_p$-representations of $\mathrm{GL}_d(k)$, via passage to highest weight vectors (see for example the appendix to [Her09]).

Each embedding $\sigma\colon K \hookrightarrow \overline{\mathbf{Q}}_p$ induces an embedding $\overline{\sigma}\colon k \hookrightarrow \overline{\mathbf{F}}_p$; if $K/\mathbf{Q}_p$ is ramified, then each $\overline{\sigma}$ corresponds to multiple embeddings σ. We say that $\underline{\lambda}$ is a lift of $\underline{k}$ if for each embedding $\overline{\sigma}\colon k \hookrightarrow \overline{\mathbf{F}}_p$, we can choose an embedding $\sigma\colon K \hookrightarrow \overline{\mathbf{Q}}_p$ lifting $\overline{\sigma}$, with the properties that:

- $\lambda_{\sigma,i} = k_{\sigma,i} + d - i$, and
- if $\sigma'\colon K \hookrightarrow \overline{\mathbf{Q}}_p$ is any other lift of $\overline{\sigma}$, then $k_{\sigma',i} = d - i$.

Lifting rings

Let $K/\mathbf{Q}_p$ be a finite extension, and let $\overline{\rho}\colon G_K \to \mathrm{GL}_d(\overline{\mathbf{F}}_p)$ be a continuous representation. Then the image of $\overline{\rho}$ is contained in $\mathrm{GL}_d(\mathbf{F})$ for any sufficiently large finite extension $\mathbf{F}/\mathbf{F}_p$. Let $\mathcal{O}$ be the ring of integers in some finite extension $E/\mathbf{Q}_p$, and suppose that the residue field of E is $\mathbf{F}$. Let $R_{\overline{\rho}}^{\square,\mathcal{O}}$ be the universal lifting $\mathcal{O}$-algebra of $\overline{\rho}$; by definition, this (pro-)represents the functor given by lifts of $\overline{\rho}$ to representations $\rho\colon G_K \to \mathrm{GL}_d(A)$, for A an Artin local $\mathcal{O}$-algebra with residue field $\mathbf{F}$. The precise choice of E is unimportant, in the sense that if $\mathcal{O}'$ is the ring of integers in a finite extension E'/E, then by [BLGGT14, Lem. 1.2.1] we have $R_{\overline{\rho}}^{\square,\mathcal{O}'} = R_{\overline{\rho}}^{\square,\mathcal{O}} \otimes_{\mathcal{O}} \mathcal{O}'$.

Fix some Hodge type $\underline{\lambda}$ and inertial type τ. If $\mathcal{O}$ is chosen large enough that the inertial type τ is defined over $E = \mathcal{O}[1/p]$, and large enough that E contains the images of all embeddings $\sigma\colon K \hookrightarrow \overline{\mathbf{Q}}_p$, then we have the usual lifting $\mathcal{O}$-algebras $R_{\overline{\rho}}^{\mathrm{crys},\underline{\lambda},\tau,\mathcal{O}}$ and $R_{\overline{\rho}}^{\mathrm{ss},\underline{\lambda},\tau,\mathcal{O}}$. By definition, these are the unique $\mathcal{O}$-flat quotients of $R_{\overline{\rho}}^{\square,\mathcal{O}}$ with the property that if B is a finite flat E-algebra, then an $\mathcal{O}$-algebra homomorphism $R_{\overline{\rho}}^{\square,\mathcal{O}} \to B$ factors through $R_{\overline{\rho}}^{\mathrm{crys},\underline{\lambda},\tau,\mathcal{O}}$ (resp. through $R_{\overline{\rho}}^{\mathrm{ss},\underline{\lambda},\tau,\mathcal{O}}$) if and only if the corresponding representation of G_K is potentially crystalline (resp. potentially semistable) of Hodge type $\underline{\lambda}$ and inertial type τ. If τ is trivial, we will sometimes omit it from the notation. By the main theorems of [Kis08], these rings are (when they are nonzero) equidimensional of dimension

$$1 + d^2 + \sum_{\sigma} \#\{1 \leq i < j \leq d | \lambda_{\sigma,i} > \lambda_{\sigma,j}\}.$$

Note that this quantity is at most $1 + d^2 + [K:\mathbf{Q}_p]d(d-1)/2$, with equality if and only if $\underline{\lambda}$ is *regular*, in the sense that $\lambda_{\sigma,i} > \lambda_{\sigma,i+1}$ for all σ and all $1 \leq i \leq d-1$. As above, we have $R_{\overline{\rho}}^{\mathrm{crys},\underline{\lambda},\tau,\mathcal{O}'} = R_{\overline{\rho}}^{\mathrm{crys},\underline{\lambda},\tau,\mathcal{O}} \otimes_{\mathcal{O}} \mathcal{O}'$, and similarly for $R_{\overline{\rho}}^{\mathrm{ss},\underline{\lambda},\tau,\mathcal{O}'}$. By [Kis08, Thm. 3.3.8] the localized rings $R_{\overline{\rho}}^{\mathrm{crys},\underline{\lambda},\tau,\mathcal{O}}[1/p]$ are regular, and thus the rings $R_{\overline{\rho}}^{\mathrm{crys},\underline{\lambda},\tau,\mathcal{O}}$ (which embed into their localizations away from p, since they are $\mathcal{O}$-flat) are reduced.

Algebra

Our conventions typically follow [Sta]. In particular, if M is an abelian topological group with a linear topology, then as in [Sta, Tag 07E7] we say that M is *complete* if the natural morphism $M \to \varprojlim_i M/U_i$ is an isomorphism, where

$\{U_i\}_{i\in I}$ is some (equivalently any) fundamental system of neighborhoods of 0 consisting of subgroups. Note that in some other references this would be referred to as being *complete and separated.*

If R is a ring, we write $D(R)$ for the (unbounded) derived category of R-modules. We say that a complex $P^\bullet$ is *good* if it is a bounded complex of finite projective R-modules; then an object $C^\bullet$ of $D(R)$ is called a *perfect complex* if there is a quasi-isomorphism $P^\bullet \to C^\bullet$ where $P^\bullet$ is good. In fact, $C^\bullet$ is perfect if and only if it is isomorphic in $D(R)$ to a good complex $P^\bullet$: if we have another complex $D^\bullet$ and quasi-isomorphisms $P^\bullet \to D^\bullet$, $C^\bullet \to D^\bullet$, then there is a quasi-isomorphism $P^\bullet \to C^\bullet$ ([Sta, Tag 064E]).

Stacks

Our conventions on algebraic stacks and formal algebraic stacks are those of [Sta] and [Eme]. We recall some terminology and results in Appendix A. Throughout the book, if A is a topological ring and $\mathcal{C}$ is a stack we write $\mathcal{C}(A)$ for $\mathcal{C}(\operatorname{Spf} A)$; if A has the discrete topology, this is equal to $\mathcal{C}(\operatorname{Spec} A)$.

Chapter Two

Rings and coefficients

In this chapter we study various rings which will be the coefficients of the φ-modules and (φ, Γ)-modules that we study in the subsequent chapters. Throughout the chapter, we fix a finite extension K of $\mathbf{Q}_p$, which we regard as a subfield of some fixed algebraic closure $\overline{\mathbf{Q}}_p$ of $\mathbf{Q}_p$.

We begin by recalling some rings used in integral p-adic Hodge theory (in particular, in defining (φ, Γ)-modules and Breuil–Kisin modules), before introducing versions of them with coefficients in certain p-adically complete $\mathbf{Z}_p$-algebras. We then prove (almost) descent results that allow us to relate φ-modules over various different rings, before introducing the (φ, Γ)-modules which we work with throughout the book.

2.1 RINGS

2.1.1 Perfectoid fields and their tilts

As usual, we let $\mathbf{C}$ denote the completion of the algebraic closure $\overline{\mathbf{Q}}_p$ of $\mathbf{Q}_p$. It is a perfectoid field, whose tilt $\mathbf{C}^\flat$ is a complete non-Archimedean valued perfect field of characteristic p. If F is a perfectoid closed subfield of $\mathbf{C}$, then its tilt $F^\flat$ is a closed, and perfect, subfield of $\mathbf{C}^\flat$.

We let $\mathcal{O}_{\mathbf{C}}$ denote the ring of integers in $\mathbf{C}$. Its tilt $\mathcal{O}_{\mathbf{C}}^\flat$ is then the ring of integers in $\mathbf{C}^\flat$. Similarly, if $\mathcal{O}_F$ denotes the ring of integers in F, then $\mathcal{O}_F^\flat$ is the ring of integers in $F^\flat$.

We may form the rings of Witt vectors $W(\mathbf{C}^\flat)$ and $W(F^\flat)$, and the rings of Witt vectors $W(\mathcal{O}_{\mathbf{C}^\flat})$ and $W(\mathcal{O}_F^\flat)$; following the standard convention, we typically denote $W(\mathcal{O}_{\mathbf{C}^\flat})$ by $\mathbf{A}_{\mathrm{inf}}$.

Each of these rings of Witt vectors is a p-adically complete ring, but we always consider them as topological rings by endowing them with a finer topology, the so-called *weak topology*, which admits the following description: If R is any of $\mathbf{C}^\flat$, $F^\flat$, $\mathcal{O}_{\mathbf{C}}^\flat$, or $\mathcal{O}_F^\flat$, endowed with its natural (valuation) topology, then there is a canonical identification (of sets),

$$W_a(R) \xrightarrow{\sim} R \times \cdots \times R \ (a \text{ factors}),$$

and we endow $W_a(R)$ with the product topology. We then endow $W(R) := \varprojlim_a W_a(R)$ with the inverse limit topology.

This topology admits the following more concrete description (in the general case of a perfectoid F; setting $F = \mathbf{C}$ recovers that particular case): If x is any element of $\mathcal{O}_F^\flat$ of positive valuation, and if $[x]$ denotes the Teichmüller lift of x, then we endow $W_a(\mathcal{O}_F^\flat)$ with the $[x]$-adic topology, so that $W(\mathcal{O}_F^\flat)$ is then endowed with the $(p, [x])$-adic topology. The topology on $W_a(F^\flat)$ is then characterized by the fact that $W_a(\mathcal{O}_F^\flat)$ is an open subring—concretely, $W_a(F^\flat) = W_a(\mathcal{O}_F^\flat)\left[\frac{1}{[x]}\right] = \varinjlim_i W_a(\mathcal{O}_F^\flat)$ (the transition maps being given by multiplication by $[x]$, and each transition map being an open and closed embedding)—while the topology on $W(F^\flat)$ is the inverse limit topology—concretely, $W(F^\flat)$ is the p-adic completion $\widehat{W(\mathcal{O}_F^\flat)\left[\frac{1}{[x]}\right]}$.

Apart from the case of $\mathbf{C}$ itself, there are two main examples of perfectoid F that will be of importance to us; see Example 2.1.7 for a justification of the claim that these fields are indeed perfectoid.

2.1.2 Example (The cyclotomic case). We write $K(\zeta_{p^\infty})$ to denote the extension of K obtained by adjoining all p-power roots of unity. It is an infinite degree Galois extension of K, whose Galois group is naturally identified with an open subgroup of $\mathbf{Z}_p^\times$. We let K_{cyc} denote the unique subextension of $K(\zeta_{p^\infty})$ whose Galois group over K is isomorphic to $\mathbf{Z}_p$ (so K_{cyc} is the "cyclotomic $\mathbf{Z}_p$-extension" of K). If we let $\widehat{K}_{\mathrm{cyc}}$ denote the closure of K_{cyc} in $\mathbf{C}$, then $\widehat{K}_{\mathrm{cyc}}$ is a perfectoid subfield of $\mathbf{C}$.

2.1.3 Example (The Kummer case). If we choose a uniformizer π of K, as well as a compatible system of p-power roots π^{1/p^n} of π (here, "compatible" has the obvious meaning, namely that $(\pi^{1/p^{n+1}})^p = \pi^{1/p^n}$), then we define $K_\infty = K(\pi^{1/p^\infty}) := \bigcup_n K(\pi^{1/p^n})$. If we let $\widehat{K}_\infty$ denote the closure of K_∞ in $\mathbf{C}$, then $\widehat{K}_\infty$ is again a perfectoid subfield of $\mathbf{C}$.

2.1.4 Remark. Let L be an algebraic extension of $\mathbf{Q}_p$, and let $\widehat{L}$ be the closure of L in $\mathbf{C}$. By Krasner's lemma (see [GR04, Prop. 9.1.16] for full details), the field $\widehat{L} \otimes_L \overline{\mathbf{Q}}_p$ is an algebraic closure of $\widehat{L}$, so that the absolute Galois groups of L and $\widehat{L}$ are canonically identified. The action of G_L on $\overline{\mathbf{Q}}_p$ extends to an action on $\mathbf{C}$, and by a theorem of Ax–Tate–Sen [Ax70], we have $\mathbf{C}^{G_L} = \widehat{L}$. We will in particular make use of these facts in the cases $L = K_{\mathrm{cyc}}$ and $L = K_\infty$.

2.1.5 Fields of norms

If L is an infinite strictly arithmetically profinite (strictly APF) extension of K in $\overline{\mathbf{Q}}_p$, then we may form the field of norms $X_K(L)$, as in [FW79] and [Win83, §2]. This is a complete discretely valued field of characteristic p, whose residue field is canonically identified with that of L. (We don't define the notion of strictly APF here, but simply refer to [Win83, Def. 1.2.1] for the definition. We will only apply these notions to Examples 2.1.2 and 2.1.3, in which case the end result of the field of norms construction can be spelled out quite explicitly.) The following theorem is well known.

2.1.6 Theorem. *Let L be an infinite strictly APF extension of K in $\overline{\mathbf{Q}}_p$.*

1. *The closure $\widehat{L}$ of L in $\mathbf{C}$ is perfectoid.*
2. *There is a canonical embedding $X_K(L) \hookrightarrow (\widehat{L})^\flat$, which identifies the target with the completion of the perfect closure of the source.*

Proof. To show that $\widehat{L}$ is perfectoid, it suffices to show that it is not discretely valued, and that the absolute Frobenius on $\mathcal{O}_L/p\mathcal{O}_L$ is surjective. This surjectivity follows from [Win83, Cor. 4.3.4]. It follows immediately from the description in [Win83, §1.4] of APF extensions as towers of elementary extensions that L, and thus $\widehat{L}$, is not discretely valued. The canonical embedding of (2) is constructed in [Win83, §4.2], and its claimed property is proved in [Win83, Cor. 4.3.4]. □

2.1.7 Example. A theorem of Sen [Sen72] shows that if the Galois group of the Galois closure of L over K is a p-adic Lie group, and the induced extension of residue fields is finite, then L is a strictly APF extension of K. (Sen's theorem shows that the Galois closure of L is strictly APF over K; see [Win83, Ex. 1.2.2]. It then follows from [Win83, Prop. 1.2.3 (iii)] that L itself is strictly APF over K.) As a consequence, we see that the theory of the field of norms, and in particular Theorem 2.1.6, applies in the cases of Examples 2.1.2 and 2.1.3.

2.1.8 Thickening fields of norms

In the context of Theorem 2.1.6, given that one may thicken $(\widehat{L})^\flat$ to the flat $\mathbf{Z}/p^a$-algebra $W_a\big((\widehat{L})^\flat\big)$, it is natural to ask whether $X_K(L)$ admits a similar thickening, such that the embedding $X_K(L) \hookrightarrow (\widehat{L})^\flat$ lifts to an embedding of the corresponding thickenings. One would furthermore like such a lifted embedding to be compatible with various auxiliary structures, such as the action of Frobenius on $W_a\big((\widehat{L})^\flat\big)$, or the action of the Galois group $\mathrm{Gal}(L/K)$, in the case when L is Galois over K (i.e., one would like the image of this embedding to be stable under these actions).

Since $X_K(L)$ is imperfect, there is no canonical thickening of $X_K(L)$ over $\mathbf{Z}/p^a$, and as far as we know, there is no simple or general answer to the question of whether such thickenings and embeddings exist with desirable extra properties, such as being Frobenius or Galois stable. However, in the cases of Examples 2.1.2 and 2.1.3, such thickenings and thickened embeddings can be constructed directly, as we now recall.

2.1.9 The cyclotomic case

The extension $K(\zeta_{p^\infty})$ of K is infinite and strictly APF, as well as being Galois over K. If we write $\widetilde{\Gamma}_K := \mathrm{Gal}(K(\zeta_{p^\infty})/K)$, then the cyclotomic character induces an embedding $\chi\colon \widetilde{\Gamma}_K \hookrightarrow \mathbf{Z}_p^\times$. Consequently, there is an isomorphism $\widetilde{\Gamma}_K \cong \Gamma_K \times \Delta$, where $\Gamma_K \cong \mathbf{Z}_p$ and Δ is finite. We have $K_{\mathrm{cyc}} = (K(\zeta_{p^\infty}))^\Delta$. Later in the book, K will typically be fixed, and we will write Γ for Γ_K.

Suppose for a moment that $K = \mathbf{Q}_p$. If we choose a compatible system of p^nth roots of 1, then these give rise in the usual way to an element $\varepsilon \in (\widehat{\mathbf{Q}_p(\zeta_{p^\infty})})^\flat$. If we identify the field of norms $X_{\mathbf{Q}_p}(\mathbf{Q}_p(\zeta_{p^\infty}))$ with a subfield of $(\widehat{\mathbf{Q}_p(\zeta_{p^\infty})})^\flat$ via the embedding $X_{\mathbf{Q}_p}(\mathbf{Q}_p(\zeta_{p^\infty})) \hookrightarrow (\widehat{\mathbf{Q}_p(\zeta_{p^\infty})})^\flat$, then $\varepsilon - 1$ is a uniformizer of $X_{\mathbf{Q}_p}(\mathbf{Q}_p(\zeta_{p^\infty}))$. As usual, let $[\varepsilon]$ denote the Teichmüller lift of ε to an element of $W(\mathcal{O}^\flat_{\widehat{\mathbf{Q}_p(\zeta_{p^\infty})}})$. There is then a continuous embedding

$$\mathbf{Z}_p[[T]] \hookrightarrow W(\mathcal{O}^\flat_{\widehat{\mathbf{Q}_p(\zeta_{p^\infty})}})$$

(the source being endowed with its (p,T)-adic topology, and the target with its weak topology), defined via $T \mapsto [\varepsilon] - 1$. We denote the image of this embedding by $(\mathbf{A}'_{\mathbf{Q}_p})^+$. This embedding extends to an embedding

$$\widehat{\mathbf{Z}_p((T))} \hookrightarrow W((\widehat{\mathbf{Q}_p(\zeta_{p^\infty})})^\flat)$$

(here the source is the p-adic completion of the Laurent series ring $\mathbf{Z}_p((T))$), whose image we denote by $\mathbf{A}'_{\mathbf{Q}_p}$.

We now return to the case of general K. We will compare this case with the case of $\mathbf{Q}_p$. Since $K\mathbf{Q}_p(\zeta_{p^\infty}) = K(\zeta_{p^\infty})$, the theory of the field of norms gives an identification $X_K(K(\zeta_{p^\infty})) = X_{\mathbf{Q}_p}(K(\zeta_{p^\infty}))$ [Win83, Rem. 2.1.4], and shows that this field is a separable extension of $X_{\mathbf{Q}_p}(\mathbf{Q}_p(\zeta_{p^\infty}))$ [Win83, Thm. 3.1.2], if we regard both as embedded in $\mathbf{C}^\flat$ via the embeddings of Theorem 2.1.6. To ease notation, from now on we write $\mathbf{E}'_K = X_K(K(\zeta_{p^\infty}))$, and $\mathbf{E}_K := (\mathbf{E}'_K)^\Delta$, so that $\mathbf{E}'_K/\mathbf{E}_K$ is a separable extension. We write $(\mathbf{E}'_K)^+$, $\mathbf{E}_K^+$ for the respective rings of integers. We write φ for the (p-power) Frobenius on E'_K, E_K, $(\mathbf{E}'_K)^+$ and $\mathbf{E}_K^+$.

We note now that $B'_{\mathbf{Q}_p} := \mathbf{A}'_{\mathbf{Q}_p}[1/p]$ is a discretely valued field admitting p as a uniformizer, with residue field $\mathbf{E}'_{\mathbf{Q}_p}$. There is then a unique finite unramified extension of $\mathbf{B}'_{\mathbf{Q}_p}$ contained in the field $W(\mathbf{C}^\flat)[1/p]$ with residue field $\mathbf{E}'_K$. We denote this extension by $\mathbf{B}'_K$; it is again a discretely valued field, admitting p as a uniformizer, and we let $\mathbf{A}'_K$ denote its ring of integers; equivalently, we have $\mathbf{A}'_K = \mathbf{B}'_K \cap W(\mathbf{C}^\flat)$. We see that $\mathbf{A}'_K$ is a discrete valuation ring, admitting p as a uniformizer, and that $\mathbf{A}'_K/p\mathbf{A}'_K = \mathbf{E}'_K$. There is a natural lift of the Frobenius φ from $\mathbf{E}'_K$ to $\mathbf{A}'_K$.

The action of $\widetilde{\Gamma}_K$ on $\mathbf{E}'_K = X_K(K(\zeta_{p^\infty}))$ induces an action of $\widetilde{\Gamma}_K$ on $\mathbf{A}'_K$, and we write $\mathbf{A}_K := (\mathbf{A}'_K)^\Delta$. This is the ring of integers in the discretely valued field $\mathbf{B}_K := \mathbf{A}_K[1/p] = (\mathbf{B}'_K)^\Delta$, and has residue field equal to $\mathbf{E}_K$. The actions of $\widetilde{\Gamma}_K$ on $\mathbf{A}'_K$ and of Γ_K on $\mathbf{A}_K$ commute with φ.

If we let T'_K denote a lift of a uniformizer of $\mathbf{E}'_K$, and let k'_∞ denote the residue field of $K(\zeta_{p^\infty})$, then there is an isomorphism $\widehat{W(k'_\infty)((T'_K))} \xrightarrow{\sim} \mathbf{A}'_K$. (Here the source denotes the p-adic completion.) Similarly if k_∞ denotes the residue field of K_{cyc}, and T_K is a lift of a uniformizer of $\mathbf{E}_K$, then there is an isomorphism $\widehat{W(k_\infty)((T_K))} \xrightarrow{\sim} \mathbf{A}_K$.

For a general extension $K/\mathbf{Q}_p$ it is hard to give an explicit formula for the actions of φ and $\widetilde{\Gamma}_K$ on $W(k'_\infty)((T'_K))^\wedge$, but in some of our arguments it is useful to reduce to a special case where we can use explicit formulae. If $K = K_0$ (that is, if $K/\mathbf{Q}_p$ is unramified) then we have such a description as follows: we have $\mathbf{E}'_{K_0} = k((\varepsilon - 1))$, and $\mathbf{A}'_{K_0} = W(k((\varepsilon - 1)))$, and we set $T'_{K_0} = [\varepsilon] - 1$, with the square brackets denoting the Teichmüller lift.

The actions of φ and $\gamma \in \widetilde{\Gamma}_{K_0}$ on $T'_{K_0} \in \mathbf{A}'_{K_0}$ are given by the explicit formulae

$$\varphi(T'_{K_0}) = (1 + T'_{K_0})^p - 1, \tag{2.1.10}$$

$$\gamma(1 + T'_{K_0}) = (1 + T'_{K_0})^{\chi(\gamma)}, \tag{2.1.11}$$

where $\chi\colon \widetilde{\Gamma}_{K_0} \to \mathbf{Z}_p^\times$ denotes the cyclotomic character. We set $(\mathbf{A}'_{K_0})^+ = W(k)[[T'_{K_0}]]$, which is visibly $(\varphi, \widetilde{\Gamma}_{K_0})$-stable. We set $T_{K_0} = \mathrm{tr}_{\mathbf{A}'_{K_0}/\mathbf{A}_{K_0}}(T'_{K_0})$ and $\mathbf{A}^+_{K_0} = W(k)[[T_{K_0}]]$; then we have $\mathbf{A}_{K_0} = W(k)((T_{K_0}))^\wedge$, and $\mathbf{A}^+_{K_0}$ is (φ, Γ_{K_0})-stable (if $p > 2$ this is [Fon90, Prop. A.3.2.3], and if $p = 2$ the same statements hold, as explained in [Her98, §1.1.2.1]). After possibly replacing T_{K_0} by $(T_{K_0} - \lambda)$ for some $\lambda \in pW(k)$, by [Ber14, Prop. 4.2, 4.3] we can and do assume that $\varphi(T_{K_0}) \in T_{K_0}\mathbf{A}^+_{K_0}$, and that $g(T_{K_0}) \in T_{K_0}\mathbf{A}^+_{K_0}$ for all $g \in \Gamma_{K_0}$.

2.1.12 Definition. We say that K is *basic* if it is contained in $K_0(\zeta_{p^\infty})$. (Although it won't play a role in what follows, we remark that this implies in particular that K is abelian over $\mathbf{Q}_p$.)

If K is basic then we have $K(\zeta_{p^\infty}) = K_0(\zeta_{p^\infty})$, so that $\mathbf{E}'_K = \mathbf{E}'_{K_0}$, and we take $\mathbf{A}'_K = \mathbf{A}'_{K_0}$, $T'_K = T'_{K_0}$, $(\mathbf{A}'_K)^+ = (\mathbf{A}'_{K_0})^+$, and similarly for $\mathbf{A}_K$, with the actions of $\widetilde{\Gamma}_K$ and Γ_K being the restrictions of the actions of $\widetilde{\Gamma}_{K_0}$ and Γ_{K_0}.

If K is not basic then, as explained above, we may still choose some T'_K so that $\mathbf{A}'_K = W(k'_\infty)((T'_K))^\wedge$, and some T_K so that $\mathbf{A}_K = W(k_\infty)((T_K))^\wedge$. Having done so, we set $(\mathbf{A}'_K)^+ = W(k'_\infty)[[T'_K]]$, $\mathbf{A}^+_K = W(k_\infty)[[T_K]]$, where the topology on $W(k_\infty)[[T]]$ is as usual the (p, T)-adic topology.

2.1.13 Remark. For an arbitrary K, it is not necessarily possible to choose T_K so that $\mathbf{A}^+_K$ is φ-stable and Γ_K-stable (and similarly for $(\mathbf{A}'_K)^+$). In fact, it is not even possible to choose T_K so that $\mathbf{A}_K^+$ is φ-stable. Indeed, by [Ber14, Prop. 4.2, 4.3], if $\mathbf{A}_K^+$ is φ-stable then it is also Γ_K-stable. As explained in [Her98, §1.1.2.2], it is possible to choose T_K so that $\mathbf{A}_K^+$ is φ and Γ_K-stable precisely when K is contained in an unramified extension of $K_0(\zeta_{p^\infty})$, i.e., when K is abelian over $\mathbf{Q}_p$.

2.1.14 The Kummer case

Set $\mathfrak{S} = W(k)[[u]]$, with a φ-semi-linear endomorphism φ determined by $\varphi(u) = u^p$, and let $\mathcal{O}_{\mathcal{E}}$ be the p-adic completion of $\mathfrak{S}[1/u]$. The extension $K_\infty = K(\pi^{1/p^\infty})$ of K is infinite and strictly APF (but not Galois). The choice of compatible system of p-power roots of π gives an element $\pi^{1/p^\infty} \in \mathcal{O}^\flat_{\widehat{K_\infty}}$, and there is a continuous φ-equivariant embedding

$$\mathfrak{S} \hookrightarrow W(\mathcal{O}^\flat_{\widehat{K_\infty}})$$

which lifts the morphism $X_K(K_\infty) \hookrightarrow (\widehat{K}_\infty)^\flat$, defined by sending $u \mapsto [\pi^{1/p^\infty}]$. This embedding extends to a continuous φ-equivariant embedding

$$\mathcal{O}_{\mathcal{E}} \hookrightarrow W((\widehat{K_\infty})^\flat).$$

2.1.15 Graded and rigid analytic techniques

In Appendix B we prove a number of results using rigid analysis and graded techniques, which we will apply in the next section in order to prove some basic facts about our coefficient rings. In order to apply these results to the various rings introduced above, we need to verify the hypotheses introduced in B.11, which are as follows. There is an Artinian local ring R, which in our present context we take to be the ring $\mathbf{Z}/p^a$. Then we work with a $\mathbf{Z}/p^a$-algebra C^+, and an element $u \in C^+$, satisfying the following properties:

(A) u is a regular element (i.e., a nonzero divisor) of C^+.
(B) C^+/u is a flat (equivalently, free) $\mathbf{Z}/p^a$-algebra.
(C) C^+/p is a rank 1 complete valuation ring, and the image of u in C^+/p (which is necessarily nonzero, by ((A))) is of positive valuation (i.e., lies in the maximal ideal of C^+/p).

The following lemma gives the examples of this construction that we will use in the next section.

2.1.16 Lemma. *The following pairs (C^+, u) satisfy axioms* (A)–(C) *above.*

1. $C^+ = W_a(\mathcal{O}_F^\flat)$, *where F is any perfectoid field, and u is any element of $W_a(\mathcal{O}_F^\flat)$ whose image in $\mathcal{O}_F^\flat$ is of positive valuation.*
2. $C^+ = \mathbf{A}_K^+/p^a$, $u = T_K$.
3. $C^+ = W_a(k)[[u]]$.

Proof. Axioms (A) and (C) are clear from the definitions in each case, so we need only verify that C^+/u is flat over $\mathbf{Z}/p^a$ in each case. This is immediate in cases (2) and (3), so we focus on case (1). Rather than checking (B) directly in this case, we instead check that $W(\mathcal{O}_F^\flat)/u$ is flat over $\mathbf{Z}_p$; base-changing to $\mathbf{Z}/p^a$ then gives us (B). For this, it suffices to observe that multiplication by u is injective on each term of the short exact sequence

$$0 \to W(\mathcal{O}_F^\flat) \xrightarrow{p\cdot} W(\mathcal{O}_F^\flat) \longrightarrow \mathcal{O}_F^\flat \to 0;$$

the snake lemma then shows that multiplication by p is injective on $W(\mathcal{O}_F^\flat)/u$, as required. $\square$

We also note the following lemma.

2.1.17 Lemma. *The actions of G_K on $W_a(\mathbf{C}^\flat)$, and of Γ_K on $\mathbf{A}_K/p^a$ and $\widetilde{\mathbf{A}}_K/p^a$, are continuous and bounded in the sense of Definition* B.32.

Proof. It suffices to show that the action of G_K on $W_a(\mathbf{C}^\flat)$ is continuous and bounded, as the other cases follow by restricting this action to the corresponding subgroup and subring. By Remark B.33, we may assume that $u=[v]$ for some $v \in \mathcal{O}_{\mathbf{C}}^\flat$ with positive valuation; since the action of G_K on $\mathcal{O}_{\mathbf{C}}^\flat$ preserves the valuation, we then have $G_K \cdot u^M W_a(\mathbf{C}^\flat) = u^M W_a(\mathbf{C}^\flat)$ for all $M \in \mathbf{Z}$, so the action is continuous and bounded. $\square$

2.2 COEFFICIENTS

Our main concern throughout this book will be families of étale (φ, Γ)-modules; an auxiliary role will also be played by families of (various flavors of) étale φ-modules. Such a family will be parameterized by an algebra of coefficients, typically denoted by A. Since we will work throughout in the context of formal algebraic stacks over Spf $\mathbf{Z}_p$ (or closely related contexts), we will always assume that our coefficient ring A is a p-adically complete $\mathbf{Z}_p$-algebra. Frequently, we will work modulo some fixed power p^a of p, and thus assume that A is actually a $\mathbf{Z}/p^a$-algebra (and sometimes we impose further conditions on A, such as that of being Noetherian, or even of finite type over $\mathbf{Z}/p^a$).

It is often convenient to introduce an auxiliary base ring for our coefficients, which we will take to be the ring of integers $\mathcal{O}$ in a finite extension E of $\mathbf{Q}_p$; in this case, we will let ϖ denote a uniformizer of $\mathcal{O}$, and A will be taken to be a p-adically complete (or, equivalently, a ϖ-adically complete) $\mathcal{O}$-algebra, or, quite frequently, an $\mathcal{O}/\varpi^a$-algebra (perhaps Noetherian, or even of finite type), for some power ϖ^a of ϖ.

For the moment, we put ourselves in the most general case; that is, we assume that A is a p-adically complete $\mathbf{Z}_p$-algebra, and define versions of the various rings considered in Section 2.1 "relative to A".

Let F be a perfectoid closed subfield of $\mathbf{C}$, and if $a \geq 1$, let v denote an element of the maximal ideal of $W_a(\mathcal{O}_F^\flat)$ whose image in $\mathcal{O}_F^\flat$ is nonzero. We then set

$$W_a(\mathcal{O}_F^\flat)_A = W_a(\mathcal{O}_F^\flat) \widehat{\otimes}_{\mathbf{Z}_p} A := \varprojlim_i \big(W_a(\mathcal{O}_F^\flat) \otimes_{\mathbf{Z}_p} A\big)/v^i$$

(so that the indicated completion is the v-adic completion). Note that any two choices of v induce the same topology on $W_a(\mathcal{O}_F^\flat) \otimes_{\mathbf{Z}_p} A$, so that $W_a(\mathcal{O}_F^\flat) \widehat{\otimes}_{\mathbf{Z}_p} A$ is well-defined independently of the choice of v. We then define

$$W_a(F^\flat)_A = W_a(F^\flat) \widehat{\otimes}_{\mathbf{Z}_p} A := W_a(\mathcal{O}_F^\flat)_A[1/v];$$

this ring is again well-defined independently of the choice of v.

There are natural reduction maps

$$W_a(\mathcal{O}_F^\flat) \widehat{\otimes}_{\mathbf{Z}_p} A \to W_b(\mathcal{O}_F^\flat) \widehat{\otimes}_{\mathbf{Z}_p} A$$

and

$$W_a(F^\flat) \widehat{\otimes}_{\mathbf{Z}_p} A \to W_b(F^\flat) \widehat{\otimes}_{\mathbf{Z}_p} A,$$

if $a \geq b$, so that we may define

$$W(\mathcal{O}_F^\flat)_A = W(\mathcal{O}_F^\flat) \widehat{\otimes}_{\mathbf{Z}_p} A := \varprojlim_a W_a(\mathcal{O}_F^\flat)_A,$$

and similarly

$$W(F^\flat)_A = W(F^\flat) \widehat{\otimes}_{\mathbf{Z}_p} A := \varprojlim_a W_a(F^\flat)_A.$$

For certain choices of F, we introduce alternative notation for the preceding constructions. In the case when F is equal to $\mathbf{C}$ itself, we write

$$\mathbf{A}_{\mathrm{inf},A} := W(\mathcal{O}_{\mathbf{C}}^\flat)_A.$$

In the case when F is equal to $\widehat{K}_{\mathrm{cyc}}$, we write

$$\widetilde{\mathbf{A}}_{K,A} := W\big((\widehat{K}_{\mathrm{cyc}})^\flat\big)_A.$$

In the case when F is equal to $\widehat{K}_\infty$, we write

$$\widetilde{\mathcal{O}}_{\mathcal{E},A} := W\big((\widehat{K}_\infty)^\flat\big)_A.$$

We next introduce various imperfect coefficient rings with A-coefficients: If T_K denotes a lift to $\mathbf{A}_K$ of a uniformizer of $\mathbf{E}_K$, defining a subring $\mathbf{A}_K^+$ of $\mathbf{A}_K$

as in Section 2.1, then we write

$$\mathbf{A}_{K,A}^{+} = \mathbf{A}_K^{+} \widehat{\otimes}_{\mathbf{Z}_p} A := \varprojlim_n \mathbf{A}_K^{+}/(p, T_K)^n \otimes_{\mathbf{Z}_p} A = \varprojlim_m (\varprojlim_n \mathbf{A}_K^{+}/(p^m, T_K^n) \otimes_{\mathbf{Z}_p} A),$$

and

$$\mathbf{A}_{K,A} = \mathbf{A}_K \widehat{\otimes}_{\mathbf{Z}_p} A := \varprojlim_m \big(\varprojlim_n (\mathbf{A}_K^{+}/(p^m, T_K^n) \otimes_{\mathbf{Z}_p} A)[1/T_K]\big).$$

Similarly, if u denotes the usual element of $\mathfrak{S}$, whose reduction modulo p is a uniformizer of $X_K(K_\infty)$, then we write

$$\mathfrak{S}_A = \mathfrak{S} \widehat{\otimes}_{\mathbf{Z}_p} A := \varprojlim_n \mathfrak{S}/(p, u)^n \otimes_{\mathbf{Z}_p} A = \varprojlim_m (\varprojlim_n \mathfrak{S}/(p^m, u^n) \otimes_{\mathbf{Z}_p} A),$$

and

$$\mathcal{O}_{\mathcal{E},A} = \mathcal{O}_{\mathcal{E}} \widehat{\otimes}_{\mathbf{Z}_p} A := \varprojlim_m \big(\varprojlim_n (\mathfrak{S}/(p^m, u^n) \otimes_{\mathbf{Z}_p} A)[1/u]\big).$$

Note that by definition the ring $\mathfrak{S}_A$ is a power series ring over $W(k) \otimes_A A$, and $\mathcal{O}_{\mathcal{E},A}$ is the p-adic completion of the corresponding Laurent series ring; and similarly for the rings $\mathbf{A}_{K,A}^{+}$ and $\mathbf{A}_{K,A}$.

2.2.1 A digression on flatness and completion

We record some results giving sufficient conditions for flatness to be preserved after passage to certain inverse limits, beginning with the following results from [FGK11].

2.2.2 Proposition. *Let R be a ring which is x-adically complete for some $x \in R$, and let S be an R-algebra which is also x-adically complete, and for which $R[1/x]$ and $S[1/x]$ are both Noetherian. If for each $k \geq 1$, the induced morphism $R/x^k \to S/x^k S$ is (faithfully) flat, then the morphism $R \to S$ is (faithfully) flat.*

Proof. The flatness claim is a special case of the discussion at the beginning of [FGK11, §5.2]. The faithful flatness claim is a special case of [FGK11, Prop. 5.2.1 (2)]. It also follows from the flatness claim together with Lemma 2.2.9 below. □

2.2.3 Remark. Under the faithful flatness hypothesis of the preceding proposition, it is proved in [FGK11, Prop. 5.2.1 (1)] that Noetherianness of $R[1/x]$ is a consequence of Noetherianness of $S[1/x]$. This is a kind of *fpqc* descent result for this property, which however we won't need in the present book.

We next present the following variation on a result of Bhatt–Morrow–Scholze.

2.2.4 Proposition. *Let R be a Noetherian ring which is x-adically complete, for some element $x \in R$, let S be an x-adically complete R-algebra, and suppose that S/x^n is (faithfully) flat over R/x^n for every $n \geq 1$. Then S is (faithfully) flat over R.*

2.2.5 Remark. In [BMS18, Rem. 4.31], the authors prove the above proposition in the particular case when R and S are flat $\mathbf{Z}_p$-algebras, and x equals p. In this case they only need to assume that S/p is flat over R/p. (Since R and S are both flat over $\mathbf{Z}_p$, it follows automatically that each S/p^n is flat over R/p^n.) Their proof makes use of the following key ingredients: that for a p-adically complete ring, any pseudo-coherent complex is derived complete; that for a complex over a p-torsion-free ring, the derived p-adic completion may be computed "naively"; and the Artin–Rees lemma for the p-adically complete and Noetherian ring R.

Our argument is identical to theirs, once we confirm that these ingredients remain available (with p-adic completions being replaced by x-adic completions). The first statement holds quite generally [Sta, Tag 0A05], and of course Artin–Rees holds for any Noetherian ring R. Thus the main point is to verify that we may compute derived x-adic completions for the (generally non-Noetherian ring) S "naively". By [Sta, Tag 0923], this is possible provided that the x-power torsion in S is bounded. So our task is to verify this boundedness (under our given hypotheses).

We begin with the following general criterion for a ring to have bounded torsion.

2.2.6 Lemma. *If x is an element of the ring R, then the following are equivalent:*

1. *$R[x^m] = R[x^\infty]$*
2. *The morphism $R[x^i] \to R/x^m$ is injective for some $i \geq 1$.*
3. *The morphism $R[x^i] \to R/x^m$ is injective for every $i \geq 1$.*

Proof. This is straightforward. Indeed, if the morphism $R[x^i] \to R/x^m$ is injective for some $i \geq 1$, then it is in particular injective if $i = 1$. Suppose then that injectivity holds for $i = 1$. If $y \in R$ is such that $x^{m+n}y = 0$ for some $n \geq 1$, then $x^{m+n-1}y$ is in the kernel of the morphism $R[x] \to R/x^m$, so $x^{m+n-1}y = 0$, and by an easy induction we have $x^m y = 0$, as required.

Conversely, if t is in the kernel of the morphism $R[x^i] \to R/x^m$ for some i, then we can write $t = x^m y$, and we have $x^i t = 0$; so $x^{m+i}y = 0$. If $R[x^m] = R[x^\infty]$, then we have $x^m y = 0$, so $t = 0$, as required. □

2.2.7 Lemma. *If x is any element of the ring R, then for any $i, n \geq 1$ we have the exact sequence*

$$R[x^i] \to R/x^n \xrightarrow{x^i} R/x^{n+i} \to R/x^i \to 0.$$

Proof. This is immediately verified. □

2.2.8 Lemma. *If x is an element of the ring R, and if $R[x^m]=R[x^\infty]$, then*

$$0 \to R[x^i] \to R/x^n \xrightarrow{x^i} R/x^{n+i} \to R/x^i \to 0$$

is an exact sequence, for any $n \geq m$ and any $i \geq 1$.

Proof. This follows from Lemmas 2.2.6 and 2.2.7. □

Proof of Proposition 2.2.4. Since R is Noetherian, we see that $R[x^m]=R[x^\infty]$ for some $m \geq 0$. Thus, by Lemma 2.2.8, for each $n \geq m$ and each $i \geq 1$, we obtain an exact sequence

$$0 \to R[x^i] \to R/x^n \xrightarrow{x^i} R/x^{n+i} \to R/x^i \to 0.$$

Tensoring with the flat R/x^{n+i}-algebra S/x^{n+i}, we obtain an exact sequence

$$0 \to S \otimes_R R[x^i] \to S/x^n \xrightarrow{x^i} S/x^{n+i} \to S/x^i \to 0.$$

Passing to the inverse limit over n (and taking into account that all the transition morphisms are surjective), we obtain an exact sequence of R/x^{n+i}-modules

$$0 \to S \otimes_R R[x^i] \to S \xrightarrow{x^i} S \to S/x^i \to 0.$$

In particular, we find that $S \otimes_R R[x^i] \xrightarrow{\sim} S[x^i]$, for every i, and so in particular $S[x^m]=S[x^\infty]$, so that S has bounded x-power torsion.

We now follow the proof of [BMS18, Rem. 4.31]. By (for example) [Sta, Tag 00M5], it is enough to show that if M is a finitely generated R-module, then $M \otimes^{\mathbf{L}}_R S$ has cohomology concentrated in degree 0. Since R is Noetherian, [Sta, Tag 066E] shows that M is pseudo-coherent when regarded as a complex of R-modules in a single degree, and so $M \otimes^{\mathbf{L}}_R S$ is a pseudo-coherent complex of S-modules, by [Sta, Tag 0650]. Also, since R and S are x-adically complete, it then follows from [Sta, Tag 0A05] that both M and $M \otimes^{\mathbf{L}}_R S$ are derived x-adically complete.

Since S has bounded x-power torsion, it follows from [Sta, Tag 0923] that we may compute derived x-adic completions naively, i.e., via $R\lim(- \otimes_S S/x^n)$. Thus

$$M \otimes^{\mathbf{L}}_R S \xrightarrow{\sim} R\lim(M \otimes^{\mathbf{L}}_R S/x^n) \xrightarrow{\sim} R\lim\big((M \otimes^{\mathbf{L}}_R R/x^n) \otimes^{\mathbf{L}}_{R/x^n} S/x^n\big).$$

Artin–Rees allows us to replace the pro-system $\{M \otimes^{\mathbf{L}}_R R/x^n\}$ with the pro-system $\{M/x^n\}$, and so we find that in fact

$$M \otimes^{\mathbf{L}}_R S \xrightarrow{\sim} R\lim(M/x^n \otimes^{\mathbf{L}}_{R/x^n} S/x^n) \\ \xrightarrow{\sim} R\lim(M \otimes_R S/x^n) \xrightarrow{\sim} \lim M \otimes_R S/x^n,$$

the penultimate isomorphism holding since S/x^n is flat over R/x^n, and the final isomorphism following from the fact that the transition morphisms in the pro-system $\{M \otimes_R S/x^n\}$ are surjective. Thus indeed $M \otimes^{\mathbf{L}}_R S$ has cohomology supported in a single degree, as required.

The claim about faithful flatness follows from the immediately following Lemma 2.2.9. □

2.2.9 Lemma. *If $R \to S$ is a flat morphism of rings, if I is an ideal in R for which R is I-adically complete, and if the morphism $R/I \to S/I$ is faithfully flat, then the morphism $R \to S$ is faithfully flat.*

Proof. Since R is I-adically complete, we see that I is contained in the Jacobson radical of R. Thus $\operatorname{Max Spec} R/I = \operatorname{Max Spec} R$, and so our assumption that $R/I \to S/I$ is faithfully flat shows that $\operatorname{Max Spec} R$ is contained in the image of $\operatorname{Spec} S/I \subseteq \operatorname{Spec} S$ in $\operatorname{Spec} R$. Since flat morphisms satisfy going down, we find that in fact $\operatorname{Spec} S \to \operatorname{Spec} R$ is surjective, as claimed. □

We also note the following lemma.

2.2.10 Lemma. *If $F \hookrightarrow F'$ is an inclusion of perfectoid fields in characteristic p, then the induced morphism $W_a(\mathcal{O}_F) \to W_a(\mathcal{O}_{F'})$ is faithfully flat, for any $a \geq 1$.*

Proof. Since $\mathcal{O}_F$ is a Bézout ring, being flat over $\mathcal{O}_F$ is the same as being torsion free, and so the inclusion $\mathcal{O}_F \hookrightarrow \mathcal{O}_{F'}$ is a flat morphism, and thus even faithfully flat, being a local morphism of local rings. A standard grading argument, applying Lemma B.8 to the p-adic filtrations on source and target, then shows that the inclusion $W_a(\mathcal{O}_F) \hookrightarrow W_a(\mathcal{O}_{F'})$ is faithfully flat, for each $a \geq 1$. □

2.2.11 Back to coefficient rings

We now return to our discussion of coefficient rings, and record that, at least if A is a finite type $\mathbf{Z}/p^a\mathbf{Z}$-algebra, the various natural maps between these rings are in fact (faithfully) flat injections. We also show that the various maps of coefficient rings induced by a (faithfully) flat morphism of finite type $\mathbf{Z}/p^a$-algebras are again (faithfully) flat.

2.2.12 Proposition. *Suppose that $A \to B$ is a flat homomorphism of finite type $\mathbf{Z}/p^a\mathbf{Z}$-algebras for some $a \geq 1$. Then all the maps in the following diagram* *are flat. Furthermore the vertical arrows are all injections, while the horizontal arrows are all faithfully flat (and so in particular also injections). If $A \to B$ is furthermore faithfully flat, then the same is true of the diagonal arrows.*

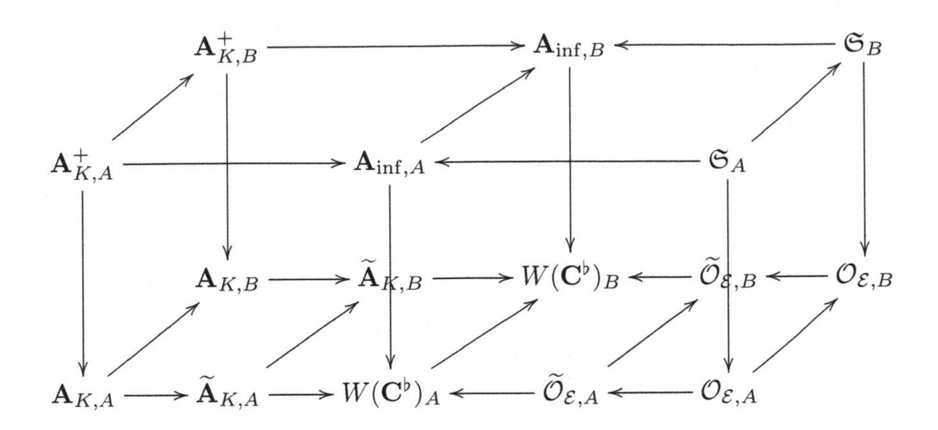

Proof. From left to right, the vertical maps are given by inverting the elements T, v, and u respectively (where as above, v is any element of the maximal ideal of $W_a(\mathcal{O}^\flat_{\mathbf{C}})$, whose image in $\mathcal{O}^\flat$ is nonzero; in particular, we can take v equal to either T or u). Since localizations are flat, to prove the proposition for these maps it is enough to note that the sources of the vertical maps are respectively T-, v-, and u-torsion free, by Lemma B.19.

We now turn to the horizontal maps, where we will make repeated use of Proposition 2.2.2. It evidently suffices to treat the maps with A-coefficients. Since each of $\mathbf{A}_{K,A}$, $\widetilde{\mathbf{A}}_{K,A}$, $W(\mathbf{C}^\flat))_A$, $\widetilde{\mathcal{O}}_{\mathcal{E},A}$, and $\mathcal{O}_{\mathcal{E},A}$ is Noetherian by Proposition B.36 (1), it follows from Proposition 2.2.2 (and the flatness of the vertical maps) that we need only show that the maps $\mathfrak{S}_A/u^i \to (W_a(\mathcal{O}^\flat_{\widehat{K_\infty}}) \otimes_{\mathbf{Z}_p} A)/u^i \to \mathbf{A}_{\mathrm{inf},A}/u^i$ and $\mathbf{A}^+_{K,A}/T^i \to (W_a(\mathcal{O}^\flat_{\widehat{K_{\mathrm{cyc}}}}) \otimes_{\mathbf{Z}_p} A)/T^i \to \mathbf{A}_{\mathrm{inf},A}/T^i$ are faithfully flat for each $i \geq 1$.

For the faithful flatness of $\mathfrak{S}_A/u^i \to (W_a(\mathcal{O}^\flat_{\widehat{K_\infty}}) \otimes_{\mathbf{Z}_p} A)/u^i$, it is enough to show that $\mathfrak{S}/p^a \to W_a(\mathcal{O}^\flat_{\widehat{K_\infty}})$ is faithfully flat. By a standard grading argument (see Lemma B.8), it suffices in turn to prove that $k[[u]] \to \mathcal{O}^\flat_{\widehat{K_\infty}}$ is faithfully flat, so in turn it is enough to check that $\mathcal{O}^\flat_{\widehat{K_\infty}}$ is u-torsion free, which is clear. To see that $(W_a(\mathcal{O}^\flat_{\widehat{K_\infty}}) \otimes_{\mathbf{Z}_p} A)/u^i \to \mathbf{A}_{\mathrm{inf},A}/u^i$ is faithfully flat, it suffices to note that $W_a(\mathcal{O}^\flat_{\widehat{K_\infty}}) \to W_a(\mathcal{O}^\flat_{\mathbf{C}})$ is faithfully flat, by Lemma 2.2.10. The faithful flatness of the remaining horizontal maps is proved in exactly the same way.

Finally, the (faithful) flatness of the diagonal maps is now immediate from another application of Proposition 2.2.2, together with the (faithful) flatness of $A \to B$. ☐

Recall that if A° is a p-adically complete $\mathcal{O}$-algebra, then A° is said to be *topologically of finite type over* $\mathcal{O}$ if it can be written as a quotient of a restricted formal power series ring in finitely many variables $\mathcal{O}\langle\langle X_1, \ldots, X_n\rangle\rangle$; equivalently, if and only if $A^\circ \otimes_{\mathcal{O}} k$ is a finite type k-algebra ([FK18, §0, Prop. 8.4.2]). In particular, if A° is a $\mathcal{O}/\varpi^a$-algebra for some $a \geq 1$, then A° is topologically of finite type over $\mathcal{O}$ if and only if it is of finite type over $\mathcal{O}/\varpi^a$. Since $\mathcal{O}$ is finite over $\mathbf{Z}_p$, it is equivalent for A to be topologically of finite type over $\mathcal{O}$ or over $\mathbf{Z}_p$, and so for the moment we consider the maximally general case of a topologically finite type $\mathbf{Z}_p$-algebra.

2.2.13 Remark. We don't know if the analogue of Proposition 2.2.12 holds when A and B are taken to be merely p-adically complete and topologically of finite type over $\mathbf{Z}_p$, rather than of finite type over $\mathbf{Z}/p^a$ for some $a \geq 1$. Each of the various coefficient rings over A and B is (by definition) formed by first forming the corresponding coefficient ring over each A/p^a or B/p^a, and then taking an inverse limit. Since the formation of inverse limits is left exact, we see that the horizontal and vertical arrows in the diagram are injective, but we don't know in general that the various arrows are flat (although we have no reason to doubt it). One can however establish some partial results, using Proposition 2.2.4. We record here one such result, which we will need later on.

2.2.14 Proposition. *If A is a p-adically complete $\mathbf{Z}_p$-algebra which is topologically of finite type, then the natural morphism $\mathfrak{S}_A \to \mathbf{A}_{\mathrm{inf},A}$ is faithfully flat.*

Proof. This follows from Propositions 2.2.4 and 2.2.12, once we note that $\mathfrak{S}_A$ is Noetherian when A is p-adically complete and topologically of finite type; indeed, since A is in particular Noetherian, so is the power series ring $W(k) \otimes_{\mathbf{Z}_p} A[[u]]$. □

We conclude this initial discussion of coefficient rings by explaining how the action of φ on the various rings $\mathfrak{S}$, $\mathbf{A}_{\mathrm{inf}}$ and so on extends to a continuous action on the corresponding rings $\mathfrak{S}_A$, $\mathbf{A}_{\mathrm{inf},A}$, etc., and similarly for the various Galois actions. For this, it is convenient for us to briefly digress, and to introduce the following situation, which will also be useful for us in Chapter 3.

2.2.15 Situation. Fix a finite extension $k/\mathbf{F}_p$ and write $\mathbf{A}^+ := W(k)[[T]]$. Write $\mathbf{A}$ for the p-adic completion of $\mathbf{A}^+[1/T]$.

If A is a p-adically complete $\mathbf{Z}_p$-algebra, we write $\mathbf{A}_A^+ := (W(k) \otimes_{\mathbf{Z}_p} A)[[T]]$; we equip $\mathbf{A}_A^+$ with its (p, T)-adic topology, so that it is a topological A-algebra (where A has the p-adic topology). Let $\mathbf{A}_A$ be the p-adic completion of $\mathbf{A}_A^+[1/T]$, which we regard as a topological A-algebra by declaring $\mathbf{A}_A^+$ to be an open subalgebra. Note that the formation of $\mathfrak{S}_A$, $\mathcal{O}_{\mathcal{E},A}$, $\mathbf{A}_{K,A}^+$ and $\mathbf{A}_{K,A}$ above are particular instances of this construction.

Let φ be a ring endomorphism of $\mathbf{A}$ which is congruent to the (p-power) Frobenius endomorphism modulo p. We say that $\mathbf{A}^+$ is φ-stable if $\varphi(\mathbf{A}^+) \subseteq \mathbf{A}^+$.

By [EG21, Lem. 5.2.2 and 5.2.5], if $\mathbf{A}^+$ is φ-stable, then φ is faithfully flat, and induces the usual Frobenius on $W(k)$; the same arguments show that this is true for φ on $\mathbf{A}$, even if $\mathbf{A}^+$ is not φ-stable.

We have the following variant of [EG21, Lem. 5.2.3].

2.2.16 Lemma. *Suppose that we are in Situation 2.2.15. Then for each integer $a \geq 1$ there is an integer $C \geq 0$ such that for all $n \geq 1$, we have $\varphi(T^n) \in T^{pn-C}\mathbf{A}^+ + p^a\mathbf{A}$. In particular, the action of φ on $\mathbf{A}$ is continuous.*

Proof. For some $h \geq 0$ we have $\varphi(T) \in T^{-h}\mathbf{A}^+ + p^a\mathbf{A}$. Write $\varphi(T) = T^p + pY$, so that $\varphi(T^n) = (T^p + pY)^n$. If we expand using the binomial theorem, then every term on the right-hand side is either divisible by p^a, or is a multiple of $T^{p(n-r)}Y^r$ for some $0 \leq r \leq a-1$. It follows that we can take $C = (a-1)(p+h)$. □

2.2.17 Lemma. *If A is a p-adically complete $\mathbf{Z}_p$-algebra, then in Situation 2.2.15, the endomorphism φ of $\mathbf{A}$ extends uniquely to an A-linear continuous endomorphism of $\mathbf{A}_A$ (which we continue to denote by φ). If $\mathbf{A}^+$ is φ-stable, then $\mathbf{A}_A^+$ is φ-stable for all A.*

Proof. Since φ is $\mathbf{Z}_p$-linear by definition, and since the topologies on $\mathbf{A}_A^+$ and $\mathbf{A}_A$ are defined by passage to the limit modulo p^a as $a \to \infty$, we are immediately reduced to the case that A is a $\mathbf{Z}/p^a\mathbf{Z}$-algebra. In this case the result is immediate from Lemma 2.2.16 and Lemma B.31. □

2.2.18 Lemma. *Let A be a p-adically complete $\mathbf{Z}_p$-algebra.*

1. *The endomorphism φ of $\mathfrak{S}$ (resp. $\mathcal{O}_{\mathcal{E}}$, $\mathbf{A}_K$, $W(\mathcal{O}_F^\flat)$, $W(F^\flat)$) extends uniquely to an A-linear continuous endomorphism of $\mathfrak{S}_A$ (resp. $\mathcal{O}_{\mathcal{E},A}$, $\mathbf{A}_{K,A}$, $W(\mathcal{O}_F^\flat)$, $W(F^\flat)_A$), which we again denote by φ. If K is furthermore basic, then the endomorphism φ of $\mathring{\mathbf{A}}_{K,A}$ preserves $\mathring{\mathbf{A}}_{K,A}^+$.*
2. *The continuous G_K-action on $W(\mathcal{O}_{\mathbf{C}}^\flat)$ and $W(\mathbf{C}^\flat)$ (resp. the continuous Γ_K-action on $\mathbf{A}_K$ and $\widetilde{\mathbf{A}}_K$) extends to a continuous A-linear action of G_K on $W(\mathcal{O}_{\mathbf{C}}^\flat)_A$ and $W(\mathbf{C}^\flat)_A$ (resp. of Γ_K on $\mathbf{A}_{K,A}$ and $\widetilde{\mathbf{A}}_{K,A}$).*

Proof. As in the proof of Lemma 2.2.17, we can immediately reduce to the case that A is a $\mathbf{Z}/p^a$-algebra. Part (1) then follows from Lemma B.31 as in the proof of Lemma 2.2.17, while part (2) follows from Lemma B.34, bearing in mind Lemmas 2.1.16 and 2.1.17. □

For our final result of this section, we compute the φ-invariants in $W(\mathbf{C}^\flat)_A$.

2.2.19 Lemma. *If A is any p-adically complete $\mathbf{Z}_p$-algebra, then for each $a \geq 1$, the natural morphism $A/p^a \to W_a(\mathbf{C}^\flat)_A$ induces an identification $A/p^a =$*

$\left(W_a(\mathbf{C}^\flat)_A\right)^{\varphi=1}$, *and (consequently) the natural morphism* $A \to W(\mathbf{C}^\flat)_A$ *induces an identification* $A = \left(W(\mathbf{C}^\flat)_A\right)^{\varphi=1}$.

Proof. Since $W(\mathbf{C}^\flat)_A = \varprojlim_a W_a(\mathbf{C}^\flat)_A$, we see that

$$\left(W(\mathbf{C}^\flat)_A\right)^{\varphi=1} = \varprojlim \left(W_a(\mathbf{C}^\flat)_A\right)^{\varphi=1},$$

and thus (as the phrasing of the lemma indicates), the claim for $W(\mathbf{C}^\flat)_A$ follows from the claims for each $W_a(\mathbf{C}^\flat)_A$. Thus for the remainder of the proof, we may and do assume that A is a $\mathbf{Z}_p/p^a$-module for some $a \geq 1$, and prove that $A = \left(W_a(\mathbf{C}^\flat)_A\right)^{\varphi=1}$. In fact, for any $\mathbf{Z}/p^a$-module M, we may define

$$W_a(\mathbf{C}^\flat)_M := \left(\varprojlim_n \left(W_a(\mathcal{O}_{\mathbf{C}}^\flat)/T^n\right) \otimes_{\mathbf{Z}/p^a} M\right)[1/T]$$

(this extends our preceding definition for $\mathbf{Z}/p^a$-algebras), and we will prove that $M = \left(W_a(\mathbf{C}^\flat)_M\right)^{\varphi=1}$. (In fact this statement for modules, which obviously implies the statement for algebras, is actually equivalent to it, as one sees by applying the statement for algebras to $\mathbf{Z}_p/p^a$-algebras of the form $\mathbf{Z}_p/p^a \oplus M$, where M is an arbitrary $\mathbf{Z}_p/p^a$-module equipped with the square zero multiplication.)

Suppose first that $a = 1$, so that M is simply an $\mathbf{F}_p$-vector space.. If we choose a basis for M, i.e., an isomorphism $\mathbf{F}_p^{\oplus I} \xrightarrow{\sim} M$, then $(\mathcal{O}_{\mathcal{C}}^\flat)_M = \prod'_{i \in I} \mathcal{O}_{\mathcal{C}}^\flat$, where the prime indicates we consider the set of tuples $(x_i)_{i \in I}$ for which $\lim_i x_i = 0$, where the limit is taken with respect to the filter of cofinite subsets of I. Thus we have, correspondingly, that $\mathcal{C}_M^\flat = \prod'_{i \in I} \mathcal{C}^\flat$ (where the prime has the same meaning). Now $(\mathcal{C}^\flat)^{\varphi=1} = \mathbf{F}_p$, and so

$$(\mathcal{C}_M^\flat)^{\varphi=1} = \prod_{i \in I}{}'(\mathcal{C}^\flat)^{\varphi=1} = \oplus_{i \in I} \mathbf{F}_p = M.$$

This establishes the case $a = 1$.

Now consider the case of general a. An application of Lemma B.20 (1) (taking $R = \mathbf{Z}/p^a$ and $C^+ = W_a(\mathcal{O}_{\mathbf{C}^\flat})$, and considering the exact sequence of $\mathbf{Z}/p^a$-modules $0 \to pM \to M \to M/pM \to 0$; see also Lemma 2.1.16) shows that the sequence

$$0 \to W_{a-1}(\mathcal{C}^\flat)_{pM} = W_a(\mathcal{C}^\flat)_{pM} \to W_a(\mathcal{C}^\flat)_M \to W_a(\mathcal{C}^\flat)_{M/pM} = \mathcal{C}_{M/pM}^\flat \to 0$$

is short exact. Suppose that $x \in \left(W_a(\mathbf{C}^\flat)_M\right)^{\varphi=1}$. The case $a = 1$ already proved then shows that we may find $m \in M$ so that $y := x - m$ lies in $W_{a-1}(\mathcal{C}^\flat)_{pM}$.

Of course $\varphi(y) = y$, and so, by induction on a, we find that $y \in pM$. Thus $x = m + y \in M$, as claimed. □

We will also use the following variant on Lemma 2.2.19. Suppose that A is a complete local Noetherian $\mathcal{O}$-algebra with finite residue field, and write $\widehat{W(\mathbf{C}^\flat)}_A$ for the $\mathfrak{m}_A$-adic completion of $W(\mathbf{C}^\flat)_A$.

2.2.20 Lemma. *Let A be a complete local Noetherian $\mathcal{O}$-algebra with finite residue field. Then $(\widehat{W(\mathbf{C}^\flat)}_A)^{\varphi=1} = A$.*

Proof. Since $\widehat{W(\mathbf{C}^\flat)}_A = \varprojlim_n W(\mathbf{C}^\flat)_{A/\mathfrak{m}_A^n}$, we have

$$\left(\widehat{W(\mathbf{C}^\flat)}_A\right)^{\varphi=1} = \varprojlim_n \left(W(\mathbf{C}^\flat)_{A/\mathfrak{m}_A^n}\right)^{\varphi=1},$$

so the result follows from Lemma 2.2.19 (applied to each $A/\mathfrak{m}_A^n$). □

2.3 ALMOST GALOIS DESCENT FOR PROFINITE GROUP ACTIONS

We will be interested in descent results for profinite group actions, and in this section we establish the key result that we will need. Our setup is slightly elaborate, but accords with the situations that will arise in practice. We begin with the case of finite group actions.

Suppose that R is a ring, and that I is an ideal in R such that $I^2 = I$. Assume further that $I \otimes_R I$ is a flat R-module; by [GR03, Prop. 2.1.7], this holds if I is a filtered union of principal ideals. In particular, these assumptions hold if R is a valuation ring with a non-discrete rank 1 valuation, and I is the maximal ideal of R.

The full subcategory of the category of R-modules whose objects are the modules annihilated by I is a Serre subcategory, and so we can form the quotient category of *almost* R-modules.

Suppose that S is an R-algebra equipped with an action of a finite group G by R-algebra automorphisms (i.e., the structural morphism $R \to S$ is equivariant with respect to the given G-action on S, and the trivial G-action on R).

2.3.1 Definition. We say that the morphism $R \to S$ makes S an *almost Galois extension of R, with Galois group G*, if the natural G-equivariant and S-linear morphism

$$S \otimes_R S \to \prod_{g \in G} S$$

(here G acts on the source through its action on the second factor, and on the target by permuting the factors, while S acts on the source through its action on the first factor and on the target through its action on each factor) defined by $s_1 \otimes s_2 \mapsto \big(s_1 g(s_2)\big)_{g\in G}$ is an *almost isomorphism* (i.e., induces an isomorphism in the category of almost R-modules).

2.3.2 Remark. Suppose that $R \to R'$ is a morphism, and define $I' = IR'$, so that $I'^2 = I'$, allowing us to also define the category of almost R'-modules. If $R \to S$ is almost Galois with Galois group G, then evidently $R' \to S' := R' \otimes_R S$ is almost Galois with Galois group G.

Conversely, if $R \to R'$ is faithfully flat, and $R' \to S'$ is almost Galois with Galois group G, then so is $R \to S$; indeed we may write $R' \otimes_R (S \otimes_R S) = (R' \otimes_R S) \otimes_{R'} (R' \otimes_R S) = S' \otimes_{R'} S'$.

2.3.3 Lemma. *Suppose that $R \to S$ makes S an almost Galois extension of S with Galois group G, and furthermore that $R \to S$ is faithfully flat. Then the morphism $R \to S^G$ is an almost isomorphism.*

Proof. Since $R \to S$ is faithfully flat, it suffices to verify that the induced morphism

$$S \to (S \otimes_R S)^G \tag{2.3.4}$$

is an almost isomorphism. By assumption the morphism $S \otimes_R S \to \prod_{g\in G} S$ is an almost isomorphism. One easily verifies that the induced morphism on invariant subrings

$$(S \otimes_R S)^G \to (\prod_{g\in G} S)^G = S$$

is then also an almost isomorphism. The composite of this map with the morphism (2.3.4) is just the identity, and so we find that (2.3.4) is indeed an almost isomorphism. □

2.3.5 Lemma. *Suppose that $R \to S$ makes S an almost Galois extension of S with Galois group G, and furthermore that $R \to S$ is faithfully flat. Then if M is an S-module equipped with a semi-linear G-action, the induced morphism $S \otimes_R M^G \to M$ is an almost isomorphism of S-modules.*

Proof. The semi-linear G-action on M can be reinterpreted as an isomorphism

$$(\prod_{g\in G} S) \otimes_S M \cong M \otimes_S (\prod_{g\in G} S)$$

(where the tensor product on the left-hand side is twisted by the G-action) of $S \otimes_R S$-modules, and hence as an almost isomorphism

$$S \otimes_R M \overset{\text{almost}}{\cong} M \otimes_R S$$

of $S \otimes_R S$-modules. The claim of the lemma now follows from faithfully flat descent in the almost category (for which see [GR03, §3.4.1]; it is in order to make this citation that we have assumed that $I \otimes_R I$ is R-flat). □

The following lemma explains our interest in almost Galois extensions.

2.3.6 Lemma. *If $F \subseteq F'$ is a finite Galois extension of perfectoid fields, with Galois group G, then the corresponding inclusion of rings of integers $\mathcal{O}_F \subseteq \mathcal{O}_{F'}$ realizes $\mathcal{O}_{F'}$ as an almost Galois extension of $\mathcal{O}_F$, with Galois group G (the "almost" structure being understood with respect to the maximal ideal of $\mathcal{O}_F$).*

Proof. There are various more or less concrete ways to see this. For example, [Sch12, Prop. 5.23] shows that the morphism $\mathcal{O}_F \to \mathcal{O}_{F'}$ is almost étale, which by definition [Sch12, Def. 4.12] means that the surjection of $\mathcal{O}_{F'}$-algebras $\mathcal{O}_{F'} \otimes_{\mathcal{O}_F} \mathcal{O}_{F'} \to \mathcal{O}_{F'}$ may be "almost" split, so that $\mathcal{O}_{F'}$ is almost a direct factor of the source. Permuting such a splitting under the action of the Galois group G then allows one to show that the morphism $\mathcal{O}_{F'} \otimes_{\mathcal{O}_F} \mathcal{O}_{F'} \to \prod_{g \in G} \mathcal{O}_{F'}$ is an almost isomorphism.

However, a more direct (but less explicit) way to prove the lemma is to use the first equivalence of categories in [Sch12, Thm. 5.2], namely the equivalence between the category of perfectoid F-algebras and the category of perfectoid almost algebras over $\mathcal{O}_F$. Under this equivalence, the morphism $\mathcal{O}_{F'} \otimes_{\mathcal{O}_F} \mathcal{O}_{F'} \to \prod_{g \in G} \mathcal{O}_{F'}$ is taken to the morphism $F' \otimes_F F' \to \prod_{g \in G} F'$. This latter morphism *is* an isomorphism, since F' is Galois over F with Galois group G, and so we conclude that the former morphism is an almost isomorphism, as required. □

In the case of perfectoid fields of characteristic p, we may extend the statement of the preceding lemma to the context of truncated rings of Witt vectors.

2.3.7 Lemma. *If $F \subseteq F'$ is a finite Galois extension of perfectoid fields in characteristic p, with Galois group G, then for each $a \geq 1$, the inclusion $W_a(\mathcal{O}_F) \hookrightarrow W_a(\mathcal{O}_{F'})$ is almost Galois, with Galois group G (the "almost" structure being understood with respect to the maximal ideal of $W_a(\mathcal{O}_F)$).*

Proof. For simplicity of notation, write $R := W_a(\mathcal{O}_F)$ and $S := W_a(\mathcal{O}_{F'})$. Each of R and S is flat over $\mathbf{Z}/p^a$, and S is also flat over R, by Lemma 2.2.10. Thus $S \otimes_R S$ is flat over S, hence over R, and hence over $\mathbf{Z}/p^a$ as well.

To see that the natural morphism $S \otimes_R S \to \prod_{g \in G} S$ is an almost isomorphism, it suffices to check the analogous condition after passing to associated graded rings for the p-adic filtration. Lemma B.8 allows us to rewrite this induced morphism on associated graded rings in the form

$$\mathbf{F}_p[T]/(T^a) \otimes_{\mathbf{F}_p} ((S/pS) \otimes_{(R/pR)} (S/pS)) \to \mathbf{F}_p[T]/(T^a) \otimes_{\mathbf{F}_p} \prod_{g \in G} (S/pS)$$

(here $\mathbf{F}_p[T]/(T^a)$ appears as the associated graded ring to $\mathbf{Z}/p^a$ with its p-adic filtration), which may be identified with the base change over $\mathbf{F}_p[T]/(T^a)$ of the natural morphism

$$(S/pS) \otimes_{(R/pR)} (S/pS) \to \prod_{g \in G} (S/pS).$$

This latter morphism is an indeed an almost isomorphism, by Lemma 2.3.6. □

We now pass to the profinite setting. We continue to suppose that we are given the ring R endowed with an idempotent ideal I. We suppose additionally that $v \in I$ is a regular element of R (i.e., a nonzero divisor) and that R is v-adically complete.

We also suppose being given a v-adically complete R-algebra S, equipped with an action of a profinite group G as R-algebra automorphisms. We assume that this action is continuous, in the sense that the action map $G \times S \to S$ is continuous when S is endowed with its v-adic topology.

We further suppose that we may write $S = \widehat{\bigcup_n S_n}$ as the v-adic completion of an increasing union of G-invariant v-adically complete R-subalgebras S_n, with $R = S_0$; that the G-action on S_n factors through a finite quotient $G_n := G/H_n$ of G, where H_n is a normal open subgroup of G; and that for each $i \geq 0$ and each $m \leq n$, the morphism $S_m/v^i \to S_n/v^i$ is faithfully flat, and realizes S_n/v^i as an almost Galois extension of S_m/v^i, having the subgroup H_m/H_n of G_n as Galois group.

2.3.8 Lemma. *For any value of m, the morphism $S_m \to S^{H_m}$ is an almost isomorphism, as is the induced morphism $S_m/v^i \to (S/v^i)^{H_m}$ for any $i \geq 0$.*

Proof. Our hypotheses, together with Lemma 2.3.3, imply that the morphism

$$S_m/v^i \to (S_n/v^i)^{H_m/H_n}$$

is an almost isomorphism for each $n \geq m$ and each $i \geq 0$. Passing to the direct limit over n gives the second claim, and then additionally passing to the inverse limit over i gives the first claim. □

2.3.9 Corollary. *The injection $R[1/v] \hookrightarrow S[1/v]^G$ is an isomorphism.*

Proof. Note that $S[1/v]^G = (S^G)[1/v]$, since v is G-invariant (being an element of R) and localization is exact. The corollary thus follows from the $m=0$ case of Lemma 2.3.8, as inverting v converts the almost isomorphism into a genuine isomorphism. $\square$

2.3.10 Definition. We say that an S-module M is *iso-projective of finite rank* if it is finitely generated and v-torsion free, and if $S[1/v] \otimes_S M$ is a projective $S[1/v]$-module.

2.3.11 Lemma. *Any iso-projective S-module of finite rank is v-adically complete.*

Proof. If M is iso-projective, then we may write $M[1/v] = eF$ where F is a finite free $S[1/v]$-module and $e \in \mathrm{End}_{S[1/v]} F$ is an idempotent. Write $F = F_0 \otimes_S S[1/v]$, where F_0 is a finite free S-module. Then, since F_0 is finitely generated, we find that eF_0 is contained in $v^{-a}F_0$ for some sufficiently large value of a; the latter S-module is v-adically separated, and thus so is the former. Since eF_0 is furthermore the image of the v-adically complete S-module F_0, it is in fact v-adically complete. Since $eF_0[1/v] = M[1/v]$, and both eF_0 and M are finitely generated S-modules, we find that $v^b eF_0 \subseteq M \subseteq v^{-b} eF_0$ for some sufficiently large value of b, and thus M is also v-adically complete, as claimed. $\square$

2.3.12 Theorem. *If M is an iso-projective S-module of finite rank, equipped with a semi-linear G-action that is continuous with respect to the v-adic topology on M, then the kernel and cokernel of the induced morphism $S \widehat{\otimes}_R M^G \to M$ (the source being the v-adically completed tensor product) are each annihilated by a power of v.*

Proof. Write $P := S[1/v] \otimes_S M$, so that (by assumption) P is a finitely presented projective module over $S[1/v]$. Since M is v-adically complete (by Lemma 2.3.11), we have an isomorphism $M \xrightarrow{\sim} \varprojlim_n M/v^n M$. This induces a corresponding isomorphism $M^G \xrightarrow{\sim} \varprojlim_n (M/v^n M)^G$. (We should also note that the topology on M^G induced by the v-adic topology on M coincides with the v-adic topology on M^G, as follows immediately from that fact that multiplication by v on M is injective (as M is iso-projective) and commutes with the G-action.) Thus, to prove the theorem, it suffices to show that the kernel and cokernel of each of the morphisms

$$S \otimes_R (M/v^n M)^G \to M/v^n M$$

are annihilated by a power of v that is bounded independently of n.

Multiplication by v^{-n} induces an isomorphism $M/v^n M \xrightarrow{\sim} v^{-n}M/M$, and we will actually prove the equivalent statement that each of the morphisms

$$S \otimes_R (v^{-n}M/M)^G \to v^{-n}M/M \tag{2.3.13}$$

has kernel and cokernel annihilated by a power of v that is bounded independently of n.

The reason for formulating our argument in terms of the quotients $v^{-n}M/M$ is that these may be conveniently regarded as submodules of the quotient P/M. Our argument will proceed by constructing, for each n, a G-invariant S-submodule Z_n of P/M, such that $v^{-n}M/M \subseteq Z_n \subseteq v^{-(n+c)}M/M$ (where c is independent of n), and such that each of the morphisms

$$S \otimes_R Z_n^G \to Z_n \tag{2.3.14}$$

has kernel and cokernel annihilated by a power of v that is bounded independently of n. This implies the corresponding statement for the morphisms (2.3.13), and thus establishes the theorem. The remainder of the argument is devoted to constructing the submodules Z_n, and proving the requisite properties of the morphisms (2.3.14).

Since P is a projective $S[1/v]$-module, we may choose a finite rank free module F over $S[1/v]$, and an idempotent $e \in \mathrm{End}_{S[1/v]}(F)$, such that $P = eF$. We choose a free R-submodule F_0 of F such that $S[1/v] \otimes_R F_0 \xrightarrow{\sim} F$. (More concretely, F_0 is simply the R-span of some chosen $S[1/v]$-basis of F.) The endomorphism e may not preserve the S-submodule $S \otimes_R F_0$ of F, but if we choose a sufficiently large, then $e' := v^a e$ will preserve $S \otimes_R F_0$.

Since M is finitely generated over S, we may and do assume that we have chosen F_0 in such a manner that $M \subseteq v^a(S \otimes_R F_0)$. (Simply replace F_0 by $v^{-b}F_0$ for some sufficiently large value of b.) Then in fact $M \subseteq e'(S \otimes_R F_0)$. Furthermore, since M and $e'(S \otimes_R F_0)$ both span P as an $S[1/v]$-module, we find that $v^c e'(S \otimes_R F_0) \subseteq M$ for some sufficiently large value of c.

For any $n \geq 0$ we have $S/v^n = \bigcup_{i \geq 0} S_i/v^n$. Thus e' mod v^n, which is an endomorphism of $(S/v^n) \otimes_R F_0$, descends to an endomorphism of $(S_i/v^n) \otimes_R F_0$ for all sufficiently large i. Replacing (S_n) by an appropriately chosen subsequence, we may and do assume that in fact for each n, e' descends to an endomorphism e'_n of $(S_n/v^n) \otimes_R F_0$.

Since G acts continuously on P and preserves M, it acts continuously on P/M. The topology on P/M is discrete, and thus any finite subset of P/M is fixed by some open subgroup of G. In particular, we find that

$$v^{-n}e'F_0/\big(v^{-n}e'F_0 \cap M\big)$$

(which we regard in the natural way as a submodule of P/M, and which we note is finitely generated over R) is pointwise fixed by some open subgroup of G. Since the H_n form a cofinal sequence of open subgroups of G, if we again replace S_n and H_n by appropriately chosen subsequences, we may and do assume that in fact $v^{-n}e'F_0/\big(v^{-n}e'F_0 \cap M\big)$ is pointwise fixed by H_n.

Since $M \supseteq v^c e'(S \otimes_R F_0)$, we see that $v^{-n}e'F_0/\big(v^{-n}e'F_0 \cap M\big)$ is annihilated by v^{n+c}, and thus that

$$G\Big(v^{-n}e'F_0/(v^{-n}e'F_0\cap M)\Big)\subseteq v^{-(n+c)}M/M\subseteq v^{-(n+c)}e'(S\otimes_R F_0)/M.$$

Since G/H_n is finite, and since H_n fixes $v^{-n}e'F_0/\big(v^{-n}e'F_0\cap M\big)$ pointwise, we see that in fact

$$G\Big(v^{-n}e'F_0/(v^{-n}e'F_0\cap M)\Big)\subseteq v^{-(n+c)}e'(T\otimes_R F_0)/M,$$

for some finitely generated R-subalgebra T of $S/v^{n+c}S$. Passing to a subsequence one more time, we may assume that T is contained in S_n.

We let Y_n, respectively Z_n, denote the S_n-submodule, respectively the S-submodule, of P/M generated by the G-translates of $v^{-n}e'F_0/\big(v^{-n}e'F_0\cap M\big)$. It remains to prove the requisite properties of Z_n. We begin by noting the inclusions

$$v^{-n}e'(S\otimes_R F_0)/M\subseteq Z_n\subseteq v^{-(n+c)}M/M\subseteq v^{-(n+c)}e'(S\otimes_R F_0)/M,$$

which imply that

$$v^cZ_n\subseteq v^{-n}e'(S\otimes_R F_0)/M\subseteq Z_n,$$

and thus that

$$v^cZ_n^{H_n}\subseteq (v^{-n}e'(S\otimes_R F_0)/M)^{H_n}\subseteq Z_n^{H_n}. \tag{2.3.15}$$

Lemma 2.3.8 shows that $S_n/v^n\to(S/v^n)^{H_n}$ is an almost isomorphism. Thus $(v^{-n}S_n/S_n)\otimes_R F_0\to(v^{-n}S/S\otimes_R F_0)^{H_n}$ is an almost isomorphism, if we declare that H_n acts trivially on F_0, and thus via its action on the first factor in the target tensor product. Since $(e')^2=v^ae'$, we find that the cokernel of the inclusion

$$e'\big((v^{-n}S\otimes_R F_0)/(S\otimes_R F_0)\big)^{H_n}\hookrightarrow\big(e'(v^{-n}S\otimes_R F_0)/(S\otimes_R F_0)\big)^{H_n}$$

is annihilated by v^a, and thus that the natural morphism

$$e'(v^{-n}S_n\otimes_R F_0)/(S_n\otimes_R F_0)\to\big(e'(v^{-n}S\otimes_R F_0)/(S\otimes_R F_0)\big)^{H_n}$$

has its kernel annihilated by I, and its cokernel annihilated by v^aI.

If we let X_n denote the image of $e'(v^{-n}S_n\otimes_R F_0)$ in $\big(e'(v^{-n}S\otimes_R F_0)/M\big)^{H_n}$, then certainly

$$X_n\subseteq Y_n\subseteq Z_n^{H_n}.$$

It follows from the conclusion of the preceding paragraph, along with the fact that $v^ce'(S_0\otimes_R F_0)\subseteq M$, that the cokernel of the inclusion

$$X_n\hookrightarrow\big(e'(v^{-n}S\otimes_R F_0)/M\big)^{H_n}$$

is annihilated by $v^{a+c}I$, while the chain of inclusions (2.3.15) shows that the cokernel of the inclusion

$$(e'(v^{-n}S \otimes_R F_0)/M)^{H_n} \subseteq Z_n^{H_n}$$

is annihilated by v^c. Thus the cokernel of the inclusion $X_n \subseteq Z_n^{H_n}$ is annihilated by $v^{a+2c}I$, and hence so is the cokernel of the inclusion $Y_n \subseteq Z_n^{H_n}$. Passing to G_n-invariants in the inclusion just mentioned, we find that there is an inclusion $Y_n^{G_n} \subseteq Z_n^G$, whose cokernel is annihilated by $v^{a+2c}I$. Extending scalars to S_n, we obtain a morphism

$$S_n \otimes_R Y_n^{G_n} \to S_n \otimes_R Z_n^G$$

(which is in fact an embedding, since S_n/v^{n+c} is flat over R/v^{n+c}, although we don't need this here), whose cokernel is annihilated by $v^{a+2c}I$. On the other hand, Lemma 2.3.5 implies that the natural morphism $S_n \otimes_R Y_n^{G_n} \to Y_n$ is an almost isomorphism.

Putting all these results together, we find that the kernel and cokernel of the natural morphism

$$S_n \otimes_R Z_n^G \to Z_n^{H_n}$$

are each annihilated by $v^{a+2c}I$. Indeed, the composite $S_n \otimes_R Y_n^{G_n} \to S_n \otimes_R Z_n^G \to Z_n^{H_n}$ factors through the almost isomorphism $S_n \otimes_R Y_n^{G_n} \to Y_n$, and we have shown that the natural morphism $Y_n \to Z_n^{H_n}$ is an injection whose cokernel is killed by $v^{a+2c}I$, while the cokernel of $S_n \otimes_R Y_n^{G_n} \to S_n \otimes_R Z_n^G$ is also killed by $v^{a+2c}I$.

Since $Y_n \subseteq Z_n^{H_n}$, and since Y_n generates Z_n over S by their very definitions, we see that the natural map

$$S \otimes_{S_n} Z_n^{H_n} \to Z_n \tag{2.3.16}$$

is surjective. We bound the exponent of its kernel as follows: The inclusion $X_n \subseteq Z_n^{H_n}$, whose cokernel we have shown above to be annihilated by $v^{a+2c}I$, induces an inclusion

$$S \otimes_{S_n} X_n \subseteq S \otimes_{S_n} Z_n^{H_n},$$

whose cokernel is again annihilated by $v^{a+2c}I$, while the defining surjection

$$e'\big((v^{-n}S_n/v^c S_n) \otimes_R F_0\big) \to X_n$$

induces a surjection

$$e'\big((v^{-n}S \otimes_R F_0)/(v^c S \otimes_R F_0)\big) \to S \otimes_{S_n} X_n.$$

Now the natural morphism

$$e'((v^{-n}S\otimes_R F_0)/(v^cS\otimes_R F_0))\to M[1/v]/M \tag{2.3.17}$$

has kernel annihilated by v^c, and so we find that the kernel of (2.3.16) is annihilated by $v^{a+3c}I$. (Indeed, if x is an element of the kernel of (2.3.16), then for any $i\in I$ we can lift $v^{a+2c}ix$ to an element of the kernel of (2.3.17).)

Putting together the results of the preceding two paragraphs, we find that the cokernel of the natural morphism

$$S\otimes_R Z_n^G\to Z_n$$

is annihilated by $v^{a+2c}I$, while its kernel is annihilated by $v^{2a+5c}I$. Recalling that $v\in I$, and noting that the powers of v just mentioned are independent of n, and also that we have the inclusions $v^{-n}M/M\subseteq Z_n\subseteq v^{-(n+c)}M/M$, we see that the proof of the theorem is completed. □

We will now deduce a descent result for projective modules over $S[1/v]$. For this, we need to make some additional hypotheses, which we now describe.

2.3.18 Hypothesis.

1. S/vS is countable.
2. $S[1/v]$ is Noetherian.
3. The morphism $R[1/v]\to S[1/v]$ is faithfully flat.

We write $S[1/v]=\varinjlim_n v^{-n}S$. If we identify $v^{-n}S$ with S via multiplication by v^n, then each of the transition maps becomes identified with the closed embedding $vS\hookrightarrow S$. Thus if we equip $S[1/v]$ with the inductive limit topology, then $S[1/v]$ becomes a topological ring, which is completely metrizable (since S is v-adically complete) and in fact Polish (since S/vS is countable, so that S and thus $S[1/v]$ are separable). Note also that vS is an open additive subgroup of $S[1/v]$ which is closed under multiplication, and consists of topologically nilpotent elements. Proposition C.6 then shows that finitely generated $S[1/v]$-modules have a canonical topology, with respect to which all $S[1/v]$-homomorphisms are continuous, with closed image.

We now establish the following descent result in this situation.

2.3.19 Theorem. *If Hypothesis* 2.3.18 *holds, and if M is a finitely generated projective $S[1/v]$-module, equipped with a continuous (when endowed with its canonical topology) semi-linear G-action, then M^G is a finitely generated and projective $R[1/v]$-module, and the natural morphism*

$$S[1/v]\otimes_{R[1/v]} M^G\to M \tag{2.3.20}$$

is an isomorphism.

Proof. Let M' denote any finitely generated S-submodule of M that generates M over $S[1/v]$; then the S-span M'' of GM' is finitely generated (note that since M' is finitely generated, there is an open subgroup H of G such that $HM' \subseteq M'$, and H has finite index in the profinite group G), so it is an iso-projective S-submodule of M which is G-invariant. Applying Theorem 2.3.12 to M'', and noting that

$$M^G = (M''[1/v])^G = (M'')^G[1/v]$$

(because localization is exact), we find that the natural morphism (2.3.20) has dense image. Since its target is finitely generated over $S[1/v]$, its image is also finitely generated (because $S[1/v]$ is Noetherian, by assumption), and thus closed in its target; combined with the density, we find that (2.3.20) is surjective.

Since $S[1/v] \otimes_{R[1/v]} M^G \to M$ is surjective, while M is finitely generated over $S[1/v]$, we see that if N is any sufficiently large finitely generated $R[1/v]$-submodule of M^G, then $S[1/v] \otimes_{R[1/v]} N \to M$ is surjective. Choose a finitely generated free $R[1/v]$-module F that surjects onto N, and let E denote the kernel of the induced surjection

$$S[1/v] \otimes_{R[1/v]} F \to S[1/v] \otimes_{R[1/v]} N \to M,$$

so that we have a short exact sequence

$$0 \to E \to S[1/v] \otimes_{R[1/v]} F \to M \to 0.$$

Passing to G-invariants, and taking into account Corollary 2.3.9, we obtain a left exact sequence

$$0 \to E^G \to F \to M^G,$$

which (by the choice of F) induces a short exact sequence

$$0 \to E^G \to F \to N \to 0.$$

Tensoring back up with $S[1/v]$ (which is flat over $R[1/v]$ by assumption), we obtain a morphism of short exact sequences

$$\begin{array}{ccccccccc} 0 & \longrightarrow & S[1/v] \otimes_{R[1/v]} E^G & \longrightarrow & S[1/v] \otimes_{R[1/v]} F & \longrightarrow & S[1/v] \otimes_{R[1/v]} N & \longrightarrow & 0 \\ & & \downarrow & & \downarrow & & \downarrow & & \\ 0 & \longrightarrow & E & \longrightarrow & S[1/v] \otimes_{R[1/v]} F & \longrightarrow & M & \longrightarrow & 0 \end{array}$$

Evidently the middle vertical arrow is the identity, and thus the natural morphism $S[1/v] \otimes_{R[1/v]} E^G \to E$ is injective. Since E is finitely generated and projective (being the kernel of a surjection from a finitely generated free module to a projective module), it follows from what we have already proved that this morphism is also surjective.

Since the left-hand side two vertical arrows are isomorphisms, so is the third, so that $S[1/v] \otimes_{R[1/v]} N \to M$ is an isomorphism. This is true for any sufficiently large choice of N, and thus we find (using the faithful flatness of $S[1/v]$ over $R[1/v]$, which holds by assumption) that all these sufficiently large choices of N coincide, implying that M^G is finitely generated and that (2.3.20) is an isomorphism. The faithful flatness of $S[1/v]$ over $R[1/v]$ then implies that M^G is projective over $R[1/v]$ [Sta, Tag 058S], as required. $\square$

The following corollary provides a convenient reformulation of the preceding theorem.

2.3.21 Corollary. *If Hypothesis* 2.3.18 *holds, then the functor*

$$M \mapsto S[1/v] \otimes_{R[1/v]} M$$

induces an equivalence between the category of finitely generated projective $R[1/v]$-modules and the category of finitely generated projective $S[1/v]$-modules endowed with a continuous semi-linear action of G. A quasi-inverse functor is given by $N \mapsto N^G$.

Proof. Theorem 2.3.19 shows that if N is a finitely generated and projective $S[1/v]$-module, endowed with a continuous semi-linear G-action, then the natural map $S[1/v] \otimes_{R[1/v]} N^G \to N$ is an isomorphism. To complete the proof of the corollary, then, it suffices to show that if M is a finitely generated and projective $R[1/v]$-module, then the natural map $M \to (S[1/v] \otimes_{R[1/v]} M)^G$ is an isomorphism. Writing M as the direct summand of a finite rank free module, we reduce to the case when M is free, which (as was already observed in the proof of Theorem 2.3.19) is established by Corollary 2.3.9. $\square$

2.4 AN APPLICATION OF ALMOST GALOIS DESCENT

Let F be a closed perfectoid subfield of $\mathbf{C}$, with tilt $F^\flat$, a closed perfectoid subfield of $\mathbf{C}^\flat$. Recall that, for any p-adically complete $\mathbf{Z}_p$-algebra A, we defined $W(\mathcal{O}_F^\flat)_A$ and $W(F^\flat)_A$ in Section 2.2.

2.4.1 Theorem. *Let A be a finite type $\mathbf{Z}/p^a$-algebra, for some $a \geq 1$. The inclusion $W(F^\flat)_A \to W(\mathbf{C}^\flat)_A$ is a faithfully flat morphism of Noetherian rings, and the functor $M \mapsto W(\mathbf{C}^\flat)_A \otimes_{W(F^\flat)_A} M$ induces an equivalence between the category of finitely generated projective $W(F^\flat)_A$-modules and the category of finitely generated projective $W(\mathbf{C}^\flat)_A$-modules endowed with a continuous semi-linear G_F-action. A quasi-inverse functor is given by $N \mapsto N^{G_F}$.*

The proof will be an application of the almost Galois descent results of Section 2.3. Thus we have to place ourselves in the framework of that section. To this end, we note that $\mathbf{C}$ may be regarded as a completion $\widehat{\overline{F}}$ of the algebraic closure $\overline{F}$ of F. We then write $\overline{F} = \bigcup_{n\geq 1} F_n$ as the increasing union of a sequence of finite Galois extensions of F; to match the notation of Section 2.3, we also write $F_0 := F$. We write $G_n := \mathrm{Gal}(F_n/F)$, $H_n := \mathrm{Gal}(\overline{F}/F_n)$, and $G := H_0 = \mathrm{Gal}(\overline{F}/F)$, so that $G \xrightarrow{\sim} \varprojlim_n G_n$ and $G/H_n \xrightarrow{\sim} G_n$. Note that each of the finite Galois extensions F_n of F is again perfectoid.

Assume now that A is a finitely generated $\mathbf{Z}/p^a$-algebra. Then $W(F^\flat)_A = W_a(F^\flat)_A$, and similarly $W(\mathbf{C}^\flat)_A = W_a(\mathbf{C}^\flat)_A$, so that from here on we may work with rings of a-truncated Witt vectors, rather than with full rings of Witt vectors themselves. To accord with the notation of Section 2.3, we also denote these rings by R and S respectively. Recall from Section 2.2 that v denotes a nonzero element of the maximal ideal of $W_a(\mathcal{O}_F^\flat)$, and that by definition

$$R := W_a(\mathcal{O}_F^\flat)_A = \varprojlim_i \bigl(W_a(\mathcal{O}_F^\flat) \otimes_{\mathbf{Z}/p^a} A\bigr)/v^i,$$

while

$$S := W_a(\mathcal{O}_{\mathbf{C}}^\flat)_A = \varprojlim_i \bigl(W_a(\mathcal{O}_{\mathbf{C}}^\flat) \otimes_{\mathbf{Z}/p^a} A\bigr)/v^i;$$

so that both R and S are v-adically complete.

We let I denote the ideal in R generated by the maximal ideal of $W_a(\mathcal{O}_F^\flat)$, and let v denote the image in I of the chosen element v of that maximal ideal (no confusion should result from this duplication of notation). We further set

$$S_n := W_a(\mathcal{O}_{F_n}^\flat)_A = \varprojlim_i \bigl(W_a(\mathcal{O}_{F_n}^\flat) \otimes_{\mathbf{Z}/p^a} A\bigr)/v^i$$

for each $n \geq 0$ (so in particular $S_0 = R$). By construction we have that S coincides with the v-adic completion of $\varinjlim_n S_n$. As we will see below, the transition morphisms $S_m \to S_n$ are in fact faithfully flat, and thus injective, and so in fact this direct limit is simply a union; thus $S = \widehat{\bigcup_n S_n}$, as required for the setup of Section 2.3. Furthermore, Lemma 2.2.18 ensures that the G-action on S is continuous.

2.4.2 Proposition. *For each $i \geq 0$, and each $m \leq n$, the morphism $S_m/v^i \to S_n/v^i$ is faithfully flat, and realizes S_n/v^i as an almost Galois extension of S_m/v^i, with Galois group H_m/H_n.*

Proof. Lemma 2.2.10 shows that each of the inclusions $W_a(\mathcal{O}_{F_m}^\flat) \hookrightarrow W_a(\mathcal{O}_{F_n}^\flat)$ is faithfully flat, thus so is the morphism $W_a(\mathcal{O}_{F_m}^\flat) \otimes_{\mathbf{Z}/p^a} A \to W_a(\mathcal{O}_{F_n}^\flat) \otimes_{\mathbf{Z}/p^a} A$, and hence so are each of the morphisms

$$S_m/v^iS_m = \big(W_a(\mathcal{O}^\flat_{F_m}) \otimes_{\mathbf{Z}/p^a} A)\big)/v^i\big(W_a(\mathcal{O}^\flat_{F_m}) \otimes_{\mathbf{Z}/p^a} A)\big)$$
$$\to \big(W_a(\mathcal{O}^\flat_{F_n}) \otimes_{\mathbf{Z}/p^a} A)\big)/v^i\big(W_a(\mathcal{O}^\flat_{F_n}) \otimes_{\mathbf{Z}/p^a} A)\big) = S_n/v^iS_n.$$

An identical argument, taking into account Lemma 2.3.7 and Remark 2.3.2, shows that $S_m/v^iS_m \to S_n/v^iS_n$ is almost Galois (with respect to the ideal in S_m generated by the maximal ideal of $W_a(\mathcal{O}^\flat_{F_m})$, and hence also with respect to IS_m, since the latter ideal is contained in the former), with Galois group H_m/H_n. □

2.4.3 Proposition.

1. *Each $S_m[1/v]$, as well as $S[1/v]$, is Noetherian. (Setting $m=0$ gives in particular that $R[1/v]$ is Noetherian.)*
2. *Each of the morphisms $S_m \to S_n$ (for $m \leq n$), as well as each morphism $S_m \to S$, is faithfully flat. In particular (setting $m=0$ and then inverting v) the morphism $R[1/v] \to S[1/v]$ is faithfully flat.*

Proof. The Noetherian claims of (1) follow from Proposition B.36 (1).

We already noted in Proposition 2.4.2 that the morphisms $S_m/v^i \to S_n/v^i$ are faithfully flat, and an identical argument shows that each morphism $S_m/v^i \to S/v^i$ is faithfully flat. Taking into account that statement of (1), we find that the claims of (2) follow from Proposition 2.2.2. □

Proof of Theorem 2.4.1. Proposition 2.4.2 verifies that the running assumptions imposed at the beginning of Section 2.3 are satisfied. Proposition 2.4.3 verifies that Hypothesis 2.3.18 is satisfied, and also establishes the Noetherian and faithful flatness claims of Theorem 2.4.1. The remainder of the theorem then follows from Corollary 2.3.21. □

2.5 ÉTALE φ-MODULES

In this section we briefly recall and generalize some definitions and results from [EG21]. Let R be a $\mathbf{Z}_p$-algebra, equipped with a ring endomorphism φ, which is congruent to the (p-power) Frobenius modulo p. If M is an R-module, we write

$$\varphi^*M := R \otimes_{R,\varphi} M.$$

2.5.1 Definition. An *étale φ-module over R* is a finite R-module M, equipped with a φ-semi-linear endomorphism $\varphi_M : M \to M$, which has the property that the induced R-linear morphism

$$\Phi_M : \varphi^*M \xrightarrow{1 \otimes \varphi_M} M$$

is an isomorphism. A morphism of étale φ-modules is a morphism of the underlying R-modules which commutes with the morphisms Φ_M. We say that M is *projective* (resp. *free*) if it is projective of constant rank (resp. free of constant rank) as an R-module.

We will typically apply this definition with R taken to be one of the coefficient rings defined in Section 2.2. Of particular interest to us will be the cases corresponding to imperfect fields of norms, as it is these cases which fit into the framework of [EG21, §5], and we will use the results of that paper to prove the basic algebraicity properties of our moduli stacks.

2.5.2 Definition. Let S be an R-algebra, and let φ_S be a ring endomorphism of S, which is congruent to Frobenius modulo p, and is compatible with φ on R. Then if M is an étale φ-module over R, the extension of scalars $S \otimes_R M$ is naturally an étale φ-module over S, with $\varphi_{S\otimes_R M} := \varphi_S \otimes \varphi_M$.

2.5.3 Multilinear algebra

We briefly recall the multilinear algebra of projective étale φ-modules. Firstly, if P is a projective étale φ-module over R, then we give its R-dual $P^\vee := \mathrm{Hom}_R(P, R)$ the structure of an étale φ-module by defining the isomorphism $\varphi^* P^\vee \to P^\vee$ to be the inverse of the transpose of Φ_P. Secondly, if M and N are projective étale φ-modules, then we endow $M \otimes_R N$ with the structure of an étale φ-module by defining $\varphi_{M\otimes N} := \varphi_M \otimes \varphi_N$.

2.5.4 Lemma. *If M, N are projective étale φ-modules over R then we have a natural identification* $\mathrm{Hom}_{R,\varphi}(M, N) = (M^\vee \otimes_R N)^{\varphi=1}$.

Proof. We have $\mathrm{Hom}_R(M, N) = M^\vee \otimes_R N$. Given $f \in \mathrm{Hom}_R(M, N)$, regarded as an element of $P := M^\vee \otimes_R N$, we have $f \in P^{\varphi=1}$ if and only if $\Phi_P(1 \otimes f) = f$, and by the definition of the φ-structure on P, this is equivalent to f intertwining Φ_M and Φ_N, as required. □

2.6 FROBENIUS DESCENT

Suppose that $\mathbf{A}$ is as in Situation 2.2.15, and that we have a continuous φ-equivariant embedding $\mathbf{A} \hookrightarrow W(\mathbf{C}^\flat)$. Let A be a finite type $\mathbf{Z}/p^a$-algebra for some $a \geq 1$. Assume that the induced map $\mathbf{A}_A \to W(\mathbf{C}^\flat)_A$ is a faithfully flat injection. Since φ is bijective on $W(\mathbf{C}^\flat)$, we have an increasing union

$$\mathbf{A}_A \subset \varphi^{-1}(\mathbf{A}_A) \subset \varphi^{-2}(\mathbf{A}_A) \subset \cdots \subset W(\mathbf{C}^\flat)_A.$$

We let $\widetilde{\mathbf{A}}_A$ be the closure of $\cup_{n\geq 0}\varphi^{-n}(\mathbf{A}_A)$ in $W(\mathbf{C}^\flat)_A$, and we set $\widetilde{\mathbf{A}}_A^+ := \widetilde{\mathbf{A}}_A \cap \mathbf{A}_{\mathrm{inf},A}$. Note that φ extends to a bijection on $\widetilde{\mathbf{A}}_A$, which induces a bijection on $\widetilde{\mathbf{A}}_A^+$.

2.6.1 Remark. We will apply the results of this section in the setting introduced in Section 2.1, taking $\mathbf{A} = \mathbf{A}_K$ or $\mathbf{A} = \mathfrak{S}$. In either case it follows from Proposition 2.2.12 that $\mathbf{A}_A \to W(\mathbf{C}^\flat)_A$ is a faithfully flat injection, and it follows easily from Theorem 2.1.6 that in the former case we have $\widetilde{\mathbf{A}}_A = \widetilde{\mathbf{A}}_{K,A}$, and in the latter case we have $\widetilde{\mathbf{A}}_A = \widetilde{\mathcal{O}}_{\mathcal{E},A}$.

2.6.2 Remark. If $\mathbf{A}_A^+$ is not φ-stable, then $\mathbf{A}_A^+$ is not a subring of $\widetilde{\mathbf{A}}_A^+$, and in particular it is not equal to $\mathbf{A}_A \cap \widetilde{\mathbf{A}}_A^+$.

We will make use of the following variant of Lemma 2.2.16; in view of Remark 2.6.2 we cannot apply Lemma 2.2.16 in the setting of $\widetilde{\mathbf{A}}_A^+$-modules.

2.6.3 Lemma. *There is an integer $r \geq 0$ such that for all $s \geq r$ we have $T^s \in \widetilde{\mathbf{A}}_A^+$, and an integer $C \geq 0$ such that for all $n \geq 1$, we have $\varphi(T^n) \in T^{pn-C}\widetilde{\mathbf{A}}_A^+$.*

Proof. Since T is topologically nilpotent in $\mathbf{A}_A^+$ it is topologically nilpotent in $\widetilde{\mathbf{A}}_A$, so in particular $T^s \in \widetilde{\mathbf{A}}_A^+$ for all s sufficiently large. For the second statement, we follow the proof of Lemma 2.2.16, and write $\varphi(T) = T^p + pY$ for some $Y \in \mathbf{A}_A$, so that $\varphi(T^n) = (T^p + pY)^n$. If we expand using the binomial theorem, and recall that $p^a = 0$, then we see that $\varphi(T^n) - T^{np}$ is a sum of terms of the form $T^{p(n-r)}Y^r$ for some $0 \leq r \leq a-1$. It therefore suffices to choose C such that $T^C T^{-pr} Y^r \in \widetilde{\mathbf{A}}_A^+$ for $0 \leq r \leq a-1$, which we can do by the topological nilpotence of T. ☐

2.6.4 Remark. As already noted in Remark 2.6.2, $\widetilde{\mathbf{A}}_A^+$ need not contain $\mathbf{A}_A^+$, and so, although $\mathbf{A}_A = \mathbf{A}_A^+[1/T]$, it doesn't make sense to write $\widetilde{\mathbf{A}}_A = \widetilde{\mathbf{A}}_A^+[1/T]$. On the other hand, Lemma 2.6.3 does ensure that $T^r \in \widetilde{\mathbf{A}}_A^+$ for some $r > 0$, and then it does make sense to write, and is true (since T is topologically nilpotent in $\widetilde{\mathbf{A}}_A$, while $\widetilde{\mathbf{A}}_A^+$ is open in $\widetilde{\mathbf{A}}_A$), that $\widetilde{\mathbf{A}}_A = \widetilde{\mathbf{A}}_A^+[1/T^r]$.

Given an étale φ-module M over $\mathbf{A}_A$, we may form its *perfection* $\widetilde{M} := \widetilde{\mathbf{A}}_A \otimes_{\mathbf{A}_A} M$, which is an étale φ-module over $\widetilde{\mathbf{A}}_A$.

2.6.5 Descending morphisms

2.6.6 Proposition. *Let M, N be projective étale φ-modules. Then the natural map $\mathrm{Hom}_{\mathbf{A}_A,\varphi}(M,N) \to \mathrm{Hom}_{\widetilde{\mathbf{A}}_A,\varphi}(\widetilde{M},\widetilde{N})$ is a bijection.*

Proof. Write $P := M^\vee \otimes N$, so that P is a projective étale φ-module over $\mathbf{A}_A$. Then by Lemma 2.5.4, we are reduced to checking that the natural morphism of A-modules $P^{\varphi=1} \to \widetilde{P}^{\varphi=1}$ is an isomorphism. It is certainly injective, so we need only show that it is surjective.

Since the formation of φ-invariants is compatible with direct sums, it follows from [EG21, Lem. 5.2.14] that it is enough to consider the case that P is in fact free. By Lemma 2.6.3 (*cf.* Remark 2.6.4) we can choose a φ-stable free $\widetilde{\mathbf{A}}_A^+$-submodule $\widetilde{\mathfrak{P}}$ of $\widetilde{P}$, which generates $\widetilde{P}$ over $\mathbf{A}_A$ (indeed, choose any basis of $\widetilde{P}$, multiply by a sufficiently large power of T, and let $\widetilde{\mathfrak{P}}$ be the submodule generated by this scaled basis).

Consider an element $x \in \widetilde{P}^{\varphi=1}$. Choose an integer $r \geq 1$ such that if $s \geq r$ then $\varphi(T^s \widetilde{\mathbf{A}}_A^+) \subseteq T^{s+1} \widetilde{\mathbf{A}}_A^+$ (such an r exists by Lemma 2.6.3). We may write $x = x_1 + x_2$ where $x_2 \in T^r \widetilde{\mathfrak{P}}$ and $x_1 \in \varphi^{-n}(\mathbf{A}_A) \otimes_{\mathbf{A}_A} P$ for some $n \geq 0$. Choose n to be minimal with this property. If $n > 0$, then since $x = \varphi(x) = \varphi(x_1) + \varphi(x_2)$, and $\varphi(x_2) \in T^r \widetilde{\mathfrak{P}}$, we see that we may replace x_1 with $\varphi(x_1) \in \varphi^{1-n}(\mathbf{A}_A) \otimes_{\mathbf{A}_A} P$, a contradiction.

Thus $x_1 \in P$, and we have $x_1 - \varphi(x_1) = \varphi(x_2) - x_2 \in T^r \widetilde{\mathfrak{P}} \cap P$. The sum $x' := x_1 + \sum_{i=0}^{\infty} \varphi^i(\varphi(x_1) - x_1)$ converges to an element of $T^r \widetilde{\mathfrak{P}}$, and the sum therefore converges in P. By definition we have $\varphi(x') = x'$. We have $x - x_1, x' - x_1 \in T^r \widetilde{\mathfrak{P}}$, so $x - x' \in T^r \widetilde{\mathfrak{P}}$; then $x - x' = \varphi(x - x') \in T^{r+1} \widetilde{\mathfrak{P}}$, and iterating gives $x = x' \in P$, as required. □

2.6.7 Descending objects

We will show that every projective étale φ-module over $\widetilde{\mathbf{A}}_A$ arises as the perfection of an étale φ-module over $\mathbf{A}_A$. We first note the following analogue of [EG21, Lem. 5.2.14] for étale φ-modules over $\widetilde{\mathbf{A}}_A$.

2.6.8 Lemma. *If M is a projective étale φ-module over $\widetilde{\mathbf{A}}_A$, then M is a direct summand of a free étale φ-module over $\widetilde{\mathbf{A}}_A$.*

Proof. This can be proved in an identical fashion to [EG21, Lem. 5.2.14]. □

2.6.9 Proposition. *Let M be a projective étale φ-module over $\widetilde{\mathbf{A}}_A$. Then there is a projective étale φ-module M_0 over $\mathbf{A}_A$ and an isomorphism $M \xrightarrow{\sim} \widetilde{M_0}$.*

Proof. Suppose firstly that M is free of some rank d as an $\widetilde{\mathbf{A}}_A$-module. Let X denote the matrix of φ with respect to a choice of basis $e_1, \ldots, e_d$. As in the proof of Proposition 2.6.6, after possibly scaling the e_i by powers of T, we may assume that X has entries in $\widetilde{\mathbf{A}}_A^+$. Since Φ_M is an isomorphism, it follows from Remark 2.6.4 that there is an integer $h \geq 0$ such that $X^{-1} \in T^{-h} M_d(\widetilde{\mathbf{A}}_A^+)$.

If we change basis via a matrix $Y \in M_d(\widetilde{\mathbf{A}}_A^+) \cap \mathrm{GL}_d(\widetilde{\mathbf{A}}_A)$, the new matrix for φ is $\varphi(Y) X Y^{-1}$. It suffices to show that we can choose Y so that this matrix has entries in $\mathbf{A}_A$ (as we can then let M_0 be the $\mathbf{A}_A$-span of this new basis).

Fix some $H > \lceil (C + h + 1)/(p-1) \rceil$, where C is as in Lemma 2.6.3, and write $X = X' + X''$ where $X' \in T^{H+h} M_d(\widetilde{\mathbf{A}}_A^+)$ and $X'' \in M_d(\varphi^{-n}(\mathbf{A}_A))$ for some sufficiently large n. (The reason for this choice of H is to be able to apply Lemma 2.6.10 below.) Taking $Y = X$, we may replace X, X', X'' by $\varphi(X), \varphi(X'), \varphi(X'')$ respectively (note that by Lemma 2.6.3, we have $\varphi(X') \in T^{H+h} M_d(\widetilde{\mathbf{A}}_A^+)$, because $p(H+h) - C \geq H + h$ by our choice of C), which has the effect of replacing n by $(n-1)$. Iterating this procedure, we can assume that $n = 0$, so $X'' \in M_d(\mathbf{A}_A)$. By Lemma 2.6.10 below, we can find Y such that $\varphi(Y) X Y^{-1} = X''$, completing the proof in the case that M is free.

We now return to the general case in which M is only assumed to be finitely generated projective, rather than free. By Lemma 2.6.8, we may write M as a direct summand of a free étale φ-module F over $\widetilde{\mathbf{A}}_A$. By the case already proved, we may write $F \xrightarrow{\sim} \widetilde{F}_0$ for some free étale φ-module F_0. By Proposition 2.6.6, the idempotent in $\mathrm{End}(F)$ corresponding to M comes from an idempotent in $\mathrm{End}(F_0)$, and we may take M_0 to be the étale φ-module corresponding to this idempotent. □

The following lemma and its proof are based on [PR09, Prop. 2.2].

2.6.10 Lemma. *Suppose that $X \in M_d(\widetilde{\mathbf{A}}_A^+) \cap \mathrm{GL}_d(\widetilde{\mathbf{A}}_A)$, and that $X^{-1} \in T^{-h} M_d(\widetilde{\mathbf{A}}_A^+)$ for some $h \geq 0$. Suppose that $X'' \in M_d(\widetilde{\mathbf{A}}_A^+) \cap \mathrm{GL}_d(\widetilde{\mathbf{A}}_A)$ is such that $X^{-1} X'' \in 1 + T^{\lceil (C+h+1)/(p-1) \rceil} M_d(\widetilde{\mathbf{A}}_A^+)$, where C is as in Lemma 2.6.3. Then there exists $Y \in M_d(\widetilde{\mathbf{A}}_A^+) \cap \mathrm{GL}_d(\widetilde{\mathbf{A}}_A)$ with $X'' = \varphi(Y) X Y^{-1}$.*

Proof. Define sequences X_i, h_i by $X_0 = X$, $h_0 = (X'')^{-1} X$, and for each $i \geq 1$, $X_i = \varphi(h_{i-1}) X_{i-1} h_{i-1}^{-1}$, $h_i = (X'')^{-1} X_i$. Then if we set $y_i = h_i h_{i-1} \cdots h_0$, we have $X'' = X_i h_i^{-1} = \varphi(y_{i-1}) X y_i^{-1}$ for each i. We claim that the y_i tend to a limit Y as $i \to \infty$; then we have $X'' = \varphi(Y) X Y^{-1}$, as required.

To see that the y_i tend to a limit, it is enough to show that $h_i \to 1$ as $i \to \infty$. To see this, suppose that $h_i \in 1 + T^s M_d(\widetilde{\mathbf{A}}_A^+)$ for some $s \geq (C+h+1)/(p-1)$. We have

$$X'' h_{i+1} = X_{i+1} = \varphi(h_i) X_i h_i^{-1} = \varphi(h_i) X'',$$

so that $h_{i+1} = (X'')^{-1} \varphi(h_i) X''$. Using Lemma 2.6.3 and the assumption that $(X'')^{-1} \in T^{-h} M_d(\widetilde{\mathbf{A}}_A^+)$, we see that $h_{i+1} \in 1 + T^{ps-C-h} M_d(\widetilde{\mathbf{A}}_A^+)$. Since $ps - C - h \geq s + 1$, we are done. □

2.6.11 An equivalence of categories

We summarize the results of this section in the following proposition.

2.6.12 Proposition. *Let A be a finite type $\mathbf{Z}/p^a$-algebra for some $a \geq 1$. Then the functor $M \mapsto \widetilde{M}$ is an equivalence of categories from the category of projective étale φ-modules over $\mathbf{A}_A$ to the category of projective étale φ-modules over $\widetilde{\mathbf{A}}_A$.*

Proof. The functor is essentially surjective by Proposition 2.6.9, and fully faithful by Proposition 2.6.6. □

2.7 (φ, Γ)–MODULES

By definition, an *étale* (φ, Γ_K)*-module* is a finitely generated $\mathbf{A}_K$-module M, equipped with

- a φ-linear morphism $\varphi\colon M \to M$ with the property that the corresponding morphism $\Phi_M : \varphi^* M \to M$ is an isomorphism (i.e., M is given the structure of an étale φ-module over $\mathbf{A}_K$), and
- a continuous semi-linear action of Γ_K that commutes with φ.

2.7.1 The relationship with Galois representations

There is an equivalence of categories between the category of continuous representations of G_K on finite $\mathbf{Z}_p$-modules and the category of étale (φ, Γ_K)-modules, which is given by functors $\mathbf{D}$ and T that are defined as follows. Let $\widehat{\mathbf{A}}_K^{\mathrm{ur}}$ denote the p-adic completion of the ring of integers of the maximal unramified extension of $\mathbf{A}_K[1/p]$ in $W(\mathbf{C}^\flat)[1/p]$; this is preserved by the natural actions of φ and G_K on $W(\mathbf{C}^\flat)[1/p]$. Then for a G_K-representation T, we define

$$\mathbf{D}(T) := (\widehat{\mathbf{A}}_K^{\mathrm{ur}} \otimes_{\mathbf{Z}_p} T)^{G_{K_{\mathrm{cyc}}}},$$

while for an étale (φ, Γ_K)-module M we define

$$T(M) := (\widehat{\mathbf{A}}_K^{\mathrm{ur}} \otimes_{\mathbf{A}_K} M)^{\varphi=1}.$$

The action of $G_{K_{\mathrm{cyc}}}$ (resp. φ) in the definition of $\mathbf{D}$ (resp. of T) is the diagonal one (with the G_K-action on M being that inflated from the action of Γ_K). There is a completely analogous theory of $(\varphi, \widetilde{\Gamma}_K)$-modules, and taking Δ_K-invariants gives an equivalence of categories between $(\varphi, \widetilde{\Gamma}_K)$-modules and (φ, Γ_K)-modules.

The functor $T(M)$ is also defined for finitely generated étale φ-modules over $\mathbf{A}_K$ (i.e., in the absence of a Γ-action on M), and in this context yields an equivalence of categories between finitely generated étale φ-modules over $\mathbf{A}_K$ and continuous $G_{K_{\mathrm{cyc}}}$-representations on finite $\mathbf{Z}_p$-modules.

Suppose now that L/K is a finite extension, and suppose further that $K = K_{\mathrm{cyc}} \cap L$ (or, equivalently, that the natural embedding $\Gamma_L \hookrightarrow \Gamma_K$ induced by the inclusion $K_{\mathrm{cyc}} \subseteq L_{\mathrm{cyc}}$ is an isomorphism). We will recall the description of the functor $\mathrm{Ind}_{G_L}^{G_K}$ in terms of (φ, Γ)-modules. (We will use this description in Section 7.3.)

By construction, there are inclusions $\mathbf{A}_K \subseteq \mathbf{A}_L \subseteq \widehat{\mathbf{A}}_K^{\mathrm{ur}}$, which are unramified embeddings of DVRs. The composite of these embeddings is G_K-equivariant (where G_K-acts on $\mathbf{A}_K$ through its quotient Γ_K), while the second is G_L-equivariant (where G_L-acts on $\mathbf{A}_L$ through its quotient Γ_L). Regarding the second of these embeddings as an $\mathbf{A}_K$-linear morphism of $\mathbf{A}_K$-modules, it induces an $\widehat{\mathbf{A}}_K^{\mathrm{ur}}$-linear surjection $\widehat{\mathbf{A}}_K^{\mathrm{ur}} \otimes_{\mathbf{A}_K} \mathbf{A}_L \to \widehat{\mathbf{A}}_K^{\mathrm{ur}}$. The source of this surjection has a diagonal action of G_K (the G_K-action on A_L being via its quotient $\Gamma_K \xrightarrow{\sim} \Gamma_L$), while the surjection itself is G_L-equivariant. Thus it induces a G_K-equivariant morphism

$$\widehat{\mathbf{A}}_K^{\mathrm{ur}} \otimes_{\mathbf{A}_K} \mathbf{A}_L \to \mathrm{Ind}_{G_L}^{G_K} \widehat{\mathbf{A}}_K^{\mathrm{ur}},$$

which is easily seen to be an isomorphism. Using the description of the induction as a tensor product, we can express this as an isomorphism

$$\widehat{\mathbf{A}}_K^{\mathrm{ur}} \otimes_{\mathbf{A}_K} \mathbf{A}_L \xrightarrow{\sim} \mathbf{Z}_p[G_K] \otimes_{\mathbf{Z}_p[G_L]} \widehat{\mathbf{A}}_K^{\mathrm{ur}}. \tag{2.7.2}$$

If M is an étale (φ, Γ)-module over $\mathbf{A}_L$, then we let M' denote M regarded as an étale (φ, Γ)-module over the subring $\mathbf{A}_K$ of $\mathbf{A}_L$. (Recall again that we are assuming $\Gamma_L \xrightarrow{\sim} \Gamma_K$.) The isomorphism (2.7.2) then yields an isomorphism

$$\begin{aligned} T(M') &:= (\widehat{\mathbf{A}}_K^{\mathrm{ur}} \otimes_{\mathbf{A}_K} M')^{\varphi=1} = \left((\widehat{\mathbf{A}}_K^{\mathrm{ur}} \otimes_{\mathbf{A}_K} \mathbf{A}_L) \otimes_{\mathbf{A}_L} M\right)^{\varphi=1} \\ &= \left(\mathbf{Z}_p[G_K] \otimes_{\mathbf{Z}_p[G_L]} \widehat{\mathbf{A}}_K^{\mathrm{ur}} \otimes_{\mathbf{A}_L} M\right)^{\varphi=1} = \mathbf{Z}_p[G_K] \otimes_{\mathbf{Z}_p[G_L]} \left(\widehat{\mathbf{A}}_K^{\mathrm{ur}} \otimes_{\mathbf{A}_L} M\right)^{\varphi=1} \\ &= \mathrm{Ind}_{G_L}^{G_K} T(M). \end{aligned} \tag{2.7.3}$$

In short, induction of Galois representations corresponds to the restriction of coefficients on the (φ, Γ)-module side.

The preceding discussion also applies in the context of étale φ-modules (in the sense of Definition 2.5.1). Our assumption that $\Gamma_L \xrightarrow{\sim} \Gamma_K$ implies that $G_{K_{\mathrm{cyc}}}/G_{L_{\mathrm{cyc}}} \to G_K/G_L$ is a bijection, so that (2.7.2) may also be interpreted as an isomorphism

$$\widehat{\mathbf{A}}_K^{\mathrm{ur}} \otimes_{\mathbf{A}_K} \mathbf{A}_L \xrightarrow{\sim} \mathbf{Z}_p[G_{K_{\mathrm{cyc}}}] \otimes_{\mathbf{Z}_p[G_{L_{\mathrm{cyc}}}]} \widehat{\mathbf{A}}_K^{\mathrm{ur}}.$$

Then, if M is a finitely generated étale φ-module over $\mathbf{A}_L$, and if M' denotes its restriction to an étale φ-module over $\mathbf{A}_K$, the preceding isomorphism induces an isomorphism

$$T(M') = \mathrm{Ind}_{G_{L_{\mathrm{cyc}}}}^{G_{K_{\mathrm{cyc}}}} T(M). \tag{2.7.4}$$

We also need a variant of the preceding theory, which again follows from the results of [Fon90], using the Kummer extension K_∞/K introduced in Example 2.1.3 and Section 2.1.14; the integral version of this theory was first studied by Breuil and Kisin; see [Kis09b, §1]. Namely, there is an equivalence of categories between the category of continuous representations of G_{K_∞} on finite $\mathbf{Z}_p$-algebras and the category of étale φ-modules over $\mathcal{O}_{\mathcal{E}}$, which is given by functors $\mathbf{D}_\infty, T_\infty$ that are defined as follows: Let $\mathcal{O}_{\widehat{\mathcal{E}^{\mathrm{ur}}}}$ denote the p-adic completion of the ring of integers in the maximal unramified extension of $\mathrm{Frac}(\mathcal{O}_{\mathcal{E}})$ in $W(\mathbf{C}^\flat)[1/p]$; this is preserved by the natural actions of φ and G_{K_∞} on $W(\mathbf{C}^\flat)[1/p]$. We define

$$\mathbf{D}_\infty(T) := (\mathcal{O}_{\widehat{\mathcal{E}^{\mathrm{ur}}}} \otimes_{\mathbf{Z}_p} T)^{G_{K_\infty}},$$

for a G_{K_∞}-representation T, and define

$$T_\infty(M) := (\mathcal{O}_{\widehat{\mathcal{E}^{\mathrm{ur}}}} \otimes_{\mathbf{A}_K} M)^{\varphi=1},$$

for an étale φ-module M.

2.7.5 Various types of φ-modules with coefficients

We now introduce coefficients.

2.7.6 Definition. Let A be a p-adically complete $\mathbf{Z}_p$-algebra. A *projective étale* (φ, Γ_K)*-module of rank* d with A-coefficients is a projective étale φ-module M of rank d over $\mathbf{A}_{K,A}$ equipped with a semi-linear action of Γ_K, which commutes with φ, and which is furthermore continuous when M is endowed with its canonical topology (i.e., the topology of Remark D.2).

If A is a finite type $\mathbf{Z}/p^a$-algebra, we can use Theorem 2.4.1 to give a useful alternative description of étale (φ, Γ_K)-modules with A-coefficients in terms of étale φ-modules with coefficients in the ring $W(\mathbf{C}^\flat)_A$, as we now explain.

2.7.7 Definition. Let A be a p-adically complete $\mathbf{Z}_p$-algebra. An étale (φ, G_K)*-module with* A*-coefficients* (resp. an étale $(\varphi, G_{K_{\mathrm{cyc}}})$*-module with* A*-coefficients*, resp. an étale (φ, G_{K_∞})*-module with* A*-coefficients*) is by definition a finitely generated $W(\mathbf{C}^\flat)_A$-module M equipped with an isomorphism

$$\varphi_M : \varphi^* M \xrightarrow{\sim} M$$

of $W(\mathbf{C}^\flat)_A$-modules, and a $W(\mathbf{C}^\flat)_A$-semi-linear action of G_K (resp. $G_{K_{\mathrm{cyc}}}$, resp. G_{K_∞}), which is continuous and commutes with φ_M. We say that M is projective if it is projective of constant rank as a $W(\mathbf{C}^\flat)_A$-module.

If A is a finite type $\mathbf{Z}/p^a$-algebra, then there is a functor from the category of finite projective étale (φ, Γ_K)-modules with A-coefficients to the category of finite projective étale (φ, G_K)-modules with A-coefficients, which takes an étale (φ, Γ_K)-module M to $W(\mathbf{C}^\flat)_A \otimes_{\mathbf{A}_{K,A}} M$, endowed with the extension of scalars of φ, and the diagonal action of G_K, with the action of G_K on M being the action inflated from Γ_K.

Similarly, there is a functor from the category of finite projective étale φ-modules over $\mathcal{O}_{\mathcal{E},A}$ to the category of finite projective étale (φ, G_{K_∞})-modules with A-coefficients, which takes an étale φ-module M to $W(\mathbf{C}^\flat)_A \otimes_{\mathcal{O}_{\mathcal{E},A}} M$, endowed with the extension of scalars of φ, and the diagonal action of G_{K_∞}, with the action of G_{K_∞} on M being the trivial action.

2.7.8 Proposition. *Let A be a finite type $\mathbf{Z}/p^a$-algebra for some $a \geq 1$.*

1. *The functor $M \mapsto W(\mathbf{C}^\flat)_A \otimes_{\mathbf{A}_{K,A}} M$ is an equivalence between the category of finite projective étale φ-modules over $\mathbf{A}_{K,A}$ and the category of finite projective $(\varphi, G_{K_{\mathrm{cyc}}})$-modules with A-coefficients.*

   It induces an equivalence of categories between the category of finite projective étale (φ, Γ_K)-modules with A-coefficients and the category of finite projective étale (φ, G_K)-modules with A-coefficients.

   A quasi-inverse functor is given by the composite of $N \mapsto N^{G_{K_{\mathrm{cyc}}}}$ and a quasi-inverse to the functor of Proposition 2.6.12.
2. *The functor $M \mapsto W(\mathbf{C}^\flat)_A \otimes_{\mathcal{O}_{\mathcal{E},A}} M$ is an equivalence of categories between the category of finite projective étale φ-modules over $\mathcal{O}_{\mathcal{E},A}$ and the category of finite projective étale (φ, G_{K_∞})-modules with A-coefficients. A quasi-inverse functor is given by $N \mapsto N^{G_{K_\infty}}$ and a quasi-inverse to the functor of Proposition* 2.6.12.

Proof. This is a formal consequence of Theorem 2.4.1 and Proposition 2.6.12. We begin with (2). Firstly, by Proposition 2.6.12 (and Remark 2.6.1) the functor $M \mapsto \widetilde{M} = \widetilde{\mathcal{O}}_{\mathcal{E},A} \otimes_{\mathcal{O}_{\mathcal{E},A}} M$ is an equivalence of categories between the category of finite projective étale φ-modules over $\mathcal{O}_{\mathcal{E},A}$ and the category of finite projective étale φ-modules over $\widetilde{\mathcal{O}}_{\mathcal{E},A}$, so it is enough to show that $\widetilde{M} \mapsto W(\mathbf{C}^\flat)_A \otimes_{\widetilde{\mathcal{O}}_{\mathcal{E},A}} \widetilde{M}$ is an equivalence of categories. By Example 2.1.3 and Remark 2.1.4, $\widehat{K}_\infty$ is a perfectoid field, and we have a canonical identification of Galois groups $G_{\widehat{K}_\infty} = G_{K_\infty}$. The result then follows from the equivalence of categories given by Theorem 2.4.1, as we can think of φ as being an isomorphism of $\widetilde{\mathcal{O}}_{\mathcal{E},A}$-modules $\varphi^* \widetilde{M} \xrightarrow{\sim} \widetilde{M}$.

The same argument shows in (1) that the functor $M \mapsto W(\mathbf{C}^\flat)_A \otimes_{\mathbf{A}_{K,A}} M$ is an equivalence of categories between the category of projective étale φ-modules over $\mathbf{A}_{K,A}$ and the category of projective étale $(\varphi, G_{K_{\mathrm{cyc}}})$-modules with A-coefficients. Since $\Gamma_K = G_K / G_{K_{\mathrm{cyc}}}$, this extends to the claimed equivalence of categories, noting that by construction the continuity of the Γ_K-action on M is equivalent to the continuity of the G_K-action on $W(\mathbf{C}^\flat)_A \otimes_{\mathbf{A}_{K,A}} M$ (since Lemma 2.2.18 shows that the G_K-action on $W(\mathbf{C}^\flat)_A$ is continuous). ☐

Chapter Three

Moduli stacks of φ-modules and (φ, Γ)-modules

In this chapter we build on the results of [EG21] (which in turn built on [PR09]) to construct our stacks of (φ, Γ)-modules, and various related stacks of φ-modules. We show in particular that our stack $\mathcal{X}_{K,d}$ of (φ, Γ)-modules is Ind-algebraic.

3.1 MODULI STACKS OF φ-MODULES

In this section we put ourselves in the context of Situation 2.2.15; we also remind the reader that the notion of étale φ-module is defined in Section 2.5. We furthermore fix a finite extension $E/\mathbf{Q}_p$ with ring of integers $\mathcal{O}$ and residue field $\mathbf{F}$; all of our coefficient rings from now on will be $\mathcal{O}$-algebras. All of our constructions are compatible with replacing E by a finite extension, and we will typically not comment on this, although see Remark 3.1.2 below.

If we fix integers $a, d \geq 1$, then we may follow [EG21, §5] and define an *fpqc* stack in groupoids $\mathcal{R}_d^a$ over $\operatorname{Spec} \mathcal{O}/\varpi^a$ as follows: For any $\mathcal{O}/\varpi^a$-algebra A, we define $\mathcal{R}_d^a(A)$ to be the groupoid of étale φ-modules over $\mathbf{A}_A$ which are projective of rank d. If $A \to B$ is a morphism of $\mathcal{O}$-algebras, and M is an object of $\mathcal{R}_d^a(A)$, then the pull-back of M to $\mathcal{R}_d^a(B)$ is defined to be the tensor product $\mathbf{A}_B \otimes_{\mathbf{A}_A} M$.

A key point is that this definition does not require $\mathbf{A}_A^+$ to be φ-stable, although (as far as we know) this hypothesis *is* required to make any deductions about $\mathcal{R}_d^a$ beyond the fact that it is an *fpqc* stack (which relies just on Drinfeld's general descent results, as described in [Dri06] and [EG21, §5.1]).

Since we are ultimately interested in questions of algebraicity or of Ind-algebraicity, from now on we regard $\mathcal{R}_d^a$ as an *fppf* stack over $\mathcal{O}/\varpi^a$. By [Sta, Tag 04WV], we may also regard the stack $\mathcal{R}_d^a$ as an *fppf* stack over $\mathcal{O}$, and as a varies, we may form the 2-colimit $\mathcal{R} := \varinjlim_a \mathcal{R}_d^a$, which is again an *fppf* stack over $\mathcal{O}$. In fact $\mathcal{R}_d$ lies over $\operatorname{Spf} \mathcal{O} := \varinjlim_a \operatorname{Spec} \mathcal{O}/\varpi^a$, the formal spectrum of $\mathcal{O}$ with respect to the ϖ-adic, or equivalently p-adic, topology.

We now fix a polynomial $F \in W(k)[T]$ which is congruent to a positive power of T modulo p (for example, an Eisenstein polynomial).

3.1.1 Definition. Suppose that $\mathbf{A}_A^+$ is φ-stable. Let h be a non-negative integer, and let A be a p-adically complete $\mathcal{O}$-algebra. A *φ-module of F-height at most h*

over $\mathbf{A}_A^+$ is a pair $(\mathfrak{M},\varphi_M)$ consisting of a finitely generated T-torsion free $\mathbf{A}_A^+$-module $\mathfrak{M}$, and a φ-semi-linear map $\varphi_{\mathfrak{M}}\colon \mathfrak{M}\to\mathfrak{M}$, with the further properties that if we write

$$\Phi_{\mathfrak{M}} := 1\otimes\varphi_{\mathfrak{M}}\colon \varphi^*\mathfrak{M}\to\mathfrak{M},$$

then $\Phi_{\mathfrak{M}}$ is injective, and the cokernel of $\Phi_{\mathfrak{M}}$ is killed by F^h.

A *φ-module of finite F-height over* $\mathbf{A}_A^+$ is a φ-module of F-height at most h for some $h\geq 0$. A morphism of φ-modules is a morphism of the underlying $\mathbf{A}_A^+$-modules which commutes with the morphisms $\Phi_{\mathfrak{M}}$.

We say that a φ-module of finite F-height is projective of rank d if it is a finitely generated projective $\mathbf{A}_A^+$-module of constant rank d.

If we maintain the assumption that $\mathbf{A}^+$ is φ-stable, and if we fix integers $a,d\geq 1$ and an integer $h\geq 0$, then we may again follow [EG21, §5] to define an *fpqc* stack in groupoids $\mathcal{C}_{d,h}^a$ over $\operatorname{Spec}\mathcal{O}/\varpi^a$ as follows: For any $\mathcal{O}/\varpi^a$-algebra A, we define $\mathcal{C}_{d,h}^a(A)$ to be the groupoid of φ-modules of F-height at most h over $\mathbf{A}_A^+$ which are projective of rank d. If $A\to B$ is a morphism of $\mathbf{Z}_p$-algebras, and $\mathfrak{M}$ is an object of $\mathcal{C}_{d,h}^a(A)$, then the pull-back of $\mathfrak{M}$ to $\mathcal{C}_{d,h}^a(B)$ is defined to be the tensor product $\mathbf{A}_B^+\otimes_{\mathbf{A}_A^+}\mathfrak{M}$.

Just as for the stack $\mathcal{R}_d^a$, we may and do also regard the stack $\mathcal{C}_{d,h}^a$ as an *fppf* stack over $\mathcal{O}$, and we then, allowing a to vary, define $\mathcal{C}_{d,h} := \varinjlim_a \mathcal{C}_{d,h}^a$, obtaining an *fppf* stack over $\mathcal{O}$ which in fact lies over $\operatorname{Spf}\mathcal{O}$. There are canonical morphisms $\mathcal{C}_{d,h}^a\to\mathcal{R}_d^a$ and $\mathcal{C}_{d,h}\to\mathcal{R}_d$ given by tensoring with $\mathbf{A}_A$ over $\mathbf{A}_A^+$.

3.1.2 Remark. If E'/E is a finite extension with ring of integers $\mathcal{O}'$, then by definition we have (with obvious notation) $\mathcal{C}_{d,h,\mathcal{O}'}=\mathcal{C}_{d,h}\times_{\mathcal{O}}\mathcal{O}'$ (in the case that $\mathbf{A}^+$ is φ-stable, so that these stacks are defined) and $\mathcal{R}_{d,\mathcal{O}'}=\mathcal{R}_{d,\mathcal{O}}\times_{\mathcal{O}}\mathcal{O}'$ (in general).

The following lemma provides a concrete interpretation of the A-valued points of the stacks we have defined, when A is a ϖ-adically complete $\mathcal{O}$-algebra (rather than just an algebra over some $\mathcal{O}/\varpi^a$).

3.1.3 Lemma. *If A is a ϖ-adically complete $\mathcal{O}$-algebra, then there is a canonical equivalence between the groupoid of morphisms* $\operatorname{Spf}A\to\mathcal{R}_d$ *and the groupoid of rank d étale φ-modules over $\mathbf{A}_A$. If $\mathbf{A}_A^+$ is furthermore φ-stable, then there is a canonical equivalence between the groupoid of morphisms* $\operatorname{Spf}A\to\mathcal{C}_{d,h}$ *and the groupoid of φ-modules of rank d and F-height at most h over $\mathbf{A}_A^+$.*

Proof. This is immediate from Lemma D.5. □

We now apply the results of [EG21, §5] to deduce various results about the stacks we have introduced. This requires the assumption that $\mathbf{A}^+$ is φ-stable.

3.1.4 Theorem. *Suppose that $\mathbf{A}^+$ is φ-stable, and let $a\geq 1$ be arbitrary.*

1. *The stack* $\mathcal{C}^a_{d,h}$ *is an algebraic stack of finite presentation over* $\operatorname{Spec} \mathcal{O}/\varpi^a$*, with affine diagonal.*
2. *The morphism* $\mathcal{C}^a_{d,h} \to \mathcal{R}^a_d$ *is representable by algebraic spaces, proper, and of finite presentation.*
3. *The diagonal morphism* $\Delta \colon \mathcal{R}^a_d \to \mathcal{R}^a_d \times_{\mathcal{O}/\varpi^a} \mathcal{R}^a_d$ *is representable by algebraic spaces, affine, and of finite presentation.*
4. $\mathcal{R}^a_d$ *is a limit preserving Ind-algebraic stack, whose diagonal is representable by algebraic spaces, affine, and of finite presentation.*

Proof. Part (1) is [EG21, Thm. 5.4.9 (1)], and parts (2), (3) and (4) are proved in [EG21, Thm. 5.4.11], except for the claim that $\mathcal{R}^a_d$ is Ind-algebraic, which is [EG21, Thm. 5.4.20]. □

3.1.5 Corollary. *Suppose that* $\mathbf{A}^+$ *is* φ*-stable.*

1. $\mathcal{C}_{d,h}$ *is a p-adic formal algebraic stack of finite presentation over* $\operatorname{Spf} \mathcal{O}$*, with affine diagonal.*
2. $\mathcal{R}_d$ *is a limit preserving Ind-algebraic stack, whose diagonal is representable by algebraic spaces, affine, and of finite presentation.*
3. *The morphism* $\mathcal{C}_{d,h} \to \mathcal{R}_d$ *is representable by algebraic spaces, proper, and of finite presentation.*
4. *The diagonal morphism* $\Delta \colon \mathcal{R}_d \to \mathcal{R}_d \times_{\operatorname{Spf} \mathcal{O}} \mathcal{R}_d$ *is representable by algebraic spaces, affine, and of finite presentation.*

Proof. The first part is immediate from Theorem 3.1.4 (1) and Proposition A.13. Everything else is immediate from Theorem 3.1.4. □

3.2 MODULI STACKS OF (φ, Γ)-MODULES

In this section we begin the study of our main objects of interest, namely the moduli stacks of étale (φ, Γ)-modules. As in Chapter 2 we fix a finite extension $K/\mathbf{Q}_p$. As in Section 3.1, we also fix a finite extension E of $\mathbf{Q}_p$ with ring of integers $\mathcal{O}$, which will serve as our ring of coefficients. As always, k denotes the residue field of the ring of integers of K, and $\mathbf{F}$ denotes the residue field of $\mathcal{O}$.

3.2.1 Definition. We let $\mathcal{X}_{K,d}$ denote the moduli stack of projective étale (φ, Γ_K)-modules of rank d. More precisely, if A is a p-adically complete $\mathcal{O}$-algebra, then we define $\mathcal{X}_{K,d}(A)$ (i.e., the groupoid of morphisms $\operatorname{Spf} A \to \mathcal{X}_{K,d}$) to be the groupoid of projective étale (φ, Γ_K)-modules of rank d with A-coefficients, in the sense of Definition 2.7.6, with morphisms given by isomorphisms. If $A \to B$ is a morphism of complete $\mathcal{O}$-algebras, and M is an object of $\mathcal{X}_{K,d}(A)$, then the pull-back of M to $\mathcal{X}_{K,d}(B)$ is defined to be the tensor product $\mathbf{A}_{K,B} \otimes_{\mathbf{A}_{K,A}} M$.

It follows from the results of [Dri06], and more specifically from [EG21, Thm. 5.1.16], that $\mathcal{X}_{K,d}$ is an *fpqc* stack over $\mathcal{O}$. As in Remark 3.1.2, the definition of $\mathcal{X}_{K,d}$ behaves naturally with respect to change of the coefficient ring $\mathcal{O}$.

One of the main results of this book is that $\mathcal{X}_{K,d}$ is a Noetherian formal algebraic stack. However, the proof of this is quite involved, and will only be fully achieved at the conclusion of Chapter 5. In this section and the two that follow it, we establish the preliminary result that $\mathcal{X}_{K,d}$ is an Ind-algebraic stack.

We begin by discussing the moduli stacks of étale φ-modules over $\mathbf{A}_{K,A}$, which will play an auxiliary role in our study of $\mathcal{X}_{K,d}$.

3.2.2 Definition. We let $\mathcal{R}_{K,d}$ denote the moduli stack of rank d projective étale φ-modules, defined as in Section 3.1, taking $\mathbf{A}$ to be $\mathbf{A}_K$.

If $\mathbf{A}_K^+$ is φ-stable, then Corollary 3.1.5 applies to $\mathcal{R}_{K,d}$; this is in particular the case if $K/\mathbf{Q}_p$ is basic in the sense of Definition 2.1.12. Our first task is to establish the same results for general K, which we will do by reducing to the basic case.

3.2.3 Definition. If $K/\mathbf{Q}_p$ is any finite extension, we set $K^{\text{basic}} := K \cap K_0(\zeta_{p^\infty})$.

By definition, K^{basic} is basic, and $(K^{\text{basic}})_0 = K_0$. Note that the natural restriction map $\widetilde{\Gamma}_K \to \widetilde{\Gamma}_{K^{\text{basic}}}$ is an isomorphism, and induces an isomorphism $\Gamma_K \to \Gamma_{K^{\text{basic}}}$. By [Win83, Thm. 3.1.2], $\mathbf{A}_K$ is a free $\mathbf{A}_{K^{\text{basic}}}$-module of rank $[K : K^{\text{basic}}] = [K(\zeta_{p^\infty}) : K_0(\zeta_{p^\infty})]$, and the inclusion $\mathbf{A}_{K^{\text{basic}}} \subset \mathbf{A}_K$ is φ-equivariant, so there is a natural morphism $\mathcal{R}_{K,d} \to \mathcal{R}_{K^{\text{basic}},d[K:K^{\text{basic}}]}$ given by forgetting the $\mathbf{A}_{K,A}$-algebra structure on an étale φ-module.

3.2.4 Remark. As noted in Remark 2.1.13, in order to ensure that $\mathbf{A}_K^+$ is φ-stable, it suffices for K to be abelian over $\mathbf{Q}_p$. Thus, rather than relating the theory for K to that for the field K^{basic} introduced above, we could just as well relate it to any other subfield K' of K which is abelian over $\mathbf{Q}_p$, and for which the natural map $\Gamma_K \to \Gamma_{K'}$ is an isomorphism; e.g., the field $K' := K \cap \mathbf{Q}_p^{\text{ab}}$ (where $\mathbf{Q}_p^{ab}$ denotes the maximal abelian extension of K). It doesn't matter (for our purposes) which particular K' we choose; K^{basic} is simply a convenient choice.

3.2.5 Lemma. *The morphism $\mathcal{R}_{K,d} \to \mathcal{R}_{K^{\text{basic}},d[K:K^{\text{basic}}]}$ is representable by algebraic spaces, affine, and of finite presentation.*

Proof. We can prove the statement after pulling back via a morphism $\operatorname{Spec} A \to \mathcal{R}_{K^{\text{basic}},d[K:K^{\text{basic}}]}$, where A is an $\mathcal{O}/\varpi^a$-algebra for some $a \geq 1$. This morphism corresponds to a projective étale φ-module over $\mathbf{A}_{K^{\text{basic}},A}$ of rank $d[K : K^{\text{basic}}]$, and we need to show that the functor on A-algebras taking B to the set of projective étale φ-modules over $\mathbf{A}_{K,B}$ of rank d, whose underlying étale φ-module over $\mathbf{A}_{K^{\text{basic}},B}$ coincides with M_B, is representable by an affine scheme of finite presentation over $\operatorname{Spec} A$.

Note that since M_B is projective and in particular flat over $\mathbf{A}_{K^{\text{basic}},B}$, we have a natural inclusion $i\colon M_B \hookrightarrow \mathbf{A}_{K,B} \otimes_{\mathbf{A}_{K^{\text{basic}},B}} M_B$. The additional structure needed to make M_B into a projective étale φ-module over $\mathbf{A}_{K,B}$ is the data of a morphism of étale φ-modules over $\mathbf{A}_{K^{\text{basic}},B}$

$$f\colon \mathbf{A}_{K,B} \otimes_{\mathbf{A}_{K^{\text{basic}},B}} M_B \to M_B$$

satisfying the conditions that

1. the composite $M_B \xrightarrow{i} \mathbf{A}_{K,B} \otimes_{\mathbf{A}_{K^{\text{basic}},B}} M_B \xrightarrow{f} M_B$ is the identity morphism, and
2. the kernel of f is $\mathbf{A}_{K,B}$-stable.

(We can then define the $\mathbf{A}_{K,B}$-module structure on M_B via the formula $\lambda \cdot m := f(\lambda \otimes m)$. The first condition guarantees that this action is compatible with the existing $\mathbf{A}_{K^{\text{basic}},B}$-module structure on M_B, and the second condition that $(\lambda_1\lambda_2)\cdot m = \lambda_1 \cdot (\lambda_2 \cdot m)$.)

By [EG21, Prop. 5.4.8], the data of a morphism of étale φ-modules $f\colon \mathbf{A}_{K,B} \otimes_{\mathbf{A}_{K^{\text{basic}},B}} M_B \to M_B$ is representable by an affine scheme of finite presentation over $\operatorname{Spec} A$, so it is enough to show that conditions (1) and (2) are closed conditions, given by finitely many equations. To see this, we follow the proof of [EG21, Prop. 5.4.8]. Exactly as in that argument, we can reduce to the case that M_B is free, and after choosing bases, any f is determined by the coefficients of finitely many powers of T in the Laurent series expansions of the entries of the matrix given by f. Condition (1) is then evidently given by finitely many equations in these coefficients.

To see that the same is true of condition (2), note that since M_B is projective and condition (1) implies in particular that f is surjective, we have a splitting $\mathbf{A}_{K,B} \otimes_{\mathbf{A}_{K^{\text{basic}},B}} M_B = M_B \oplus \ker(f)$. The projection onto $\ker(f)$ is given by $(1 - i \circ f)$, so the condition that $\ker(f)$ is $\mathbf{A}_{K,B}$-stable is the condition that for any λ in $\mathbf{A}_{K,B}$, and any $m \in M_B$, we have

$$f(\lambda(m - i(f(m)))) = 0.$$

This is evidently a closed condition, and since $\mathbf{A}_{K,B}$ and M_B are both finitely generated $\mathbf{A}_{K^{\text{basic}},B}$-modules, it is determined by finitely many equations, as required. □

3.2.6 Corollary. *The stack $\mathcal{R}_{K,d}$ is a limit preserving Ind-algebraic stack, whose diagonal is representable by algebraic spaces, affine, and of finite presentation.*

Proof. This follows from Lemma 3.2.5, Corollary 3.1.5 (which establishes the claimed properties for $\mathcal{R}_{K^{\text{basic}},d[K:K^{\text{basic}}]}$), and Corollary 3.2.9 below. □

The following series of results concerning morphisms of stacks culminates in Corollary 3.2.9, which was used in the proof of Corollary 3.2.6.

3.2.7 Lemma. *Let $\mathcal{X} \to \mathcal{Y}$ be a morphism of stacks over a base scheme S, which is representable by algebraic spaces. As usual, let $\Delta_f : \mathcal{X} \to \mathcal{X} \times_{\mathcal{Y}} \mathcal{X}$ denote the diagonal of f. If f is of finite type and quasi-separated, then Δ_f is of finite presentation.*

Proof. This can be checked after pulling back along an arbitrary morphism $T \to \mathcal{Y}$, where T is a scheme, and hence reduced to the case of a morphism from an algebraic space to a scheme. In this case, the claim of the lemma is proved in [Sta, Tag 084P]. □

3.2.8 Lemma. *Let $f : \mathcal{X} \to \mathcal{Y}$ be a morphism of stacks over a base scheme S which is representable by algebraic spaces, has affine diagonal, and is of finite type. Suppose that the diagonal of $\mathcal{Y}$ is representable by algebraic spaces, affine, and of finite presentation. Then the diagonal of $\mathcal{X}$ is also representable by algebraic spaces, affine, and of finite presentation.*

Proof. We may factor the diagonal of $\mathcal{X}$ as

$$\mathcal{X} \to \mathcal{X} \times_{\mathcal{Y}} \mathcal{X} \to \mathcal{X} \times_S \mathcal{X}.$$

Since the morphism $\mathcal{X} \times_{\mathcal{Y}} \mathcal{X} \to \mathcal{X} \times_S \mathcal{X}$ is pulled back from the diagonal $\mathcal{Y} \to \mathcal{Y} \times_S \mathcal{Y}$, it is enough to show that the relative diagonal $\Delta_f : \mathcal{X} \to \mathcal{X} \times_{\mathcal{Y}} \mathcal{X}$ is representable by algebraic spaces, affine, and of finite presentation.

Now Δ_f is representable by algebraic spaces (since f is), and affine (by assumption). Since affine morphisms are quasi-compact, we see that f is quasi-separated, as well as being of finite type (by assumption); thus Δ_f is of finite presentation (by Lemma 3.2.7). □

3.2.9 Corollary. *Let $\mathcal{X} \to \mathcal{Y}$ be a morphism of stacks which is representable by algebraic spaces, affine, and of finite presentation. If $\mathcal{Y}$ is a limit preserving Ind-algebraic stack, whose diagonal is representable by algebraic spaces, affine, and of finite presentation, then the same is true of $\mathcal{X}$.*

Proof. The claimed limit preserving property of $\mathcal{X}$ follows, by [EG21, Lem. 2.3.20 (3)], from that of $\mathcal{Y}$ and the fact that $\mathcal{X} \to \mathcal{Y}$ is of finite presentation, while the claimed Ind-algebraic property of $\mathcal{X}$ follows from that of $\mathcal{Y}$ and the fact that $\mathcal{X} \to \mathcal{Y}$ is representable by algebraic spaces. Finally, the claimed properties of the diagonal of $\mathcal{X}$ follow from those of $\mathcal{Y}$ by Lemma 3.2.8. (Note that by assumption the hypotheses of Lemma 3.2.8 hold, since morphisms of finite presentation are in particular of finite type, while affine morphisms are separated, and thus have their diagonals being closed immersions, which in particular are again affine.) □

We now give a concrete description of $\mathcal{R}^a_{K,d}$ as an Ind-algebraic stack which will be useful in Chapter 7. In order to state it, we introduce the notation $\mathcal{C}^a_{K^{\mathrm{basic}},d[K:K^{\mathrm{basic}}],h}$ for the stack over $\operatorname{Spec}\mathcal{O}/\varpi^a$ classifying rank d projective φ-modules over $\mathbf{A}^+_{K^{\mathrm{basic}},A}$ of T-height at most h; by Theorem 3.1.4, $\mathcal{C}^a_{K^{\mathrm{basic}},d[K:K^{\mathrm{basic}}],h}$ is an algebraic stack of finite presentation over $\operatorname{Spec}\mathcal{O}/\varpi^a$.

We begin with a lemma which is a variant of one of the steps appearing in the proof of [EG21, Thm. 5.4.20].

3.2.10 Lemma. *If $T \to \mathcal{R}^a_{K,d}$ is a morphism whose source is a Noetherian scheme, then there is a Noetherian scheme Z and a scheme-theoretically dominant and surjective morphism $Z \to T$ such that the composite $Z \to T \to \mathcal{R}^a_{K,d} \to \mathcal{R}^a_{K^{\mathrm{basic}},d[K:K^{\mathrm{basic}}]}$ can be lifted to a morphism $Z \to \mathcal{C}^a_{K^{\mathrm{basic}},d[K:K^{\mathrm{basic}}],h}$ for some sufficiently large value of h.*

Proof. Since T is Noetherian, it is quasi-compact, and so admits a scheme-theoretically dominant surjection from a Noetherian affine scheme (e.g., the disjoint union of the members of a finite cover of T by affine open subsets). Thus we are reduced to the affine case.

Since T is affine, by [EG21, Prop. 5.4.7], we may find a scheme-theoretically dominant surjection $Z = \operatorname{Spec} A \to T$ such that the étale φ-module M over A corresponding to the composite $Z \to T \to \mathcal{R}_{K,d}$ is free (of rank d) over $\mathbf{A}_{K,A}$ (rather than merely projective). Thus M is also free (of rank $d[K:K^{\mathrm{basic}}]$) over $\mathbf{A}_{K^{\mathrm{basic}},A}$. We may then choose a φ-stable $\mathbf{A}_{K^{\mathrm{basic}},A}$-basis of M. If we let $\mathfrak{M}$ denote the $\mathbf{A}^+_{K^{\mathrm{basic}},A}$-module spanned by this basis, then $\mathfrak{M}$ is φ-invariant, and has some height h. Thus $\mathfrak{M}$ induces the desired morphism $Z \to \mathcal{C}^a_{K^{\mathrm{basic}},d[K:K^{\mathrm{basic}}],h}$. □

We now give the promised description of the Ind-algebraic stack structure on $\mathcal{R}_{K,d}$.

3.2.11 Lemma. *Fix $a \geq 1$, and for each $h \geq 0$, let $\mathcal{R}^a_{K,d,K^{\mathrm{basic}},h}$ denote the scheme-theoretic image of the base-changed morphism*

$$\mathcal{C}^a_{d[K:K^{\mathrm{basic}}],h} \times_{\mathcal{R}_{K^{\mathrm{basic}},d[K:K^{\mathrm{basic}}]}} \mathcal{R}_{K,d} \to \mathcal{R}_{K,d},$$

so that $\mathcal{R}^a_{K,d,K^{\mathrm{basic}},h}$ is a closed algebraic substack of $\mathcal{R}^a_{K,d}$. Then the canonical morphism $\varinjlim_h \mathcal{R}^a_{K,d,K^{\mathrm{basic}},h} \to \mathcal{R}^a_{K,d}$ is an isomorphism.

Proof. We have to show that any morphism $T \to \mathcal{R}^a_{K,d}$ whose source is a scheme factors through the inductive limit. Since $\mathcal{R}^a_{K,d}$ is limit preserving, we may assume that T is of finite type over $\mathcal{O}/\varpi^a$. Let $Z \to T$ be as in the statement of Lemma 3.2.10. Then the composite

$$Z \to T \to \mathcal{R}^a_{K,d} \tag{3.2.12}$$

lifts to a morphism $Z \to \mathcal{C}^a_{d[K:K^{\mathrm{basic}}],h} \times_{\mathcal{R}_{K^{\mathrm{basic}},d[K:K^{\mathrm{basic}}]}} \mathcal{R}_{K,d}$, and hence the morphism (3.2.12) factors through $\mathcal{R}^a_{K,d,K^{\mathrm{basic}},h}$. Since $Z \to T$ is scheme-theoretically dominant, the original morphism $T \to \mathcal{R}^a_{K,d}$ also factors through $\mathcal{R}^a_{K,d,K^{\mathrm{basic}},h}$, and hence through the inductive limit. Thus the lemma is proved. (We remark that this argument is essentially identical to that used to prove [EG21, Thm. 5.4.20].) □

We now turn to studying Γ_K-actions on our φ-modules. To ease notation, write $\Gamma = \Gamma_K$ from now on. We choose a topological generator γ of Γ, and let $\Gamma_{\mathrm{disc}} := \langle \gamma \rangle$; so $\Gamma_{\mathrm{disc}} \cong \mathbf{Z}$.

3.2.13 Remark. Since Γ_{disc} is dense in Γ, while a projective étale φ-module M is complete with respect to its canonical topology, in order to endow M with the structure of an étale (φ,Γ)-module, it suffices to equip M with a continuous action of Γ_{disc} (where we equip Γ_{disc} with the topology induced on it by Γ).

In order to study the properties of $\mathcal{X}_{K,d}$ we will take advantage of Remark 3.2.13. Accordingly, we now consider the moduli stack of projective étale φ-modules of rank d equipped with a semi-linear action of Γ_{disc}. We don't introduce particular notation for this stack, since the following proposition identifies it with a fixed point stack $\mathcal{R}_{K,d}^{\Gamma_{\mathrm{disc}}}$, which we now define.

There is a canonical action of Γ_{disc} on $\mathcal{R}_d$ (that is, a canonical morphism $\gamma\colon \mathcal{R}_d \to \mathcal{R}_d$): if M is an object of $\mathcal{R}_d(A)$, then $\gamma(M)$ is given by $\gamma^* M := \mathbf{A}_{K,A} \otimes_{\gamma,\mathbf{A}_{K,A}} M$. (Note that this is naturally a φ-module, because the action of γ on $\mathbf{A}_{K,A}$ commutes with φ.) Then we set

$$\mathcal{R}_d^{\Gamma_{\mathrm{disc}}} := \mathcal{R}_d \underset{\Delta,\mathcal{R}_d \times \mathcal{R}_d,\Gamma_\gamma}{\times} \mathcal{R}_d,$$

where Δ is the diagonal of $\mathcal{R}_d$ and Γ_γ is the graph of γ, so that $\Gamma_\gamma(x) = (x, \gamma(x))$.

3.2.14 Proposition. *The moduli stack of projective étale φ-modules of rank d equipped with a semi-linear action of Γ_{disc} is isomorphic to the fixed point stack $\mathcal{R}_d^{\Gamma_{\mathrm{disc}}}$.*

Proof. Using the usual construction of the 2-fiber product, we see that $\mathcal{R}_d^{\Gamma_{\mathrm{disc}}}$ consists of tuples (x,y,α,β), with x,y being objects of $\mathcal{R}_d$, and $\alpha\colon x \xrightarrow{\sim} y$ and $\beta\colon \gamma(x) \xrightarrow{\sim} y$ being isomorphisms. This is equivalent to the category fibred in groupoids given by pairs (x,ι) consisting of an object x of $\mathcal{R}_d$ and an isomorphism $\iota\colon \gamma(x) \xrightarrow{\sim} x$. Thus an object of $\mathcal{R}_d^{\Gamma_{\mathrm{disc}}}(A)$ is a projective étale φ-module of rank d with A-coefficients M, together with an isomorphism of φ-modules $\iota\colon \gamma^* M \xrightarrow{\sim} M$; but this isomorphism is precisely the data of a semi-linear action of $\Gamma_{\mathrm{disc}} = \langle \gamma \rangle$ on M, as required. □

Since $\mathcal{R}_d$ is an Ind-algebraic stack, so is $\mathcal{R}_d^{\Gamma_{\mathrm{disc}}}$. More precisely, we have the following lemma.

3.2.15 Lemma. *$\mathcal{R}_d^{\Gamma_{\mathrm{disc}}}$ is a limit preserving Ind-algebraic stack, whose diagonal is representable by algebraic spaces, affine, and of finite presentation.*

Proof. The description of $\mathcal{R}_d^{\Gamma_{\mathrm{disc}}}$ as a fiber product shows that the forgetful morphism $\mathcal{R}_d^{\Gamma_{\mathrm{disc}}} \to \mathcal{R}_d$ (given by forgetting the Γ_{disc}-action) is representable by algebraic spaces, affine, and of finite presentation, since these properties hold for the diagonal of $\mathcal{R}_d$, by Corollary 3.2.6. The claim of the lemma is now seen to follow via another application of Corollary 3.2.6, together with Corollary 3.2.9. □

Restricting the Γ-action on an étale (φ, Γ)-module to Γ_{disc} (and taking into account Proposition 3.2.14), we obtain a morphism $\mathcal{X}_{K,d} \to \mathcal{R}_d^{\Gamma_{\mathrm{disc}}}$, which by Remark 3.2.13 is fully faithful. Thus $\mathcal{X}_{K,d}$ may be regarded as a substack of $\mathcal{R}_d^{\Gamma_{\mathrm{disc}}}$.

As already noted, the first step in proving that $\mathcal{X}_{K,d}$ is a Noetherian formal algebraic stack is to show that it is an Ind-algebraic stack. Although $\mathcal{X}_{K,d}$ is a substack of the Ind-algebraic stack $\mathcal{R}_d^{\Gamma_{\mathrm{disc}}}$, it is not a closed substack, but should rather be thought of as a certain formal neighborhood of $\mathcal{X}_{d,\mathrm{red}}$ in $\mathcal{R}_d^{\Gamma_{\mathrm{disc}}}$ (see Remark 7.2.18), and so even this statement will require additional work to prove. We begin with the following lemma, which allows us to reduce to the case that K is basic.

3.2.16 Lemma. *We have a 2-Cartesian diagram*

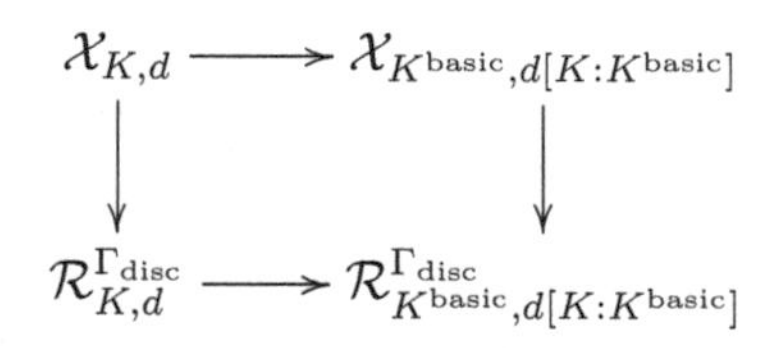

$$\begin{array}{ccc} \mathcal{X}_{K,d} & \longrightarrow & \mathcal{X}_{K^{\mathrm{basic}},d[K:K^{\mathrm{basic}}]} \\ \downarrow & & \downarrow \\ \mathcal{R}_{K,d}^{\Gamma_{\mathrm{disc}}} & \longrightarrow & \mathcal{R}_{K^{\mathrm{basic}},d[K:K^{\mathrm{basic}}]}^{\Gamma_{\mathrm{disc}}} \end{array}$$

where the horizontal arrows are the natural maps (forgetting the $\mathbf{A}_K$-module structure), and the vertical arrows are the monomorphisms given by restricting the Γ-action to Γ_{disc}.

Proof. Unwinding the definitions, we need to show that if A is an $\mathcal{O}/\varpi^a$-algebra for some $a \geq 1$, and M is an étale φ-module over $\mathbf{A}_{K,A}$ with a semi-linear action of Γ_{disc}, then the action of Γ_{disc} extends to a continuous action of Γ if and only if the same is true of M regarded as a module $\mathbf{A}_{K^{\mathrm{basic}},A}$. Since $\mathbf{A}_{K,A}$ is free of finite rank over $\mathbf{A}_{K^{\mathrm{basic}},A}$, this is clear (for example, because it follows from Lemma D.7 that the set of lattices in M regarded as an $\mathbf{A}_{K,A}$-module is cofinal in the set of lattices in M regarded as an $\mathbf{A}_{K^{\mathrm{basic}},A}$-module). □

As a consequence of Lemmas 3.2.5 and 3.2.16, we can deduce some properties of $\mathcal{X}_d$ from the corresponding properties of $\mathcal{R}_d$.

3.2.17 Proposition

1. The morphism $\mathcal{X}_{K,d} \to \mathcal{X}_{K^{\text{basic}},d[K:K^{\text{basic}}]}$ is representable by algebraic spaces, affine, and of finite presentation.
2. The diagonal of $\mathcal{X}_{K,d}$ is representable by algebraic spaces, affine, and of finite presentation.

Proof. To make the proof easier to read, we write $\mathcal{R}_K$ for $\mathcal{R}_{K,d}$ and $\mathcal{R}_{K^{\text{basic}}}$ for $\mathcal{R}_{K^{\text{basic}},d[K:K^{\text{basic}}]}$. We factor the morphism $\mathcal{R}_K^{\Gamma_{\text{disc}}} \to \mathcal{R}_{K^{\text{basic}}}^{\Gamma_{\text{disc}}}$ as

$$\mathcal{R}_K^{\Gamma_{\text{disc}}} := \mathcal{R}_K \underset{\Delta, \mathcal{R}_K \times \mathcal{R}_K, \Gamma_\gamma}{\times} \mathcal{R}_K \to \mathcal{R}_K \underset{\Delta, \mathcal{R}_{K^{\text{basic}}} \times \mathcal{R}_{K^{\text{basic}}}, \Gamma_\gamma}{\times} \mathcal{R}_K$$
$$\to \mathcal{R}_{K^{\text{basic}}} \underset{\Delta, \mathcal{R}_{K^{\text{basic}}} \times \mathcal{R}_{K^{\text{basic}}}, \Gamma_\gamma}{\times} \mathcal{R}_{K^{\text{basic}}} =: \mathcal{R}_{K^{\text{basic}}}^{\Gamma_{\text{disc}}}.$$

By Lemma 3.2.5, the morphism $\mathcal{R}_K \to \mathcal{R}_{K^{\text{basic}}}$ is representable by algebraic spaces, affine, and of finite presentation. Thus so is the second arrow in the preceding displayed expression. The first arrow is a base change of a product of two copies of the diagonal

$$\mathcal{R}_K \to \mathcal{R}_K \times_{\mathcal{R}_{K^{\text{basic}}}} \mathcal{R}_K.$$

As in the proof of Lemma 3.2.8, we deduce from Lemma 3.2.5 that this diagonal is representable by algebraic spaces, affine, and of finite presentation. Thus so is the first arrow in the expression, and hence so is the morphism $\mathcal{R}_K^{\Gamma_{\text{disc}}} \to \mathcal{R}_{K^{\text{basic}}}^{\Gamma_{\text{disc}}}$.

It follows from Lemma 3.2.16 that $\mathcal{X}_{K,d} \to \mathcal{X}_{K^{\text{basic}},d[K:K^{\text{basic}}]}$ is then also representable by algebraic spaces, affine, and of finite presentation. The claimed properties of the diagonal of $\mathcal{X}_{K,d}$ follow from the corresponding properties of the diagonal of $\mathcal{R}_{K,d}^{\Gamma_{\text{disc}}}$ proved in Lemma 3.2.15, together with the fact that $\mathcal{X}_{K,d} \to \mathcal{R}_{K,d}^{\Gamma_{\text{disc}}}$ is a monomorphism. □

From now on we will typically drop K from the notation, simply writing $\mathcal{X}_d$, $\mathcal{R}_d$ and so on. We conclude this section by showing that $\mathcal{X}_d$ is limit preserving, using the material on lattices and continuity developed in Appendix D.

3.2.18 Lemma. *We have* $\gamma(T) - T \in p\mathbf{A}_{K,A} + T^2\mathbf{A}_{K,A}^+$. *If K is basic then* $\gamma(T) - T \in (p, T)T\mathbf{A}_{K,A}^+$.

Proof. Given the definitions of $\mathbf{A}_{K,A}$ and $\mathbf{A}_{K,A}^+$ in terms of $\mathbf{A}_K$ and $\mathbf{A}_K^+$, it suffices to prove the lemma for these latter rings (i.e., in the case when $A = \mathbf{Z}_p$). We then reduce modulo p, and write $\gamma(T) = \sum_{i=n}^{\infty} a_i T^i$, with $a_i \in k_\infty$ and $a_n \neq 0$. Since the action of γ is continuous for the T-adic topology, we see in particular that for $m > 0$ sufficiently large, we have $\gamma(T)^m = \gamma(T^m) \in Tk_\infty[[T]]$.

Since $\gamma(T)^m = a_n^m T^{nm} + \dots$, we see that $n > 0$. Since $\gamma^{p^s} \to 1$ as $s \to \infty$, we see that $n = 1$ and that $a_1^{p^s} = 1$ for all sufficiently large s, which implies that $a_1 = 1$.

If K is basic then we have chosen T so that $\gamma(T) \in T\mathbf{A}^+_{K,A}$, so that $\gamma(T) - T \in (p,T)T\mathbf{A}^+_{K,A}$ by the above. □

If K is basic, it follows from Lemma 3.2.18 that (D.24) holds, so that by Lemma D.27 we can use the material on T-quasi-linear morphisms developed in Appendix D to study the action of γ.

3.2.19 Lemma. *$\mathcal{X}_d$ is limit preserving.*

Proof. By Proposition 3.2.17 it suffices to prove this in the case that K is basic (since a morphism which is of finite presentation is in particular limit preserving). Since $\mathcal{X}_d \hookrightarrow \mathcal{R}_d^{\Gamma_{\text{disc}}}$ is fully faithful, and $\mathcal{R}_d^{\Gamma_{\text{disc}}}$ is limit preserving by Corollary 3.1.5, it suffices to prove that $\mathcal{X}_d$ is limit preserving on objects. Since $\mathcal{R}_d^{\Gamma_{\text{disc}}}$ is limit preserving on objects, we are reduced to showing that if $T = \varprojlim_i T_i$ is a limit of affine schemes, and $T_{i_0} \to \mathcal{R}_d^{\Gamma_{\text{disc}}}$ is a morphism with the property that the composite

$$T \to T_{i_0} \to \mathcal{R}_d^{\Gamma_{\text{disc}}}$$

factors through $\mathcal{X}_d$, then for some $i \geq i_0$, the composite

$$T_i \to T_{i_0} \to \mathcal{R}_d^{\Gamma_{\text{disc}}}$$

factors through $\mathcal{X}_d$. This follows from the equivalence of conditions (1) and (7) of Lemma D.28, together with Lemma D.31. □

3.3 WEAK WACH MODULES

In this section we introduce the notion of a weak Wach module of height at most h and level at most s. These will play a purely technical auxiliary role for us, and will be used only in order to show that $\mathcal{X}_d$ is an Ind-algebraic stack.

We suppose throughout this section that K is basic.

3.3.1 Remark. By Lemma D.28, if A is an $\mathcal{O}/\varpi^a$-algebra for some $a \geq 1$, and $\mathfrak{M}$ is a rank d projective φ-module of T-height $\leq h$ over A, such that $\mathfrak{M}[1/T]$ is equipped with a semi-linear action of Γ_{disc}, then this action extends to a continuous action of Γ if and only if for some $s \geq 0$ we have $(\gamma^{p^s} - 1)(\mathfrak{M}) \subseteq T\mathfrak{M}$.

3.3.2 Definition. Suppose that K is basic. A *rank d projective weak Wach module of T-height $\leq h$ with A-coefficients* is a rank d projective φ-module $\mathfrak{M}$ over $\mathbf{A}^+_{K,A}$, which is of T-height $\leq h$, such that $\mathfrak{M}[1/T]$ is equipped with a semi-linear action of Γ which is furthermore continuous when $\mathfrak{M}[1/T]$ is endowed with its canonical topology (see Remark D.2).

If $s \geq 0$, then we say that $\mathfrak{M}$ has *level* $\leq s$ if $(\gamma^{p^s}-1)(\mathfrak{M}) \subseteq T\mathfrak{M}$. Remark 3.3.1 shows that any projective weak Wach module is of level $\leq s$ for some $s \geq 0$.

A special role in the classical theory of (φ,Γ)-modules is played by the weak Wach modules of level 0. However, this will not be important for us.

3.3.3 Definition. We let $\mathcal{W}_{d,h}$ denote the moduli stack of rank d projective weak Wach modules of T-height $\leq h$. (That this is an *fpqc* stack follows as in Definition 3.2.1.) We let $\mathcal{W}_{d,h,s}$ denote the substack of rank d projective weak Wach modules of T-height $\leq h$ and level $\leq s$.

We recall from Section 3.1 that there is a p-adic formal algebraic stack $\mathcal{C}_{d,h}$ classifying rank d projective φ-modules of T-height at most h. We consider the fiber product $\mathcal{R}_d^{\Gamma_{\rm disc}} \times_{\mathcal{R}_d} \mathcal{C}_{d,h}$, where the map $\mathcal{R}_d^{\Gamma_{\rm disc}} \to \mathcal{R}_d$ is the canonical morphism given by forgetting the $\Gamma_{\rm disc}$ action. By Proposition 3.2.14, this is the moduli stack of rank d projective φ-modules $\mathfrak{M}$ of T-height at most h, equipped with a semi-linear action of $\Gamma_{\rm disc}$ on $\mathfrak{M}[1/T]$.

3.3.4 Lemma. *$\mathcal{R}_d^{\Gamma_{\rm disc}} \times_{\mathcal{R}_d} \mathcal{C}_{d,h}$ is a p-adic formal algebraic stack of finite presentation over* Spf $\mathcal{O}$.

Proof. The map $\mathcal{R}_d^{\Gamma_{\rm disc}} \to \mathcal{R}_d$ is representable by algebraic spaces and of finite presentation, being a base change of the diagonal of $\mathcal{R}_d$, which is representable by algebraic spaces and of finite presentation by Corollary 3.1.5. Again by Corollary 3.1.5, $\mathcal{C}_{d,h}$ is a p-adic formal algebraic stack of finite presentation, and thus $\mathcal{R}_d^{\Gamma_{\rm disc}} \times_{\mathcal{R}_d} \mathcal{C}_{d,h}$ is a p-adic formal algebraic stack of finite presentation, as claimed. ☐

Restricting the Γ-action on a weak Wach module to $\Gamma_{\rm disc}$, we obtain a morphism $\mathcal{W}_{d,h} \to \mathcal{R}_d^{\Gamma_{\rm disc}} \times_{\mathcal{R}_d} \mathcal{C}_{d,h}$, which, by the evident analogue for weak Wach modules of Remark 3.2.13, is fully faithful. Thus $\mathcal{W}_{d,h}$ may be regarded as a substack of $\mathcal{R}_d^{\Gamma_{\rm disc}} \times_{\mathcal{R}_d} \mathcal{C}_{d,h}$. The following proposition records the basic properties of the stacks $\mathcal{W}_{d,h}$ and $\mathcal{W}_{d,h,s}$.

3.3.5 Proposition

1. For $s \geq 0$, the morphism

$$\mathcal{W}_{d,h,s} \longrightarrow \mathcal{R}_d^{\Gamma_{\rm disc}} \times_{\mathcal{R}_d} \mathcal{C}_{d,h}$$

is a closed immersion of finite presentation. In particular, each of the stacks $\mathcal{W}_{d,h,s}$ is a p-adic formal algebraic stack of finite presentation over Spf $\mathcal{O}$.

2. If $s' \geq s$, then the canonical monomorphism $\mathcal{W}_{d,h,s} \hookrightarrow \mathcal{W}_{d,h,s'}$ is a closed immersion of finite presentation.
3. The canonical morphism $\varinjlim_s \mathcal{W}_{d,h,s} \to \mathcal{W}_{d,h}$ is an isomorphism. In particular, $\mathcal{W}_{d,h}$ is an Ind-algebraic stack.

Proof. Since $\mathcal{R}_d^{\Gamma_{\mathrm{disc}}} \times_{\mathcal{R}_d} \mathcal{C}_{d,h}$ classifies φ-modules $\mathfrak{M}$ that are projective of rank d and of T-height $\leq h$, which are endowed with a Γ_{disc}-action on the underlying étale φ-module, in order to prove (1), we must show that the condition $(\gamma^{p^s} - 1)(\mathfrak{M}) \subseteq T\mathfrak{M}$ is a closed condition, and is determined by finitely many equations. It suffices to check this after pulling back via an arbitrary morphism $\operatorname{Spec} A \to \mathcal{R}_d^{\Gamma_{\mathrm{disc}}} \times_{\mathcal{R}_d} \mathcal{C}_{d,h}$, where A is an $\mathcal{O}/\varpi^a$-algebra for some $a \geq 1$.

This morphism gives rise to a projective $\mathbf{A}_{K,A}^+$ φ-module $\mathfrak{M}$, such that $M := \mathfrak{M}[1/T]$ is étale, and is furthermore endowed with a semi-linear Γ_{disc}-action. As already stated above, we must now check that the condition $(\gamma^{p^s} - 1)(\mathfrak{M}) \subseteq T\mathfrak{M}$ is a finitely presented closed condition (in the sense that it holds after replacing $\mathfrak{M}$ by $\mathfrak{M}_B := \mathbf{A}_{K,B}^+ \otimes_{\mathbf{A}_{K,A}^+} \mathfrak{M}$ for an A-algebra B if and only if $\operatorname{Spec} B \to \operatorname{Spec} A$ factors through a certain finitely presented closed subscheme of $\operatorname{Spec} A$).

Note firstly that for n sufficiently large, we have $(\gamma^{p^s} - 1)(T^n\mathfrak{M}) \subseteq T\mathfrak{M}$. Indeed, writing

$$\gamma^{p^s} - 1 = ((\gamma - 1) + 1)^{p^s} - 1$$

and expanding via the binomial theorem, it suffices to show that for n sufficiently large, we have $(\gamma - 1)^m(T^n\mathfrak{M}) \subseteq T\mathfrak{M}$ for all $0 \leq m \leq p^s$; and this is immediate from Lemma D.19.

We next choose a finitely generated projective $\mathbf{A}_{K,A}^+$-module $\mathfrak{N}$ such that $\mathfrak{F} := \mathfrak{M} \oplus \mathfrak{N}$ is free, and write $N := \mathfrak{N}[1/T]$, $F := \mathfrak{F}[1/T]$. Extend the morphism $\gamma^{p^s} - 1\colon M \to M$ to the morphism $f\colon F \to F$ given by $(\gamma^{p^s} - 1, 0)\colon M \oplus N \to M \oplus N$, so that the condition that $(\gamma^{p^s} - 1)(\mathfrak{M}) \subseteq T\mathfrak{M}$ is equivalent to asking that $f(\mathfrak{F}) \subseteq T\mathfrak{F}$. By our choice of n above, the morphism $\mathfrak{F} \to F/T\mathfrak{F}$ induced by f factors through a morphism $\mathfrak{F}/T^n\mathfrak{F} \to F/T\mathfrak{F}$; since $\mathfrak{F}$ is finitely generated, after enlarging n if necessary, it factors through $\mathfrak{F}/T^n\mathfrak{F} \to T^{-n}\mathfrak{F}/T\mathfrak{F}$. The vanishing of this morphism is obviously a closed condition; indeed it is given by the vanishing of finitely many matrix entries, so is closed and of finite presentation, as required.

Since $\mathcal{R}_d^{\Gamma_{\mathrm{disc}}} \times_{\mathcal{R}_d} \mathcal{C}_{d,h}$ is a p-adic formal algebraic stack of finite presentation over $\operatorname{Spf} \mathcal{O}$, it follows that so is each $\mathcal{W}_{d,h,s}$. That $\mathcal{W}_{d,h,s} \hookrightarrow \mathcal{W}_{d,h,s'}$ is a closed immersion of finite presentation follows immediately from a consideration of the composite

$$\mathcal{W}_{d,h,s} \hookrightarrow \mathcal{W}_{d,h,s'} \hookrightarrow \mathcal{R}_d^{\Gamma_{\mathrm{disc}}} \times_{\mathcal{R}_d} \mathcal{C}_{d,h},$$

as we have just shown that both the composite and the second morphism are closed immersions of finite presentation. Thus we have proved (1) and (2).

For (3), we need to show that every morphism $\operatorname{Spec} A \to \mathcal{W}_{d,h}$, where A is an $\mathcal{O}/\varpi^a$-algebra for some $a \geq 1$, factors through $\mathcal{W}_{d,h,s}$ for s sufficiently large.

As we already noted in Definition 3.3.2, this follows from Remark 3.3.1. Since each $\mathcal{W}_{d,h,s}$ is an Ind-algebraic stack, so is $\mathcal{W}_{d,h}$. □

3.4 $\mathcal{X}_d$ IS AN IND-ALGEBRAIC STACK

Continue to assume that K is basic. By definition, we have a 2-Cartesian diagram

$$\begin{array}{ccc} \mathcal{W}_{d,h} & \longrightarrow & \mathcal{R}_d^{\Gamma_{\mathrm{disc}}} \times_{\mathcal{R}_d} \mathcal{C}_{d,h} \\ \downarrow & & \downarrow \\ \mathcal{X}_d & \longrightarrow & \mathcal{R}_d^{\Gamma_{\mathrm{disc}}} \end{array} \tag{3.4.1}$$

If $h' \geq h$ then the closed immersion $\mathcal{C}_{d,h} \hookrightarrow \mathcal{C}_{d,h'}$ is compatible with the morphisms from each of its source and target to $\mathcal{R}_d$, and so we obtain a closed immersion

$$\mathcal{W}_{d,h} \hookrightarrow \mathcal{W}_{d,h'}. \tag{3.4.2}$$

By construction, the morphisms $\mathcal{W}_{d,h} \to \mathcal{X}_d$ are compatible, as h varies, with the closed immersions (3.4.2). Thus we also obtain a morphism

$$\varinjlim_h \mathcal{W}_{d,h} \to \mathcal{X}_d. \tag{3.4.3}$$

Roughly speaking, we will prove that $\mathcal{X}_d$ is an Ind-algebraic stack by showing that it is the "scheme-theoretic image" of the morphism $\varinjlim_h \mathcal{W}_{d,h} \to \mathcal{R}_d^{\Gamma_{\mathrm{disc}}}$ induced by (3.4.3). More precisely, choose $s \geq 0$, and consider the composite

$$\mathcal{W}_{d,h,s} \to \mathcal{W}_{d,h} \to \mathcal{X}_d \to \mathcal{R}_d^{\Gamma_{\mathrm{disc}}}. \tag{3.4.4}$$

This admits the alternative factorization

$$\mathcal{W}_{d,h,s} \to \mathcal{W}_{d,h} \to \mathcal{R}_d^{\Gamma_{\mathrm{disc}}} \times_{\mathcal{R}_d} \mathcal{C}_{d,h} \to \mathcal{R}_d^{\Gamma_{\mathrm{disc}}}.$$

Proposition 3.3.5 shows that the composite of the first two arrows is a closed embedding of finite presentation, while Corollary 3.1.5 shows that the third arrow is representable by algebraic spaces, proper, and of finite presentation. Thus (3.4.4) is representable by algebraic spaces, proper, and of finite presentation.

Fix an integer $a \geq 1$, and write $\mathcal{W}_{d,h,s}^a := \mathcal{W}_{d,h,s} \times_{\mathrm{Spf}\,\mathcal{O}} \operatorname{Spec} \mathcal{O}/\varpi^a$. Proposition 3.3.5 shows that $\mathcal{W}_{d,h,s}$ is a p-adic formal algebraic stack of finite presentation over $\mathrm{Spf}\,\mathcal{O}$, and so $\mathcal{W}_{d,h,s}^a$ is an algebraic stack, and a closed substack of $\mathcal{W}_{d,h,s}$.

3.4.5 Definition. We let $\mathcal{X}^a_{d,h,s}$ denote the scheme-theoretic image of the composite

$$\mathcal{W}^a_{d,h,s} \hookrightarrow \mathcal{W}_{d,h,s} \xrightarrow{(3.4.4)} \mathcal{R}_d^{\Gamma_{\text{disc}}}, \tag{3.4.6}$$

defined via the formalism of scheme-theoretic images for morphisms of Ind-algebraic stacks developed in Appendix A.

More concretely, since $\mathcal{R}_d^{\Gamma_{\text{disc}}}$ is an Ind-algebraic stack, constructed as the 2-colimit of a directed system of algebraic stacks whose transition morphisms are closed immersions, the morphism (3.4.6), which is representable by algebraic spaces, proper, and of finite presentation, factors through a closed algebraic substack $\mathcal{Z}$ of $\mathcal{R}_d^{\Gamma_{\text{disc}}}$. We then define $\mathcal{X}^a_{d,h,s}$ to be the scheme-theoretic image of $\mathcal{W}^a_{d,h,s}$ in $\mathcal{Z}$. Note that $\mathcal{X}^a_{d,h,s}$ is a closed algebraic substack of $\mathcal{R}_d^{\Gamma_{\text{disc}}}$, and is independent of the choice of $\mathcal{Z}$.

3.4.7 Remark. As already observed above, the morphism (3.4.6) factors through $\mathcal{X}_d$. However, since at this point we don't know that $\mathcal{X}_d$ is Ind-algebraic, we can't directly define a scheme-theoretic image of $\mathcal{W}^a_{d,h,s}$ in $\mathcal{X}_d$. It might be possible to do this using the formalism of [EG21]; since (3.4.6) is proper, this scheme-theoretic image would then coincide with $\mathcal{X}^a_{d,h,s}$. We don't do this here; but we do prove somewhat more directly, in Lemma 3.4.9 below, that $\mathcal{X}^a_{d,h,s}$ is a substack of $\mathcal{X}_d$.

As in Definition D.6, a *lattice* $\mathfrak{M}$ in a projective étale φ-module M is a finitely generated $\mathbf{A}^+_{K,A}$-submodule of M whose $\mathbf{A}_{K,A}$-span is M. Note that $\mathfrak{M}$ is not assumed to be projective.

3.4.8 Lemma. *Suppose that M is a projective étale φ-module of rank d over a finite type Artinian $\mathcal{O}/\varpi^a$-algebra A, and that M is endowed with an action of Γ_{disc}, such that the corresponding morphism* $\operatorname{Spec} A \to \mathcal{R}_d^{\Gamma_{\text{disc}}}$ *factors through $\mathcal{X}^a_{d,h,s}$. Then M contains a φ-invariant lattice $\mathfrak{M}$ of T-height $\leq h$, such that $(\gamma^{p^s} - 1)(\mathfrak{M}) \subseteq T\mathfrak{M}$.*

Proof. Since an Artinian ring is a direct product of Artinian local rings, it suffices to treat the case that A is local. Let the residue field of A be $\mathbf{F}'$, a finite extension of $\mathbf{F}$, and write $\mathcal{O}'$ for the ring of integers in the compositum of E and $W(\mathbf{F}')[1/p]$, so that $\mathcal{O}'$ has residue field $\mathbf{F}'$; note that A is naturally an $\mathcal{O}'$-algebra.

The projective étale φ-module M is in fact free of rank d, since $\mathbf{A}_{K,A}$ is again a local ring, and we fix an (ordered) basis for M as an $\mathbf{A}_{K,A}$-module. Write $M_{\mathbf{F}'} := \mathbf{A}_{K,\mathbf{F}'} \otimes_{\mathbf{A}_{K,A}} M$; the $\mathbf{A}_{K,A}$-basis of M gives rise to a corresponding $\mathbf{A}_{K,\mathbf{F}'}$-basis of $M_{\mathbf{F}'}$. Let $\mathcal{C}_{\mathcal{O}'}$ be the category of Artinian local $\mathcal{O}'/\varpi^a$-algebras for which the structure map induces an isomorphism on residue fields. By a *lifting* of $M_{\mathbf{F}'}$ to an object Λ of $\mathcal{C}_{\mathcal{O}'}$, we mean a triple consisting of an étale

φ-module M_Λ which is free of rank d, a choice of (ordered) $\mathbf{A}_{K,\Lambda}$-basis of M_Λ, and an isomorphism $M_\Lambda \otimes_\Lambda \mathbf{F}' \cong M_{\mathbf{F}'}$ of étale φ-modules which takes the chosen basis of M_Λ to the fixed basis of $M_{\mathbf{F}'}$. Let D be the functor $\mathcal{C}_{\mathcal{O}'} \to \underline{Sets}$ taking Λ to the set of isomorphism classes of liftings of $M_{\mathbf{F}'}$ to an étale φ-module M_Λ with Λ-coefficients, endowed with an action of Γ_{disc} lifting that on $M_{\mathbf{F}'}$. Note that A is an object of $\mathcal{C}_{\mathcal{O}'}$, and that our originally chosen basis for M is classified by a continuous morphism $R \to A$.

The functor D is pro-representable by an object R of pro-$\mathcal{C}_{\mathcal{O}'}$, by the same argument as in the proof of [EG21, Prop. 5.3.6]: namely, by Grothendieck's representability theorem, it suffices to prove the compatibility of D with fiber products in $\mathcal{C}_{\mathcal{O}'}$, which is obvious. The universal lifting M_R gives a morphism $\operatorname{Spf} R \to \mathcal{R}_d^{\Gamma_{\text{disc}}}$. The composite $\operatorname{Spec} A \to \operatorname{Spf} R \to \mathcal{R}_d^{\Gamma_{\text{disc}}}$ is of course just the morphism that classifies M.

We let D' denote the subfunctor of D consisting of those lifts for which there is a lattice $\mathfrak{M}_\Lambda$ of F-height $\leq h$ and level $\leq s$, with $\mathfrak{M}_\Lambda[1/T] = M$. (More precisely, we require that $T^h \mathfrak{M}_\Lambda \subseteq (1 \otimes \varphi_{M_\Lambda})(\varphi^* \mathfrak{M}_\Lambda)$ and $(\gamma^{p^s} - 1)(\mathfrak{M}_\Lambda) \subseteq T\mathfrak{M}_\Lambda$.) We claim that the functor D' is pro-representable by a quotient S of R. To see this, it is again enough to show that D' preserves fiber products, and in turn it is enough to show that the property of an étale φ-module M with an action of Γ_{disc} admitting a φ-stable lattice $\mathfrak{M}$ of T-height $\leq h$, and such that $(\gamma^{p^s} - 1)(\mathfrak{M}) \subseteq T\mathfrak{M}$, is stable under taking direct sums and subquotients. This is obvious for direct sums, and the case of subquotients follows as in [EG21, Lem. 5.3.10] (which is the same result without the conditions on Γ_{disc}). More precisely, once checks that if we have a short exact sequence $0 \to M' \to M \to M'' \to 0$ of étale φ-modules with Λ-coefficients and actions of Γ_{disc}, and $\mathfrak{M}$ is a lattice of the appropriate kind in M, then the kernel and image of the map $\mathfrak{M} \to M''$ give the appropriate lattices $\mathfrak{M}'$, $\mathfrak{M}''$ in M', M'' respectively. (The properties that $(\gamma^{p^s} - 1)(\mathfrak{M}') \subseteq T\mathfrak{M}'$ and $(\gamma^{p^s} - 1)(\mathfrak{M}'') \subseteq T\mathfrak{M}''$ follow from a short diagram chase, using that the composite $\mathfrak{M} \xrightarrow{(\gamma^{p^s}-1)} M \to M/T\mathfrak{M}$ vanishes by hypothesis.)

By the definition of D', the statement of the lemma is equivalent to the statement that the map $R \to A$ factors through S. Let $X = \mathcal{W}_{d,h,s} \times_{\mathcal{R}_d^{\Gamma_{\text{disc}}}} \operatorname{Spf} R$, a formal algebraic space, and let $\operatorname{Spf} T$ be the scheme-theoretic image of the morphism $X \to \operatorname{Spf} R$. Since the morphism $\operatorname{Spec} A \to \mathcal{R}_d^{\Gamma_{\text{disc}}}$ factors through $\mathcal{X}^a_{d,h,s}$ by hypothesis, the morphism $\operatorname{Spec} A \to \operatorname{Spf} R$ factors through $\operatorname{Spf} T$, so it is enough to show that $\operatorname{Spf} T$ is a closed formal subscheme of $\operatorname{Spf} S$. By Lemma A.32, it in turn suffices to show that if A' is a finite type Artinian local R-algebra for which $R \to A'$ factors through a discrete quotient of R, and for which the morphism $\mathcal{W}_{d,h,s} \times_{\mathcal{R}_d^{\Gamma_{\text{disc}}}} \operatorname{Spec} A' \to \operatorname{Spec} A'$ admits a section, then $R \to A'$ factors through S.

By replacing R with $R \otimes_{W(\mathbf{F}')} \kappa(A')$ we can reduce to the case that A' has residue field $\mathbf{F}'$ (*cf.* [EG21, Cor. 5.3.18]), so that $R \to A'$ factors through S if and only if the étale φ-module corresponding to $\operatorname{Spec} A' \to \mathcal{R}_d^{\Gamma_{\text{disc}}}$ admits a φ-stable lattice $\mathfrak{M}_{A'}$ of T-height $\leq h$ for which $(\gamma^{p^s} - 1)(\mathfrak{M}_{A'}) \subseteq T\mathfrak{M}_{A'}$. But a section to the morphism $\mathcal{W}_{d,h,s} \times_{\mathcal{R}_d^{\Gamma_{\mathrm{disc}}}} \operatorname{Spec} A' \to \operatorname{Spec} A'$ gives us a morphism $\operatorname{Spec} A' \to \mathcal{W}_{d,h,s}$, and the corresponding weak Wach module provides the required lattice. □

3.4.9 Lemma. *Each $\mathcal{X}^a_{d,h,s}$ is a closed substack of $\mathcal{X}_d$.*

Proof. Since $\mathcal{X}^a_{d,h,s}$ is a closed substack of $\mathcal{R}_d^{\Gamma_{\mathrm{disc}}}$, if it is a substack of $\mathcal{X}_d$, it will in fact be a closed substack. Thus it suffices to show that $\mathcal{X}^a_{d,h,s}$ is indeed a substack of $\mathcal{X}_d$. Since $\mathcal{X}^a_{d,h,s}$ is limit preserving, it is enough to check that if A is a finite type $\mathcal{O}/\varpi^a$-algebra, then for any morphism $\operatorname{Spec} A \to \mathcal{X}^a_{d,h,s}$, the composite morphism $\operatorname{Spec} A \to \mathcal{X}^a_{d,h,s} \to \mathcal{R}_d^{\Gamma_{\mathrm{disc}}}$ factors through $\mathcal{X}_d$. Equivalently, if M denotes the étale φ-module over A, endowed with a Γ_{disc}-action, associated to the given point $\operatorname{Spec} A \to \mathcal{R}_d^{\Gamma_{\mathrm{disc}}}$, then we must show that the Γ_{disc}-action on M is continuous.

By Lemma D.28, to see that the Γ_{disc}-action on M is continuous, it suffices to produce a lattice $\mathfrak{M} \subseteq M$ such that $(\gamma^{p^s} - 1)(\mathfrak{M}) \subseteq T\mathfrak{M}$. We will produce such a lattice by reduction to the Artinian case, as follows. Let $\{A_i\}_{i\in I}$ be the directed system of Artinian quotients of A. Since A is of finite type over $\mathcal{O}/\varpi^a$, and so Noetherian, the natural map $A \to B := \prod_i A_i$ is injective. Lemma 3.4.8 shows that each base-changed module M_{A_i} admits a φ-invariant lattice $\mathfrak{M}_i$ of T-height $\le h$ satisfying the condition $(\gamma^{p^s} - 1)(\mathfrak{M}_i) \subseteq T\mathfrak{M}_i$.

Write $\mathfrak{M}_B := \prod_i \mathfrak{M}_i \subset M_B$, and set $\mathfrak{M} := M \cap \mathfrak{M}_B \subset M_B$. Since $(\gamma^{p^s} - 1)(\mathfrak{M}_B) \subseteq T\mathfrak{M}_B$, we have $(\gamma^{p^s} - 1)(\mathfrak{M}) \subseteq T\mathfrak{M}$, so it only remains to check that $\mathfrak{M}$ is a lattice. To see this, let $\mathfrak{M}'$ be a finite height lattice in M (i.e., a lattice in M which is φ-stable, and for which the cokernel of Φ is killed by a power of T; such a lattice exists by [EG21, Lem. 5.2.15]). By Lemma D.7 (3), it suffices to prove that there is an integer $l \ge 0$ such that $T^l\mathfrak{M}' \subseteq \mathfrak{M} \subseteq T^{-l}\mathfrak{M}$. Accordingly, if for each i we let $\mathfrak{M}'_i$ denoted the image of $\mathfrak{M}_i \otimes_A A_i$ in M_{A_i}, we need to show that we have $T^l\mathfrak{M}'_i \subseteq \mathfrak{M}_i \subseteq T^{-l}\mathfrak{M}'_i$.

The existence of such an l is established in the course of the proof of [EG21, Lem. 5.3.14], although this is not explicitly recorded there. For the convenience of the reader, we recall the argument in our present setting. Increasing h if necessary, we may assume that $\mathfrak{M}'$ has T-height at most h. It follows that $\mathfrak{M}'_i$ and $\mathfrak{M}_i$ are each φ-stable lattices of height at most h in M_{A_i}. We claim that we may take $l = \lfloor (h+ap)/(p-1) \rfloor$.

To see this, let j be minimal such that $T^j\mathfrak{M}_i \subseteq \mathfrak{M}'_i$; then we have

$$\Phi_{\mathfrak{M}_i}(\varphi^*\mathfrak{M}_i) \subseteq \mathfrak{M}_i \subseteq T^{-j}\mathfrak{M}'_i \subseteq T^{-h-j}\Phi_{\mathfrak{M}'_i}(\varphi^*\mathfrak{M}'_i),$$

so that $T^{h+j}\varphi^*\mathfrak{M}_i \subseteq \varphi^*\mathfrak{M}'_i$. It follows from [EG21, Lem. 5.3.13] that $h+j > (j-a)p$, so that $j < (h+ap)/(p-1)$, and $j \le l$ by definition. Thus $T^l\mathfrak{M}_i \subseteq \mathfrak{M}'_i$, as claimed. Reversing the roles of $\mathfrak{M}_i$ and $\mathfrak{M}'_i$, we have $T^l\mathfrak{M}'_i \subseteq \mathfrak{M}_i$, and we are done. □

We now prove our first key structural result for $\mathcal{X}_d$, in the case when K is basic.

3.4.10 Proposition. *If K is basic, then the canonical morphism $\varinjlim \mathcal{X}^a_{d,h,s} \to \mathcal{X}_d$ is an isomorphism. Thus $\mathcal{X}_d$ is an Ind-algebraic stack, and may in fact be written as the inductive limit of algebraic stacks of finite presentation, with the transition maps being closed immersions.*

Proof. Note firstly that if $a' \geq a$, $h' \geq h$ and $s' \geq s$ then the canonical morphism $\mathcal{X}^a_{d,h,s} \to \mathcal{X}^{a'}_{d,h',s'}$ is a closed immersion by construction. By Lemma 3.4.9, each $\mathcal{X}^a_{d,h,s}$ is a closed substack of $\mathcal{X}_d$, so it remains to show that any morphism $T \to \mathcal{X}_d$ whose source is an affine scheme factors through some $\mathcal{X}^a_{d,h,s}$, or equivalently, that the closed immersion

$$\mathcal{X}^a_{d,h,s} \times_{\mathcal{X}_d} T \to T \tag{3.4.11}$$

is an isomorphism, for some choice of h and s.

Since $\mathcal{X}_d$ is limit preserving, by Lemma 3.2.19, we can reduce to the case where $T = \operatorname{Spec} A$ for a Noetherian $\mathcal{O}/\varpi^a$-algebra A. If M denotes the étale (φ, Γ)-module corresponding to the morphism $\operatorname{Spec} A \to \mathcal{X}_d$, then an application of [EG21, Prop. 5.4.7] shows that we may find a scheme-theoretically dominant morphism $\operatorname{Spec} B \to \operatorname{Spec} A$ such that M_B is free of rank d. If we show that the composite $\operatorname{Spec} B \to \operatorname{Spec} A \to \mathcal{X}_d$ factors through $\mathcal{X}^a_{d,h,s}$ for some h and s, then we see that the morphism $\operatorname{Spec} B \to \operatorname{Spec} A$ factors through the closed subscheme $\mathcal{X}^a_{d,h,s} \times_{\mathcal{X}_d} \operatorname{Spec} A$ of $\operatorname{Spec} A$. Since $\operatorname{Spec} B \to \operatorname{Spec} A$ is scheme-theoretically dominant, this implies that (3.4.11) is indeed an isomorphism, as required.

Since M_B is free, we may choose a φ-invariant free lattice $\mathfrak{M} \subseteq M_B$, of height $\leq h$ for some sufficiently large value of h. Since the Γ_{disc}-action on M, and hence on M_B, is continuous by assumption, Lemma D.28 shows that $(\gamma^{p^s} - 1)(\mathfrak{M}) \subseteq T\mathfrak{M}$ for some sufficiently large value of s. Then $\mathfrak{M}$ gives rise to a B-valued point of $\mathcal{W}^a_{d,h,s}$, whose image in $\mathcal{R}_d^{\Gamma_{\text{disc}}}$ is equal to the étale φ-module M_B. Thus the morphism $B \to \mathcal{X}_d$ corresponding to M_B does indeed factor through $\mathcal{X}^a_{d,h,s}$. □

Finally, we drop our assumption that K is basic.

3.4.12 Proposition. *Let K be an arbitrary finite extension of $\mathbf{Q}_p$. Then $\mathcal{X}_d$ is an Ind-algebraic stack, and may in fact be written as the inductive limit of algebraic stacks of finite presentation over $\operatorname{Spec} \mathcal{O}$, with the transition maps being closed immersions. Furthermore the diagonal of $\mathcal{X}_d$ is representable by algebraic spaces, affine, and of finite presentation.*

Proof. This is immediate from Propositions 3.2.17 and 3.4.10, together with Corollary 3.2.9. □

3.5 CANONICAL ACTIONS AND WEAK WACH MODULES

We now explain an alternative perspective on some of the above results, which gives more information about the moduli stacks of weak Wach modules. The results of this section are only used in Chapter 7, where we use them to establish a concrete description of $\mathcal{X}_d$ in the case $d=1$. For each $s \geq 0$, we write $K_{\text{cyc},s}$ for the unique subfield of K_{cyc} which is cyclic over K of degree p^s. The following lemma can be proved in exactly the same way as Lemma 4.3.3 below.

3.5.1 Lemma. *Assume that K is basic. For any fixed a, h, there is a constant $N(a,h)$ such that if $N \geq N(a,h)$, there is a positive integer $s(a,h,N)$ with the property that for any finite type $\mathcal{O}/\varpi^a$-algebra A, any finite projective φ-module $\mathfrak{M}$ over $\mathbf{A}^+_{K,A}$ of T-height at most h, and any $s \geq s(a,h,N)$, there is a unique continuous action of $G_{K_{\text{cyc},s}}$ on $\mathfrak{M}^{\text{inf}} := \mathbf{A}_{\text{inf},A} \otimes_{\mathbf{A}^+_{K,A}} \mathfrak{M}$ which commutes with φ and is semi-linear with respect to the natural action of $G_{K_{\text{cyc},s}}$ on $\mathbf{A}_{\text{inf},A}$, with the additional property that for all $g \in G_{K_{\text{cyc},s}}$ we have $(g-1)(\mathfrak{M}) \subset T^N \mathfrak{M}^{\text{inf}}$.*

It is possible to prove the following result purely in the world of (φ,Γ)-modules (by following the proof of Lemma 4.3.3, interpreting the existence of a semi-linear action of $\Gamma_{K_{\text{cyc},s}}$ in terms of linear maps between twists of φ-modules, satisfying certain compatibilities), but we have found it more straightforward to argue with Proposition 2.7.8.

3.5.2 Corollary. *Assume that K is basic. For any fixed a, h, there is a constant $N(a,h)$ such that if $N \geq N(a,h)$, there is a positive integer $s(a,h,N)$ with the property that for any finite type $\mathcal{O}/\varpi^a$-algebra A, any finite projective φ-module $\mathfrak{M}$ over $\mathbf{A}^+_{K,A}$ of T-height at most h, and any $s \geq s(a,h,N)$, there is a unique semi-linear action of $\langle \gamma^{p^s} \rangle$ on $\mathfrak{M}$ which commutes with φ, with the additional property that $(\gamma^{p^s}-1)(\mathfrak{M}) \subseteq T^N \mathfrak{M}$.*

In particular, this action gives $\mathfrak{M}[1/T]$ the structure of a projective étale $(\varphi, \Gamma_{K_{\text{cyc},s}})$-module.

Proof. By Lemma 3.5.1, there is a a unique continuous semi-linear action of $G_{K_{\text{cyc},s}}$ on $\mathfrak{M}^{\text{inf}} := \mathbf{A}_{\text{inf},A} \otimes_{\mathbf{A}^+_{K,A}} \mathfrak{M}$ which commutes with φ, with the additional property that for all $g \in G_{K_{\text{cyc},s}}$ we have $(g-1)(\mathfrak{M}) \subset T^N \mathfrak{M}^{\text{inf}}$. Write $M := \mathbf{A}_{K,A} \otimes_{\mathbf{A}^+_{K,A}} \mathfrak{M}$, $\widetilde{M} := W(\mathbf{C}^\flat)_A \otimes_{\mathbf{A}_{\text{inf},A}} \mathfrak{M}^{\text{inf}}$. By Proposition 2.7.8, the $G_{K_{\text{cyc},s}}$-action on $\widetilde{M}$ endows M with the structure of a projective $(\varphi, \Gamma_{K_{\text{cyc},s}})$-module with A-coefficients, and $\widetilde{M}$ with its $G_{K_{\text{cyc},s}}$-action is recovered as

$$\widetilde{M} = W(\mathbf{C}^\flat)_A \otimes_{\mathbf{A}_{K,A}} M.$$

Since $\Gamma_{K_{\mathrm{cyc},s}}$ is topologically generated by γ^{p^s}, and since $(g-1)(\mathfrak{M}) \subset T^N \mathfrak{M}^{\mathrm{inf}}$ for $g \in G_{K_{\mathrm{cyc},s}}$, we have $(\gamma^{p^s}-1)(\mathfrak{M}) \subset T^N \mathfrak{M}^{\mathrm{inf}}$. We also have $(\gamma^{p^s}-1)(\mathfrak{M}) \subseteq M$, and since $M \cap \mathfrak{M}^{\mathrm{inf}} = \mathfrak{M}$ (as is easily checked, by reducing to the case that $\mathfrak{M}$ is free, and then to the case that it is free of rank 1), we have $(\gamma^{p^s}-1)(\mathfrak{M}) \subset T^N \mathfrak{M}$, as required. For the uniqueness, note that if we have two such actions, then taking their difference gives a nonzero φ-linear morphism $(\gamma^{p^s})^* \mathfrak{M} \to T^N \mathfrak{M}$, which is impossible for N sufficiently large (see, e.g., the first part of the proof of Lemma 4.3.2 below).

Finally, the claim that this gives $\mathfrak{M}[1/T]$ the structure of a $(\varphi, \Gamma_{K_{\mathrm{cyc},s}})$-module is immediate from Remark 3.3.1. □

We continue to assume that K is basic. For any $s \geq 1$, we write $\Gamma_{s,\mathrm{disc}}$ to denote the subgroup of Γ_{disc} generated by γ^{p^s}. We write $\mathcal{R}_d^{\Gamma_{s,\mathrm{disc}}}$ in obvious analogy to the notation $\mathcal{R}_d^{\Gamma_{\mathrm{disc}}}$ introduced above. The result of Corollary 3.5.2 may be interpreted as constructing a morphism

$$\mathcal{C}_{d,h}^a \to \mathcal{R}_d^{\Gamma_{s,\mathrm{disc}}} \tag{3.5.3}$$

for sufficiently large values of s (depending on a and h), lying over the morphism $\mathcal{C}_{d,h}^a \to \mathcal{R}_d$. Of course there is also a morphism $\mathcal{R}_d^{\Gamma_{\mathrm{disc}}} \to \mathcal{R}_d^{\Gamma_{s,\mathrm{disc}}}$ given by restricting the action of Γ_{disc} to $\Gamma_{s,\mathrm{disc}}$.

Since the morphism $\mathcal{R}_d^{\Gamma_{s,\mathrm{disc}}} \to \mathcal{R}_d$ given by forgetting the $\Gamma_{s,\mathrm{disc}}$-action is representable by algebraic spaces and separated (indeed, even affine—it is a base change of the morphism $\mathcal{R}_d \to \mathcal{R}_d \times_{\mathcal{O}} \mathcal{R}_d$ giving the graph of the action of γ^{p^s}, and this latter morphism is representable by algebraic spaces and affine, since the diagonal morphism of $\mathcal{R}_d$ is so), the diagonal morphism

$$\mathcal{R}_d^{\Gamma_{s,\mathrm{disc}}} \to \mathcal{R}_d^{\Gamma_{s,\mathrm{disc}}} \times_{\mathcal{R}_d} \mathcal{R}_d^{\Gamma_{s,\mathrm{disc}}}$$

is a closed immersion. We may thus define a closed substack $\mathcal{Z}_{d,h,s}^a$ of $\mathcal{R}_d^{\Gamma_{\mathrm{disc}}} \times_{\mathcal{R}_d} \mathcal{C}_{d,h}^a$ via the following 2-Cartesian diagram:

$$\begin{array}{ccc} \mathcal{Z}_{d,h,s}^a & \longrightarrow & \mathcal{R}_d^{\Gamma_{\mathrm{disc}}} \times_{\mathcal{R}_d} \mathcal{C}_{d,h}^a \\ \downarrow & & \downarrow \\ \mathcal{R}_d^{\Gamma_{s,\mathrm{disc}}} & \longrightarrow & \mathcal{R}_d^{\Gamma_{s,\mathrm{disc}}} \times_{\mathcal{R}_d} \mathcal{R}_d^{\Gamma_{s,\mathrm{disc}}} \end{array}$$

in which the lower horizontal arrow is the diagonal, and the right-hand side vertical arrow is the product of the restriction morphism and the morphism (3.5.3). In less formal language, the stack $\mathcal{Z}_{d,h,s}^a$ parameterizes projective rank d φ-modules $\mathfrak{M}$ over $\mathbf{A}_{K,A}^+$ of T-height at most h, for which $\mathfrak{M}[1/T]$ is endowed with a Γ_{disc}-action extending the canonical action of $\Gamma_{s,\mathrm{disc}}$ given by Corollary 3.5.2.

3.5.4 Proposition. *Assume that K is basic. Then each $\mathcal{Z}^a_{d,h,s}$ is contained (as a closed algebraic substack) in $\mathcal{X}_d \times_{\mathcal{R}_d} \mathcal{C}_{d,h}$, and the natural morphism $\varinjlim_{a,s} \mathcal{Z}^a_{d,h,s} \to \mathcal{X}_d \times_{\mathcal{R}_d} \mathcal{C}_{d,h}$ is an isomorphism.*

Proof. That $\mathcal{Z}^a_{d,h,s}$ is a substack of $\mathcal{X}_d \times_{\mathcal{R}_d} \mathcal{C}_{d,h}$ is immediate from Remark 3.3.1; indeed, if we are given $\mathfrak{M}$ over $\mathbf{A}^+_{K,A}$ of T-height at most h, for which $\mathfrak{M}[1/T]$ is endowed with a Γ_{disc}-action extending the canonical action of $\Gamma_{s,\text{disc}}$ given by Corollary 3.5.2, then we have $(\gamma^{p^s} - 1)(\mathfrak{M}) \subseteq T^N\mathfrak{M} \subseteq T\mathfrak{M}$. Since $\mathcal{Z}^a_{d,h,s}$ is a closed substack of $\mathcal{R}_d^{\Gamma_{\text{disc}}} \times_{\mathcal{R}_d} \mathcal{C}^a_{d,h}$, it is a closed substack of $\mathcal{X}_d \times_{\mathcal{R}_d} \mathcal{C}_{d,h}$.

To see that the natural morphism $\varinjlim_{a,s} \mathcal{Z}^a_{d,h,s} \to \mathcal{X}_d \times_{\mathcal{R}_d} \mathcal{C}_{d,h}$ is an isomorphism, we need to show that it is surjective, and so we need to show that for any finite type $\mathcal{O}/\varpi^a$-algebra A, and any projective φ-module $\mathfrak{M}$ over $\mathbf{A}^+_{K,A}$ of T-height at most h, for which $\mathfrak{M}[1/T]$ is endowed with a continuous Γ_{disc}-action, there is some $s \geq 1$ such that the restriction of this action to $\gamma_{s,\text{disc}}$ agrees with the canonical action of $\Gamma_{s,\text{disc}}$ given by Corollary 3.5.2.

By Remark 3.3.1, there is some $s' \geq 1$ such that $(\gamma^{p^{s'}} - 1)(\mathfrak{M}) \subseteq T\mathfrak{M}$. Let $N = N(a,h)$ be as in the statement of Lemma 3.5.1. By the equivalence of conditions (3) and (4) of Lemma D.28, there is some $s \geq s'$ such that $(\gamma^{p^s} - 1)(\mathfrak{M}) \subseteq T^N\mathfrak{M}$, as required. □

For arbitrary K (not necessarily basic) we can deduce the following description of $\mathcal{X}^a_d$ as an Ind-algebraic stack, in the style of Lemma 3.2.11.

3.5.5 Lemma. *Fix $a \geq 1$. For each h, and each sufficiently large (depending on a and h) value of s, let $\mathcal{X}^a_{K,d,K^{\text{basic}},h,s}$ denote the scheme-theoretic image of the base-changed morphism*

$$\mathcal{Z}^a_{d[K:K^{\text{basic}}],h,s} \times_{\mathcal{X}_{K^{\text{basic}},d[K:K^{\text{basic}}]}} \mathcal{X}^a_{K,d} \to \mathcal{X}^a_{K,d},$$

so that $\mathcal{X}^a_{K,d,K^{\text{basic}},h,s}$ is a closed algebraic substack of $\mathcal{X}_{K,d}$. Then the natural morphism $\varinjlim_{h,s} \mathcal{X}^a_{K,d,K^{\text{basic}},h,s} \to \mathcal{X}^a_{K,d}$ is an isomorphism.

Proof. This is proved in the same way as Lemma 3.2.11, bearing in mind Proposition 3.5.4. □

3.6 THE CONNECTION WITH GALOIS REPRESENTATIONS

In Section 2.7.1 we recalled the relationship between étale (φ,Γ)-modules without coefficients (which is to say, with coefficients in $\mathbf{Z}_p$) and p-adic representations of G_K and the similar relationship between étale φ-modules over $\mathcal{O}_{\mathcal{E}}$ and p-adic representations of G_{K_∞}. In this section, we revisit those topics in the context of étale (φ,Γ)-modules (resp. étale φ-modules) and G_K-representations (resp. G_{K_∞}-modules) with coefficients.

3.6.1 Galois representations with coefficients

For a general p-adically complete $\mathbf{Z}_p$-algebra A, a projective étale (φ,Γ)-module with A-coefficients need not correspond to a family of G_K-representations. It will be useful, though, to have a version of such a correspondence in the case that A is complete local Noetherian with finite residue field. In this case, following [Dee01] we let $\widehat{\mathbf{A}}_{K,A}$ denote the $\mathfrak{m}_A$-adic completion of $\mathbf{A}_{K,A}$, and we define a *formal étale (φ,Γ)-module* with A-coefficients to be an étale (φ,Γ)-module over $\widehat{\mathbf{A}}_{K,A}$ in the obvious sense.

3.6.2 Remark. If A is a complete local Noetherian $\mathcal{O}$-algebra, with finite residue field, then the groupoid of formal étale (φ,Γ)-modules is equivalent to the groupoid $\mathcal{X}_d(A)$ (which we remind the reader refers to the groupoid of morphisms $\operatorname{Spf} A \to \mathcal{X}_d$; here $\operatorname{Spf} A$ is taken with respect to the $\mathfrak{m}_A$-adic topology on A). Indeed, by definition, the latter groupoid may be identified with the 2-limit $\varprojlim_i \mathcal{X}_d(A/\mathfrak{m}_A^i)$, which is easily seen to be equivalent to the groupoid of formal étale (φ,Γ)-modules, via an application of [GD71, Prop. 0.7.2.10(ii)].

We let $\widehat{\mathbf{A}}^{\mathrm{ur}}_{K,A}$ denote $\widehat{\mathbf{A}}^{\mathrm{ur}}_K \widehat{\otimes}_{\mathbf{Z}_p} A$, where the completed tensor product is with respect to the usual topology on $\widehat{\mathbf{A}}^{\mathrm{ur}}_K$, and the $\mathfrak{m}_A$-adic topology on A. The functors T_A, $\mathbf{D}_A$ defined by

$$\mathbf{D}_A(T) = (\widehat{\mathbf{A}}^{\mathrm{ur}}_{K,A} \otimes_A T)^{G_{K_{\mathrm{cyc}}}},$$

$$T_A(M) = (\widehat{\mathbf{A}}^{\mathrm{ur}}_{K,A} \otimes_{\widehat{\mathbf{A}}_{K,A}} M)^{\varphi=1}$$

then give equivalences of categories between the category of finite projective formal étale (φ,Γ)-modules with A-coefficients and the category of finite free A-modules with a continuous action of G_K. (In fact the results of [Dee01] do not require the (φ,Γ)-modules to be projective, but we will for the most part only consider projective modules in this book.)

Continuing to assume that A is complete local Noetherian with finite residue field, we let $\widehat{\mathcal{O}}_{\mathcal{E},A}$ denote the $\mathfrak{m}_A$-adic completion of $\mathcal{O}_{\mathcal{E},A}$, and we define $\widehat{\mathcal{O}}_{\widehat{\mathcal{E}^{\mathrm{ur}}},A} := \mathcal{O}_{\widehat{\mathcal{E}^{\mathrm{ur}}}} \widehat{\otimes}_{\mathbf{Z}_p} A$, where the completed tensor product is with respect to the usual topology on $\mathcal{O}_{\widehat{\mathcal{E}^{\mathrm{ur}}}}$, and the $\mathfrak{m}_A$-adic topology on A. Then the analogous statements to those of the previous paragraph, relating representations of G_{K_∞} on finite free A-modules and étale φ-modules over $\widehat{\mathcal{O}}_{\mathcal{E},A}$[1] via the functors

$$\mathbf{D}_{\infty,A}(T) = (\widehat{\mathcal{O}}_{\widehat{\mathcal{E}^{\mathrm{ur}}},A} \otimes_A T)^{G_{K_\infty}},$$
$$T_{\infty,A}(M) = (\widehat{\mathcal{O}}_{\widehat{\mathcal{E}^{\mathrm{ur}}},A} \otimes_{\widehat{\mathcal{O}}_{\mathcal{E},A}} M)^{\varphi=1}$$

[1]One might refer to an étale φ-module over $\widehat{\mathcal{O}}_{\mathcal{E},A}$ as a "formal étale φ-module over A", but since there are several different species of étale φ-modules under consideration throughout the book, we avoid using this potentially ambiguous terminology.

can be proved in exactly the same way as in [Dee01] (by passage to the limit over $A/\mathfrak{m}_A^n$).

We will also occasionally apply these statements in the case $A = \overline{\mathbf{F}}_p$. Their validity in this case follows from their validity in the case when A is a finite extension of $\mathbf{F}_p$, and the fact that both Galois representations and projective étale (φ, Γ)-modules over $\overline{\mathbf{F}}_p$ arise as base changes from such finite contexts; for Galois representations, this follows from the compactness of G_K and G_{K_∞}, and for étale (φ, Γ)-modules, it is an immediate consequence of Lemma 3.2.19.

We can use the equivalence between Galois representations and (φ, Γ)-modules as a tool to deduce facts about the finite type points of $\mathcal{X}_d$. More precisely, if $\mathbf{F}'/\mathbf{F}$ is a finite extension, then the groupoid of points $x \in \mathcal{X}_d(\mathbf{F}')$ is canonically equivalent to the groupoid of Galois representations $\overline{\rho}\colon G_K \to \mathrm{GL}_d(\mathbf{F}')$. For this reason, we will often denote such a point x simply by the corresponding Galois representation $\overline{\rho}$.

Suppose now that $\mathbf{F}'/\mathbf{F}$ is a finite extension, and that $x\colon \operatorname{Spec} \mathbf{F}' \to \mathcal{X}_d$ is a finite type point, with corresponding Galois representation $\overline{\rho}\colon G_K \to \mathrm{GL}_d(\mathbf{F}')$. Let $\widehat{(\mathcal{X}_d)}_x$ be the category of Definition A.25 (with $\mathcal{F}$ there being $\mathcal{X}_d$). Let $\mathcal{O}' = \mathcal{O} \otimes_{W(\mathbf{F})} W(\mathbf{F}')$ be the ring of integers in the finite extension $E' = W(\mathbf{F}')E$.

3.6.3 Proposition. *There is a morphism* $\operatorname{Spf} R_{\overline{\rho}}^{\square, \mathcal{O}'} \to \mathcal{X}_d$ *which is versal at the point x corresponding to $\overline{\rho}$, an isomorphism* $\operatorname{Spf} R_{\overline{\rho}}^{\square, \mathcal{O}'} \times_{\mathcal{X}_d} \operatorname{Spf} R_{\overline{\rho}}^{\square, \mathcal{O}'} \xrightarrow{\sim} \widehat{\mathrm{GL}}_{d, R_{\overline{\rho}}^{\square, \mathcal{O}'}, Z(\overline{\rho})}$, *where* $\widehat{\mathrm{GL}}_{d, R_{\overline{\rho}}^{\square, \mathcal{O}'}, Z(\overline{\rho})}$ *denotes the completion of* $(\mathrm{GL}_d)_{R_{\overline{\rho}}^{\square, \mathcal{O}'}}$ *along the closed subgroup of* $(\mathrm{GL}_d)_{\mathbf{F}'}$ *given by the centralizer of $\overline{\rho}$, and an isomorphism* $\operatorname{Spf} R_{\overline{\rho}}^{\square, \mathcal{O}'} \times_{\widehat{(\mathcal{X}_d)}_x} \operatorname{Spf} R_{\overline{\rho}}^{\square, \mathcal{O}'} \xrightarrow{\sim} \widehat{\mathrm{GL}}_{d, R_{\overline{\rho}}^{\square, \mathcal{O}'}, 1}$, *where* $\widehat{\mathrm{GL}}_{d, R_{\overline{\rho}}^{\square, \mathcal{O}'}, 1}$ *denotes the completion of* $(\mathrm{GL}_d)_{R_{\overline{\rho}}^{\square, \mathcal{O}'}}$ *along the identity of* $(\mathrm{GL}_d)_{\mathbf{F}'}$.

Proof. The existence of the morphism follows from the theorem of Dee recalled above, and the descriptions of the two fiber products are clear from its very definition. To see that this morphism is versal, it suffices to show that if $\rho\colon G_K \to \mathrm{GL}_d(A)$ is a representation with A a finite Artinian $\mathcal{O}$-algebra, and if $\rho_B : G_K \to \mathrm{GL}_d(B)$ is a second representation, with B a finite Artinian $\mathcal{O}$-algebra admitting a surjection onto A, such that the base change ρ_A of ρ_B to A is isomorphic to ρ (more concretely, so that there exists $M \in \mathrm{GL}_d(A)$ with $\rho = M \rho_A M^{-1}$), then we may find $\rho'\colon G_K \to \mathrm{GL}_d(B)$ which lifts ρ, and is isomorphic to ρ_B. But this is clear: the natural morphism $\mathrm{GL}_d(B) \to \mathrm{GL}_d(A)$ is surjective, and so if M' is any lift of M to an element of $\mathrm{GL}_d(B)$, then we may set $\rho' = M' \rho_B (M')^{-1}$. □

Because of the equivalence between (φ, Γ)-modules and Galois representations with $\mathbf{Z}_p$-coefficients, and because of the traditional notation ρ for Galois representations, we will often denote a family of rank d projective étale

(φ, Γ)-modules over T, i.e., a morphism $T \to \mathcal{X}_d$, by ρ_T, or some similar notation. We caution the reader that this notation is chosen purely for its psychological suggestiveness; a family of (φ, Γ)-modules over a general base T does not admit a literal interpretation in terms of Galois representations.

3.6.4 Galois representations associated to (φ, G_K)-modules

At times it will be convenient to consider the Galois representations associated to étale (φ, G_K)-modules. Let A be a finite $\mathcal{O}/\varpi^a$-algebra for some $a \geq 1$, and let M be a finite projective étale (φ, G_K)-module (resp. étale (φ, G_{K_∞})-module) with A-coefficients. Then we may apply the equivalence of Proposition 2.7.8 to obtain an étale (φ, Γ)-module (resp. an étale φ-module) from M; applying the functor T_A (resp. $T_{\infty,A}$) of Section 3.6.1 to this latter object yields a G_K-representation (resp. G_{K_∞}-representation) on a finite free A-module, which we denote simply by $T_A(M)$ (resp. $T_{\infty,A}(M)$).

By passage to the limit over a, we can extend these functors to the case where A is a finite $\mathcal{O}$-algebra, and in particular to the case that $A = \mathcal{O}'$ is the ring of integers in a finite extension E'/E. As in Section 3.6.1, we can and do then further extend these functors to the case $A = \overline{\mathbf{F}}_p$.

While we have defined the functors $T_A(M)$ (resp. $T_{\infty,A}(M)$) via our equivalences of categories, we note that both also admit a more direct description: namely $T_A(M) = M^{\varphi=1}$ (resp. $T_{\infty,A}(M) = M^{\varphi=1}$). Similarly, the composites of the functors $\mathbf{D}_A$ and $\mathbf{D}_{\infty,A}$ with the equivalences of Proposition 2.7.8 admit the following simple description: if T_A is a finite free A-module with a continuous action of G_K (resp. G_{K_∞}), then the corresponding étale (φ, G_K)-module (resp. (φ, G_{K_∞})-module) is given by $W(\mathbf{C}^\flat)_A \otimes_A T_A$ (with φ acting on the first factor in the tensor product, and G_K (resp. G_{K_∞}) acting diagonally).

3.6.5 Galois representations over certain fields

There is one more context in which we will need to consider the correspondence between Galois representations and étale (φ, Γ)-modules. Namely, suppose that A is a complete Noetherian local $\mathbf{Z}_p$-algebra which is a domain, and which has finite residue field. Let $\mathcal{K}$ denote the fraction field of A, and as usual, let $\mathfrak{m}$ denote the maximal ideal of A. We say that a representation $\rho\colon G_K \to \mathrm{GL}_d(\mathcal{K})$ is *continuous* if there exists a finitely generated A-submodule L of $\mathcal{K}^d$ which spans $\mathcal{K}^d$ over $\mathcal{K}$ (a "lattice"), such that G_K preserves L and acts continuously (when L is given its $\mathfrak{m}$-adic topology).

We write $\widehat{\mathbf{A}}_{K,\mathcal{K}} := \widehat{\mathbf{A}}_{K,A} \otimes_A \mathcal{K}$, and also write $\widehat{\mathbf{A}}^{\mathrm{ur}}_{K,\mathcal{K}} := \widehat{\mathbf{A}}^{\mathrm{ur}}_{K,A} \otimes_A \mathcal{K}$. We have the obvious notion of a projective (φ, Γ)-module with coefficients in $\widehat{\mathbf{A}}_{K,\mathcal{K}}$. We say that such a (φ, Γ)-module D is *étale* if there exists a (not necessarily projective) étale (φ, Γ)-module D_A over $\widehat{\mathbf{A}}_{K,A}$ contained in D such that the evident morphism $\widehat{\mathbf{A}}_{K,\mathcal{K}} \otimes_{\widehat{\mathbf{A}}_{K,A}} D_A \to D$ is an isomorphism.

Then, analogously to the case of A itself, we have functors $T_{\mathcal{K}}$, $\mathbf{D}_{\mathcal{K}}$ defined by

$$\mathbf{D}_{\mathcal{K}}(T) = (\widehat{\mathbf{A}}^{\mathrm{ur}}_{K,\mathcal{K}} \otimes_{\mathcal{K}} T)^{G_{K_{\mathrm{cyc}}}},$$
$$T_{\mathcal{K}}(M) = (\widehat{\mathbf{A}}^{\mathrm{ur}}_{K,\mathcal{K}} \otimes_{\widehat{\mathbf{A}}_{K,\mathcal{K}}} M)^{\varphi=1},$$

which give equivalences of categories between the category of finite projective formal étale (φ,Γ)-modules with $\mathcal{K}$-coefficients and the category of finite dimensional $\mathcal{K}$-vector spaces with a continuous action of G_K. (This is proved by passing to lattices on each side, and using the results of [Dee01]; note that the lattices involved need not be projective in general, and so we apply those results in their full generality.)

This formalism is most often applied in the literature in the case when $A = \mathbf{Z}_p$, so that $\mathcal{K} = \mathbf{Q}_p$. However, we will not consider étale (φ,Γ)-modules with $\mathbf{Q}_p$-coefficients in this book. Rather we will apply the preceding formalism only once, namely in our analysis of the closed $\overline{\mathbf{F}}_p$-points of $\mathcal{X}_d$, which we make in Section 6.6; and in this application, we will take A to be a complete local domain of characteristic p.

3.7 (φ, G_K)-MODULES AND RESTRICTION

Our goal in this section is to define certain morphisms of stacks which are the analogues, for families of (φ,Γ)-modules, of the restriction functors on Galois representations $\rho \mapsto \rho_{|G_{K_\infty}}$ and $\rho \mapsto \rho_{|G_L}$ (for any finite extension L of K).

3.7.1 Definition. We let $\mathcal{R}_{\mathrm{BK},d}$ denote the moduli stack of rank d projective étale φ-modules over $\mathcal{O}_{\mathcal{E},A}$; that is, the stack $\mathcal{R}_d$ constructed in Section 3.1 in the case when $\mathbf{A}_A = \mathcal{O}_{\mathcal{E},A}$.

Recall that we also have the stack $\mathcal{X}_d$ of étale (φ,Γ)-modules of Definition 3.2.1. We now construct a morphism $\mathcal{X}_d \to \mathcal{R}_{\mathrm{BK},d}$ which corresponds to the restriction of Galois representations from G_K to G_{K_∞}.

3.7.2 Proposition. *There is a canonical morphism $\mathcal{X}_d \to \mathcal{R}_{\mathrm{BK},d}$. If A is a complete local Noetherian $\mathcal{O}$-algebra with finite residue field, or equals $\overline{\mathbf{F}}_p$, then the morphism $\mathcal{X}_d(A) \to \mathcal{R}_{\mathrm{BK},d}(A)$ is given by restriction of the corresponding representation of G_K to G_{K_∞}.*

Proof. If A is a finite type $\mathcal{O}/\varpi^a$-algebra, for some $a \geq 1$, then we define a morphism of groupoids $\mathcal{X}_d(A) \to \mathcal{R}_{\mathrm{BK},d}(A)$ via the equivalences of categories of Proposition 2.7.8; that is, given a finite projective étale (φ,Γ)-module with A-coefficients, we form the corresponding finite projective étale (φ, G_K)-module, which yields a finite projective étale (φ, G_{K_∞})-module by restricting the G_K-action to G_{K_∞}, and thus gives an étale φ-module over $\mathcal{O}_{\mathcal{E},A}$. Since both $\mathcal{X}_d$

and $\mathcal{R}_{\mathrm{BK},d}$ are limit preserving (by Corollary 3.1.5 and Lemma 3.2.19), it follows from [EG21, Lem. 2.5.4, Lem. 2.5.5 (1)] that this construction determines a morphism $\mathcal{X}_d \to \mathcal{R}_{\mathrm{BK},d}$.

Suppose now that A is a complete local Noetherian $\mathcal{O}$-algebra with finite residue field. Let M be a formal étale (φ,Γ)-module corresponding to a morphism $\operatorname{Spf} A \to \mathcal{X}_d$, and M_∞ be the étale φ-module over $\widehat{\mathcal{O}}_{\mathcal{E},A}$ corresponding to the composite $\operatorname{Spf} A \to \mathcal{X}_d \to \mathcal{R}_{\mathrm{BK},d}$. The relationship between M and M_∞ is expressed as an isomorphism

$$\widehat{W(\mathbf{C}^\flat)_A} \otimes_{\widehat{\mathbf{A}}_{K,A}} M \xrightarrow{\sim} \widehat{W(\mathbf{C}^\flat)_A} \otimes_{\widehat{\mathcal{O}}_{\mathcal{E},A}} M_\infty. \tag{3.7.3}$$

Recall that the Galois representation associated to M is defined via $T_A(M) := (\widehat{\mathbf{A}}^{\mathrm{ur}}_{K,A} \otimes_{\widehat{\mathbf{A}}_{K,A}} M)^{\varphi=1}$, and that the evident (φ, G_K)-equivariant $\widehat{\mathbf{A}}^{\mathrm{ur}}_{K,A}$-linear morphism

$$\widehat{\mathbf{A}}^{\mathrm{ur}}_{K,A} \otimes_A T_A(M) \to \widehat{\mathbf{A}}^{\mathrm{ur}}_{K,A} \otimes_{\widehat{\mathbf{A}}_{K,A}} M$$

is then an isomorphism (see for example [Dee01, Prop. 2.1.26]). Thus there is an induced natural (φ, G_K)-equivariant isomorphism

$$\widehat{W(\mathbf{C}^\flat)_A} \otimes_A T_A(M) \xrightarrow{\sim} \widehat{W(\mathbf{C}^\flat)_A} \otimes_{\widehat{\mathbf{A}}_{K,A}} M,$$

where we recall that we write $\widehat{W(\mathbf{C}^\flat)_A}$ for the $\mathfrak{m}_A$-adic completion of $W(\mathbf{C}^\flat)_A$. Similarly, we obtain a natural (φ, G_{K_∞})-equivariant isomorphism

$$\widehat{W(\mathbf{C}^\flat)_A} \otimes_A T_{\infty,A}(M_\infty) \xrightarrow{\sim} \widehat{W(\mathbf{C}^\flat)_A} \otimes_{\widehat{\mathcal{O}}_{\mathcal{E},A}} M_\infty.$$

Combining these two isomorphisms with (3.7.3), we obtain a natural (φ, G_{K_∞})-equivariant isomorphism

$$\widehat{W(\mathbf{C}^\flat)_A} \otimes_A T_A(M) \xrightarrow{\sim} \widehat{W(\mathbf{C}^\flat)_A} \otimes_A T_{\infty,A}(M_\infty).$$

Each of $T_A(M)$ and $T_{\infty,A}(M_\infty)$ is a free A-module with trivial φ-action, and so if we pass to φ-invariants in this isomorphism, and take into account Lemma 2.2.20, we obtain an isomorphism $T_A(M)|_{G_{K_\infty}} \xrightarrow{\sim} T_{\infty,A}(M_\infty)$, which is what we wanted to show. Finally, if $A = \overline{\mathbf{F}}_p$, the result follows by taking direct limits. □

3.7.4 Proposition. *The diagonal morphism $\Delta\colon \mathcal{X}_d \to \mathcal{X}_d \times_{\mathcal{R}_{\mathrm{BK},d}} \mathcal{X}_d$ induced by the morphism of Proposition* 3.7.2 *is a closed immersion.*

Proof. The product $\mathcal{X}_d \times_{\mathcal{R}_{\mathrm{BK}}} \mathcal{X}_d$ can be described explicitly as follows: its A-valued points are pairs (M_1, M_2) of étale (φ, G_K)-modules with a G_{K_∞}-equivariant isomorphism between them. The diagonal is defined by $M \mapsto (M, M)$, with the isomorphism being the identity. To see that this defines a closed immersion, we have to check that if we are given a pair (M_1, M_2) of (φ, G_K)-modules with A-coefficients, together with a G_{K_∞}-equivariant isomorphism $f\colon M_1 \xrightarrow{\sim} M_2$, then the locus where f becomes G_K-equivariant is closed. That this is so follows from Corollary B.27. (That corollary shows that, for each $g \in G_K$, there is an ideal $J_g \subseteq A$ which cuts out the condition for the given isomorphism to commute with g. The ideal $J := \sum_g J_g$ then cuts out the condition for the given isomorphism to be G_K-equivariant.) □

We now construct the restriction maps corresponding to finite extensions of K.

3.7.5 Lemma. *Let L/K be a finite extension. There is a canonical morphism $\mathcal{X}_{K,d} \to \mathcal{X}_{L,d}$, which is representable by algebraic spaces, affine, and of finite presentation. If A is a complete local Noetherian $\mathcal{O}$-algebra with finite residue field, or equals $\overline{\mathbf{F}}_p$, then the morphism $\mathcal{X}_{K,d}(A) \to \mathcal{X}_{L,d}(A)$ is given by restriction of the corresponding representation of G_K to G_L.*

Proof. As in the case of Proposition 3.7.2, the morphism is defined first in the case of $\mathcal{O}$-algebras that are of finite type over $\mathcal{O}/\varpi^a$ for some a by applying the equivalences of categories of Proposition 2.7.8: we send an étale (φ, G_K)-module M to an étale (φ, G_L)-module M_L via restricting the continuous G_K-action to a continuous G_L-action. Since $\mathcal{X}_{K,d}$ and $\mathcal{X}_{L,d}$ are limit preserving (by Lemma 3.2.19), it follows from [EG21, Lem. 2.5.4, Lem. 2.5.5 (1)] that this construction determines a morphism $\mathcal{X}_{K,d} \to \mathcal{X}_{L,d}$.

Before verifying the various claimed properties of this morphism, we confirm that it has the claimed effect on associated Galois representations. To this end, we note that if A is a complete local Noetherian $\mathcal{O}$-algebra, then, as in the proof of Proposition 3.7.2, it follows from the definition that we have a natural (φ, G_L)-equivariant isomorphism

$$\widehat{W(\mathbf{C}^\flat)}_A \otimes_A T_A(M) \xrightarrow{\sim} \widehat{W(\mathbf{C}^\flat)}_A \otimes_A T_A(M_L),$$

and by Lemma 2.2.20, taking φ-invariants induces an isomorphism $T_A(M)|_{G_L} \xrightarrow{\sim} T_A(M_L)$. We deduce the same statement if $A = \overline{\mathbf{F}}_p$ by taking direct limits.

We now verify the claimed properties of the restriction morphism $\mathcal{X}_{K,d} \to \mathcal{X}_{L,d}$. Suppose firstly that L/K is Galois, and let $\{g_i\}$ be a set of (finitely many) coset representatives for G_L in G_K. For each i, we may give $g_i^* M$ the structure of a (φ, G_L)-module by letting each $h \in G_L$ act as $g_i^{-1} h g_i$ acts on M; then to extend the action of G_L to an action of G_K is to give an isomorphism of (φ, G_L)-modules $g_i^* M \to M$ for each index i, satisfying a slew of compatibilities. (Since G_L is open in G_K, the continuity of the G_K-action is automatic, given that the G_L-action that it is extending is continuous.)

We let $\mathcal{Y}_i$ denote the stack classifying objects M_A of $\mathcal{X}_{L,d}(A)$, endowed with an isomorphism $g_i^* M_A \xrightarrow{\sim} M_A$. If we regard g_i^* as an automorphism of $\mathcal{X}_{L,d}(A)$, then we may form its graph Γ_i, and we then have an isomorphism of stacks

$$\mathcal{Y}_i \xrightarrow{\sim} \mathcal{X}_{L,d}(A) \times_{\Delta, \mathcal{X}_{L,d}(A) \times \mathcal{X}_{L,d}(A), \Gamma_i} \mathcal{X}_{L,d}(A).$$

(Here Δ denotes the diagonal of $\mathcal{X}_{L,d}$; cf. the definition of $\mathcal{R}_d^{\Gamma_{\mathrm{disc}}}$ in Section 3.2 above, together with Proposition 3.2.14.) The projection onto the second factor $\mathcal{Y}_i \to \mathcal{X}_{L,d}(A)$, which corresponds to forgetting the isomorphism, is a base-change of the diagonal Δ, and so is representable by algebraic spaces, affine and of finite presentation by Proposition 3.2.17.

We can rephrase our interpretation of objects of $\mathcal{X}_{K,d}(A)$ as objects of $\mathcal{X}_{L,d}(A)$ endowed with isomorphisms $g_i^* M_A \xrightarrow{\sim} M_A$ satisfying certain compatibilities as the existence of a closed immersion

$$\mathcal{X}_{K,d} \hookrightarrow \mathcal{Y}_1 \times_{\mathcal{X}_{L,d}} \times \cdots \times_{\mathcal{X}_{L,d}} \mathcal{Y}_n$$

(the point being that the compatibilities arise as base changes of the double diagonal, and so impose closed conditions). Since the morphisms $\mathcal{Y}_i \to \mathcal{X}_{L,d}$ are representable by algebraic spaces, affine and of finite presentation, it follows that the same is true of the morphism $\mathcal{X}_{K,d} \to \mathcal{X}_{L,d}$, as claimed.

Finally, let L/K be a general finite extension. Let M/K be the normal closure of L/K, so that we have morphisms

$$\mathcal{X}_{K,d} \to \mathcal{X}_{L,d} \to \mathcal{X}_{M,d}.$$

By what we have already proved, both the second arrow and the composite of the two arrows are representable by algebraic spaces and affine, and of finite presentation. An affine morphism is separated, and thus has affine and finite type diagonal. One sees (e.g., by Lemma 3.2.7) that the diagonal of an affine morphism of finite type (and so, in particular, the diagonal of an affine morphism of finite presentation) is furthermore of finite presentation. A standard graph argument, in which we factor the first morphism as

$$\mathcal{X}_{K,d} \hookrightarrow \mathcal{X}_{K,d} \times_{\mathcal{X}_{M,d}} \mathcal{X}_{L,d} \to \mathcal{X}_{L,d}$$

(the first arrow in this factorization being the graph of the first morphism, which is a base change of the diagonal $\mathcal{X}_{L,d} \to \mathcal{X}_{L,d} \times_{\mathcal{X}_{M,d}} \mathcal{X}_{L,d}$, and the second arrow being the projection, which is a base change of the composite $\mathcal{X}_{K,d} \to \mathcal{X}_{L,d} \to \mathcal{X}_{M,d}$), then shows that the first morphism is also representable by algebraic spaces and affine, and of finite presentation. $\square$

3.8 TENSOR PRODUCTS AND DUALITY

If M is a projective étale (φ, Γ)-module with A-coefficients, we define the dual (φ, Γ)-module by

$$M^\vee := \mathrm{Hom}_{\mathbf{A}_{K,A}}(M, \mathbf{A}_{K,A});$$

if A is a finite $\mathcal{O}$-module, then there is a natural isomorphism $T(M^\vee) \cong T(M)^\vee$. Similarly, for each M, we let $M(1) := M \otimes_{\mathbf{A}_{K,A}} \mathbf{A}_{K,A}(1)$ denote the Tate twist, where $\mathbf{A}_{K,A}(1)$ denotes the free (φ, Γ)-module of rank 1 with a generator v on which Γ acts via the cyclotomic character and $\varphi(v) = v$. (Note that if $A = \mathcal{O}$ then $T(\mathbf{A}_{K,A}(1)) = \mathcal{O}(1)$, the usual Tate twist on the Galois side.) We also define the Cartier dual $M^* := M^\vee(1)$, which again in the case that A is a finite $\mathcal{O}$-algebra is compatible with the usual notion for Galois modules. If $T \to \mathcal{X}_d$ corresponds to a family ρ_T, then we adopt the natural convention of writing $\rho_T(1)$, $\rho_T^\vee$ and ρ_T^* for the corresponding families.

Given two projective étale (φ, Γ)-modules M_1, M_2 with A-coefficients of respective ranks d_1, d_2, we may form the tensor product (φ, Γ)-module $M_1 \otimes M_2$ (given by the tensor product on underlying $\mathbf{A}_{K,A}$-modules, and the tensor products of the actions of each of φ and Γ). If A is a finite $\mathcal{O}$-module, then we have a natural isomorphism

$$T(M_1 \otimes M_2) \cong T(M_1) \otimes T(M_2).$$

The tensor product induces a natural morphism

$$\mathcal{X}_{d_1} \times_{\mathcal{O}} \mathcal{X}_{d_2} \to \mathcal{X}_{d_1 d_2};$$

in particular, twisting by (the (φ, Γ)-module corresponding to) any character $G_K \to \mathcal{O}^\times$ gives an automorphism of each $\mathcal{X}_d$. Given morphisms $S \to \mathcal{X}_{d_1}$, $T \to \mathcal{X}_{d_2}$, denoted by ρ_S and ρ'_T, we write $\rho_S \boxtimes \rho'_T$ for the family corresponding to the composite

$$S \times_{\mathcal{O}} T \to \mathcal{X}_{d_1} \times_{\mathcal{O}} \mathcal{X}_{d_2} \to \mathcal{X}_{d_1 d_2}.$$

If $S = T$ then we write $\rho_T \otimes \rho'_T$ for the composite

$$T \to T \times_{\mathcal{O}} T \to \mathcal{X}_{d_1} \times_{\mathcal{O}} \mathcal{X}_{d_2} \to \mathcal{X}_{d_1 d_2},$$

where the first map is the diagonal embedding.

Chapter Four

Crystalline and semistable moduli stacks

In this chapter we build on the results of Appendix F to define the potentially semistable and potentially crystalline substacks of $\mathcal{X}_{K,d}$. Using the results of the earlier chapters, and Kisin's results on potentially semistable lifting rings [Kis08], we show that these stacks are p-adic formal algebraic stacks, and compute the dimensions of their special fibres.

4.1 NOTATION

Recall from Section 2.1.14 that the embedding $\mathfrak{S}_A \hookrightarrow \mathbf{A}_{\mathrm{inf},A}$ depends on the choice of a uniformizer π of K, and of a compatible family π^{1/p^∞} of p-power roots of unity in $\overline{K}$. In this chapter we will want to consider all possible choices of π and π^{1/p^∞}, and consequently we introduce some notation to do so. We write $\pi^\flat \in \mathcal{O}_{\mathbf{C}}^\flat$ for the element determined by π^{1/p^∞}. For each $s \geq 0$ we write $K_{\pi^\flat,s}$ for $K(\pi^{1/p^s})$, and $K_{\pi^\flat,\infty}$ for $\cup_s K_{\pi^\flat,s}$. We write $\mathfrak{S}_{\pi^\flat,A}$ for the image of $\mathfrak{S}_A$ in $\mathbf{A}_{\mathrm{inf},A}$ via the homomorphism determined by $u \mapsto [\pi^\flat]$, and $\mathcal{O}_{\mathcal{E},\pi^\flat,A}$ for the image of $\mathcal{O}_{\mathcal{E},A}$ in $W(\mathbf{C}^\flat)_A$ under the corresponding homomorphism. When $\pi^\flat$ is fixed, we will often write u for $[\pi^\flat]$ when discussing $\mathfrak{S}_{\pi^\flat,A}$. We write E_π for the Eisenstein polynomial corresponding to π. We also have a natural map $\theta\colon \mathbf{A}_{\mathrm{inf},A} \to \mathcal{O}_{\mathbf{C},A}$ (where the target is the tensor product $\mathcal{O}_{\mathbf{C}} \widehat{\otimes} A$, which is the quotient of the source by the principal ideal generated by the kernel of the usual map $\theta\colon \mathbf{A}_{\mathrm{inf}} \to \mathcal{O}_{\mathbf{C}}$).

4.1.1 Remark. In contrast to many papers in the literature (e.g., [BMS18, Liu12]) we do not twist the embedding $\mathfrak{S} \to \mathbf{A}_{\mathrm{inf}}$ by φ. While such a twist is important in applications involving comparisons to Fontaine's period rings, it is more convenient for us to use the embedding we have given here. Since φ is an automorphism of $\mathbf{A}_{\mathrm{inf}}$, it is in any case straightforward to pass back and forth between these two conventions.

4.2 BREUIL–KISIN MODULES AND BREUIL–KISIN–FARGUES MODULES

4.2.1 Definition. Fix a choice of $\pi^\flat$. Let A be a p-adically complete $\mathcal{O}$-algebra which is topologically of finite type. We define a *projective Breuil–Kisin module* (resp. a *projective Breuil–Kisin–Fargues module*) *of height at most h with A-coefficients* to be a finitely generated projective $\mathfrak{S}_{\pi^\flat,A}$-module (resp. $\mathbf{A}_{\mathrm{inf},A}$-module) $\mathfrak{M}$, equipped with a φ-semi-linear morphism $\varphi\colon \mathfrak{M}\to\mathfrak{M}$, with the property that the corresponding morphism $\Phi_{\mathfrak{M}}\colon \varphi^*\mathfrak{M}\to\mathfrak{M}$ is injective, with cokernel killed by E_π^h. If $\mathfrak{M}$ is a Breuil–Kisin module then $\mathbf{A}_{\mathrm{inf},A}\otimes_{\mathfrak{S}_{\pi^\flat,A}}\mathfrak{M}$ is a Breuil–Kisin–Fargues module.

4.2.2 Remark. Note that our definition of a Breuil–Kisin–Fargues module is less general than the definition made in [BMS18], in that we require φ to take $\mathfrak{M}$ to itself; this corresponds to only considering Galois representations with non-negative Hodge–Tate weights. This definition is convenient for us, as it allows us to make direct reference to the literature on Breuil–Kisin modules. The restriction to non-negative Hodge–Tate weights is harmless in our main results, as we can reduce to this case by twisting by a large enough power of the cyclotomic character (the interpretation of which on Breuil–Kisin–Fargues modules is explained in [BMS18, Ex. 4.24]).

4.2.3 Definition. Let A be a p-adically complete $\mathcal{O}$-algebra which is topologically of finite type. A Breuil–Kisin–Fargues G_K-module with A-coefficients is a Breuil–Kisin–Fargues module with A-coefficients, equipped with a continuous semi-linear action of G_K which commutes with φ.

Note that if $\mathfrak{M}^{\mathrm{inf}}$ is a Breuil–Kisin–Fargues G_K-module with A-coefficients, then $W(\mathbf{C}^\flat)_A\otimes_{\mathbf{A}_{\mathrm{inf},A}}\mathfrak{M}^{\mathrm{inf}}$ is naturally an étale (φ,G_K)-module in the sense of Definition 2.7.7. The following definition is motivated by Theorem F.11 and Corollary F.23.

4.2.4 Definition. Let A be a p-adically complete $\mathcal{O}$-algebra which is topologically of finite type over $\mathcal{O}$ (and recall then that $\mathbf{A}_{\mathrm{inf},A}$ is faithfully flat over $\mathfrak{S}_{\pi^\flat,A}$, by Proposition 2.2.14). We say that a Breuil–Kisin–Fargues G_K-module with A-coefficients $\mathfrak{M}^{\mathrm{inf}}$ *descends for $\pi^\flat$* or *descends to $\mathfrak{S}_{\pi^\flat,A}$* if there is a Breuil–Kisin module $\mathfrak{M}_{\pi^\flat}$ with $\mathfrak{M}_{\pi^\flat}\subseteq(\mathfrak{M}^{\mathrm{inf}})^{G_{K_{\pi^\flat,\infty}}}$ for which the natural map $\mathbf{A}_{\mathrm{inf},A}\otimes_{\mathfrak{S}_{\pi^\flat,A}}\mathfrak{M}_{\pi^\flat}\to\mathfrak{M}^{\mathrm{inf}}$ is an isomorphism.

We say that $\mathfrak{M}^{\mathrm{inf}}$ *admits all descents* if it descends for every choice of $\pi^\flat$ (for every choice of π), and if furthermore

1. The $W(k)\otimes_{\mathbf{Z}_p}A$-submodule $\mathfrak{M}_{\pi^\flat}/[\pi^\flat]\mathfrak{M}_{\pi^\flat}$ of $W(\overline{k})_A\otimes_{\mathbf{A}_{\mathrm{inf},A}}\mathfrak{M}^{\mathrm{inf}}$ is independent of the choice of π and $\pi^\flat$.
2. The $\mathcal{O}_K\otimes_{\mathbf{Z}_p}A$-submodule $\varphi^*\mathfrak{M}_{\pi^\flat}/E_{\pi^\flat}\varphi^*\mathfrak{M}_{\pi^\flat}$ of $\mathcal{O}_{\mathbf{C},A}\otimes_{\mathbf{A}_{\mathrm{inf},A},\theta}\varphi^*\mathfrak{M}^{\mathrm{inf}}$ is independent of the choice of π and $\pi^\flat$.

If $\mathfrak{M}^{\mathrm{inf}}$ admits all descents, then we say that it is *crystalline* if for each $\pi^\flat$, and each $g\in G_K$, we have

$$(g-1)(\mathfrak{M}_{\pi^\flat})\subseteq\varphi^{-1}(\mu)[\pi^\flat]\mathfrak{M}^{\mathrm{inf}} \tag{4.2.5}$$

(where $\mu = [\varepsilon] - 1$ for some compatible choice of roots of unity $\varepsilon = (1, \zeta_p, \zeta_{p^2}, \dots) \in \mathcal{O}_{\mathbf{C}}^\flat$).

If L/K is a finite Galois extension, then we say that $\mathfrak{M}^{\mathrm{inf}}$ *admits all descents over* L if the corresponding Breuil–Kisin–Fargues G_L-module (obtained by restricting the G_K-action to G_L) admits all descents.

4.2.6 Remark. There is considerable redundancy and rigidity in Definition 4.2.4. Note in particular that if for some $\pi^\flat$ there exists a $\mathfrak{S}_{\pi^\flat,A}$-module $\mathfrak{M}_{\pi^\flat}$ with $\mathfrak{M}^{\mathrm{inf}} = \mathbf{A}_{\mathrm{inf},A} \otimes_{\mathfrak{S}_{\pi^\flat,A}} \mathfrak{M}_{\pi^\flat}$, then since $\mathfrak{S}_{\pi^\flat,A} \to \mathbf{A}_{\mathrm{inf},A}$ is faithfully flat, the $\mathfrak{S}_{\pi^\flat,A}$-module $\mathfrak{M}_{\pi^\flat}$ is automatically finite projective (by [Sta, Tag 058S]).

We will find the following basic lemmas useful.

4.2.7 Lemma. *Fix a choice of π. Suppose that $\mathfrak{M}^{\mathrm{inf}}$ is a Breuil–Kisin–Fargues G_K-module with A coefficients, where A is a p-adically complete $\mathcal{O}$-algebra which is topologically of finite type over $\mathcal{O}$. Then $\mathfrak{M}^{\mathrm{inf}}$ descends for some choice of $\pi^\flat$ (for our fixed π) if and only if it descends for all such choices.*

Proof. We can choose an element $g \in G_K$ with $g(\pi^\flat) = (\pi^\flat)'$, so that $\mathfrak{S}_{(\pi^\flat)',A} = g(\mathfrak{S}_{\pi^\flat,A})$. Then if $\mathfrak{M}_\pi$ is a descent to $\mathfrak{S}_{\pi,A}$, $g(\mathfrak{M}_\pi)$ is a descent to $\mathfrak{S}_{(\pi^\flat)',A}$. ☐

4.2.8 Lemma. *Let A be a p-adically complete $\mathcal{O}$-algebra which is topologically of finite type over $\mathcal{O}$. Suppose that $\mathfrak{M}^{\mathrm{inf}}$ is a Breuil–Kisin–Fargues G_K-module. Then if $\mathfrak{M}^{\mathrm{inf}}$ descends for $\pi^\flat$, the Breuil–Kisin module $\mathfrak{M}_{\pi^\flat}$ is uniquely determined.*

Proof. Assume to begin with that A is an $\mathcal{O}/\varpi^a$-algebra of finite type, for some $a \geq 1$. By Proposition 2.7.8, the φ-module $(W(\mathbf{C}^\flat)_A \otimes_{\mathbf{A}_{\mathrm{inf},A}} \mathfrak{M}^{\mathrm{inf}})^{G_{K_{\pi^\flat,\infty}}}$ uniquely descends to an étale φ-module M over $\mathcal{O}_{\mathcal{E},\pi^\flat,A}$. Thus if $\mathfrak{M}_1$, $\mathfrak{M}_2$ are two descents of $\mathfrak{M}^{\mathrm{inf}}$ to $\mathfrak{S}_{\pi^\flat,A}$, then $\mathcal{O}_{\mathcal{E},\pi^\flat,A} \otimes_{\mathfrak{S}_{\pi^\flat,A}} \mathfrak{M}_1$ and $\mathcal{O}_{\mathcal{E},\pi^\flat,A} \otimes_{\mathfrak{S}_{\pi^\flat,A}} \mathfrak{M}_2$ coincide; they are both equal to M.

This allows us to show that $\mathfrak{M}_1 + \mathfrak{M}_2$ is also a descent of $\mathfrak{M}^{\mathrm{inf}}$. For this, we have to show that the natural map $\mathbf{A}_{\mathrm{inf},A} \otimes_{\mathfrak{S}_{\pi^\flat,A}} (\mathfrak{M}_1 + \mathfrak{M}_2) \to \mathfrak{M}^{\mathrm{inf}}$ is an isomorphism. (Note that it will then follow automatically that $\mathfrak{M}_1 + \mathfrak{M}_2$ is projective over $\mathfrak{S}_{\pi^\flat,A}$, by Remark 4.2.6.) That it is a surjection is clear, since it is already a surjection when restricted to each summand. To see that it is an injection, we may check that its composite with the injection $\mathfrak{M}^{\mathrm{inf}} \hookrightarrow W(\mathbf{C}^\flat)_A \otimes_{\mathbf{A}_{\mathrm{inf},A}} \mathfrak{M}^{\mathrm{inf}}$ is injective. But this composite may be factored as the composite of the embedding

$$\mathbf{A}_{\mathrm{inf},A} \otimes_{\mathfrak{S}_{\pi^\flat,A}} (\mathfrak{M}_1 + \mathfrak{M}_2) \hookrightarrow \mathbf{A}_{\mathrm{inf},A} \otimes_{\mathfrak{S}_{\pi^\flat,A}} M$$

(obtained by tensoring the embedding $\mathfrak{M}_1 + \mathfrak{M}_2 \hookrightarrow M$ with the extension $\mathbf{A}_{\mathrm{inf},A}$ of $\mathfrak{S}_{\pi^\flat,A}$, which is flat by Proposition 2.2.14) and the isomorphism

$$\mathbf{A}_{\mathrm{inf},A} \otimes_{\mathfrak{S}_{\pi^\flat,A}} M \cong \left(\mathbf{A}_{\mathrm{inf},A} \otimes_{\mathfrak{S}_{\pi^\flat,A}} \mathcal{O}_{\mathcal{E},\pi^\flat,A}\right) \otimes_{\mathcal{O}_{\mathcal{E},\pi^\flat,A}} M$$
$$\cong W(\mathbf{C}^\flat)_A \otimes_{\mathcal{O}_{\mathcal{E},\pi^\flat,A}} M \cong W(\mathbf{C}^\flat)_A \otimes_{\mathbf{A}_{\mathrm{inf},A}} \mathfrak{M}^{\mathrm{inf}}.$$

Since $\mathfrak{M}_1 + \mathfrak{M}_2$ is also a descent, we may replace $\mathfrak{M}_2$ by $\mathfrak{M}_1 + \mathfrak{M}_2$, and therefore assume that $\mathfrak{M}_1 \subseteq \mathfrak{M}_2$. Since $\mathbf{A}_{\mathrm{inf},A} \otimes_{\mathfrak{S}_{\pi^\flat,A}} (\mathfrak{M}_2/\mathfrak{M}_1) = 0$, and as $\mathfrak{S}_{\pi^\flat,A} \to \mathbf{A}_{\mathrm{inf},A}$ is faithfully flat (again by Proposition 2.2.14), we see that $\mathfrak{M}_2/\mathfrak{M}_1 = 0$, as required.

Suppose now that A is p-adically complete and topologically of finite type. Let $\mathfrak{M}_1$ and $\mathfrak{M}_2$ be two descents of $\mathfrak{M}^{\mathrm{inf}}$. Since $\mathfrak{S}_A \to \mathbf{A}_{\mathrm{inf},A}$ is faithfully flat (again by Proposition 2.2.14), we find that $\mathfrak{M}_1/\varpi^a$ and $\mathfrak{M}_1/\varpi^a$ both embed into $\mathfrak{M}^{\mathrm{inf}}/\varpi^a$, for any $a \geq 1$, and both provide descents for $\pi^\flat$ of $\mathfrak{M}^{\mathrm{inf}}/\varpi^a$. Applying the uniqueness result we've already proved (our coefficients now being A/ϖ^a), we find that $\mathfrak{M}_1/\varpi^a$ and $\mathfrak{M}_2/\varpi^a$ coincide as submodules of $\mathfrak{M}^{\mathrm{inf}}/\varpi^a$. Passing to the inverse limit over a, we find that $\mathfrak{M}_1$ and $\mathfrak{M}_2$ coincide as submodules of $\mathfrak{M}^{\mathrm{inf}}$, proving the desired uniqueness in general. □

4.3 CANONICAL EXTENSIONS OF G_{K_∞}-ACTIONS

In this section we will consider a fixed choice of $\pi^\flat$, which we will accordingly drop from the notation. For each $s \geq 1$, write $K_s = K(\pi^{1/p^s})$. We now use a variant of the arguments of [CL11], which show that the trivial action of G_{K_∞} on a Breuil–Kisin module with $\mathcal{O}/\varpi^a$-coefficients can be extended to some G_{K_s}. We begin with some preliminary lemmas, the first of which is an analogue for Breuil–Kisin modules of [EG21, Lem. 5.2.14], and is proved in a similar way.

4.3.1 Lemma. *Let A be a p-adically complete $\mathcal{O}$-algebra which is topologically of finite type, and let $\mathfrak{M}$ be a finite projective Breuil–Kisin module with A-coefficients of height at most h. Then $\mathfrak{M}$ is a direct summand of a finite free Breuil–Kisin module with A-coefficients of height at most h.*

Proof. Since the map $\Phi_{\mathfrak{M}} : \varphi^*\mathfrak{M} \to \mathfrak{M}$ is an injection with cokernel killed by E^h, there is a map $\Psi_{\mathfrak{M}} : \mathfrak{M} \to \varphi^*\mathfrak{M}$ with $\Psi_{\mathfrak{M}} \circ \Phi_{\mathfrak{M}}$ and $\Phi_{\mathfrak{M}} \circ \Psi_{\mathfrak{M}}$ both being given by multiplication by E^h. Let $\mathfrak{P}$ be a finite projective $\mathfrak{S}_A$-module such that $\mathfrak{F} := \mathfrak{M} \oplus \mathfrak{P}$ is a finite free $\mathfrak{S}_A$ module. Set $\mathfrak{Q} := \mathfrak{P} \oplus \mathfrak{M} \oplus \mathfrak{P}$, a finite projective $\mathfrak{S}_A$-module.

Note that $\mathfrak{M} \oplus \mathfrak{Q} \cong \mathfrak{F} \oplus \mathfrak{F}$ is a finite free $\mathfrak{S}_A$-module, so it suffices to show that $\mathfrak{Q}$ can be endowed with the structure of a Breuil–Kisin module of height at most h. Since $\mathfrak{F}$ is free, we can choose an isomorphism $\Phi_{\mathfrak{F}} : \varphi^*\mathfrak{F} \xrightarrow{\sim} \mathfrak{F}$, and we then define $\Phi_{\mathfrak{Q}} : \varphi^*\mathfrak{Q} \to \mathfrak{Q}$ as the composite

$$\begin{aligned}\varphi^*\mathfrak{Q} &= \varphi^*\mathfrak{P} \oplus \varphi^*\mathfrak{M} \oplus \varphi^*\mathfrak{P} \\ &\xrightarrow{\sim} \varphi^*\mathfrak{M} \oplus \varphi^*\mathfrak{P} \oplus \varphi^*\mathfrak{P} \\ &= \varphi^*\mathfrak{F} \oplus \varphi^*\mathfrak{P} \\ &\xrightarrow{\Phi_{\mathfrak{F}}} \mathfrak{F} \oplus \varphi^*\mathfrak{P} \\ &\xrightarrow{\sim} \mathfrak{P} \oplus \mathfrak{M} \oplus \varphi^*\mathfrak{P} \\ &\xrightarrow{\Psi_{\mathfrak{M}}} \mathfrak{P} \oplus \varphi^*\mathfrak{M} \oplus \varphi^*\mathfrak{P} \\ &= \mathfrak{P} \oplus \varphi^*\mathfrak{F} \\ &\xrightarrow{\Phi_{\mathfrak{F}}} \mathfrak{P} \oplus \mathfrak{F} = \mathfrak{Q}.\end{aligned}$$

Every morphism in this composite other than $\Psi_{\mathfrak{M}}$ is an isomorphism. Since $\Psi_{\mathfrak{M}} \circ \Phi_{\mathfrak{M}} = \Phi_{\mathfrak{M}} \circ \Psi_{\mathfrak{M}} = E^h$, we see that $\Psi_{\mathfrak{M}}$ is an injection with cokernel killed by E^h, so the same is true of $\Phi_{\mathfrak{Q}}$, as required. □

The following lemma is proved by a standard Frobenius amplification argument, which is in particular almost identical to the proof of [CEGS19, Lem. 4.1.9].

4.3.2 Lemma. *Let $\mathfrak{M}^{\inf}$, $\mathfrak{N}^{\inf}$ be projective Breuil–Kisin–Fargues modules of height at most h, where A is a finite type $\mathcal{O}/\varpi^a$-algebra. Suppose that $N \geq e(a+h)/(p-1)$. Let*

$$f\colon \mathfrak{M}^{\inf} \to \mathfrak{N}^{\inf}$$

be an $\mathbf{A}_{\inf,A}$-linear map, and suppose that for all $m \in \mathfrak{M}^{\inf}$, $(f \circ \Phi_{\mathfrak{M}^{\inf}} - \Phi_{\mathfrak{N}^{\inf}} \circ \varphi^* f)(m) \in u^{pN}\mathfrak{N}^{\inf}$.

Then there is a unique $\mathbf{A}_{\inf,A}$-linear, φ-linear morphism $f'\colon \mathfrak{M}^{\inf} \to \mathfrak{N}^{\inf}$ with the property that for all $m \in \mathfrak{M}^{\inf}$, $(f'-f)(m) \in u^N\mathfrak{N}^{\inf}$; in fact, we have $(f'-f)(m) \in u^{N+1}\mathfrak{N}^{\inf}$.

Proof. We firstly prove uniqueness. Suppose that f'' also satisfies the properties that f' does, and write $g = f' - f''$, so that by assumption we have that $\operatorname{im} g \subseteq u^N\mathfrak{N}^{\inf}$ and $g \circ \Phi_{\mathfrak{M}^{\inf}} = \Phi_{\mathfrak{N}^{\inf}} \circ \varphi^* g$. We claim that this implies that $\operatorname{im} g \subseteq u^{N+1}\mathfrak{N}^{\inf}$; if this is the case, then by induction on N we have $\operatorname{im} g \subseteq u^N\mathfrak{N}^{\inf}$ for all N, and so $g = 0$ (note that as $\mathfrak{N}^{\inf}$ is a finite projective $\mathbf{A}_{\inf,A}$-module, it is u-adically complete and separated).

To prove the claim, note that since $\operatorname{im} g \subseteq u^N\mathfrak{N}^{\inf}$ we have $\operatorname{im} \varphi^* g \subseteq u^{pN}\mathfrak{N}^{\inf}$. By [EG21, Lem. 5.2.6] (and its proof), and the assumption that $\mathfrak{M}^{\inf}$ has height at most h, we have $\operatorname{im} \Phi_{\mathfrak{M}^{\inf}} \supseteq u^{e(a+h-1)}\mathfrak{M}^{\inf}$. Since $g \circ \Phi_{\mathfrak{M}^{\inf}} = \Phi_{\mathfrak{N}^{\inf}} \circ \varphi^* g$, it follows that

$$u^{e(a+h-1)} \operatorname{im} g \subseteq \operatorname{im} g \circ \Phi_{\mathfrak{M}^{\inf}} = \operatorname{im} \Phi_{\mathfrak{N}^{\inf}} \circ \varphi^* g \subseteq u^{pN}\mathfrak{N}^{\inf},$$

and since $\mathfrak{N}^{\inf}$ is u-torsion free, we have $\operatorname{im} g \subseteq u^{pN-e(a+h-1)}\mathfrak{N}^{\inf}$. Our assumption on N implies that $pN - e(a+h-1) \geq N+1$, as required.

We now prove the existence of f'. For any $\mathbf{A}_{\inf,A}$-linear map $h\colon \mathfrak{M}^{\inf} \to \mathfrak{N}^{\inf}$ we set $\delta(h) := \Phi_{\mathfrak{N}^{\inf}} \circ \varphi^* h - h \circ \Phi_{\mathfrak{M}^{\inf}}\colon \varphi^*\mathfrak{M}^{\inf} \to \mathfrak{N}^{\inf}$. We claim that for any $s\colon \varphi^*\mathfrak{M}^{\inf} \to u^{pN}\mathfrak{N}^{\inf}$, we can find $t\colon \mathfrak{M}^{\inf} \to u^{N+1}\mathfrak{N}^{\inf}$ with $\delta(t) = s$. Given this claim, the existence of f' is immediate, taking $s = \delta(f)$ and $f' := f - t$, so that $\delta(f') = 0$.

To prove the claim, we first prove the weaker claim that for any s as above we can find $h\colon \mathfrak{M}^{\inf} \to u^{N+1}\mathfrak{N}^{\inf}$ with $\operatorname{im}(\delta(h) - s) \subseteq u^{p(N+1)}\mathfrak{N}^{\inf}$. Admitting this second claim, we prove the first claim by successive approximation: we set $t_0 = h$, so that $\operatorname{im}(\delta(t_0) - s) \subseteq u^{p(N+1)}\mathfrak{N}^{\inf}$. Applying the second claim again with N replaced by $N+1$, and s replaced by $s - \delta(t_0)$, we find $h\colon \mathfrak{M}^{\inf} \to u^{N+2}\mathfrak{N}^{\inf}$ with $\operatorname{im}(\delta(h) - (s - \delta(t_0))) \subseteq u^{p(N+2)}\mathfrak{N}^{\inf}$. Setting $t_1 = t_0 + h$, and proceeding inductively, we obtain a Cauchy sequence converging (in the u-adically complete finite $\mathbf{A}_{\inf,A}$-module $\operatorname{Hom}_{\mathbf{A}_{\inf,A}}(\mathfrak{M}^{\inf}, \mathfrak{N}^{\inf})$) to the required t.

Finally, we prove this second claim. If $h\colon \mathfrak{M}^{\inf} \to u^{N+1}\mathfrak{N}^{\inf}$ then $\operatorname{im} \varphi^* h \subseteq u^{p(N+1)}\mathfrak{N}^{\inf}$, so it suffices to show that we can find h such that $h \circ \Phi_{\mathfrak{M}^{\inf}} = -s$; but this is immediate, because the cokernel of $\Phi_{\mathfrak{M}^{\inf}}$ is killed by $u^{e(a+h-1)}$, and $pN - e(a+h-1) \geq N+1$. ∎

The following lemma is proved by a reinterpretation of some of the arguments of [CL11, §2].

4.3.3 Lemma. *For any fixed a, h, and any $N \geq e(a+h)/(p-1)$, there is a positive integer $s(a,h,N)$ with the property that for any finite type $\mathcal{O}/\varpi^a$-algebra A, any projective Breuil–Kisin module $\mathfrak{M}$ of height at most h, and any $s \geq s(a,h,N)$, there is a unique continuous action of G_{K_s} on $\mathfrak{M}^{\inf} := \mathbf{A}_{\inf,A} \otimes_{\mathfrak{S}_A} \mathfrak{M}$ which commutes with φ and is semi-linear with respect to the natural action of G_{K_s} on $\mathbf{A}_{\inf,A}$, with the additional property that for all $g \in G_{K_s}$ we have $(g-1)(\mathfrak{M}) \subset u^N\mathfrak{M}^{\inf}$.*

Proof. By definition, to give a semi-linear action of G_{K_s} on $\mathfrak{M}^{\inf}$ is to give for each $g \in G_{K_s}$ a morphism of Breuil–Kisin–Fargues modules

$$\beta_g\colon g^*\mathfrak{M}^{\inf} \to \mathfrak{M}^{\inf}$$

with the property that for all $g, h \in G_{K_s}$, we have $\beta_{gh} = \beta_h \circ h^*\beta_g$. Now, the requirement that $(g-1)(\mathfrak{M}) \subseteq u^N\mathfrak{M}^{\inf}$ implies that if we have two such morphisms β_g, β'_g then $(\beta_g - \beta'_g)(g^*\mathfrak{M}^{\inf}) \subseteq u^N\mathfrak{M}^{\inf}$, so that $\beta'_g = \beta_g$ by the uniqueness assertion of Lemma 4.3.2.

We now use the existence part of Lemma 4.3.2 to construct the required action. Since the homomorphism $\mathfrak{S}_A \hookrightarrow \mathbf{A}_{\inf,A}$ takes u to the Teichmüller lift of $(\pi^{1/p^n})_n$, we can and do choose $s(a,h,N)$ sufficiently large that for all $s \geq$

$s(a,h,N)$ and $g \in G_{K_s}$, we have $g(u) - u \in u^{pN}\mathbf{A}_{\mathrm{inf},A}$; thus for all $\lambda \in \mathfrak{S}_A$, we have $g(\lambda) - \lambda \in u^{pN}\mathbf{A}_{\mathrm{inf},A}$.

We claim that for each $g \in G_{K_s}$, we may define an $\mathbf{A}_{\mathrm{inf},A}$-linear map

$$\alpha_g \colon g^*\mathfrak{M}^{\mathrm{inf}} \to \mathfrak{M}^{\mathrm{inf}}$$

with the following two properties:

- $(\alpha_g \circ \Phi_{g^*\mathfrak{M}^{\mathrm{inf}}} - \Phi_{\mathfrak{M}^{\mathrm{inf}}} \circ \varphi^*\alpha_g)(\varphi^* g^*\mathfrak{M}^{\mathrm{inf}}) \subseteq u^{pN}\mathfrak{M}^{\mathrm{inf}}$.
- For each $m \in \mathfrak{M}$, we have $\alpha_g(1 \otimes m) - m \in u^N\mathfrak{M}^{\mathrm{inf}}$.

It suffices to construct such maps in the case when $\mathfrak{M}$ is free; indeed, by Lemma 4.3.1, we may write $\mathfrak{F} = \mathfrak{M} \oplus \mathfrak{P}$ where $\mathfrak{F}, \mathfrak{P}$ are respectively free and projective Breuil–Kisin modules of height at most h, and given a morphism $\alpha_g \colon g^*\mathfrak{F}^{\mathrm{inf}} \to \mathfrak{F}^{\mathrm{inf}}$ satisfying our requirements (where we write $\mathfrak{F}^{\mathrm{inf}} = \mathbf{A}_{\mathrm{inf},A} \otimes_{\mathfrak{S}_A} \mathfrak{F}$), we may obtain the desired morphism $\alpha_g \colon g^*\mathfrak{M}^{\mathrm{inf}} \to \mathfrak{M}^{\mathrm{inf}}$ by projection onto the corresponding direct summands.

If then $\mathfrak{F}$ is a free Breuil–Kisin module of height at most h, we choose a basis $f_1, \ldots, f_d$ for $\mathfrak{F}$, and then, for each $g \in G_{K_s}$, we define an $\mathbf{A}_{\mathrm{inf},A}$-linear map

$$\alpha_g \colon g^*\mathfrak{F}^{\mathrm{inf}} \to \mathfrak{F}^{\mathrm{inf}}$$

(which depends on our choice of basis) by

$$\alpha_g\Big(\sum_i \lambda_i \otimes f_i\Big) = \sum_i \lambda_i f_i.$$

We now check that α_g has the claimed properties. If $\Phi_{\mathfrak{F}}(1 \otimes f_i) = \sum_j \theta_{i,j} f_j$ (so that $\theta_{i,j} \in \mathfrak{S}_A$), then for any $\sum_i \lambda_i \otimes f_i \in \varphi^* g^*\mathfrak{F}^{\mathrm{inf}}$, we compute that

$$(\alpha_g \circ \Phi_{g^*\mathfrak{F}^{\mathrm{inf}}} - \Phi_{\mathfrak{F}^{\mathrm{inf}}} \circ \varphi^*\alpha_g)\Big(\sum_i \lambda_i \otimes f_i\Big) = \sum_{i,j} \lambda_i (g(\theta_{i,j}) - \theta_{i,j}) f_j \in u^{pN}\mathfrak{F}^{\mathrm{inf}},$$

while for any $\sum_i \lambda_i f_i \in \mathfrak{F}$ (note then that each $\lambda_i \in \mathfrak{S}_A$), we compute that

$$\begin{aligned}\alpha_g\Big(1 \otimes \Big(\sum_i \lambda_i f_i\Big)\Big) - \sum_i \lambda_i f_i \\ = \alpha_g\Big(\sum_i g(\lambda_i) \otimes g_i\Big) - \sum_i \lambda_i g_i = \sum_i (g(\lambda_i) - \lambda_i) g_i \in u^N\mathfrak{F}^{\mathrm{inf}};\end{aligned}$$

in both computations we have used the fact that, by our choice of s, if $\lambda \in \mathfrak{S}_A$, then $g(\lambda) - \lambda \in u^{pN}\mathfrak{F}^{\mathrm{inf}} \subseteq u^N\mathfrak{F}^{\mathrm{inf}}$.

Returning to the general case of a projective Breuil–Kisin module $\mathfrak{M}$ of height at most h, it follows from Lemma 4.3.2 that there is a unique morphism of Breuil–Kisin–Fargues modules

$$\beta_g : g^*\mathfrak{M}^{\mathrm{inf}} \to \mathfrak{M}^{\mathrm{inf}}$$

with $\mathrm{im}(\beta_g - \alpha_g) \subseteq u^N \mathbf{A}_{\mathrm{inf},A} \otimes_{\mathfrak{S}_A} \mathfrak{M}$. Since we have $\alpha_{gh} = \alpha_h \circ h^*\alpha_g$, it follows from this uniqueness that $\beta_{gh} = \beta_h \circ h^*\beta_g$, so we have constructed an action of G_{K_s} on $\mathfrak{M}^{\mathrm{inf}}$. It remains to check that $(g-1)(\mathfrak{M}) \subseteq u^N\mathfrak{M}^{\mathrm{inf}}$; for this, note that, for any $m \in \mathfrak{M}$, we have

$$(g-1)m = \beta_g(1 \otimes m) - m = (\beta_g - \alpha_g)(1 \otimes m) + \alpha_g(1 \otimes m) - m \in u^N\mathfrak{M}^{\mathrm{inf}};$$

here we have used the fact $\mathrm{im}(\beta_g - \alpha_g) \subseteq u^N \mathbf{A}_{\mathrm{inf},A} \otimes_{\mathfrak{S}_A} \mathfrak{M}$, together with the second of the properties satisfied by the maps α_g. □

4.4 BREUIL–KISIN–FARGUES G_K-MODULES AND CANONICAL ACTIONS

In an earlier draft of this book, we claimed to show that for Breuil–Kisin–Fargues G_K-modules admitting descents for all $\pi^\flat$, the actions of the $G_{K_{\pi^\flat,s}} \subseteq G_K$ (for sufficiently large values of s) are necessarily the canonical actions considered in Section 4.3. We then used this claim to establish finiteness properties of the stacks of Breuil–Kisin–Fargues G_K-modules that we consider later in this chapter. We are grateful to Dat Pham for pointing out a mistake in our argument, and we do not know whether this is in fact automatic. However, the following result of Caruso–Liu (which motivated our consideration of canonical actions in the first place) shows that the Breuil–Kisin–Fargues G_K-modules coming from potentially semistable G_K-representations do necessarily have this property; so we can (and do) impose this as an extra condition in the definition of our stacks.

The following lemma is essentially the content of [CL11, §3] (although they are concerned with getting precise bounds on the size of s, which is unimportant for us). Recall that, by Theorem F.11, to any $\mathcal{O}$-lattice in a semistable E-representation of G_K with Hodge–Tate weights in $[0, h]$ (for some integer $h \geq 0$), there corresponds a Breuil–Kisin–Fargues G_K-module $\mathfrak{M}^{\mathrm{inf}}$ of height at most h which admits all descents.

4.4.1 Proposition. *For any fixed K, a, h and $N \geq e(a+h)/(p-1)$ there is a constant $s'(K, a, h, N) \geq s(a, h, N)$ (where $s(a, h, N)$ is as in Lemma 4.3.3) with the following property: for any Breuil–Kisin–Fargues G_K-module $\mathfrak{M}^{\mathrm{inf}}$ corresponding as above to a semistable G_K representation with Hodge–Tate weights in $[0, h]$, and for any choice of $\pi^\flat$ with corresponding descent $\mathfrak{M}_{\pi^\flat}$, if $s \geq s'(K, a, h, N)$, the restriction to $G_{K_{\pi^\flat,s}}$ of the action of G_K on $\mathfrak{M}^{\mathrm{inf}} \otimes_{\mathcal{O}} \mathcal{O}/\varpi^a$ agrees with the action obtained from $\mathfrak{M}_{\pi^\flat}$ by Lemma 4.3.3.*

Proof. This is essentially the content of [CL11, Prop. 3.3.1]. By the statement of Lemma 4.3.3, it is enough to check that if s is sufficiently large, then for all $g \in G_{K_{\pi^\flat,s}}$ we have $(g-1)(\mathfrak{M}_{\pi^\flat} \otimes_{\mathcal{O}} \mathcal{O}/\varpi^a) \subset u^N \mathbf{A}_{\mathrm{inf}} \otimes_{\mathfrak{S}} \mathfrak{M}_{\pi^\flat} \otimes_{\mathcal{O}} \mathcal{O}/\varpi^a$, where the action of g is via the given action of G_K on $\mathfrak{M}^{\mathrm{inf}}$. This is immediate from [CL11, Lem. 3.2.1] (taking $\mathfrak{L} = \mathfrak{M}_{\pi^\flat}$, and noting that in their notation, $\widehat{\mathfrak{L}}$ is an extension of scalars of $\mathfrak{L}$ to a subring of $\mathbf{A}_{\mathrm{inf},\mathcal{O}/\varpi^a}$; this extension of scalars involves a twist by φ, but since φ is a continuous automorphism of $\mathbf{A}_{\mathrm{inf}}$, this is harmless). □

4.5 STACKS OF SEMISTABLE AND CRYSTALLINE BREUIL–KISIN–FARGUES MODULES

We now define moduli stacks parameterizing the Breuil–Kisin–Fargues G_K-modules admitting all descents that were introduced in Definition 4.2.4. As discussed in Section 4.4, we also impose a condition that the G_K-actions agree with the canonical actions introduced in Section 4.3. To this end, we make the following definition.

4.5.1 Definition. For each K, a, h and $N \geq e(a+h)/(p-1)$, we fix a constant $s'(K,a,h,N)$ as in the statement of Proposition 4.4.1. Let A be a p-adically complete $\mathcal{O}$-algebra which is topologically of finite type, and let $\mathfrak{M}^{\mathrm{inf}}$ be a projective Breuil–Kisin–Fargues G_K-module with A-coefficients which admits all descents. Then we say that the G_K-action on $\mathfrak{M}^{\mathrm{inf}}$ is *canonical* if for any $a \geq 1$, any $\pi^\flat$, and any $s \geq s'(K,a,h,N)$, the restriction to $G_{K_{\pi^\flat,s}}$ of the action of G_K on $\mathfrak{M}^{\mathrm{inf}} \otimes_{\mathcal{O}} \mathcal{O}/\varpi^a$ agrees with the action obtained from $\mathfrak{M}_{\pi^\flat}$ by Lemma 4.3.3.

4.5.2 Remark. The point of this definition is that it is immediate from Proposition 4.4.1 that the Breuil–Kisin–Fargues modules coming from lattices in semistable G_K-representations have canonical G_K-actions.

We now define our moduli stacks of Breuil–Kisin–Fargues G_K-modules.

4.5.3 Definition. For any $h \geq 0$ we let $\mathcal{C}^a_{d,\mathrm{ss},h}$ denote the limit preserving category of groupoids over $\operatorname{Spec} \mathcal{O}/\varpi^a$ determined by decreeing, for any finite type $\mathcal{O}/\varpi^a$-algebra A, that $\mathcal{C}^a_{d,\mathrm{ss},h}(A)$ is the groupoid of rank d projective Breuil–Kisin–Fargues G_K-modules with A-coefficients, which are of height at most h, which admit all descents, and whose G_K-action is canonical.

We let $\mathcal{C}^a_{d,\mathrm{crys},h}$ denote the limit preserving subcategory of groupoids of $\mathcal{C}^a_{d,\mathrm{ss},h}$ consisting of those Breuil–Kisin–Fargues G_K-modules $\mathfrak{M}^{\mathrm{inf}}$ which are furthermore crystalline. We let $\mathcal{C}_{d,\mathrm{ss},h} := \varinjlim_a \mathcal{C}^a_{d,\mathrm{ss},h}$, and let $\mathcal{C}_{d,\mathrm{crys},h} := \varinjlim_a \mathcal{C}^a_{d,\mathrm{crys},h}$.

4.5.4 Remark. This definition uniquely determines limit preserving categories fibred in groupoids over $\mathcal{O}$ by [EG21, Lem. 2.5.4]. We will see shortly that the inductive limits $\mathcal{C}_{d,\mathrm{ss},h}$ and $\mathcal{C}_{d,\mathrm{crys},h}$ are in fact p-adic formal algebraic stacks (Theorem 4.5.20).

If A is a p-adically complete $\mathcal{O}$-algebra, then as usual we write $\mathcal{C}_{d,\mathrm{ss},h}(A)$ (resp. $\mathcal{C}_{d,\mathrm{crys},h}(A)$) to denote the groupoid of morphisms from $\operatorname{Spf} A$ (defined using the p-adic topology on A) to $\mathcal{C}_{d,\mathrm{ss},h}$ (resp. $\mathcal{C}_{d,\mathrm{crys},h}$). More concretely, we have

$$\mathcal{C}_{d,\mathrm{ss},h}(A) = \varprojlim_{a} \mathcal{C}^a_{d,\mathrm{ss},h}(A/\varpi^a),$$

and similarly for $\mathcal{C}_{d,\mathrm{crys},h}(A)$. We will typically apply this convention in the case when A is furthermore topologically of finite type over $\mathcal{O}$. In this case each of the quotients A/ϖ^a is of finite type over $\mathcal{O}/\varpi^a$ (by the definition of topologically of finite type), and so the objects of $\mathcal{C}^a_{d,\mathrm{ss},h}(A/\varpi^a)$ admit a concrete interpretation as Breuil–Kisin–Fargues G_K-modules over A. The objects of $\mathcal{C}_{d,\mathrm{ss},h}(A)$ and of $\mathcal{C}_{d,\mathrm{crys},h}(A)$ then similarly admit a concrete interpretation.

4.5.5 Lemma. *Let A be a p-adically complete $\mathcal{O}$-algebra which is topologically of finite type. Then $\mathcal{C}_{d,\mathrm{ss},h}(A)$ is the groupoid of rank d projective Breuil–Kisin–Fargues G_K-modules with A-coefficients, which are of height at most h, which admit all descents, and whose G_K-action is canonical. In addition, $\mathcal{C}_{d,\mathrm{crys},h}(A)$ is the subgroupoid consisting of those Breuil–Kisin–Fargues G_K-modules $\mathfrak{M}^{\mathrm{inf}}$ which are furthermore crystalline.*

Proof. This follows from Lemma D.5. □

The following lemma is crucial: it shows that Breuil–Kisin–Fargues modules which admit all descents give rise to (φ,Γ)-modules.

4.5.6 Lemma. *There is a natural morphism $\mathcal{C}_{d,\mathrm{ss},h} \to \mathcal{X}_d$, which on* $\operatorname{Spf} A$*-points (for p-adically complete $\mathcal{O}$-algebras A that are topologically of finite type) is given by extending scalars to $W(\mathbf{C}^\flat)_A$.*

Proof. As indicated in the statement of the lemma, this morphism is defined, for finite type $\mathcal{O}/\varpi^a$-algebras, via $\mathfrak{M}^{\mathrm{inf}} \mapsto W(\mathbf{C}^\flat)_A \otimes_{\mathbf{A}_{\mathrm{inf},A}} \mathfrak{M}^{\mathrm{inf}}$, with the target object being regarded as an A-valued point of $\mathcal{X}_d$ via the equivalence of Proposition 2.7.8. Since both $\mathcal{C}_{d,\mathrm{ss},h}$ and $\mathcal{X}_d$ are limit preserving (the former by its very construction, and the latter by Lemma 3.2.19), it follows from [EG21, Lem. 2.5.4, Lem. 2.5.5 (1)] that this construction determines a morphism $\mathcal{C}_{d,\mathrm{ss},h} \to \mathcal{X}_d$. □

4.5.7 Remark. The proof of Lemma 4.5.6 illustrates a general principle (which was also applied in the proofs of the results in Section 3.7), namely that to construct a morphism between limit preserving stacks over $\operatorname{Spec} \mathcal{O}/\varpi^a$, or over $\operatorname{Spf} \mathcal{O}$, it suffices to define the intended morphism on finite type $\mathcal{O}/\varpi^a$-algebras (with a either fixed or varying, depending on which case we are in).

Similarly, if P is any property of such a morphism that is preserved under base change, and that is tested by pulling back over A-valued points, then to test for the property P, it suffices to consider such pull-backs in the case when A is of finite type over $\mathcal{O}/\varpi^a$.

We will apply these principles consistently throughout the remainder of this section.

For any $h \geq 0$ and any choice of $\pi^\flat$, we write $\mathcal{C}_{\pi^\flat,d,h}$ for the moduli stack of rank d projective Breuil–Kisin modules over $\mathfrak{S}_{\pi^\flat,A}$ of height at most h, and $\mathcal{R}_{\pi^\flat,d}$ for the corresponding stack of rank d projective étale φ-modules. We also write $\mathcal{C}^a_{\pi^\flat,d,h}$ and $\mathcal{R}^a_{\pi^\flat,d}$ for their base change over $\mathcal{O}/\varpi^a$, for any $a \geq 1$.

Recall that, by Proposition 3.7.2, we have a natural morphism $\mathcal{X}_{K,d} \to \mathcal{R}_{\pi^\flat,d}$, which for each $s \geq 1$ can be factored, via Lemma 3.7.5, as

$$\mathcal{X}_{K,d} \to \mathcal{X}_{K_{\pi^\flat,s},d} \to \mathcal{R}_{\pi^\flat,d}.$$

4.5.8 Proposition. *For any fixed a, h, and any N and $s(a,h,N)$ as in Lemma 4.3.3, for any $s \geq s(a,h,N)$ there is a canonical morphism $\mathcal{C}^a_{\pi^\flat,d,h} \to \mathcal{X}^a_{K_{\pi^\flat,s},d}$ obtained from the canonical action of Lemma 4.3.3. This morphism fits into a commutative triangle*

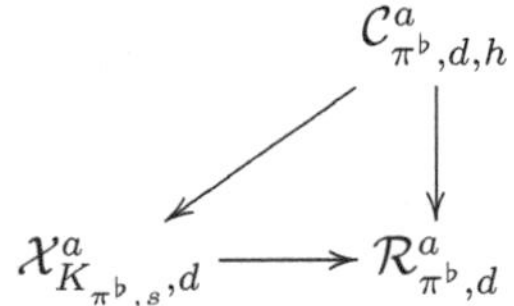

Proof. As discussed in Remark 4.5.7, it is enough to show that if A is a finite type $\mathcal{O}/\varpi^a$-algebra, and $\mathfrak{M}$ is a finite projective étale Breuil–Kisin module with A-coefficients, then we can canonically extend the action of $G_{K_{\pi^\flat,\infty}}$ on $W(\mathbf{C}^\flat)_A \otimes_{\mathfrak{S}_{\pi^\flat,A}} \mathfrak{M}$ to an action of $G_{K_{\pi^\flat,s}}$, if $s \geq s(a,h,N)$. Such an extension is provided by Lemma 4.3.3. ☐

4.5.9 Lemma. *The morphism $\mathcal{C}^a_{\pi^\flat,d,h} \to \mathcal{X}^a_{K_{\pi^\flat,s},d}$ of Proposition 4.5.8 is representable by algebraic spaces, proper, and of finite presentation.*

Proof. This follows by a standard graph argument. Namely, we write this morphism as the composite

$$\mathcal{C}^a_{\pi^\flat,d,h} \to \mathcal{C}^a_{\pi^\flat,d,h} \times_{\mathcal{R}^a_{\pi^\flat,d}} \mathcal{X}^a_{K_{\pi^\flat,s},d} \to \mathcal{X}^a_{K_{\pi^\flat,s},d}.$$

The first arrow is a closed immersion, being the base change of the diagonal morphism $\mathcal{X}_d \to \mathcal{X}_d \times_{\mathcal{R}_{\pi^\flat,d}} \mathcal{X}_d$, which is a closed immersion by Proposition 3.7.4. The second arrow is representable by algebraic spaces and proper, because it is the base change of the morphism $\mathcal{C}^a_{\pi^\flat,d,h} \to \mathcal{R}^a_{\pi^\flat,d}$ of Theorem 3.1.4. The composite morphism is thus representable by algebraic spaces and proper. Such a morphism is in particular of finite type and separated (and so also quasi-separated), and hence Lemma 4.5.10 below allows us to conclude that it is also of finite presentation (once we take into account Lemma 3.2.19, which shows that its target is limit preserving). □

4.5.10 Lemma. *Let $\mathcal{Z} \hookrightarrow \mathcal{X}$ be a morphism of stacks over a locally Noetherian base scheme S which is representable by algebraic spaces, of finite type, and quasi-separated. If $\mathcal{X}$ is limit preserving, then this morphism is furthermore of finite presentation.*

Proof. It is enough to show, for any morphism $T \to \mathcal{X}$ whose source is an affine S-scheme T, that the pulled-back morphism (of algebraic spaces) $T \times_{\mathcal{X}} \mathcal{Z} \to T$ is of finite presentation. Since $\mathcal{X}$ is limit preserving, the morphism $T \to \mathcal{X}$ factors through a morphism of S-schemes $T \to T'$, where T' is affine and of finite presentation over S. Thus, replacing T by T', we may assume that T is of finite presentation over S, and consequently Noetherian.

The base-changed morphism $T \times_{\mathcal{X}} \mathcal{Z} \to T$ is thus a finite type and quasi-separated morphism from an algebraic space to a Noetherian scheme, and hence is of finite presentation by [Sta, Tag 06G4]. This proves the lemma. □

4.5.11 Definition. The morphisms of Lemma 3.7.5 and Proposition 4.5.8 allow us to define a stack $\mathcal{C}^a_{\pi^\flat,s,d,h}$ by the requirement that it fits into a 2-Cartesian diagram

$$\begin{array}{ccc} \mathcal{C}^a_{\pi^\flat,s,d,h} & \longrightarrow & \mathcal{C}^a_{\pi^\flat,d,h} \\ \downarrow & & \downarrow \\ \mathcal{X}^a_{K,d} & \longrightarrow & \mathcal{X}^a_{K_{\pi^\flat,s},d} \end{array} \tag{4.5.12}$$

The stack $\mathcal{C}^a_{\pi^\flat,s,d,h}$ classifies rank d projective Breuil–Kisin modules $\mathfrak{M}$ of height at most h over $\mathfrak{S}_{\pi^\flat,A}$ equipped with an extension of the canonical $G_{K_{\pi^\flat,s}}$-action on $W(\mathbf{C}^\flat)_A \otimes_{\mathfrak{S}_{\pi^\flat,A}} \mathfrak{M}$ to an action of G_K (making it an étale (φ, G_K)-module).

Since the lower horizontal arrow in (4.5.12) is representable by algebraic spaces and of finite presentation, by Lemma 3.7.5, and since $\mathcal{C}^a_{\pi^\flat,d,h}$ is an algebraic stack of finite presentation over $\mathcal{O}/\varpi^a$, by Theorem 3.1.4, we see that $\mathcal{C}^a_{\pi^\flat,s,d,h}$ is also an algebraic stack of finite presentation over $\mathcal{O}/\varpi^a$. Since the right-hand side vertical arrow in this diagram is representable by algebraic spaces, proper, and of finite presentation, by Lemma 4.5.9, so is the left-hand side vertical arrow.

4.5.13 Lemma. *The diagonal of $\mathcal{C}^a_{\pi^\flat,s,d,h}$ is affine and of finite presentation.*

Proof. As we already noted, the morphism $\mathcal{C}^a_{\pi^\flat,s,d,h} \to \mathcal{X}_{K,d}$ is representable by algebraic spaces and proper (and so in particular is both separated and of finite type); the claim of the lemma thus follows from Proposition 3.2.17, together with Lemma 4.5.14 below. □

4.5.14 Lemma. *Suppose that $f\colon \mathcal{X} \to \mathcal{Y}$ is a morphism of stacks over a base scheme S which is representable by algebraic spaces, separated, and of finite type. Suppose also that the diagonal $\Delta_{\mathcal{Y}}\colon \mathcal{Y} \to \mathcal{Y} \times_S \mathcal{Y}$ is affine and of finite presentation. Then the diagonal $\Delta_{\mathcal{X}}\colon \mathcal{X} \to \mathcal{X} \times_S \mathcal{X}$ is affine and of finite presentation.*

Proof. The diagonal morphism

$$\mathcal{X} \to \mathcal{X} \times_S \mathcal{X}$$

may be factored as the composite of the relative diagonal

$$\mathcal{X} \to \mathcal{X} \times_{\mathcal{Y}} \mathcal{X} \tag{4.5.15}$$

and the morphism

$$\mathcal{X} \times_{\mathcal{Y}} \mathcal{X} \to \mathcal{X} \times_S \mathcal{X}, \tag{4.5.16}$$

which is a base change of the diagonal morphism $\Delta_{\mathcal{Y}}$. Our assumption on $\Delta_{\mathcal{Y}}$, then, implies that the morphism (4.5.16) is affine and of finite presentation. Since $\mathcal{X} \to \mathcal{Y}$ is representable by algebraic spaces and separated, its diagonal (4.5.15) is a closed immersion, and thus affine. Since it is furthermore of finite type, the morphism (4.5.15) is also of finite presentation, by Lemma 3.2.7. Thus $\Delta_{\mathcal{X}}$ is the composite of morphisms that are affine and of finite presentation, and the lemma follows. □

4.5.17 Proposition. *For each $\pi^\flat$ and each $s \geq s'(K,a,h,N)$, there are natural closed immersions $\mathcal{C}^a_{d,\mathrm{crys},h} \to \mathcal{C}^a_{d,\mathrm{ss},h} \to \mathcal{C}^a_{\pi^\flat,s,d,h}$. In particular, $\mathcal{C}^a_{d,\mathrm{ss},h}$ and $\mathcal{C}^a_{d,\mathrm{crys},h}$ are algebraic stacks of finite presentation over $\mathcal{O}/\varpi^a$, and have affine diagonals.*

Proof. We have already observed that $\mathcal{C}^a_{\pi^\flat,s,d,h}$ is of finite presentation over $\mathcal{O}/\varpi^a$, and has affine diagonal by Lemma 4.5.13. Once we construct the claimed closed immersions, the claims of finite presentation will follow from Lemma 4.5.10; and since closed immersions are in particular monomorphisms, the diagonals of $\mathcal{C}_{d,\mathrm{ss},h}$ and $\mathcal{C}_{d,\mathrm{crys},h}$ are then obtained via base change from the diagonal of $\mathcal{C}^a_{\pi^\flat,s,d,h}$, and are in particular affine. We thus focus on constructing these closed immersions. Throughout the proof, we take into account Remark 4.5.7, which allows us to restrict our attention to the points of the various stacks in question that are defined over finite type $\mathcal{O}/\varpi^a$-algebras.

Lemma 4.5.6 constructs a morphism $\mathcal{C}^a_{d,\mathrm{ss},h} \to \mathcal{X}^a_{K,d}$, given by extending scalars to $W(\mathbf{C}^\flat)$, and it follows from Lemma 4.2.8 that there is a natural morphism $\mathcal{C}^a_{d,\mathrm{ss},h} \to \mathcal{C}^a_{\pi^\flat,d,h}$, defined via $\mathfrak{M}^{\mathrm{inf}} \mapsto \mathfrak{M}_{\pi^\flat}$. The composite morphisms $\mathcal{C}^a_{d,\mathrm{ss},h} \to \mathcal{C}^a_{\pi^\flat,d,h} \to \mathcal{X}^a_{K_{\pi^\flat,s},d}$ and $\mathcal{C}^a_{d,\mathrm{ss},h} \to \mathcal{X}^a_{K,d} \to \mathcal{X}^a_{K_{\pi^\flat,s},d}$ coincide by definition (see Definitions 4.5.1 and 4.5.3). Thus these morphisms induce a morphism

$$\mathcal{C}^a_{d,\mathrm{ss},h} \to \mathcal{C}^a_{\pi^\flat,s,d,h}. \tag{4.5.18}$$

To see that (4.5.18) is a monomorphism, it is enough to note that if A is any finite type $\mathcal{O}/\varpi^a$-algebra A, and $\mathfrak{M}^{\mathrm{inf}}$ is any Breuil–Kisin–Fargues module over A which admits all descents and whose G_K-action is canonical, corresponding to an A-valued point of $\mathcal{C}^a_{d,\mathrm{ss},h}$, then $\mathfrak{M}^{\mathrm{inf}} = \mathbf{A}_{\mathrm{inf},A} \otimes_{\mathfrak{S}_{\pi^\flat,A}} \mathfrak{M}_{\pi^\flat}$ is determined by $\mathfrak{M}_{\pi^\flat}$, and the G_K-action on $\mathfrak{M}^{\mathrm{inf}}$ is determined by the G_K-action on $W(\mathbf{C}^\flat)_A \otimes_{\mathbf{A}_{\mathrm{inf},A}} \mathfrak{M}^{\mathrm{inf}}$.

We now show that (4.5.18) is a closed immersion. It is enough to do this after pulling back to some finite type $\mathcal{O}/\varpi^a$-algebra A, where we need to show that the conditions that $\mathfrak{M}^{\mathrm{inf}} = \mathbf{A}_{\mathrm{inf},A} \otimes_{\mathfrak{S}_{\pi^\flat,A}} \mathfrak{M}_{\pi^\flat}$ is G_K-stable and admits all descents, and that the G_K-action is canonical, are closed conditions. We begin with the first of these. It is enough to show that for each $g \in G_K$, the condition that $g(\mathfrak{M}^{\mathrm{inf}}) \subset \mathfrak{M}^{\mathrm{inf}}$ is closed. This condition is equivalent to the vanishing of the composite morphism

$$g^*\mathfrak{M}^{\mathrm{inf}} \to \mathfrak{M}^{\mathrm{inf}}[1/u] \to \mathfrak{M}^{\mathrm{inf}}[1/u]/\mathfrak{M}^{\mathrm{inf}},$$

and this is a closed condition by Lemma B.28. Let $\mathcal{C}^a_{d,G_K}$ denote the closed substack of $\mathcal{C}^a_{\pi^\flat,s,d,h}$ for which $\mathfrak{M}^{\mathrm{inf}}$ is G_K-stable.

We next show that for each choice of uniformizer π', and each $(\pi')^\flat$, the condition that $\mathfrak{M}^{\mathrm{inf}}$ admits a descent to $\mathfrak{S}_{A,(\pi')^\flat}$ is a closed condition. To this end, note that by Proposition 2.7.8, $\mathfrak{M}^{\mathrm{inf}}[1/u]$ descends uniquely to $\mathfrak{S}_{(\pi')^\flat,A}[1/u]$, so we have a morphism $\mathcal{C}^a_{d,G_K} \to \mathcal{R}^a_{(\pi')^\flat,d}$. Then the fiber product $\mathcal{C}^a_{d,G_K} \times_{\mathcal{R}^a_{(\pi')^\flat,d}} \mathcal{C}^a_{(\pi')^\flat,d,h}$ is the moduli space of pairs $(\mathfrak{M}, \mathfrak{M}')$ where $\mathfrak{M} := \mathfrak{M}_{\pi^\flat}$ is as above (an object classified by $\mathcal{C}^a_{d,G,K}$), and $\mathfrak{M}'$ is a Breuil–Kisin module for $\mathfrak{S}_{(\pi')^\flat,A}$ of height at most h which satisfies

$$W(\mathbf{C}^\flat)_A \otimes_{\mathfrak{S}_{(\pi')^\flat,A}} \mathfrak{M}' = W(\mathbf{C}^\flat)_A \otimes_{\mathfrak{S}_{\pi^\flat,A}} \mathfrak{M}.$$

Consider the substack $\mathcal{C}^a_{d,G_K,(\pi')^\flat}$ of this fiber product satisfying the condition that

$$\mathbf{A}_{\mathrm{inf},A} \otimes_{\mathfrak{S}_{(\pi')^\flat,A}} \mathfrak{M}' = \mathfrak{M}^{\mathrm{inf}}. \tag{4.5.19}$$

Note that this is exactly the condition that $\mathfrak{M}^{\mathrm{inf}}$ admits a descent for $(\pi')^\flat$, and that $\mathfrak{M}'$ is this descent (which is uniquely determined by Lemma 4.2.8), so the projection $\mathcal{C}^a_{d,G_K,(\pi')^\flat} \to \mathcal{C}^a_{d,G_K}$ is a monomorphism, and we need to show that it is a closed immersion. Since the projection $\mathcal{C}^a_{d,G_K} \times_{\mathcal{R}^a_{(\pi')^\flat,d}} \mathcal{C}^a_{(\pi')^\flat,d,h} \to \mathcal{C}^a_{d,G_K}$ is proper (being a base change of the proper morphism $\mathcal{C}^a_{(\pi')^\flat,d,h} \to \mathcal{R}^a_{(\pi')^\flat,d}$), and since proper monomorphisms are closed immersions, it is enough to show that $\mathcal{C}^a_{d,G_K,(\pi')^\flat}$ is a closed substack of $\mathcal{C}^a_{d,G_K} \times_{\mathcal{R}^a_{(\pi')^\flat,d}} \mathcal{C}^a_{(\pi')^\flat,d,h}$; that is, we must show that (4.5.19) is a closed condition. This again follows from Lemma B.28.

We now consider the closed substack of $\mathcal{C}^a_{\pi^\flat,s,d,h}$ for which $\mathfrak{M}^{\mathrm{inf}}$ is G_K-stable and admits a descent for each $(\pi')^\flat$. We need to show that for each $(\pi')^\flat$, the further conditions

$$\mathfrak{M}_{\pi^\flat}/[\pi^\flat]\mathfrak{M}_{\pi^\flat} = \mathfrak{M}_{(\pi')^\flat}/[(\pi')^\flat]\mathfrak{M}_{(\pi')^\flat}$$

and

$$\varphi^*\mathfrak{M}_{\pi^\flat}/E_{\pi^\flat}\varphi^*\mathfrak{M}_{\pi^\flat} = \varphi^*\mathfrak{M}_{(\pi')^\flat}/E_{(\pi')^\flat}\varphi^*\mathfrak{M}_{(\pi')^\flat}$$

are closed conditions.

The arguments in both cases are very similar (and in turn are similar to the proof of Lemma B.28), so we only give the argument in the second case, leaving the first to the reader. Each side is a projective $\mathcal{O}_K \otimes_{\mathbf{Z}_p} A$-submodule of the projective $\mathcal{O}_{\mathbf{C}} \otimes_{\mathbf{Z}_p} A$-module $\mathcal{O}_{\mathbf{C}} \otimes_{\mathbf{A}_{\mathrm{inf},A},\theta} \varphi^*\mathfrak{M}^{\mathrm{inf}}$, and indeed spans this module after extension of scalars to $\mathcal{O}_{\mathbf{C}}$. By symmetry, it is enough to show that for each element $m \in \varphi^*\mathfrak{M}_{(\pi')^\flat}/E_{(\pi')^\flat}\varphi^*\mathfrak{M}_{(\pi')^\flat}$, the condition that $m \in \varphi^*\mathfrak{M}_{\pi^\flat}/E_{\pi^\flat}\varphi^*\mathfrak{M}_{\pi^\flat}$ is closed.

Write $P := \varphi^*\mathfrak{M}_{\pi^\flat}/E_{\pi^\flat}\varphi^*\mathfrak{M}_{\pi^\flat}$, and choose a finite projective $\mathcal{O}_K \otimes_{\mathbf{Z}_p} A$-module Q such that $F := P \oplus Q$ is free. We can think of m as an element of $\mathcal{O}_{\mathbf{C}} \otimes_{\mathcal{O}_K} P$ and thus as an element of $\mathcal{O}_{\mathbf{C}} \otimes_{\mathcal{O}_K} F$, and it is enough to check that the condition that $m \in F$ is closed. Choosing a basis for F, we reduce to the case that F is one-dimensional, and thus to showing that the condition that an element of $\mathcal{O}_{\mathbf{C}} \otimes_{\mathbf{Z}_p} A$ lies in $\mathcal{O}_K \otimes_{\mathbf{Z}_p} A$ is closed. Since $\mathcal{O}_{\mathbf{C}}$ is a torsion-free and thus flat $\mathcal{O}_K$-module, $\mathcal{O}_{\mathbf{C}}/\varpi^a$ is flat and thus free as a module for the Artinian ring $\mathcal{O}_K/\varpi^a$. Thus $\mathcal{O}_{\mathbf{C}} \otimes_{\mathbf{Z}_p} A$ is a free $\mathcal{O}_K \otimes_{\mathbf{Z}_p} A$-module, and the result follows by choosing a basis.

We now need to show that the condition that the G_K-action is canonical is a closed condition. Explicitly, by Lemma 4.3.3, we need to show that for each $b \le a$, each $N \ge e(b+h)/(p-1)$, each $s \ge s'(K,b,h,N)$, each $\pi^\flat$, and each $g \in G_{K_{\pi^\flat,s}}$, the condition that

$$(g-1)(\mathfrak{M}_{\pi^\flat} \otimes_{\mathcal{O}} \mathcal{O}/\varpi^b) \subset u^N(\mathfrak{M}^{\mathrm{inf}} \otimes_{\mathcal{O}} \mathcal{O}/\varpi^b)$$

is closed. This follows from Lemma B.29.

Finally, it remains to show that the monomorphism $\mathcal{C}^a_{d,\mathrm{crys},h} \to \mathcal{C}^a_{d,\mathrm{ss},h}$ is a closed immersion; that is, for each $g \in G_K$ and each $\pi^\flat$ we need to show that the condition that

$$(g-1)(\mathfrak{M}_{\pi^\flat}) \subseteq \varphi^{-1}(\mu)[\pi^\flat]\mathfrak{M}^{\text{inf}}$$

is a closed condition. It is enough to show that for each $m \in \mathfrak{M}_{\pi^\flat}$, the condition that $(g-1)(m) \in \varphi^{-1}(\mu)[\pi^\flat]\mathfrak{M}^{\text{inf}}$ is closed. Since $\varphi^{-1}(\mu)[\pi^\flat]\mathfrak{M}^{\text{inf}}$ is a finite projective $\mathbf{A}_{\text{inf},A}$-module, this follows from another application of Lemma B.29. □

4.5.20 Theorem. *$\mathcal{C}_{d,\text{ss},h}$ is a p-adic formal algebraic stack of finite presentation, as is its closed substack $\mathcal{C}_{d,\text{crys},h}$. In addition, both of these stacks have affine diagonal, and each of the morphisms $\mathcal{C}_{d,\text{ss},h} \to \mathcal{X}_{K,d}$ and $\mathcal{C}_{d,\text{crys},h} \to \mathcal{X}_{K,d}$ is representable by algebraic spaces, proper, and of finite presentation.*

Proof. Since $\mathcal{C}_{d,\text{crys},h}$ is a closed substack of $\mathcal{C}_{d,\text{ss},h}$, it suffices to prove the statements of the lemma for $\mathcal{C}_{d,\text{ss},h}$. (Here we use the fact that a closed immersion is a monomorphism to see that the diagonal of $\mathcal{C}_{d,\text{crys},h}$ is obtained via base change from the diagonal of $\mathcal{C}_{d,\text{ss},h}$, while we use Lemma 4.5.10 to transfer finite presentation properties from $\mathcal{C}_{d,\text{ss},h}$ to $\mathcal{C}_{d,\text{crys},h}$).

By Proposition A.13, to see that $\mathcal{C}_{d,\text{ss},h}$ is a p-adic formal algebraic stack of finite presentation, it is enough to show that each $\mathcal{C}^a_{d,\text{ss},h}$ is an algebraic stack of finite presentation over $\mathcal{O}/\varpi^a$, which was proved in Proposition 4.5.17. Since each $\mathcal{C}^a_{d,\text{ss},h}$ has affine diagonal, by the same proposition, we see as well that $\mathcal{C}_{d,\text{ss},h}$ has affine diagonal.

That $\mathcal{C}_{d,\text{ss},h} \to \mathcal{X}_{K,d}$ is representable by algebraic spaces, proper, and of finite presentation also follows from Proposition 4.5.17. Indeed, for each a the morphism $\mathcal{C}^a_{d,\text{ss},h} \to \mathcal{X}_{K,d}$ factors as $\mathcal{C}^a_{d,\text{ss},h} \to \mathcal{C}^a_{\pi^\flat,s,d,h} \to \mathcal{X}^a_{K,d}$, where the first morphism is a closed immersion of finite type algebraic stacks, so it is enough to prove the same properties for the morphism $\mathcal{C}^a_{\pi^\flat,s,d,h} \to \mathcal{X}^a_{K,d}$; as explained in Definition 4.5.11, this follows from Lemma 4.5.9. □

4.5.21 Potentially crystalline and potentially semistable stacks

We now consider the corresponding potentially semistable and potentially crystalline versions of these moduli stacks of Breuil–Kisin–Fargues modules.

4.5.22 Definition. For a Galois extension L/K, define stacks $\mathcal{C}^{L/K}_{d,\text{ss},h}$ and $\mathcal{C}^{L/K}_{d,\text{crys},h}$ as follows: for each $a \geq 1$ we let $\mathcal{C}^{L/K,a}_{d,\text{ss},h}$ denote the limit preserving category of groupoids over $\operatorname{Spec} \mathcal{O}/\varpi^a$ determined by decreeing, for any finite type $\mathcal{O}/\varpi^a$-algebra A, that $\mathcal{C}^{L/K,a}_{d,\text{ss},h}(A)$ is the groupoid of Breuil–Kisin–Fargues G_K-modules with A-coefficients, which are of height at most h, which admit all descents over L, and whose G_L-actions are canonical.

We let $\mathcal{C}^{L/K,a}_{d,\mathrm{crys},h}$ denote the limit preserving subcategory of groupoids of $\mathcal{C}^{a}_{d,\mathrm{ss},h}$ consisting of those Breuil–Kisin–Fargues G_K-modules $\mathfrak{M}^{\mathrm{inf}}$ for which the action of G_L is crystalline.

We let $\mathcal{C}^{L/K}_{d,\mathrm{ss},h} := \varinjlim_a \mathcal{C}^{L/K,a}_{d,\mathrm{ss},h}$, and let $\mathcal{C}^{L/K}_{d,\mathrm{crys},h} := \varinjlim_a \mathcal{C}^{L/K,a}_{d,\mathrm{crys},h}$.

The following lemma is proved in exactly the same way as Lemma 4.5.5.

4.5.23 Lemma. *Let A be a p-adically complete $\mathcal{O}$-algebra which is topologically of finite type over $\mathcal{O}$. Then $\mathcal{C}^{L/K}_{d,\mathrm{ss},h}(A)$ is the groupoid of Breuil–Kisin–Fargues G_K-modules with A-coefficients, which are of height at most h, and which admit all descents over L, and whose G_L-actions are canonical; and $\mathcal{C}^{L/K}_{d,\mathrm{crys},h}(A)$ is the subgroupoid of those Breuil–Kisin–Fargues G_K-modules $\mathfrak{M}^{\mathrm{inf}}$ which are furthermore crystalline.*

4.5.24 Proposition. *For any Galois extension L/K, any d, and any h, $\mathcal{C}^{L/K}_{d,\mathrm{ss},h}$ is a p-adic formal algebraic stack of finite presentation, and $\mathcal{C}^{L/K}_{d,\mathrm{crys},h}$ is a closed substack, which is thus again a p-adic formal algebraic stack of finite presentation. Both $\mathcal{C}^{L/K}_{d,\mathrm{ss},h}$ and $\mathcal{C}^{L/K}_{d,\mathrm{crys},h}$ have affine diagonal, and the natural morphisms $\mathcal{C}^{L/K}_{d,\mathrm{ss},h} \to \mathcal{X}_{K,d}$ and $\mathcal{C}^{L/K}_{d,\mathrm{crys},h} \to \mathcal{X}_{K,d}$ are representable by algebraic spaces, proper, and of finite presentation.*

Proof. We have a natural morphism $\mathcal{C}^{L/K}_{d,\mathrm{ss},h} \to \mathcal{C}^{L/L}_{d,\mathrm{ss},h}$ (given by restricting the action of G_K to G_L; the target is of course just the stack denoted $\mathcal{C}_{d,\mathrm{ss},h}$ above, but for L instead of K), and a natural morphism $\mathcal{C}^{L/K}_{d,\mathrm{ss},h} \to \mathcal{X}_{K,d}$, and thus a natural morphism

$$\mathcal{C}^{L/K}_{d,\mathrm{ss},h} \to \mathcal{C}^{L/L}_{d,\mathrm{ss},h} \times_{\mathcal{X}_{L,d}} \mathcal{X}_{K,d}. \tag{4.5.25}$$

We claim that (4.5.25) is a closed immersion. Given this, the proposition follows. Indeed, by Theorem 4.5.20, $\mathcal{C}^{L/L}_{d,\mathrm{ss},h}$ is a p-adic formal algebraic stack of finite presentation, and the morphism $\mathcal{C}^{L/L}_{d,\mathrm{ss},h} \to \mathcal{X}_{L,d}$ is representable by algebraic spaces, proper, and of finite presentation. Furthermore, $\mathcal{X}_{K,d} \to \mathcal{X}_{L,d}$ is representable by algebraic spaces and of finite presentation by Lemma 3.7.5. That $\mathcal{C}^{L/K}_{d,\mathrm{crys},h}$ is a closed substack of $\mathcal{C}^{L/K}_{d,\mathrm{ss},h}$ follows from Proposition 4.5.17. The claims on the diagonals are proved by appeal to Lemma 4.5.14 (cf. the proof of Lemma 4.5.13).

It remains to prove the claim. It suffices to prove this after pulling back via a morphism $\operatorname{Spec} A \to \mathcal{X}_{L,d}$ for some finite type $\mathcal{O}/\varpi^a$-algebra A. Unwinding the definitions, we have to show that if $\mathfrak{M}^{\mathrm{inf}}$ is a Breuil–Kisin–Fargues G_L-module with A-coefficients, with a compatible action of G_K on $W(\mathbf{C}^\flat)_A \otimes_{\mathbf{A}_{\mathrm{inf},A}} \mathfrak{M}^{\mathrm{inf}}$, then the condition that $\mathfrak{M}^{\mathrm{inf}}$ is G_K-stable is a closed condition. It suffices to show that for each $g \in G_K$, the condition that $g(\mathfrak{M}^{\mathrm{inf}}) \subseteq \mathfrak{M}^{\mathrm{inf}}$ is a closed condition;

as in the proof of Proposition 4.5.17, this follows from Lemma B.28, applied to the composite morphism

$$g^*\mathfrak{M}^{\mathrm{inf}} \to \mathfrak{M}^{\mathrm{inf}}[1/u] \to \mathfrak{M}^{\mathrm{inf}}[1/u]/\mathfrak{M}^{\mathrm{inf}}. \qquad \square$$

We end this section with the following lemma, which will be used in the proof of Proposition 4.8.10.

4.5.26 Lemma. *Let R be a complete local Noetherian $\mathcal{O}$-algebra with residue field $\mathbf{F}$, together with a morphism $\operatorname{Spf} R \to \mathcal{X}_{K,d}$, and let $\widehat{C} := \mathcal{C}^{L/K}_{d,\mathrm{ss},h} \times_{\mathcal{X}_{K,d}} \operatorname{Spf} R$. Then there is a projective morphism of schemes $C \to \operatorname{Spec} R$ whose $\mathfrak{m}_R$-adic completion is isomorphic to $\widehat{C}$.*

Proof. Note that since $\mathcal{C}^{L/K}_{d,\mathrm{ss},h} \to \mathcal{X}_{K,d}$ is proper and representable by algebraic spaces, we see that $\widehat{C}$ is a formal algebraic space, and the morphism $\widehat{C} \to \operatorname{Spf} R$ is proper and representable by algebraic spaces. In particular, $\widehat{C}$ is a proper $\mathfrak{m}_R$-adic formal algebraic space over $\operatorname{Spf} R$.

To show that we can algebraize $\widehat{C}$, we will use an argument of Kisin—see [Kis09b, Prop. 2.1.10]—which proves a similar statement for moduli of Breuil–Kisin modules. In order to use this, we show that we can realize $\mathcal{C}^{L/K}_{d,\mathrm{ss},h}$ as a closed substack of a product of moduli stacks of Breuil–Kisin modules.

Let π, π' be two choices of uniformizers of L, chosen such that π/π' is not a pth power, and choose compatible systems of p-power roots $\pi^\flat, (\pi')^\flat$. Note that the closure of the subgroup of G_L generated by $G_{L_{\pi^\flat,\infty}}$ and $G_{L_{(\pi')^\flat,\infty}}$ is just G_L, because its fixed field is contained in $L_{\pi^\flat,\infty} \cap L_{(\pi')^\flat,\infty} = L$, so that a continuous action of G_L is determined by its restrictions to $G_{L_{\pi^\flat,\infty}}$ and $G_{L_{(\pi')^\flat,\infty}}$.

Suppose firstly that $L = K$. We have a natural morphism

$$\mathcal{C}_{d,\mathrm{ss},h} \to \mathcal{C}_{\pi^\flat,d,h} \times_{\operatorname{Spf}\mathcal{O}} \mathcal{C}_{(\pi')^\flat,d,h}, \tag{4.5.27}$$

which we claim is a monomorphism. To see this, it is enough to note that $\mathfrak{M}^{\mathrm{inf}} = \mathbf{A}_{\mathrm{inf},A} \otimes_{\mathfrak{S}_{\pi^\flat,A}} \mathfrak{M}_{\pi^\flat}$ is determined by $\mathfrak{M}_{\pi^\flat}$, while the G_K-action on $\mathfrak{M}^{\mathrm{inf}}$ is determined by the actions of $G_{K_{\pi^\flat,\infty}}$ and $G_{K_{(\pi')^\flat,\infty}}$, which are in turn determined by the conditions that they act trivially on $\mathfrak{M}_{\pi^\flat}$ and $\mathfrak{M}_{(\pi')^\flat}$ respectively.

We may factor (4.5.27) as the composite

$$\begin{aligned}\mathcal{C}_{d,\mathrm{ss},h} &\to (\mathcal{C}_{\pi^\flat,d,h} \times_{\operatorname{Spf}\mathcal{O}} \mathcal{C}_{(\pi')^\flat,d,h}) \times_{(\mathcal{R}_{\pi^\flat,d} \times_{\operatorname{Spf}\mathcal{O}} \mathcal{R}_{(\pi')^\flat,d})} \mathcal{X}_{K,d} \\ &\to \mathcal{C}_{\pi^\flat,d,h} \times_{\operatorname{Spf}\mathcal{O}} \mathcal{C}_{(\pi')^\flat,d,h}.\end{aligned}$$

Since the composite is a monomorphism, the first morphism is a monomorphism. This first morphism is also proper (since the morphism $\mathcal{C}_{d,\mathrm{ss},h} \to \mathcal{X}_{K,d}$ obtained by composing it with the projection to $\mathcal{X}_{K,d}$ is proper, by Theorem 4.5.20, while this projection is separated, being a base change of the product of the morphisms $\mathcal{C}_{(\pi)^\flat,d,h} \to \mathcal{R}_{\pi^\flat,d}$ and $\mathcal{C}_{(\pi')^\flat,d,h} \to \mathcal{R}_{(\pi')^\flat,d}$, each of which is proper, and so in particular separated, by Theorem 3.1.4), and so it is in fact a closed immersion. Thus $\widehat{C}$ is a closed algebraic subspace of

$$(\mathcal{C}_{\pi^\flat,d,h} \times_{\mathcal{R}_{\pi^\flat,d}} \operatorname{Spf} R) \times_{\operatorname{Spf} R} (\mathcal{C}_{(\pi')^\flat,d,h} \times_{\mathcal{R}_{(\pi')^\flat,d}} \operatorname{Spf} R).$$

By the Grothendieck Existence theorem for algebraic spaces [Knu71, Thm. V.6.3] (and the symmetry between $\pi^\flat$ and $(\pi')^\flat$), we are therefore reduced to showing that the morphism $\mathcal{C}_{\pi^\flat,d,h} \times_{\mathcal{R}_{\pi^\flat,d}} \operatorname{Spf} R \to \operatorname{Spf} R$ can be algebraized to a projective morphism. In the case $h=1$, this is [Kis09b, Prop. 2.1.10] (bearing in mind the main result of [BL95], as in the proof of [Kis09b, Prop. 2.1.7]), and the case of general h can be proved in exactly the same way, as explained in the proof of [Kis10, Prop. 1.3]; the key point is that $\mathcal{C}_{\pi^\flat,d,h} \times_{\mathcal{R}_{\pi^\flat,d}} \operatorname{Spf} R$ inherits a natural very ample (formal) line bundle from the affine Grassmannian.

We now consider the case of a general finite Galois extension L/K, where we argue as in the proof of Proposition 3.7.5. Let $\{g_i\}_{i=1,\dots,n}$ be a set of coset representatives for G_L in G_K. By definition, to give an object of $\mathcal{C}^{L/K}_{d,\mathrm{ss},h}(A)$ (for some A) is the same as giving an object of $\mathcal{C}^{L/L}_{d,\mathrm{ss},h}(A)$, together with an extension of the action of G_L on the underlying Breuil–Kisin–Fargues module $\mathfrak{M}^{\mathrm{inf}}_A$ to an action of G_K. Now, for each i we can give $g_i^*\mathfrak{M}^{\mathrm{inf}}_A$ the structure of a Breuil–Kisin–Fargues module by letting $h \in G_L$ act as $g_i^{-1}hg_i$ acts on $\mathfrak{M}^{\mathrm{inf}}_A$; it follows from the definitions that $g_i^*\mathfrak{M}^{\mathrm{inf}}_A$ is also an object of $\mathcal{C}^{L/L}_{d,\mathrm{ss},h}(A)$. Then to give an extension of the action of G_L on $\mathfrak{M}^{\mathrm{inf}}_A$ to an action of G_K is to give for each i an isomorphism $g_i^*\mathfrak{M}^{\mathrm{inf}}_A \xrightarrow{\sim} \mathfrak{M}^{\mathrm{inf}}_A$ of objects of $\mathcal{C}^{L/L}_{d,\mathrm{ss},h}(A)$, satisfying a slew of compatibilities.

We let $\mathcal{Y}_i$ denote the stack classifying objects $\mathfrak{M}^{\mathrm{inf}}_A$ of $\mathcal{C}^{L/L}_{d,\mathrm{ss},h}(A)$, endowed with an isomorphism $g_i^*\mathfrak{M}^{\mathrm{inf}}_A \to \mathfrak{M}^{\mathrm{inf}}_A$. If we regard g_i^* as an automorphism of $\mathcal{C}^{L/L}_{d,\mathrm{ss},h}(A)$, then we may form its graph Γ_i, and we then have an isomorphism of stacks

$$\mathcal{Y}_i \xrightarrow{\sim} \mathcal{C}^{L/L}_{d,\mathrm{ss},h}(A) \times_{\Delta,\mathcal{C}^{L/L}_{d,\mathrm{ss},h}(A)\times\mathcal{C}^{L/L}_{d,\mathrm{ss},h}(A),\Gamma_i} \mathcal{C}^{L/L}_{d,\mathrm{ss},h}(A).$$

(Here Δ denotes the diagonal; cf. the definition of $\mathcal{R}_d^{\Gamma_{\mathrm{disc}}}$ in Section 3.2 above, together with Proposition 3.2.14.) The projection onto the second factor $\mathcal{Y}_i \to \mathcal{C}^{L/L}_{d,\mathrm{ss},h}(A)$, which corresponds to forgetting the isomorphism, is a base change of the diagonal Δ, and so is affine, since Δ is affine by Theorem 4.5.20. In particular, $\mathcal{Y}_i \times_{\mathcal{X}_{L,d}} \operatorname{Spf} R$ admits a natural ample formal line bundle, pulled back from the natural very ample formal line bundle on $\mathcal{C}^{L/L}_{d,\mathrm{ss},h} \times_{\mathcal{X}_{L,d}} \operatorname{Spf} R$, whose existence we established above.

We can rephrase our interpretation of objects of $\mathcal{C}^{L/K}_{d,\mathrm{ss},h}(A)$ as objects of $\mathcal{C}^{L/L}_{d,\mathrm{ss},h}(A)$ endowed with isomorphisms $g_i^*\mathfrak{M}^{\mathrm{inf}}_A \to \mathfrak{M}^{\mathrm{inf}}_A$ satisfying compatibilities as the existence of a closed immersion

$$\mathcal{C}^{L/K}_{d,\mathrm{ss},h} \hookrightarrow \mathcal{Y}_1 \times_{\mathcal{C}^{L/L}_{d,\mathrm{ss},h}} \times \cdots \times_{\mathcal{C}^{L/L}_{d,\mathrm{ss},h}} \mathcal{Y}_n$$

(the compatibilities that cut out $\mathcal{C}^{L/K}_{d,\mathrm{ss},h}$ arise as base changes of the double diagonal, and so impose closed conditions). It follows that $\mathcal{C}^{L/K}_{d,\mathrm{ss},h} \times_{\mathcal{X}_{L,d}} \operatorname{Spf} R$ inherits a natural ample formal line bundle. As noted in the proof of Lemma 3.7.5, since the morphism $\mathcal{X}_{K,d} \to \mathcal{X}_{L,d}$ is affine, it has affine diagonal, so that the natural morphism

$$\mathcal{C}^{L/K}_{d,\mathrm{ss},h} \times_{\mathcal{X}_{K,d}} \operatorname{Spf} R \to \mathcal{C}^{L/K}_{d,\mathrm{ss},h} \times_{\mathcal{X}_{L,d}} \operatorname{Spf} R$$

is affine. It follows that $\mathcal{C}^{L/K}_{d,\mathrm{ss},h} \times_{\mathcal{X}_{K,d}} \operatorname{Spf} R$ in turn inherits a natural ample formal line bundle (see, e.g., [Sta, Tag 0892]), and the result then follows from another application of [Knu71, Thm. V.6.3]. □

4.6 INERTIAL TYPES

In this section we examine how to extract the inertial type of the Galois representation associated to a Breuil–Kisin–Fargues G_K-module; in Section 4.7 we study the same problem for Hodge–Tate weights, and in Section 4.8 we use these results to define our moduli stacks of potentially semistable and potentially crystalline representations of fixed inertial and Hodge type.

In contrast to the rest of the book, in this and the following sections we will need to consider Kisin modules with coefficients with p inverted. To this end, we will write A° and B° for p-adically complete flat $\mathcal{O}$-algebras which are topologically of finite type over $\mathcal{O}$, and write $A = A^\circ[1/p]$, $B = B^\circ[1/p]$. We hope that this change of notation will not cause any confusion. We will freely use that if A° is topologically of finite type over $\mathcal{O}$, then $A := A^\circ[1/p]$ is Noetherian and Jacobson ([FK18, §0, Prop. 9.3.2, 9.3.10]), and the residue fields of the maximal ideals of A are finite extensions of K ([FK18, §0, Cor. 9.3.7]).

Let L/K be a finite Galois extension with inertia group $I_{L/K}$, and suppose now that E is large enough that it contains the images of all embeddings $L \hookrightarrow \overline{\mathbf{Q}}_p$, that all irreducible E-representations of $I_{L/K}$ are absolutely irreducible, and that every irreducible $\overline{\mathbf{Q}}_p$-representation of $I_{L/K}$ is defined over E. Write l for the residue field of L, and write $L_0 = W(l)[1/p]$.

Let $\mathfrak{M}_{A^\circ}$ be a Breuil–Kisin–Fargues G_K-module with A°-coefficients which admits all descents over L, and write $\overline{\mathfrak{M}}_{A^\circ} := \mathfrak{M}_{A^\circ,\pi^\flat}/[\pi^\flat]\mathfrak{M}_{A^\circ,\pi^\flat}$ for the module considered in Definition 4.2.4 (1) (for some choice of $\pi^\flat$, with π a uniformizer of L; note that by the definition of $\mathfrak{M}_{A^\circ}$ admitting all descents over L, the quotient $\overline{\mathfrak{M}}_{A^\circ}$ is actually well-defined as a $W(l) \otimes_{\mathbf{Z}_p} A^\circ$-submodule of $W(\overline{k})_A \otimes_{\mathbf{A}_{\mathrm{inf},A}} \mathfrak{M}_{A^\circ}$, independently of the choice π or $\pi^\flat$). Then $\overline{\mathfrak{M}}_{A^\circ}$ has a natural $W(l) \otimes_{\mathbf{Z}_p} A$-semi-linear action of $\mathrm{Gal}(L/K)$, which is defined as follows: if $g \in \mathrm{Gal}(L/K)$, then $g(\mathfrak{M}_{A^\circ,\pi^\flat}) = \mathfrak{M}_{A^\circ,g(\pi^\flat)}$ (see the proof of Lemma 4.2.7), so the morphism $g\colon \mathfrak{M}_{A^\circ,\pi^\flat} \to g(\mathfrak{M}_{A^\circ,\pi^\flat}) = \mathfrak{M}_{A^\circ,g(\pi^\flat)}$ induces a morphism

$$g\colon \mathfrak{M}_{A^\circ,\pi^\flat}/[\pi^\flat]\mathfrak{M}_{A^\circ,\pi^\flat} \to \mathfrak{M}_{A^\circ,g(\pi^\flat)}/[g(\pi^\flat)]\mathfrak{M}_{A^\circ,g(\pi^\flat)},$$

and the source and target are both canonically identified with $\overline{\mathfrak{M}}_{A^\circ}$.

This action of $\mathrm{Gal}(L/K)$ on $\overline{\mathfrak{M}}_{A^\circ}$ induces an $L_0 \otimes_{\mathbf{Q}_p} A$-linear action of $I_{L/K}$ on the projective $L_0 \otimes_{\mathbf{Q}_p} A$-module $\overline{\mathfrak{M}}_{A^\circ} \otimes_{A^\circ} A$. Fix a choice of embedding $\sigma\colon L_0 \hookrightarrow E$, and let $e_\sigma \in L_0 \otimes_{\mathbf{Q}_p} E$ be the corresponding idempotent. Then $e_\sigma(\overline{\mathfrak{M}}_{A^\circ} \otimes_{A^\circ} A)$ is a projective A-module of rank d, with an A-linear action of $I_{L/K}$. (Indeed, since $\mathfrak{M}_{A^\circ}$ is a finite projective $\mathfrak{S}_A$-module of rank d, $\overline{\mathfrak{M}}_{A^\circ}$ is a finite projective $W(k) \otimes_{\mathbf{Z}_p} A^\circ$-module of rank d, and $\overline{\mathfrak{M}}_{A^\circ} \otimes_{A^\circ} A$ is a finite projective $L_0 \otimes_{\mathbf{Q}_p} A$-module of rank d. It follows that $e_\sigma(\overline{\mathfrak{M}}_{A^\circ} \otimes_{A^\circ} A)$ is a finite projective A-module of rank d.)

Note that up to canonical isomorphism, this module does not depend on the choice of e_σ: indeed the induced action of φ commutes with $I_{L/K}$ and induces isomorphisms between the $e_\sigma(\overline{\mathfrak{M}}_{A^\circ} \otimes_{A^\circ} A)$ (with σ varying), because the cokernel of φ is killed by $E_{\pi^\flat}$, which is a unit in our setting (in which $[\pi^\flat]=0$ and p is a unit).

4.6.1 Definition. Let $\mathfrak{M}^{\mathrm{inf}}_{A^\circ}$ be as above. Then we write

$$\mathrm{WD}(\mathfrak{M}^{\mathrm{inf}}_{A^\circ}) := e_\sigma(\overline{\mathfrak{M}}_{A^\circ} \otimes_{A^\circ} A),$$

a projective A-module of rank d with an A-linear action of $I_{L/K}$.

4.6.2 Remark. The point of this definition is that if A° is a finite flat $\mathcal{O}$-algebra, then it computes inertial types in the following sense: writing $A := A^\circ[1/p]$, and $M_{A^\circ} := W(\mathbf{C}^\flat)_{A^\circ} \otimes_{\mathbf{A}_{\mathrm{inf},A^\circ}} \mathfrak{M}^{\mathrm{inf}}_{A^\circ}$, we have a potentially semistable representation of G_K on a free A-module given by $V_A(M_{A^\circ}) := T_{A^\circ}(M_{A^\circ}) \otimes_{A^\circ} A$. As explained in Section F.24, the inertial type $D_{\mathrm{pst}}(V_A(M_{A^\circ}))|_{I_K}$ is then given by $\mathrm{WD}(\mathfrak{M}^{\mathrm{inf}}_{A^\circ})$ with its action of $I_{L/K}$.

4.6.3 Proposition. *Let A° be a p-adically complete flat $\mathcal{O}$-algebra which is topologically of finite type over $\mathcal{O}$, and write $A = A^\circ[1/p]$. Let $\mathfrak{M}^{\mathrm{inf}}_{A^\circ}$ and $\mathfrak{M}_{A^\circ}$ be as above, and fix $\sigma\colon L_0 \hookrightarrow E$. Then $\mathrm{WD}(\mathfrak{M}^{\mathrm{inf}}_{A^\circ})$ is a finite projective A-module of rank d with an action of I_K, whose formation is compatible with base changes $A^\circ \to B^\circ$ of p-adically complete flat $\mathcal{O}$-algebras which are topologically of finite type over $\mathcal{O}$.*

Proof. Writing u for $[\pi^\flat]$, the compatibility of formation with base change reduces to the observation that the natural map $\mathfrak{S}_A \otimes_A B \to \mathfrak{S}_B$ induces an isomorphism

$$\mathfrak{S}_B/u\mathfrak{S}_B \cong (\mathfrak{S}_A/u\mathfrak{S}_A) \otimes_A B. \qquad \square$$

Let τ be a d-dimensional E-representation of $I_{L/K}$.

4.6.4 Definition. In the setting of Proposition 4.6.3, we say that $\mathfrak{M}^{\mathrm{inf}}_{A^\circ}$ has inertial type τ if Zariski locally on $\operatorname{Spec} A$, $\mathrm{WD}(\mathfrak{M}^{\mathrm{inf}}_{A^\circ})$ is isomorphic to the base change of τ to A.

4.6.5 Corollary. *In the setting of Proposition* 4.6.3, *we can decompose* $\operatorname{Spec} A$ *as the disjoint union of open and closed subschemes* $\operatorname{Spec} A^\tau$, *where* $\operatorname{Spec} A^\tau$ *is the locus over which* $\mathfrak{M}^{\mathrm{inf}}_{A^\circ}$ *has inertial type* τ. *Furthermore, the formation of this decomposition is compatible with base changes* $A^\circ \to B^\circ$ *of* p*-adically complete flat* $\mathcal{O}$*-algebras which are topologically of finite type over* $\mathcal{O}$.

Proof. By Proposition 4.6.3 (more precisely, by the compatibility with base change), we can define $\operatorname{Spec} A^\tau$ to be the locus over which $\mathfrak{M}^{\mathrm{inf}}_{A^\circ}$ has inertial type τ. That $\operatorname{Spec} A$ is actually the disjoint union of the $\operatorname{Spec} A^\tau$ follows easily from our assumptions on E, and standard facts about the representation theory of finite groups in characteristic 0. For lack of a convenient reference, we sketch a proof as follows.

Since E has characteristic 0, the representation $P := \oplus_r r$ is a projective generator of the category of $E[I_{L/K}]$-modules, where r runs over a set of representatives for the isomorphism classes of irreducible E-representations of $I_{L/K}$. Our assumption that E is large enough that each r is absolutely irreducible furthermore shows that $\operatorname{End}_{I_{L/K}}(r) = E$ for each r, so that $\operatorname{End}_{I_{L/K}}(P) = \prod_r E$.

Standard Morita theory then shows that the functor $M \mapsto \operatorname{Hom}_{I_{L/K}}(P, M)$ induces an equivalence between the category of $E[I_{L/K}]$-modules and the category of $\prod_r E$-modules. Of course, a $\prod_r E$-module is just given by a tuple $(N_r)_r$ of E-vector spaces, and in this optic, the functor $\operatorname{Hom}_{I_{L/K}}(P, -)$ can be written as $M \mapsto \left(\operatorname{Hom}_{I_{L/K}}(r, M)\right)_r$, with a quasi-inverse functor being given by $(N_r) \mapsto \bigoplus_r r \otimes_E N_r$. It is easily seen (just using the fact that $\operatorname{Hom}_{I_{L/K}}(P, -)$ induces an equivalence of categories) that M is a finitely generated projective A-module, for some E-algebra A, if and only if each $\operatorname{Hom}_{I_{L/K}}(r, M)$ is a finitely generated projective A-module. Writing the various τ in the form $\oplus_r r^{n_r}$, we are done. □

4.7 HODGE–TATE WEIGHTS

Let A° be a p-adically complete flat $\mathcal{O}$-algebra which is topologically of finite type over $\mathcal{O}$, and let $\mathfrak{M}^{\mathrm{inf}}_{A^\circ}$ be a Breuil–Kisin–Fargues G_K-module of height at most h with A°-coefficients, which admits all descents. We now explain how to interpret the condition that $\mathfrak{M}^{\mathrm{inf}}_{A^\circ}$ has a fixed Hodge type, following [Kis08, Lem. 2.6.1, Cor. 2.6.2] (but bearing in mind the corrections to these results explained in [Kis09a, A.4]). We would like to thank Mark Kisin for a helpful conversation about these results, and for some suggestions regarding the proof of Proposition 4.7.2.

Fix some choice of $\pi^\flat$, and write $\mathfrak{M}_{A^\circ}$ for $\mathfrak{M}_{\pi^\flat,A^\circ}$, and u for $[\pi^\flat]$. For each $0 \leq i \leq h$ we define $\operatorname{Fil}^i \varphi^*\mathfrak{M}_{A^\circ} = \Phi_{\mathfrak{M}_{A^\circ}}^{-1}(E(u)^i\mathfrak{M}_{A^\circ})$, and we set $\operatorname{Fil}^i \varphi^*\mathfrak{M}_{A^\circ} = \varphi^*\mathfrak{M}_{A^\circ}$ for $i < 0$.

4.7.1 Remark. The point of this definition is that it captures the Hodge filtration. Indeed, if A° is a finite flat $\mathcal{O}$-algebra, then writing $A = A^\circ[1/p]$, and $M_{A^\circ} := W(\mathbf{C}^\flat)_{A^\circ} \otimes_{\mathbf{A}_{\mathrm{inf},A^\circ}} \mathfrak{M}^{\mathrm{inf}}_{A^\circ}$, we have a representation of G_K on a free A-module given by $V_A(M_{A^\circ}) := T_{A^\circ}(M_{A^\circ}) \otimes_{A^\circ} A$. As explained in Section F.24 (or see the proof of [Kis08, Cor. 2.6.2]), there is a natural identification of $D_{\mathrm{dR}}(V_A(M_{A^\circ}))$ with $(\varphi^*\mathfrak{M}_{A^\circ}/E(u)\varphi^*\mathfrak{M}) \otimes_{A^\circ} A$, under which $\operatorname{Fil}^i D_{\mathrm{dR}}(V_A(M_{A^\circ}))$ is identified with

$$(\operatorname{Fil}^i \varphi^*\mathfrak{M}_{A^\circ}/E(u)\operatorname{Fil}^{i-1}\varphi^*\mathfrak{M}_{A^\circ}) \otimes_{A^\circ} A;$$

in particular, this latter module is a finite projective $K \otimes_{\mathbf{Q}_p} A$-module.

The following proposition shows that the final conclusion of the preceding remark holds for more general choices of A°; the proof uses that particular case (i.e., the case that A° is a finite flat $\mathcal{O}$-algebra) as an input.

4.7.2 Proposition. *Let A° be a p-adically complete flat $\mathcal{O}$-algebra which is topologically of finite type over $\mathcal{O}$, and write $A = A^\circ[1/p]$. Let $\mathfrak{M}^{\mathrm{inf}}_{A^\circ}$ and $\mathfrak{M}_{A^\circ}$ be as above. For each $0 \leq i \leq h$,*

$$(\operatorname{Fil}^i \varphi^*\mathfrak{M}_{A^\circ}/E(u)\operatorname{Fil}^{i-1}\varphi^*\mathfrak{M}_{A^\circ}) \otimes_{A^\circ} A$$

is a finite projective $K \otimes_{\mathbf{Q}_p} A$-module, whose formation is compatible with base changes $A^\circ \to B^\circ$ of p-adically complete flat $\mathcal{O}$-algebras which are topologically of finite type over $\mathcal{O}$.

Proof. For any morphism of $\mathcal{O}$-algebras $A^\circ \to B^\circ$, we write $\mathfrak{M}_{B^\circ} := \mathfrak{S}_{B^\circ} \otimes_{\mathfrak{S}_{A^\circ}} \mathfrak{M}_{A^\circ}$. We begin by showing (following the proof of [Kis08, Lem. 2.6.1]) that for any p-adically complete $\mathcal{O}$-algebra A° which is topologically of finite type over $\mathcal{O}$, both $\mathfrak{M}_{A^\circ}/\operatorname{im}\Phi_{\mathfrak{M}_{A^\circ}}$ and $\varphi^*\mathfrak{M}_{A^\circ}/\operatorname{Fil}^h\varphi^*\mathfrak{M}_{A^\circ}$ are finite projective $\mathcal{O}_K \otimes_{\mathbf{Z}_p} A^\circ$-modules, whose formation is compatible with arbitrary base changes $A^\circ \to B^\circ$ (with B° also topologically of finite type).

Indeed, that $\mathfrak{M}_{A^\circ}/\operatorname{im}\Phi_{\mathfrak{M}_{A^\circ}}$ is a finite projective $\mathcal{O}_K \otimes_{\mathbf{Z}_p} A^\circ$-module whose formation is compatible with base change follows as in the proof of [Kis08, Lem. 2.6.1] from the fact that $\Phi_{\mathfrak{M}}$ remains injective after any base change (because $E(u)$ remains a nonzero divisor after any base change), and the result for $\varphi^*\mathfrak{M}_{A^\circ}/\operatorname{Fil}^h\varphi^*\mathfrak{M}_{A^\circ}$ follows from this and the short exact sequence

$$0 \to \varphi^*\mathfrak{M}_{A^\circ}/\operatorname{Fil}^h\varphi^*\mathfrak{M}_{A^\circ} \to \mathfrak{M}_{A^\circ}/E(u)^h\mathfrak{M}_{A^\circ} \to \mathfrak{M}_{A^\circ}/\operatorname{im}\Phi_{\mathfrak{M}_{A^\circ}} \to 0.$$

We claim that for each $0 \leq i \leq h$, $(\operatorname{Fil}^h \varphi^*\mathfrak{M}_{A^\circ}/E(u)^i\operatorname{Fil}^{h-i}\varphi^*\mathfrak{M}_{A^\circ}) \otimes_{A^\circ} A$ is a finite projective $K \otimes_{\mathbf{Q}_p} A$-module whose formation is compatible with base change. Admitting the claim, the proposition follows from the short exact sequence

$$\begin{aligned}0 \longrightarrow \mathrm{Fil}^i\, \varphi^*\mathfrak{M}_{A^\circ}/E(u)\,\mathrm{Fil}^{i-1}\, \varphi^*\mathfrak{M}_{A^\circ} \\ \xrightarrow{E(u)^{h-i}} \mathrm{Fil}^h\, \varphi^*\mathfrak{M}_{A^\circ}/E(u)^{h-i+1}\,\mathrm{Fil}^{i-1}\, \varphi^*\mathfrak{M}_{A^\circ} \\ \longrightarrow \mathrm{Fil}^h\, \varphi^*\mathfrak{M}_{A^\circ}/E(u)^{h-i}\,\mathrm{Fil}^i\, \varphi^*\mathfrak{M}_{A^\circ} \longrightarrow 0 \quad (4.7.3)\end{aligned}$$

(because after tensoring with A, the second and third terms are projective and compatible with base change, so that the sequence splits, and the first term is also projective and compatible with base change).

To prove the claim, we firstly consider for each $0 \le i \le h$ the finite $\mathcal{O}_K \otimes_{\mathbf{Z}_p} A^\circ$-module

$$\varphi^*\mathfrak{M}_{A^\circ}/(E(u)^i\varphi^*\mathfrak{M}_{A^\circ} + \mathrm{Fil}^h\, \varphi^*\mathfrak{M}_{A^\circ}).$$

Since this is the cokernel of the morphism

$$E(u)^i\varphi^*\mathfrak{M}_{A^\circ} \to \varphi^*\mathfrak{M}_{A^\circ}/\,\mathrm{Fil}^h\, \varphi^*\mathfrak{M}_{A^\circ},$$

we see that its formation is compatible with base change. We have a short exact sequence of finite type A°-modules

$$\begin{aligned}0 \to \mathrm{Fil}^h\, \varphi^*\mathfrak{M}_{A^\circ}/E(u)^i\,\mathrm{Fil}^{h-i}\, \varphi^*\mathfrak{M}_{A^\circ} \to \varphi^*\mathfrak{M}_{A^\circ}/E(u)^i\varphi^*\mathfrak{M}_{A^\circ} \\ \to \varphi^*\mathfrak{M}_{A^\circ}/(E(u)^i\varphi^*\mathfrak{M}_{A^\circ} + \mathrm{Fil}^h\, \varphi^*\mathfrak{M}_{A^\circ}) \to 0,\end{aligned} \quad (4.7.4)$$

in which the second and third terms are compatible with base change, and we need to prove that after inverting p, the first term is projective and is of formation compatible with base change. To this end, note firstly that, as discussed above, if A° is a finite flat $\mathcal{O}$-algebra, then each

$$(\mathrm{Fil}^i\, \varphi^*\mathfrak{M}_{A^\circ}/E(u)\,\mathrm{Fil}^{i-1}\, \varphi^*\mathfrak{M}_{A^\circ}) \otimes_{A^\circ} A$$

is a finite projective $K \otimes_{\mathbf{Q}_p} A$-module, and it follows from (4.7.3) and an easy induction that the same is true of $(\mathrm{Fil}^h\, \varphi^*\mathfrak{M}_{A^\circ}/E(u)^i\,\mathrm{Fil}^{h-i}\, \varphi^*\mathfrak{M}_{A^\circ}) \otimes_{A^\circ} A$.

To simplify notation, we write (4.7.4) as

$$0 \to K(A^\circ) \to M_1 \to M_2 \to 0,$$

and for any A°-algebra C, we write $K(C)$ for the kernel of the surjection $M_1 \otimes_{A^\circ} C \to M_2 \otimes_{A^\circ} C$. In particular for B° as in the statement of the proposition, if as usual we write $B := B^\circ[1/p]$, then we have

$$K(B) = (\mathrm{Fil}^h\, \varphi^*\mathfrak{M}_{B^\circ}/E(u)^i\,\mathrm{Fil}^{h-i}\, \varphi^*\mathfrak{M}_{B^\circ}) \otimes_{B^\circ} B;$$

so we need to show that $K(A)$ is projective, and that $K(B)=K(A)\otimes_A B$.

Let $\mathfrak{m}$ be a maximal ideal of A, so that $A/\mathfrak{m}^i$ is a finite E-algebra for each i. Then since $\widehat{A}_{\mathfrak{m}}$ is a flat A-algebra, since the Artin–Rees Lemma shows that tensoring with $\widehat{A}_{\mathfrak{m}}$ coincides with $\mathfrak{m}$-adic completion for finite type modules, and since the formation of kernels commutes with limits, we see that

$$K(A)\otimes_{A^\circ}\widehat{A}_{\mathfrak{m}}=K(\widehat{A}_{\mathfrak{m}})=\varprojlim_i K(A/\mathfrak{m}^i).$$

Since $A/\mathfrak{m}^i$ is a finite E-algebra, we see that $K(A/\mathfrak{m}^i)$ is a finite projective $(K\otimes_{\mathbf{Q}_p}A)/\mathfrak{m}^i$-module of rank bounded independently of i. It follows that $K(A)\otimes_{A^\circ}\widehat{A}_{\mathfrak{m}}$ is a finite projective $\mathcal{O}_K\otimes_{\mathbf{Z}_p}\widehat{A}_{\mathfrak{m}}$-module. Since $A_{\mathfrak{m}}$ is Noetherian, $\widehat{A}_{\mathfrak{m}}$ is a faithfully flat $A_{\mathfrak{m}}$-algebra, so that $K(A)\otimes_{A^\circ}A_{\mathfrak{m}}$ is a finite projective $\mathcal{O}_K\otimes_{\mathbf{Z}_p}A_{\mathfrak{m}}$-module. Since this holds for all $\mathfrak{m}$, we see that $K(A)$ is a finite projective $\mathcal{O}_K\otimes_{\mathbf{Z}_p}A$-module, as claimed.

It remains to show that we have $K(B)=K(A)\otimes_A B$. We have a natural surjective morphism of finite projective $K\otimes_{\mathbf{Q}_p}B$-modules

$$K(A)\otimes_A B\to K(B), \tag{4.7.5}$$

which we need to show is an isomorphism. Note firstly that if $B=A/\mathfrak{m}$ for some maximal ideal $\mathfrak{m}$ of A, then this follows from the previous paragraph. Now suppose that B is general. Since the kernel of (4.7.5) is in particular a finite projective B-module, it is enough to prove that for any maximal ideal $\mathfrak{m}_B$ of B, (4.7.5) becomes an isomorphism after tensoring with $B/\mathfrak{m}_B$. Suppose that $\mathfrak{m}_B$ lies over a maximal ideal $\mathfrak{m}$ of A, so that $B/\mathfrak{m}_B$ is a finite field extension of $A/\mathfrak{m}$; we need to show that the induced surjection

$$K(A/\mathfrak{m})\otimes_{A/\mathfrak{m}}B/\mathfrak{m}_B\to K(B/\mathfrak{m}_B)$$

is an isomorphism. Each side is determined by the de Rham filtration on the corresponding G_K-representation, which is compatible with the extension of scalars, so we are done. □

In what follows, we will work in the relative setting of an extension L/K. To this end, we let A° be a p-adically complete flat $\mathcal{O}$-algebra, with $A:=A^\circ[1/p]$, let L/K be a finite Galois extension, and let $\mathfrak{M}^{\mathrm{inf}}_{A^\circ}$ be a Breuil–Kisin–Fargues G_K-module of rank d and height at most h with A°-coefficients, which admits all descents over L. Fix some choice of $\pi^\flat$ a uniformizer of L, write $\mathfrak{M}_{A^\circ}$ for $\mathfrak{M}_{\pi^\flat,A^\circ}$, and u for $[\pi^\flat]$. Applying Proposition 4.7.2, with L in place of K, we obtain a projective $L\otimes_{\mathbf{Q}_p}A$-module $(\varphi^*\mathfrak{M}_{A^\circ}/E(u)\varphi^*\mathfrak{M}_{A^\circ})\otimes_{A^\circ}A$, which is filtered by projective submodules. This filtered module has a natural action of $\mathrm{Gal}(L/K)$, which is semi-linear with respect to the action of $\mathrm{Gal}(L/K)$ on $L\otimes_{\mathbf{Q}_p}A$ induced by its action on the first factor. Since L/K is a Galois extension, the tensor product $L \otimes_{\mathbf{Q}_p} A$ is an étale $\mathrm{Gal}(L/K)$-extension of $K \otimes_{\mathbf{Q}_p} A$, and so étale descent allows us to descend $(\varphi^*\mathfrak{M}_{A^\circ}/E(u)\varphi^*\mathfrak{M}_{A^\circ}) \otimes_{A^\circ} A$ to a filtered module over $K \otimes_{\mathbf{Q}_p} A$; concretely, this descent is achieved by taking $\mathrm{Gal}(L/K)$-invariants. This leads to the following definition.

4.7.6 Definition. In the preceding situation, we write

$$D_{\mathrm{dR}}(\mathfrak{M}^{\mathrm{inf}}_{A^\circ}) := \big((\varphi^*\mathfrak{M}_{A^\circ}/E(u)\varphi^*\mathfrak{M}_{A^\circ}) \otimes_{A^\circ} A\big)^{\mathrm{Gal}(L/K)},$$

and more generally, for each $i \geq 0$, we write

$$\mathrm{Fil}^i D_{\mathrm{dR}}(\mathfrak{M}^{\mathrm{inf}}_{A^\circ}) := \big((\mathrm{Fil}^i \varphi^*\mathfrak{M}_{A^\circ}/E(u)\,\mathrm{Fil}^{i-1} \varphi^*\mathfrak{M}_{A^\circ}) \otimes_{A^\circ} A\big)^{\mathrm{Gal}(L/K)}$$

(and for $i < 0$, we write $\mathrm{Fil}^i D_{\mathrm{dR}}(\mathfrak{M}^{\mathrm{inf}}_{A^\circ}) := D_{\mathrm{dR}}(\mathfrak{M}^{\mathrm{inf}}_{A^\circ})$). The property of being a finite rank projective module is preserved under étale descent, and so we find that $D_{\mathrm{dR}}(\mathfrak{M}^{\mathrm{inf}}_{A^\circ})$ is a rank d projective $K \otimes_{\mathbf{Q}_p} A$-module, filtered by projective submodules.

Since A is an E-algebra, we have the product decomposition $K \otimes_{\mathbf{Q}_p} A \xrightarrow{\sim} \prod_{\sigma\colon K \hookrightarrow E} A$, and so, if we write e_σ for the idempotent corresponding to the factor labeled by σ in this decomposition, we find that

$$D_{\mathrm{dR}}(\mathfrak{M}^{\mathrm{inf}}_{A^\circ}) = \prod_{\sigma\colon K \hookrightarrow E} e_\sigma D_{\mathrm{dR}}(\mathfrak{M}^{\mathrm{inf}}_{A^\circ}),$$

where each $e_\sigma D_{\mathrm{dR}}(\mathfrak{M}^{\mathrm{inf}}_{A^\circ})$ is a projective A-module of rank d. For each i, we write

$$\mathrm{Fil}^i e_\sigma D_{\mathrm{dR}}(\mathfrak{M}^{\mathrm{inf}}_{A^\circ}) = e_\sigma \mathrm{Fil}^i D_{\mathrm{dR}}(\mathfrak{M}^{\mathrm{inf}}_{A^\circ}).$$

Each $\mathrm{Fil}^i e_\sigma D_{\mathrm{dR}}(\mathfrak{M}^{\mathrm{inf}}_{A^\circ})$ is again a projective A-module.

The base change property proved in Proposition 4.7.2 shows that the various quotients $\mathrm{Fil}^i e_\sigma D_{\mathrm{dR}}(\mathfrak{M}^{\mathrm{inf}}_{A^\circ})/\mathrm{Fil}^{i+1} e_\sigma D_{\mathrm{dR}}(\mathfrak{M}^{\mathrm{inf}}_{A^\circ})$ are again projective A-modules. Phrased more geometrically, then, we see that each $e_\sigma D_{\mathrm{dR}}(\mathfrak{M}^{\mathrm{inf}}_{A^\circ})$ gives rise to a vector bundle over $\mathrm{Spec}\, A$ which is endowed with a filtration by subbundles. We may thus decompose $\mathrm{Spec}\, A$ into a disjoint union of open and closed subschemes over which the ranks of the various subbundles $\mathrm{Fil}^i e_\sigma D_{\mathrm{dR}}(\mathfrak{M}^{\mathrm{inf}}_{A^\circ})$ (or equivalently, the ranks of the various constituents

$$\mathrm{Fil}^i e_\sigma D_{\mathrm{dR}}(\mathfrak{M}^{\mathrm{inf}}_{A^\circ})/\mathrm{Fil}^{i+1} e_\sigma D_{\mathrm{dR}}(\mathfrak{M}^{\mathrm{inf}}_{A^\circ})$$

of the associated graded bundle) are constant.

To encode this rank data, and the corresponding decomposition of $\mathrm{Spec}\, A$, it is traditional to use the terminology of Hodge types, which we now recall.

4.7.7 Definition. A Hodge type $\underline{\lambda}$ of rank d is by definition a set of tuples of integers $\{\lambda_{\sigma,j}\}_{\sigma\colon K\hookrightarrow\overline{\mathbf{Q}}_p,1\leq j\leq d}$ with $\lambda_{\sigma,j}\geq\lambda_{\sigma,j+1}$ for all σ and all $1\leq j\leq d-1$.

If $\underline{D}:=(D_\sigma)_{\sigma\colon K\hookrightarrow E}$ is a collection of rank d vector bundles over $\operatorname{Spec} A$, labeled (as indicated) by the embeddings $\sigma\colon K\hookrightarrow E$, then we say that $\underline{D}$ has Hodge type $\underline{\lambda}$ if $\operatorname{Fil}^i D_\sigma$ has constant rank equal to $\#\{j\mid\lambda_{\sigma,j}\geq i\}$.

4.7.8 Corollary. *In the preceding context, if $\underline{\lambda}$ is a Hodge type of rank d, then we let* $\operatorname{Spec} A^{\underline{\lambda}}$ *denote the open and closed subscheme of* $\operatorname{Spec} A$ *over which the tuple $\left(e_\sigma D_{\mathrm{dR}}(\mathfrak{M}^{\mathrm{inf}}_{A^\circ})\right)$ of filtered vector bundles is of Hodge type $\underline{\lambda}$. We have a corresponding decomposition*

$$\operatorname{Spec} A=\coprod_{\underline{\lambda}}\operatorname{Spec} A^{\underline{\lambda}},\tag{4.7.9}$$

labeled by the set of Hodge types $\underline{\lambda}$ of rank d. This decomposition is compatible with base changes $A^\circ\to B^\circ$ of p-adically complete flat $\mathcal{O}$-algebras which are topologically of finite type over $\mathcal{O}$.

Proof. This follows from the preceding discussion. □

4.7.10 Remark. A priori, the tuple of filtered vector bundles $\left(e_\sigma D_{\mathrm{dR}}(\mathfrak{M}^{\mathrm{inf}}_{A^\circ})\right)_{\sigma\colon K\hookrightarrow E}$ on $\operatorname{Spec} A$, and hence the decomposition (4.7.9) of $\operatorname{Spec} A$, depends on the descent $\mathfrak{M}_{\pi^\flat,A^\circ}$, and hence on the choice of $\pi^\flat$. However, Theorem 4.7.13 below implies in fact that the decomposition (4.7.9) is independent of the choice of $\pi^\flat$; see Remark 4.7.14 for a more detailed explanation of this.

As in Section 4.6, suppose now that E is large enough that it contains the images of all embeddings $L\hookrightarrow\overline{\mathbf{Q}}_p$, that all irreducible E-representations of $I_{L/K}$ are absolutely irreducible, and that every irreducible $\overline{\mathbf{Q}}_p$-representation of $I_{L/K}$ is defined over E. We can then immediately combine Corollaries 4.6.5 and 4.7.8, obtaining the following result.

4.7.11 Corollary. *In the preceding situation, we have a decomposition*

$$\operatorname{Spec} A=\coprod_{\underline{\lambda},\tau}\operatorname{Spec} A^{\underline{\lambda},\tau},\tag{4.7.12}$$

labeled by the set of Hodge types $\underline{\lambda}$ of rank d, and the set of inertial types τ of $I_{L/K}$. This decomposition is compatible with base changes $A^\circ\to B^\circ$ of p-adically complete flat $\mathcal{O}$-algebras which are topologically of finite type over $\mathcal{O}$. The Breuil–Kisin–Fargues G_K-module $\mathfrak{M}^{\mathrm{inf}}_{A^\circ}$ is of Hodge type $\underline{\lambda}$ and inertial type τ if and only if $A^{\underline{\lambda},\tau}=A$.

The following key theorem relates the constructions of this section, and the previous one, to the p-adic Hodge theory of G_K-representations. If B is a finite E-algebra, then we say that a sub-$\mathcal{O}$-algebra $B^\circ \subseteq B$ is an *order of* B if $B^\circ[1/p] = B$, and B° is a finite $\mathcal{O}$-algebra.

4.7.13 Theorem. *Suppose that A° is a finite flat $\mathcal{O}$-algebra, let M be a projective étale (φ, G_K)-module with A°-coefficients, and write $V_A(M) = T_{A^\circ}(M)[1/p]$. Let L/K be a finite Galois extension. Then $V_A(M)|_{G_L}$ is semistable with Hodge–Tate weights in $[0,h]$ if and only if there is an order $(A^\circ)'$ of $A := A^\circ[1/p]$ that contains A°, and a Breuil–Kisin–Fargues G_K-module $\mathfrak{M}^{\inf}_{(A^\circ)'}$ with $(A^\circ)'$-coefficients, which is of height at most h, which admits all descents over L, whose G_L-action is canonical, and which satisfies $M_{(A^\circ)'} = W(\mathbf{C}^\flat)_{(A^\circ)'} \otimes_{\mathbf{A}_{\inf,(A^\circ)'}} \mathfrak{M}^{\inf}_{(A^\circ)'}$.*

Furthermore $V_A(M)|_{G_L}$ is crystalline if and only if $\mathfrak{M}^{\inf}_{(A^\circ)'}$ is crystalline as a Breuil–Kisin–Fargues G_L-module with $(A^\circ)'$ coefficients.

In either case, the Hodge type is determined by applying Definition 4.7.7 *to the tuple $\big(e_\sigma D_{\mathrm{dR}}(\mathfrak{M}^{\inf}_{(A^\circ)'})\big)_{\sigma:\, K \hookrightarrow E}$ arising from Definition* 4.7.6, *and the inertial type of $V_A(M)$ is given by Definition* 4.6.4.

Proof. Suppose firstly that $(A^\circ)'$ and $\mathfrak{M}^{\inf}_{(A^\circ)'}$ exist. Then if we simply forget the $(A^\circ)'$-coefficients and consider $\mathfrak{M}^{\inf}_{(A^\circ)'}$ as a Breuil–Kisin–Fargues G_K-module with $\mathbf{Z}_p$-coefficients, the theorem follows from Corollaries F.23 and F.25. Conversely, suppose that $V_A(M)|_{G_L}$ is semistable with Hodge–Tate weights in $[0,h]$. If we show that $(A^\circ)'$ and $\mathfrak{M}^{\inf}_{(A^\circ)'}$ exist, then (as discussed in Remarks 4.6.2 and 4.7.1) it again follows from Corollary F.25 that the inertial and Hodge types are given by the claimed recipes; so it suffices to prove this existence. We begin with a preliminary reduction.

The ring A is a product of Artinian local E-algebras, say $A = \prod_i A_i$; then if A_i° denotes the image of A° in A_i, the product $\prod_i A_i^\circ$ is an order of A containing A°. Thus it is no loss of generality to replace A° by this product, and hence, by working one factor at a time, to assume that A is local. The residue field E' of A is then a finite extension of E, and A is naturally an E'-algebra. The compositum $\mathcal{O}_{E'}A^\circ$ is an order in A containing A° which is furthermore an $\mathcal{O}_{E'}$-algebra. Replacing A° by this compositum, we may thus assume that A° is an $\mathcal{O}_{E'}$-algebra. Then, relabelling E' as E if necessary, we can and do assume that $E' = E$. Thus, reformulating the preceding discussion slightly, we have $A_{\mathrm{red}} = E$ and $A^\circ_{\mathrm{red}} = \mathcal{O}_E$.

By Corollary F.23 there is a unique Breuil–Kisin–Fargues G_K-module $\mathfrak{M}^{\inf}$ with $\mathbf{Z}_p$-coefficients which is of height at most h and admits all descents over L, and satisfies

$$M = W(\mathbf{C}^\flat) \otimes_{\mathbf{A}_{\inf}} \mathfrak{M}^{\inf}.$$

For each $\pi^\flat$ we have by definition a descent $\mathfrak{M}_{\pi^\flat}$ of $\mathfrak{M}^{\inf}$. Fix some choice of $\pi^\flat$, write u for $[\pi^\flat]$, and write $\mathfrak{S}$ for $\mathfrak{S}_{\pi^\flat}$ and $\mathfrak{M}$ for $\mathfrak{M}_{\pi^\flat}$.

We now follow the proof of [Kis08, Prop. 1.6.4]. In particular, we work for the most part with the Breuil–Kisin module $\mathfrak{M}$, rather than the Breuil–Kisin–Fargues module $\mathfrak{M}^{\mathrm{inf}}$. It is possible that by using [BMS18, Prop. 4.13], we could make our arguments with $\mathfrak{M}^{\mathrm{inf}}$ itself, but since this does not seem likely to significantly simplify the proof, and would make it harder for the reader to compare to the arguments of [Kis08], we have not attempted to do this.

Note firstly that $\mathfrak{M} \subset \mathfrak{M}^{\mathrm{inf}} \subset M$ are stable under the action of A° on M. Indeed, since A° is local, it is enough to check stability under the action of $(A^\circ)^\times$. In the case of $\mathfrak{M}^{\mathrm{inf}}$ it is immediate from the unicity of $\mathfrak{M}^{\mathrm{inf}}$ that we have $\mathfrak{M}^{\mathrm{inf}} = a\mathfrak{M}^{\mathrm{inf}}$ for any $a \in (A^\circ)^\times$, and similarly for $\mathfrak{M}$ it follows from Lemma 4.2.8.

In particular, $\mathfrak{M}$ is naturally a finite $\mathfrak{S}_{A^\circ}$-module, which is projective as an $\mathfrak{S}$-module. While $\mathfrak{M}$ need not be a projective $\mathfrak{S}_{A^\circ}$-module, $\mathcal{O}_{\mathcal{E},A^\circ} \otimes_{\mathfrak{S}_{A^\circ}} \mathfrak{M}$ is a projective $\mathcal{O}_{\mathcal{E},A^\circ}$-module, because after the faithfully flat base extension $\mathcal{O}_{\mathcal{E},A^\circ} \hookrightarrow W(\mathbf{C}^\flat)_{A^\circ}$ it is identified with M. It follows from [Kis08, Lem. 1.6.1] that $\mathfrak{M}[1/p]$ is a projective $\mathfrak{S}_{A^\circ}[1/p]$-module, necessarily of rank d, and that $\mathfrak{M}[1/u]$ is a projective $\mathfrak{S}_{A^\circ}[1/u]$-module, again of rank d.

In fact, it will be useful to note that $\mathfrak{M}[1/p]$ is actually free of rank d. To see this, it suffices to prove it after base-changing to

$$(\mathfrak{S}_{A^\circ})[1/p]_{\mathrm{red}} = (\mathfrak{S}_{A^\circ_{\mathrm{red}}})[1/p] = \mathfrak{S}_{\mathcal{O}}[1/p] \cong \prod_{\sigma : W(k) \hookrightarrow \mathcal{O}} \mathcal{O}[[u]][1/p].$$

Since $\mathcal{O}[[u]][1/p]$ is a PID (by the Weierstrass preparation theorem), any finitely generated projective module over $(\mathfrak{S}_{A^\circ_{\mathrm{red}}})[1/p]$ of constant rank is necessarily free. For later use, we note that since $(\mathcal{O}_{\mathcal{E},A^\circ})_{\mathrm{red}} = \mathcal{O} \otimes_{\mathbf{Z}_p} \mathcal{O}_{\mathcal{E}} = \prod_{\sigma : W(k) \hookrightarrow \mathcal{O}} \mathcal{O}_{\mathcal{E}}$, and $\mathcal{O}_{\mathcal{E}}$ is a local ring (indeed a discrete valuation ring, with uniformizer p), any projective module of constant rank over $\mathcal{O}_{\mathcal{E},A^\circ}$ is also necessarily free.

Now let $\mathfrak{M}'_{\mathcal{O}}$ denote the image of $\mathfrak{M}$ under the projection $\mathfrak{M}[1/u] \to \mathfrak{M}[1/u] \otimes_A E$ (the map $A \to E$ being the projection from A to its residue field), and let $\mathfrak{M}'_{\mathcal{O}} \subset \mathfrak{M}_{\mathcal{O}}$ be the canonical inclusion (with finite cokernel) of $\mathfrak{M}'_{\mathcal{O}}$ into the corresponding finite projective $\mathfrak{S}_{\mathcal{O}}$-module. (The existence of $\mathfrak{M}_{\mathcal{O}}$ follows from the structure theory of $\mathfrak{S}$-modules; see [BMS18, Prop. 4.3]. Concretely, we have $\mathfrak{M}_{\mathcal{O}} = \mathfrak{M}'_{\mathcal{O}}[1/u] \cap \mathfrak{M}'_{\mathcal{O}}[1/p]$.) The module $\mathfrak{M}_{\mathcal{O}}$ is again a Breuil–Kisin module of height at most h.

Choose an $\mathfrak{S}_{\mathcal{O}}$-basis for $\mathfrak{M}_{\mathcal{O}}$, and lift it to an $\mathfrak{S}_{A^\circ}[1/p]$-basis of $\mathfrak{M}[1/p]$, and let $(A^\circ)''$ be the A°-subalgebra of A generated by these matrix coefficients. These coefficients have bounded powers of p in their denominators (when we think of A as equalling $A^\circ[1/p]$) and lie in $\mathcal{O}$ after passing to the quotient of A by its maximal ideal (equivalently its nilradical); in other words these entries lie in $A^\circ + p^{-N}\,\mathrm{nil}(A^\circ)$ for some $N \geq 0$, and this A°-submodule of A generates a finite A°-subalgebra of A. Thus $(A^\circ)''$ is an order in A. Let $\mathfrak{M}_{(A^\circ)''}$ be the $\mathfrak{S}_{(A^\circ)''}$-submodule of $\mathfrak{M}[1/p]$ spanned by this basis. Then $\mathfrak{M}_{(A^\circ)''}$ is φ-stable by construction. Again by construction we have that

$$\mathfrak{M}_{(A^\circ)''}[1/p] = (A^\circ)'' \otimes_{A^\circ} \mathfrak{M}[1/p],$$

and that

$$\mathcal{O} \otimes_{(A^\circ)''} \mathcal{O}_{\mathcal{E},(A^\circ)''} \otimes_{\mathfrak{S}_{(A^\circ)''}} \mathfrak{M}_{(A^\circ)''} = \mathcal{O}_{\mathcal{E}} \otimes_{\mathfrak{S}} \mathfrak{M}_{\mathcal{O}} = \mathcal{O}_{\mathcal{E}} \otimes_{\mathfrak{S}} \mathfrak{M}'_{\mathcal{O}}$$

(where the first tensor product is with respect to the surjection $(A^\circ)'' \to (A^\circ)''_{\mathrm{red}} = \mathcal{O}$). Thus $\mathcal{O}_{\mathcal{E},(A^\circ)''} \otimes_{\mathfrak{S}_{(A^\circ)''}} \mathfrak{M}_{(A^\circ)''}$ and $\mathcal{O}_{\mathcal{E},(A^\circ)''} \otimes_{\mathfrak{S}_{A^\circ}} \mathfrak{M}$ are two rank d projective $\mathcal{O}_{\mathcal{E},(A^\circ)''}$-modules which coincide after inverting p, and also after reducing modulo $\mathrm{nil}\big((A^\circ)''\big)$. In fact, they are both free of rank d, by our observation above that a projective $\mathcal{O}_{\mathcal{E},A^\circ}$-module of constant rank is necessarily free (applied now with $(A^\circ)''$ in place of A°). If we choose bases for each of them which coincide modulo $\mathrm{nil}\big((A^\circ)''\big)$, then we may regard each of these bases as a basis of $\mathcal{O}_{\mathcal{E},(A^\circ)''} \otimes_{\mathfrak{S}_{A^\circ}} \mathfrak{M}[1/p]$ over $\mathcal{O}_{\mathcal{E},(A^\circ)''}[1/p]$, and so they differ by a change-of-basis matrix lying in $1 + \mathrm{nil}\big((A^\circ)''\big) M_d\big(\mathcal{O}_{\mathcal{E},(A^\circ)''}[1/p]\big)$. Just as we argued above in the construction of $(A^\circ)''$, we see that we may enlarge $(A^\circ)''$ to an order $(A^\circ)'$ such that the entries of this change-of-basis matrix lie in $(A^\circ)'$. Consequently, if we set $\mathfrak{M}_{(A^\circ)'} := \mathfrak{S}_{(A^\circ)'}) \otimes_{\mathfrak{S}_{(A^\circ)''}} \mathfrak{M}_{(A^\circ)''}$, then we find that

$$\mathcal{O}_{\mathcal{E},(A^\circ)'} \otimes_{\mathfrak{S}_{A^\circ}} \mathfrak{M} = \mathcal{O}_{\mathcal{E},(A^\circ)'} \otimes_{\mathfrak{S}_{(A^\circ)''}} \mathfrak{M}_{(A^\circ)'}.$$

Extending scalars further, we find that

$$\begin{aligned} M_{(A^\circ)'} &:= W(\mathbf{C}^\flat)_{(A^\circ)'} \otimes_{W(\mathbf{C}^\flat)_{A^\circ}} M \\ &= W(\mathbf{C}^\flat)_{(A^\circ)'} \otimes_{\mathfrak{S}_{A^\circ}} \mathfrak{M} = W(\mathbf{C}^\flat)_{(A^\circ)''} \otimes_{\mathfrak{S}_{(A^\circ)'}} \mathfrak{M}_{(A^\circ)'}. \end{aligned}$$

We claim that $\mathfrak{M}_{(A^\circ)'}$ is a Breuil–Kisin module of height at most h, i.e., that the cokernel of $\Phi_{\mathfrak{M}_{(A^\circ)'}}$ is killed by $E(u)^h$. This cokernel is, as an $\mathfrak{S}$-module, a successive extension of copies of the cokernel of $\Phi_{\mathfrak{M}_{\mathcal{O}}}$ (the injectivity of Φ implies that the formation of the cokernel of Φ is exact), and is in particular p-torsion free (by [Kis09b, Lem. 1.2.2 (2)]). After inverting p the cokernel is a base change of the cokernel of $\Phi_{\mathfrak{M}}$, and is therefore killed by $E(u)^h$, as required.

We now set $\mathfrak{M}^{\mathrm{inf}}_{(A^\circ)'} := \mathbf{A}_{\mathrm{inf},(A^\circ)'} \otimes_{\mathfrak{S}_{(A^\circ)'}} \mathfrak{M}_{(A^\circ)'}$, a Breuil–Kisin–Fargues G_K-module of height at most h with $(A^\circ)'$-coefficients which by construction satisfies $M_{(A^\circ)'} = W(\mathbf{C}^\flat)_{(A^\circ)'} \otimes_{\mathbf{A}_{\mathrm{inf},(A^\circ)'}} \mathfrak{M}^{\mathrm{inf}}_{(A^\circ)'}$. It remains to prove that $\mathfrak{M}^{\mathrm{inf}}_{(A^\circ)'}$ admits all descents over L and that the G_L-action is canonical. Applying Corollary F.23 again to $M_{(A^\circ)'}$ regarded as an étale (φ, G_K)-module with $\mathbf{Z}_p$-coefficients, we see that there is a Breuil–Kisin–Fargues G_K-module $(\mathfrak{M}^{\mathrm{inf}})'$ of height at most h with $\mathbf{Z}_p$-coefficients which admits all descents over L and satisfies $M_{(A^\circ)'} = W(\mathbf{C}^\flat) \otimes_{\mathbf{A}_{\mathrm{inf}}} (\mathfrak{M}^{\mathrm{inf}})'$. Write $(\mathfrak{M})'$ for the descent of $(\mathfrak{M}^{\mathrm{inf}})'$, for our particular choice of $\pi^\flat$; then by the uniqueness of Breuil–Kisin modules with $\mathbf{Z}_p$-coefficients ([Kis06, Prop. 2.1.12]) we have $(\mathfrak{M})' = \mathfrak{M}_{(A^\circ)'}$, so that in fact $(\mathfrak{M}^{\mathrm{inf}})' = \mathfrak{M}^{\mathrm{inf}}_{(A^\circ)'}$.

Thus $\mathfrak{M}^{\inf}_{(A^\circ)'}$ admits all descents over L to Breuil–Kisin modules with $\mathbf{Z}_p$-coefficients, and by another application of Lemma 4.2.8, these Breuil–Kisin modules are $(A^\circ)'$-stable. Finally, the G_L-action is canonical by Proposition 4.4.1, and we are done. □

4.7.14 Remark. It follows from Theorem 4.7.13 that the decomposition (4.7.9) is independent of the choice of $\pi^\flat$. Indeed, the condition that $\mathrm{Fil}^i\, e_\sigma D_{\mathrm{dR}}(\mathfrak{M}^{\inf}_{A^\circ})$ has given constant rank can be checked after base changing to all $A/\mathfrak{m}$, for $\mathfrak{m}$ a maximal ideal of A, so it follows from Proposition 4.7.2 and Theorem 4.7.13 that the direct factor $A^{\underline{\lambda}}$ of A is characterized as follows: for any A°-algebra B° which is finite and flat as an $\mathcal{O}$-algebra, the canonical morphism $A \to B^\circ[1/p]$ factors through $A^{\underline{\lambda}}$ if and only if the G_K-representation $V_B(\mathfrak{M}^{\inf}_{B^\circ})$ has Hodge type $\underline{\lambda}$.

4.8 MODULI STACKS OF POTENTIALLY SEMISTABLE REPRESENTATIONS

We are now in a position to define the main objects of interest in this chapter, which are the closed substacks of $\mathcal{X}_d$ classifying representations which are potentially crystalline or potentially semistable of fixed Hodge and inertial types.

In this section L/K will denote a finite Galois extension with inertia group $I_{L/K}$; we will always assume (without specifically recalling this) that E is large enough that all irreducible E-representations of $I_{L/K}$ are absolutely irreducible, and that every irreducible $\overline{\mathbf{Q}}_p$-representation of $I_{L/K}$ is defined over E. For any such L/K we have the stacks $\mathcal{C}^{L/K}_{d,\mathrm{ss},h}$ and $\mathcal{C}^{L/K}_{d,\mathrm{crys},h}$ that we defined in Definition 4.5.22, and we write $\mathcal{C}^{L/K,\mathrm{fl}}_{d,\mathrm{ss},h}$ and $\mathcal{C}^{L/K,\mathrm{fl}}_{d,\mathrm{crys},h}$ respectively for their flat parts (in the sense recalled in Appendix A).

4.8.1 Definition. We say that a Hodge type $\underline{\lambda}$ is *effective* if $\lambda_{\sigma,i} \geq 0$ for each σ and i, and that $\underline{\lambda}$ is *bounded by* h if $\lambda_{\sigma,i} \in [0,h]$ for each σ and i.

4.8.2 Proposition. *Let L/K be a finite Galois extension. Then the stacks $\mathcal{C}^{L/K,\mathrm{fl}}_{d,\mathrm{ss},h}$ and $\mathcal{C}^{L/K,\mathrm{fl}}_{d,\mathrm{crys},h}$ are scheme-theoretic unions of closed substacks $\mathcal{C}^{L/K,\mathrm{fl},\underline{\lambda},\tau}_{d,\mathrm{ss},h}$ and $\mathcal{C}^{L/K,\mathrm{fl},\underline{\lambda},\tau}_{d,\mathrm{crys},h}$, where $\underline{\lambda}$ runs over all effective Hodge types that are bounded by h, and τ runs over all d-dimensional E-representations of $I_{L/K}$. These latter closed substacks are uniquely characterized by the following property: if A° is a finite flat $\mathcal{O}$-algebra, then an A°-point of $\mathcal{C}^{L/K,\mathrm{fl}}_{d,\mathrm{ss},h}$ (resp. $\mathcal{C}^{L/K,\mathrm{fl}}_{d,\mathrm{crys},h}$) is a point of $\mathcal{C}^{L/K,\mathrm{fl},\underline{\lambda},\tau}_{d,\mathrm{ss},h}$ (resp. $\mathcal{C}^{L/K,\mathrm{fl},\underline{\lambda},\tau}_{d,\mathrm{crys},h}$) if and only if the corresponding Breuil–Kisin–Fargues module $\mathfrak{M}^{\inf}_{A^\circ}$ has Hodge type $\underline{\lambda}$ in the sense given by applying Definition 4.7.7 to the tuple $\left(e_\sigma D_{\mathrm{dR}}(\mathfrak{M}^{\inf}_{A^\circ})\right)_{\sigma\colon K\hookrightarrow E}$ arising from Definition 4.7.6, and inertial type τ*

in the sense of Definition 4.6.4; *or, more succinctly (and writing* $A := A^\circ[1/p]$, *as usual), if and only if* $A^{\underline{\lambda},\tau} = A$.

Finally, each stack $\mathcal{C}_{d,\mathrm{ss},h}^{L/K,\mathrm{fl},\underline{\lambda},\tau}$ *or* $\mathcal{C}_{d,\mathrm{crys},h}^{L/K,\mathrm{fl},\underline{\lambda},\tau}$ *is a p-adic formal algebraic stack of finite presentation which is flat over* $\operatorname{Spf}\mathcal{O}$ *whose diagonal is affine, and the natural morphisms to* $\mathcal{X}_d$ *are representable by algebraic spaces, proper, and of finite presentation.*

Proof. We give the proof for $\mathcal{C}_{d,\mathrm{ss},h}^{L/K,\mathrm{fl}}$, the crystalline case being formally identical. Since $\mathcal{C}_{d,\mathrm{ss},h}^{L/K,\mathrm{fl}}$ is a flat p-adic formal algebraic stack of finite presentation over $\operatorname{Spf}\mathcal{O}$, we can choose a smooth surjection

$$\operatorname{Spf} B^\circ \to \mathcal{C}_{d,\mathrm{ss},h}^{L/K,\mathrm{fl}} \tag{4.8.3}$$

where B° is topologically of finite type over $\mathcal{O}$.

As usual, we write $B := B^\circ[1/p]$. By Corollary 4.7.11 we may write $\operatorname{Spec} B$ as a disjoint union $\operatorname{Spec} B = \coprod_{\underline{\lambda},\tau} B^{\underline{\lambda},\tau}$, and so correspondingly factor B as a product $B = \prod_{\underline{\lambda},\tau} B^{\underline{\lambda},\tau}$. If we let $B^{\underline{\lambda},\tau,\circ}$ denote the image of B° in $B^{\underline{\lambda},\tau}$, then we obtain an induced injection $B^\circ \hookrightarrow \prod_{\underline{\lambda},\tau} B^{\underline{\lambda},\tau,\circ}$, which induces a scheme-theoretically dominant morphism

$$\coprod_{\underline{\lambda},\tau} \operatorname{Spf} B^{\underline{\lambda},\tau,\circ} \to \operatorname{Spf} B^\circ. \tag{4.8.4}$$

Write $R = \operatorname{Spf} B^\circ \times_{\mathcal{C}_{d,\mathrm{ss},h}^{L/K,\mathrm{fl}}} \operatorname{Spf} B^\circ$, so that $[\operatorname{Spf} B^\circ/R] \xrightarrow{\sim} \mathcal{C}_{d,\mathrm{ss},h}^{L/K,\mathrm{fl}}$. Then we claim that each $\operatorname{Spf} B^{\underline{\lambda},\tau,\circ}$ is R-invariant. Granting this, if we write $R^{\underline{\lambda},\tau}$ for the restriction of R to $\operatorname{Spf} B^{\underline{\lambda},\tau,\circ}$, and then define $\mathcal{C}_{d,\mathrm{ss},h}^{L/K,\mathrm{fl},\underline{\lambda},\tau} := [\operatorname{Spf} B^{\underline{\lambda},\tau,\circ}/R^{\underline{\lambda},\tau}]$, it follows from the discussion of closed substacks in Appendix A that $\mathcal{C}_{d,\mathrm{ss},h}^{L/K,\mathrm{fl},\underline{\lambda},\tau}$ embeds as a closed sub-formal algebraic stack of $\mathcal{C}_{d,\mathrm{ss},h}^{L/K,\mathrm{fl}}$. Furthermore, the induced morphism

$$\coprod_{\underline{\lambda},\tau} \mathcal{C}_{d,\mathrm{ss},h}^{L/K,\mathrm{fl},\underline{\lambda},\tau} \to \mathcal{C}_{d,\mathrm{ss},h}^{L/K,\mathrm{fl}} \tag{4.8.5}$$

induces the morphism (4.8.4) after pull-back via the morphism (4.8.3), which is representable by algebraic spaces, smooth, and surjective. Since this latter morphism is scheme-theoretically dominant, so is (4.8.5).

We now verify that $\operatorname{Spf} B^{\underline{\lambda},\tau,\circ}$ is R-invariant, i.e., that $R_0 := \operatorname{Spf} B^{\underline{\lambda},\tau,\circ} \times_{\mathcal{C}_{d,\mathrm{ss},h}^{L/K,\mathrm{fl}}} \operatorname{Spf} B^\circ$ and $R_1 := \operatorname{Spf} B^\circ \times_{\mathcal{C}_{d,\mathrm{ss},h}^{L/K,\mathrm{fl}}} \operatorname{Spf} B^{\underline{\lambda},\tau,\circ}$ coincide as closed sub-formal algebraic spaces of $R = \operatorname{Spf} B^\circ \times_{\mathcal{C}_{d,\mathrm{ss},h}^{L/K,\mathrm{fl}}} \operatorname{Spf} B^\circ$ (and thus that both coincide with $R^{\underline{\lambda},\tau} := \operatorname{Spf} B^{\underline{\lambda},\tau,\circ} \times_{\mathcal{C}_{d,\mathrm{ss},h}^{L/K,\mathrm{fl}}} \operatorname{Spf} B^{\underline{\lambda},\tau,\circ}$).

Since $\operatorname{Spf} B^\circ \to \mathcal{C}_{d,\mathrm{ss},h}^{L/K,\mathrm{fl}}$ is representable by algebraic spaces and smooth, it is in particular representable by algebraic spaces, flat, and locally of finite type, and thus so are each of the projections $R_0 \to \operatorname{Spf} B^{\underline{\lambda},\tau,\circ}$ and $R_1 \to \operatorname{Spf} B^{\underline{\lambda},\tau,\circ}$. It follows from Lemmas A.3 and A.4 that any affine formal algebraic space which is étale over either of R_0 or R_1 is of the form $\operatorname{Spf} A^\circ$, where A° is ϖ-adically complete, topologically of finite type, and flat over $\mathcal{O}$. Thus, to show that R_0 and R_1 coincide as subspaces of R, it suffices to show that if A° is any ϖ-adically complete, topologically of finite type, and flat $\mathcal{O}$-algebra endowed with a morphism $\operatorname{Spf} A^\circ \to R$, then this morphism factors through R_0 if and only if it factors through R_1. Unwinding the definitions of R, R_0, and R_1 as fiber products, this amounts to showing that if we are given a pair of morphism $\operatorname{Spf} A^\circ \rightrightarrows \operatorname{Spf} B^\circ$ which induce the same morphism to $\mathcal{C}_{d,\mathrm{ss},h}^{L/K,\mathrm{fl}}$, then one factors through $\operatorname{Spf} B^{\underline{\lambda},\tau,\circ}$ if and only if the other does.

The pair of morphisms $\operatorname{Spf} A^\circ \rightrightarrows \operatorname{Spf} B^\circ$ correspond to a pair of morphisms $f_0, f_1 \colon B^\circ \rightrightarrows A^\circ$. The fact that these morphisms induce the same morphism to $\mathcal{C}_{d,\mathrm{ss},h}^{L/K,\mathrm{fl}}$ may be rephrased as saying that $f_0^*\mathfrak{M}^{\mathrm{inf}}$ and $f_1^*\mathfrak{M}^{\mathrm{inf}}$ are isomorphic Breuil–Kisin–Fargues modules over A°. Now, applying the base change statements from Propositions 4.6.3 and 4.7.2, we find that both $f_0^*\mathrm{WD}(\mathfrak{M}^{\mathrm{inf}}_{B^\circ})$ and $f_1^*\mathrm{WD}(\mathfrak{M}^{\mathrm{inf}}_{B^\circ})$ are isomorphic, as I_K-representations, to $\mathrm{WD}(\mathfrak{M}^{\mathrm{inf}}_{A^\circ})$, and that both $f_0^*D_{\mathrm{dR}}(\mathfrak{M}^{\mathrm{inf}}_{B^\circ})$ and $f_1^*D_{\mathrm{dR}}(\mathfrak{M}^{\mathrm{inf}}_{B^\circ})$ are isomorphic, as filtered $K\otimes_{\mathbf{Q}_p} B$-modules, to $D_{\mathrm{dR}}(\mathfrak{M}^{\mathrm{inf}}_{A^\circ})$. Thus either, and hence both, of $f_0^*\mathrm{WD}(\mathfrak{M}^{\mathrm{inf}}_{B^\circ})$ and $f_1^*\mathrm{WD}(\mathfrak{M}^{\mathrm{inf}}_{B^\circ})$ are of inertial type τ if and only if $\mathrm{WD}(\mathfrak{M}^{\mathrm{inf}}_{A^\circ})$ is of inertial type τ, while either, and hence both, of $f_0^*D_{\mathrm{dR}}(\mathfrak{M}^{\mathrm{inf}}_{B^\circ})$ and $f_1^*D_{\mathrm{dR}}(\mathfrak{M}^{\mathrm{inf}}_{B^\circ})$ are of Hodge type $\underline{\lambda}$ if and only if $D_{\mathrm{dR}}(\mathfrak{M}^{\mathrm{inf}}_{A^\circ})$ is of Hodge type $\underline{\lambda}$. Consequently f_0 factors through $\operatorname{Spf} B^{\underline{\lambda},\tau,\circ}$ if and only f_1 does. This shows that $\operatorname{Spf} B^{\underline{\lambda},\tau,\circ}$ is indeed R-invariant.

We now show that $\mathcal{C}_{d,\mathrm{ss},h}^{L/K,\mathrm{fl},\underline{\lambda},\tau} := [\operatorname{Spf} B^{\underline{\lambda},\tau,\circ}/R^{\underline{\lambda},\tau}]$ satisfies the required property, i.e., that for any finite flat $\mathcal{O}$-algebra A°, a morphism $\operatorname{Spf} A^\circ \to \mathcal{C}_{d,\mathrm{ss},h}^{L/K,\mathrm{fl}}$ factors through $\mathcal{C}_{d,\mathrm{ss},h}^{L/K,\mathrm{fl},\underline{\lambda},\tau}$ if and only if $A^{\underline{\lambda},\tau} = A$. By construction, such a factorization occurs if and only if the induced morphism $\operatorname{Spf} A^\circ \times_{\mathcal{C}_{d,\mathrm{ss},h}^{L/K,\mathrm{fl}}} \operatorname{Spf} B^\circ \to \operatorname{Spf} B^\circ$ factors through $\operatorname{Spf} B^{\underline{\lambda},\tau,\circ}$.

Since the surjection $\operatorname{Spf} B^\circ \to \mathcal{C}_{d,\mathrm{ss},h}^{L/K,\mathrm{fl}}$ is representable by algebraic spaces, smooth, and surjective, the fiber product $\operatorname{Spf} A^\circ \times_{\mathcal{C}_{d,\mathrm{ss},h}^{L/K,\mathrm{fl}}} \operatorname{Spf} B^\circ$ is a formal algebraic space, and the surjection $\operatorname{Spf} A^\circ \times_{\mathcal{C}_{d,\mathrm{ss},h}^{L/K,\mathrm{fl}}} \operatorname{Spf} B^\circ \to \operatorname{Spf} A^\circ$ is representable by algebraic spaces, smooth, and surjective. We may thus find an open cover of $\operatorname{Spf} A^\circ \times_{\mathcal{C}_{d,\mathrm{ss},h}^{L/K,\mathrm{fl}}} \operatorname{Spf} B^\circ$ by affine formal algebraic spaces $\operatorname{Spf} C^\circ$, with C° being a faithfully flat ϖ-adically complete A°-algebra which is topologically of finite type. (We are again applying Lemmas A.3 and A.4.) If we write $C := C^\circ[1/p]$, then the base change property of Corollary 4.7.11 shows that the hypothesized factorization occurs if and only if $C = C^{\underline{\lambda},\tau}$, and also that $C^{\underline{\lambda},\tau} = A^{\underline{\lambda},\tau} \otimes_A C$, so that (since C is faithfully flat over A) $C^{\underline{\lambda},\tau} = C$ if and only if $A^{\underline{\lambda},\tau} = A$. In conclusion, we have shown that $\operatorname{Spf} A^\circ \to \mathcal{C}_{d,\mathrm{ss},h}^{L/K,\mathrm{fl}}$ factors through $\mathcal{C}_{d,\mathrm{ss},h}^{L/K,\mathrm{fl},\underline{\lambda},\tau}$ if and only if $A^{\underline{\lambda},\tau} = A$, as required.

Finally, the uniqueness of $\mathcal{C}_{d,\mathrm{ss},h}^{L/K,\mathrm{fl},\underline{\lambda},\tau}$ is immediate from Proposition 4.8.6 below; the properties of being of finite presentation, of having affine diagonal, and of the morphism $\mathcal{C}_{d,\mathrm{ss},h}^{L/K,\mathrm{fl},\underline{\lambda},\tau} \to \mathcal{X}_d$ being representable by algebraic spaces, proper, and of finite presentation, are immediate from Proposition 4.5.24 and Lemma 4.5.10. □

4.8.6 Proposition. *Suppose that $\mathcal{Y}$ and $\mathcal{Z}$ are two Noetherian formal algebraic stacks, both lying over* $\operatorname{Spf}\mathcal{O}$ *and both flat over $\mathcal{O}$, and both embedded as closed substacks of a stack $\mathcal{X}$ over* $\operatorname{Spec}\mathcal{O}$ *(in the usual sense that the inclusions of substacks $\mathcal{Y},\mathcal{Z}\hookrightarrow\mathcal{X}$ are representable by algebraic spaces, and induce closed immersions when pulled back over any scheme-valued point of $\mathcal{X}$). Suppose also that each of $\mathcal{Y}_{\mathrm{red}}$ and $\mathcal{Z}_{\mathrm{red}}$ are locally of finite type over* $\mathbf{F}$*, and suppose further that, for any finite flat $\mathcal{O}$-algebra A°, a morphism* $\operatorname{Spf} A^\circ \to \mathcal{X}$ *factors through $\mathcal{Y}$ if and only if it factors through $\mathcal{Z}$. Then $\mathcal{Y}$ and $\mathcal{Z}$ coincide.*

Proof. The $\mathcal{O}$-flat part $\mathcal{W} := (\mathcal{Y}\times_{\mathcal{X}}\mathcal{Z})_{\mathrm{fl}}$ of the 2-fiber product $\mathcal{Y}\times_{\mathcal{X}}\mathcal{Z}$ embeds as a closed substack of each of $\mathcal{Y}$ and $\mathcal{Z}$, and it suffices to show that each of those embeddings are isomorphisms. The stack $\mathcal{W}$ inherits all of the hypotheses shared by $\mathcal{Y}$ and $\mathcal{Z}$, including the fact that, for any finite flat $\mathcal{O}$-algebra A°, the A°-valued points of $\mathcal{W}$ coincide with the A°-valued points of each of $\mathcal{Y}$ and $\mathcal{Z}$. Thus, replacing the pair $\mathcal{Y}$, $\mathcal{Z}$ by the pairs $\mathcal{W},\mathcal{Y}$ and $\mathcal{W},\mathcal{Z}$ in turn, we may reduce to the case when $\mathcal{Y}$ is actually a closed substack of $\mathcal{Z}$.

Since $\mathcal{Z}$ is Noetherian, lies over $\operatorname{Spf}\mathcal{O}$, and is flat over $\mathcal{O}$, we can find a smooth surjection $U\to\mathcal{Z}$, whose source is a disjoint union of affine formal algebraic spaces $\operatorname{Spf} C^\circ$, where C° is a flat I-adically complete Noetherian $\mathcal{O}$-algebra, for some ideal I which contains ϖ and defines the topology on C°.

The assumption on $\mathcal{Z}_{\mathrm{red}}$ further implies that C°/I is a finite type $\mathbf{F}$-algebra. Then $\operatorname{Spf} C^\circ\times_{\mathcal{Z}}\mathcal{Y}$ is a closed sub-formal algebraic space of $\operatorname{Spf} C^\circ$, and (since it is also $\mathcal{O}$-flat) it follows from Lemmas A.3 and A.4 that it is of the form $\operatorname{Spf} B^\circ$, where B° is an $\mathcal{O}$-flat quotient of C°, again endowed with the I-adic topology. Since the property of being an isomorphism can be checked *fppf* locally on the target, we may replace the closed embedding $\mathcal{Y}\hookrightarrow\mathcal{Z}$ by $\operatorname{Spf} B^\circ\hookrightarrow\operatorname{Spf} C^\circ$. In summary, we have a surjection $C^\circ\to B^\circ$ of flat, I-adically complete $\mathcal{O}$-algebras, with C°/I (and hence also B°/I) being of finite type over $\mathbf{F}$, and with the further property that any morphism $C^\circ\to A^\circ$ of $\mathcal{O}$-algebras in which A° is finite flat over $\mathcal{O}$ factors through B°; we then have to prove that this surjection is an isomorphism.

Write $B=B^\circ[1/p]$, $C=C^\circ[1/p]$; since B° and C° are flat over $\mathcal{O}$, it is enough to check that the surjection $C\to B$ is also injective. Now, we can embed C into the product of its localizations at all maximal ideals, and since C is Noetherian, it in fact embeds into the product of the completions of these localizations. It therefore embeds into the product of all of its local Artinian quotients A. We claim that any such quotient is in fact obtained from a morphism $C^\circ\to A^\circ$, where A° is a finite flat $\mathcal{O}$-algebra; since any such morphism $C^\circ\to A^\circ$ factors through B° by assumption, it follows that any such A is a quotient of B, so that the surjection $C \to B$ is indeed injective.

It remains to prove the claim. Let A° be the image of the composite $C^\circ \to C \to A$; we need to show that A° is a finite $\mathcal{O}$-algebra. This follows from Lemma 4.8.7 below. □

The following lemma is no doubt standard, but we recall a proof for the sake of completeness.

4.8.7 Lemma. *If $R \to A$ is a morphism of $\mathcal{O}$-algebras, with R being p-adically complete and Noetherian (e.g., a complete local $\mathcal{O}$-algebra with finite residue field), and A being a finite-dimensional E-algebra, then the image of R in A is finite over $\mathcal{O}$.*

Proof. Since R is Noetherian and p-adically complete, the Artin–Rees lemma shows that the same is true of its image in A (see, e.g., [AM69, Prop. 10.13]). Thus, by replacing R by its image in A and A itself by the E-span of this image, we are reduced to checking the following statement: if M is a p-adically complete and separated torsion-free $\mathcal{O}$-module, and if $M[1/p]$ is finite-dimensional, then M is finite over $\mathcal{O}$. There are many ways to see this, of course; here is one.

Since $V := M[1/p]$ is finite dimensional, we may find a finite spanning set for this vector space contained in M; if L denotes its $\mathcal{O}$-span, then L is a lattice in V. Since L is compact with respect to its p-adic topology, the p-adic topology on M induces the p-adic topology on L, and in particular L is closed in the p-adic topology on M. Thus M/L is a p-adically separated submodule of V/L. Note that V/L consists entirely of p-power torsion elements, and thus that the same is true of M/L.

This implies that $\bigcap_n p^n\big((M/L)[p^{n+1}]\big) \subseteq \bigcap_n p^n(M/L) = 0$. On the other hand, $p^n\big((M/L)[p^{n+1}]\big) \subseteq (M/L)[p] \subseteq (V/L)[p] \xrightarrow{\sim} L/pL$, and so $\{p^n\big((M/L)[p^{n+1}]\big)$ is a decreasing sequence of finite sets, which thus eventually stabilizes. We conclude that $p^n\big((M/L)[p^{n+1}]\big) = 0$ for some sufficiently large value of n; equivalently, $(M/L)[p^n] = (M/L)[p^{n+1}]$, and thus in fact $(M/L)[p^n] = (M/L)[p^\infty] = M/L$. Consequently $L \subseteq M \subseteq p^{-n}L$, showing that M itself is finite over $\mathcal{O}$, as claimed. □

4.8.8 Definition. Let τ be an inertial type, and let $\underline{\lambda}$ be a Hodge type. To begin with, assume in addition that $\underline{\lambda}$ is effective, and choose $h \geq 0$ such that $\underline{\lambda}$ is bounded by h, as well as a finite Galois extension L/K for which the kernel of τ contains I_L. We then define $\mathcal{X}_{K,d}^{\mathrm{crys},\lambda,\tau}$ to be the scheme-theoretic image of the morphism $\mathcal{C}_{d,\mathrm{crys},h}^{L/K,\mathrm{fl},\underline{\lambda},\tau} \to \mathcal{X}_{K,d}$, and $\mathcal{X}_{K,d}^{\mathrm{ss},\underline{\lambda},\tau}$ to be the scheme-theoretic image of the morphism $\mathcal{C}_{d,\mathrm{ss},h}^{L/K,\mathrm{fl},\underline{\lambda},\tau} \to \mathcal{X}_{K,d}$. (The characterization of these closed substacks of $\mathcal{X}_{K,d}$ given in Theorem 4.8.12 will show that they are well-defined substacks of $\mathcal{X}_{K,d}$, independently of the auxiliary choices of h and L/K that were used in their definition.)

In general (i.e., if $\underline{\lambda}$ is not effective), we choose an integer h' so that $\lambda'_{\sigma,i} := \lambda_{\sigma,i} + h' \geq 0$ for all σ and all i, and define $\mathcal{X}_{K,d}^{\mathrm{ss},\underline{\lambda},\tau}$ to be the unique substack of $\mathcal{X}_{K,d}$ with the property that if B is a $\mathcal{O}/\varpi^a$-algebra for some $a \geq 1$, then a morphism ρ: $\operatorname{Spec} B \to \mathcal{X}_{K,d}$ factors through $\mathcal{X}_{K,d}^{\mathrm{ss},\underline{\lambda},\tau}$ if and only if the morphism $\rho \otimes \epsilon^{h'}$: $\operatorname{Spec} B \to \mathcal{X}_{K,d}$ factors through $\mathcal{X}_{K,d}^{\mathrm{ss},\underline{\lambda}',\tau}$. (Here we are using the notation of Section 3.8; more precisely, we are twisting by the étale (φ,Γ)-module corresponding to $\epsilon^{h'}$.) We define $\mathcal{X}_{K,d}^{\mathrm{crys},\underline{\lambda},\tau}$ in the same way. The point of this definition is that if A° is a finite flat $\mathcal{O}$-algebra, then a representation ρ: $G_K \to \mathrm{GL}_d(A^\circ[1/p])$ is potentially semistable (resp. crystalline) of Hodge type $\underline{\lambda}$ and inertial type τ if and only if $\rho \otimes \epsilon^{h'}$ is potentially semistable (resp. crystalline) of Hodge type $\underline{\lambda}'$ and inertial type τ; again, Theorem 4.8.12 below shows that these stacks are defined independently of the choices made in this definition, and in particular independently of the choice of h'.

4.8.9 Remark. Rather than introducing the twist by $\epsilon^{h'}$, it would perhaps be more natural to work throughout with Breuil–Kisin and Breuil–Kisin–Fargues modules which are not necessarily φ-stable, as in [BMS18]. However, it would still frequently be useful to make the corresponding Tate twists (see for example the proof of [BMS18, Lem. 4.26]), and in particular we would need to make such twists in order to use the results of [PR09, EG21], so we do not see any advantage in doing so.

The rest of this section is devoted to proving some fundamental properties of the stacks $\mathcal{X}_{K,d}^{\mathrm{crys},\lambda,\tau}$ and $\mathcal{X}_{K,d}^{\mathrm{ss},\underline{\lambda},\tau}$. We begin with an analysis of versal rings. To this end, fix a point $\operatorname{Spec} \mathbf{F}' \to \mathcal{X}_{K,d}$ for some finite extension $\mathbf{F}'$ of $\mathbf{F}$, giving rise to a continuous representation $\overline{\rho}$: $G_K \to \mathrm{GL}_d(\mathbf{F}')$. Let $\mathcal{O}' = W(\mathbf{F}') \otimes_{W(\mathbf{F})} \mathcal{O}$, the ring of integers in a finite extension E' of E having residue field $\mathbf{F}'$. Then the versal morphism $\operatorname{Spf} R_{\overline{\rho}}^{\square,\mathcal{O}'} \to \mathcal{X}_{K,d}$ of Proposition 3.6.3 induces morphisms $\operatorname{Spf} R_{\overline{\rho}}^{\mathrm{crys},\underline{\lambda},\tau,\mathcal{O}'} \to \mathcal{X}_{K,d}$ and $\operatorname{Spf} R_{\overline{\rho}}^{\mathrm{ss},\underline{\lambda},\tau,\mathcal{O}'} \to \mathcal{X}_{K,d}$ (where these deformation rings are as in Section 1.12).

4.8.10 Proposition. *The morphisms* $\operatorname{Spf} R_{\overline{\rho}}^{\mathrm{crys},\underline{\lambda},\tau,\mathcal{O}'} \to \mathcal{X}_{K,d}$ *and* $\operatorname{Spf} R_{\overline{\rho}}^{\mathrm{ss},\underline{\lambda},\tau,\mathcal{O}'} \to \mathcal{X}_{K,d}$ *factor through versal morphisms* $\operatorname{Spf} R_{\overline{\rho}}^{\mathrm{crys},\underline{\lambda},\tau,\mathcal{O}'} \to \mathcal{X}_{K,d}^{\mathrm{crys},\underline{\lambda},\tau}$ *and* $\operatorname{Spf} R_{\overline{\rho}}^{\mathrm{ss},\underline{\lambda},\tau,\mathcal{O}'} \to \mathcal{X}_{K,d}^{\mathrm{ss},\underline{\lambda},\tau}$ *respectively.*

Proof. By definition, we can and do assume that $\underline{\lambda}$ is effective. We give the proof for $\operatorname{Spf} R_{\overline{\rho}}^{\mathrm{crys},\underline{\lambda},\tau,\mathcal{O}'}$, the argument in the semistable case being identical. We begin by introducing the fiber product $\widehat{C} := \mathcal{C}_{d,\mathrm{crys},h}^{L/K,\mathrm{fl},\underline{\lambda},\tau} \times_{\mathcal{X}_{K,d}} \operatorname{Spf} R_{\overline{\rho}}^{\square,\mathcal{O}'}$. Since $\mathcal{C}_{d,\mathrm{crys},h}^{L/K,\mathrm{fl},\underline{\lambda},\tau} \to \mathcal{X}_{K,d}$ is proper and representable by algebraic spaces, we see that

$\widehat{C}$ is a formal algebraic space, and that the morphism

$$\widehat{C} \to \operatorname{Spf} R_{\overline{\rho}}^{\square,\mathcal{O}'} \tag{4.8.11}$$

is proper and representable by algebraic spaces. The latter condition can be reexpressed by saying that (4.8.11) is an *adic* morphism; thus we see that $\widehat{C}$ is a proper $\mathfrak{m}_{R_{\overline{\rho}}^{\square,\mathcal{O}'}}$-adic formal algebraic space over $\operatorname{Spf} R_{\overline{\rho}}^{\square,\mathcal{O}'}$.

If we let $\operatorname{Spf} R$ denote the scheme-theoretic image of the morphism (4.8.11), then by Propositions 3.4.12 and 4.8.2 and Lemma A.30, the versal morphism $\operatorname{Spf} R_{\overline{\rho}}^{\square,\mathcal{O}'} \to \mathcal{X}_{K,d}$ induces a versal morphism $\operatorname{Spf} R \to \mathcal{X}_{K,d}^{\mathrm{crys},\underline{\lambda},\tau}$, and so the assertion of the proposition may be rephrased as the claim that R and $R_{\overline{\rho}}^{\mathrm{crys},\underline{\lambda},\tau,\mathcal{O}'}$ coincide as quotients of $R_{\overline{\rho}}^{\square,\mathcal{O}'}$.

Since $\mathcal{C}_{d,\mathrm{crys},h}^{L/K,\mathrm{fl}}$ is flat over $\operatorname{Spf} \mathcal{O}$, it follows from Lemma A.31 that R is an $\mathcal{O}$-flat quotient of $R_{\overline{\rho}}^{\square,\mathcal{O}'}$. Since $R_{\overline{\rho}}^{\mathrm{crys},\underline{\lambda},\tau,\mathcal{O}'}$ is also an $\mathcal{O}$-flat quotient of $R_{\overline{\rho}}^{\square,\mathcal{O}'}$, we are reduced by Proposition 4.8.6 to showing that if A° is any finite flat $\mathcal{O}'$-algebra, then an $\mathcal{O}'$-homomorphism $R_{\overline{\rho}}^{\square,\mathcal{O}'} \to A^\circ$ factors through R if and only if it factors through $R_{\overline{\rho}}^{\mathrm{crys},\underline{\lambda},\tau,\mathcal{O}'}$.

To this end, note that if the morphism factors through $R_{\overline{\rho}}^{\mathrm{crys},\underline{\lambda},\tau,\mathcal{O}'}$, then by Theorem 4.7.13 there is an order $(A^\circ)' \supseteq A^\circ$ in $A^\circ[1/p]$ such that the induced morphism $\operatorname{Spf}(A^\circ)' \to \operatorname{Spf} R_{\overline{\rho}}^{\square,\mathcal{O}'}$ lifts to $\widehat{C}$. Consequently the morphism $\operatorname{Spf}(A^\circ)' \to \operatorname{Spf} R_{\overline{\rho}}^{\square,\mathcal{O}'}$ factors through $\operatorname{Spf} R$, and since $\operatorname{Spf}(A^\circ)' \to \operatorname{Spf} A^\circ$ is scheme-theoretically dominant, the morphism $\operatorname{Spf} A^\circ \to \operatorname{Spf} R_{\overline{\rho}}^{\square,\mathcal{O}'}$ also factors through $\operatorname{Spf} R$.

It remains to prove the converse, namely that if $\operatorname{Spf} A^\circ \to \operatorname{Spf} R_{\overline{\rho}}^{\square,\mathcal{O}'}$ factors through $\operatorname{Spf} R$, then it factors through $R_{\overline{\rho}}^{\mathrm{crys},\underline{\lambda},\tau,\mathcal{O}'}$. If we write $S := R \otimes_{R_{\overline{\rho}}^{\square,\mathcal{O}'}} R_{\overline{\rho}}^{\mathrm{crys},\underline{\lambda},\tau,\mathcal{O}'}$ (a quotient of R), then equivalently, we wish to show that any morphism $\operatorname{Spf} A^\circ \to \operatorname{Spf} R$ in fact factors through $\operatorname{Spf} S$.

To do this, we will apply Lemma A.32, not in the context of $\mathcal{O}'$-algebras, but rather in the context of E'-algebras. That is, we will invert p, and work with the rings $R[1/p]$ and $S[1/p]$ (or rather on certain completions of these rings). We would also like to invert p on the object $\widehat{C}$, but since the latter is a formal algebraic space, doing so directly would lead us into considerations of rigid analytic geometry that we prefer to avoid. Thus, we begin by observing that by Lemma 4.5.26 (together with [Knu71, Thm. V.6.3], to pass from $\mathcal{C}_{d,\mathrm{ss},h}^{L/K}$ to its closed substack $\mathcal{C}_{d,\mathrm{crys},h}^{L/K,\mathrm{fl}}$), $\widehat{C}$ can be promoted to a projective scheme over $\operatorname{Spec} R$. That is, there is a projective morphism of schemes $C \to \operatorname{Spec} R_{\overline{\rho}}^{\square,\mathcal{O}'}$, with scheme-theoretic image equal to $\operatorname{Spec} R$, whose $\mathfrak{m}_{R_{\overline{\rho}}^{\square,\mathcal{O}'}}$-adic completion is isomorphic to $\widehat{C}$. Write $C[1/p] := C \otimes_{\mathcal{O}'} E'$. Since scheme-theoretic dominance is preserved by flat base change, the morphism $C[1/p] \to \operatorname{Spec} R[1/p]$ is proper and scheme-theoretically dominant.

Returning to our argument, we want to show, for any finite flat $\mathcal{O}'$-algebra A°, that a morphism $R \to A^\circ$ necessarily factors through S, or, equivalently, that the induced morphism $R[1/p] \to A := A^\circ[1/p]$ factors through $S[1/p]$. Since A is a product of Artinian local E-algebras, we can check this assertion factor by factor; replacing A by one of these factors (and A° by the image of A° in this factor) we may thus assume that A is local. We then let $\widehat{R[1/p]}$ denote the completion of R at the kernel of its map to A, and we let $\widehat{S[1/p]}$ denote the corresponding completion of $S[1/p]$; by Artin–Rees, we also have that $\widehat{S[1/p]} = \widehat{R[1/p]} \otimes_{R[1/p]} S[1/p]$, and so the map $R[1/p] \to A$ factors through $S[1/p]$ if and only if the induced morphism $\widehat{R[1/p]} \to A$ factors through $\widehat{S[1/p]}$.

To show the desired factorization, we see from Lemma A.32 that it suffices to show that if $\widehat{R[1/p]} \to B$ is a morphism to a finite Artinian E-algebra for which $C_B \to \operatorname{Spec} B$ admits a section, then this morphism factors through $\widehat{S[1/p]}$.

Write B° to denote the image of the composite $R_{\overline{\rho}}^{\square,\mathcal{O}'} \to \widehat{R[1/p]} \to B$; by Lemma 4.8.7, we see that B° is an order in B. Let $Z \hookrightarrow C_{B^\circ}$ denote the scheme-theoretic closure of the section $\operatorname{Spec} B \to C_B$ in C_{B°. This is proper over $\operatorname{Spec} B^\circ$, flat over $\mathcal{O}$, and irreducible of dimension 1. Thus Z is finite over $\operatorname{Spec} B^\circ$, and hence $Z = \operatorname{Spec}(B^\circ)'$ for some order $(B^\circ)'$ in B containing B°. Thinking of Z as a section of C over $(B^\circ)'$, we find that the induced morphism $\operatorname{Spf}(B^\circ)' \to \mathcal{X}_{K,d}$ lifts to a morphism $\operatorname{Spf}(B^\circ)' \to \mathcal{C}_{d,\mathrm{crys},h}^{L/K,\mathrm{fl},\underline{\lambda},\tau}$. Theorem 4.7.13 then implies that the morphism $R_{\overline{\rho}}^{\square,\mathcal{O}'} \to B^\circ$ factors through $R_{\overline{\rho}}^{\mathrm{crys},\underline{\lambda},\tau,\mathcal{O}'}$, and thus the morphism $\widehat{R[1/p]} \to B$ does indeed factor through $\widehat{S[1/p]}$, as required. □

4.8.12 Theorem. *Let τ be an inertial type, and let $\underline{\lambda}$ be a Hodge type. Then the closed substacks $\mathcal{X}_{K,d}^{\mathrm{crys},\underline{\lambda},\tau}$ and $\mathcal{X}_{K,d}^{\mathrm{ss},\underline{\lambda},\tau}$ of $\mathcal{X}_{K,d}$ are p-adic formal algebraic stacks which are of finite type and flat over $\operatorname{Spf}\mathcal{O}$, and are uniquely determined as $\mathcal{O}$-flat closed substacks of $\mathcal{X}_{K,d}$ by the following property: if A° is a finite flat $\mathcal{O}$-algebra, then $\mathcal{X}_{K,d}^{\mathrm{ss},\lambda,\tau}(A^\circ)$ (resp. $\mathcal{X}_{K,d}^{\mathrm{crys},\lambda,\tau}(A^\circ)$) is precisely the subgroupoid of $\mathcal{X}_{K,d}(A^\circ)$ consisting of G_K-representations which become potentially semistable (resp. potentially crystalline) of Hodge type $\underline{\lambda}$ and inertia type τ after inverting p.*

Proof. We can and do assume that $\underline{\lambda}$ is effective. By Propositions 3.4.12, 4.8.2 and A.21, $\mathcal{X}_{K,d}^{\mathrm{crys},\underline{\lambda},\tau}$ and $\mathcal{X}_{K,d}^{\mathrm{ss},\underline{\lambda},\tau}$ of $\mathcal{X}_{K,d}$ are p-adic formal algebraic stacks which are of finite type and flat over $\operatorname{Spf}\mathcal{O}$.

We now verify the claimed description of the A°-valued points of $\mathcal{X}_{K,d}^{\mathrm{ss},\lambda,\tau}(A^\circ)$ and $\mathcal{X}_{K,d}^{\mathrm{crys},\lambda,\tau}(A^\circ)$, for finite flat $\mathcal{O}$-algebras A°. Since the argument is identical in either case, we give it in the semistable case. Any such algebra A° is a product of finitely many finite flat local $\mathcal{O}$-algebras, and so we immediately reduce to verifying the claim in the case when A° is furthermore local. Since A° is then a complete local finite flat $\mathcal{O}$-algebra, the claimed description follows from Proposition 4.8.10. Indeed, the residue field $\mathbf{F}'$ is a finite extension of $\mathbf{F}$, and if as above we set $\mathcal{O}' = W(\mathbf{F}') \otimes_{W(\mathbf{F})} \mathcal{O}$, then the morphism $\operatorname{Spf} A^\circ \to \mathcal{X}_{K,d}$ factors through $\mathcal{X}_{K,d}^{\mathrm{ss},\underline{\lambda},\tau}$ if and only if it factors through the versal morphism $\operatorname{Spf} R_{\overline{\rho}}^{\mathrm{ss},\underline{\lambda},\tau,\mathcal{O}'} \to \mathcal{X}_{K,d}^{\mathrm{ss},\underline{\lambda},\tau}$ of Proposition 4.8.10. In turn, this happens if and only if the corresponding G_K-representation is potentially semistable of Hodge type $\underline{\lambda}$ and inertial type τ, by the defining property of $R_{\overline{\rho}}^{\mathrm{ss},\underline{\lambda},\tau,\mathcal{O}'}$.

It remains to show the claimed uniqueness statement. We prove in Corollary 5.5.18 below that $\mathcal{X}_{K,d}$ is a Noetherian formal algebraic stack, and in Theorem 5.5.12 below that $\mathcal{X}_{d,\mathrm{red}}$ is of finite presentation over $\mathbf{F}$. The reader can easily confirm that Chapter 5 contains no citation to the present chapter, and thus that those results are independent of the present ones; in particular, it is safe to invoke them here. These results imply that any closed substack $\mathcal{Y}$ of $\mathcal{X}_{K,d}$ is Noetherian (so that it makes sense to speak of $\mathcal{Y}$ being flat over $\mathcal{O}$), and that $\mathcal{Y}_{\mathrm{red}}$ is of finite type over $\mathcal{F}$. The claimed uniqueness then follows from Proposition 4.8.6. □

We will make use of the following corollary in Chapter 6.

4.8.13 Corollary. *For any Hodge type $\underline{\lambda}$, and for any $a \geq 1$, the corresponding morphism $\operatorname{Spf} R_{\overline{\rho}}^{\mathrm{crys},\underline{\lambda},\mathcal{O}'}/\varpi^a \to \mathcal{X}_{K,d}^a$ is effective, i.e., is induced by a morphism $\operatorname{Spec} R_{\overline{\rho}}^{\mathrm{crys},\underline{\lambda},\mathcal{O}'}/\varpi^a \to \mathcal{X}_{K,d}^a$.*

Proof. By Proposition 4.8.10, the morphism $\operatorname{Spf} R_{\overline{\rho}}^{\mathrm{crys},\underline{\lambda},\mathcal{O}'}/\varpi^a \to \mathcal{X}_{K,d}^a$ factors through a versal morphism $\operatorname{Spf} R_{\overline{\rho}}^{\mathrm{crys},\underline{\lambda},\mathcal{O}'}/\varpi^a \to \mathcal{X}_{K,d}^{\mathrm{crys},\lambda} \times_{\operatorname{Spf}\mathcal{O}} \operatorname{Spec} \mathcal{O}/\varpi^a$. Since $\mathcal{X}_{K,d}^{\mathrm{crys},\lambda}$ is a p-adic formal algebraic stack (by Theorem 4.8.12) the base-change $\mathcal{X}_{K,d}^{\mathrm{crys},\lambda} \times_{\operatorname{Spf}\mathcal{O}} \operatorname{Spec} \mathcal{O}/\varpi^a$ is an algebraic stack, and so the result follows from [Sta, Tag 07X8]. □

Finally, we can compute the dimensions of our potentially crystalline and semistable stacks. It is presumably possible to develop the dimension theory of p-adic formal algebraic stacks in some generality, but we do not need to do so, as we can instead work with the special fibres, which are algebraic stacks.

4.8.14 Theorem. *The algebraic stacks $\overline{\mathcal{X}}_{K,d}^{\mathrm{crys},\underline{\lambda},\tau} := \mathcal{X}_{K,d}^{\mathrm{crys},\underline{\lambda},\tau} \times_{\operatorname{Spf}\mathcal{O}} \operatorname{Spec}\mathbf{F}$ and $\overline{\mathcal{X}}_{K,d}^{\mathrm{ss},\underline{\lambda},\tau} := \mathcal{X}_{K,d}^{\mathrm{ss},\underline{\lambda},\tau} \times_{\operatorname{Spf}\mathcal{O}} \operatorname{Spec}\mathbf{F}$ are equidimensional of dimension*

$$\sum_{\sigma} \#\{1 \leq i < j \leq d \mid \lambda_{\sigma,i} > \lambda_{\sigma,j}\}.$$

In particular, if $\underline{\lambda}$ *is regular, then the algebraic stacks* $\overline{\mathcal{X}}_{K,d}^{\mathrm{crys},\underline{\lambda},\tau}$ *and* $\overline{\mathcal{X}}_{K,d}^{\mathrm{ss},\underline{\lambda},\tau}$ *are equidimensional of dimension* $[K:\mathbf{Q}_p]d(d-1)/2$.

Proof. Once again, we give the argument in the crystalline case, the semistable case being identical. Write $d_{\underline{\lambda}} := \sum_\sigma \#\{1 \le i < j \le d | \lambda_{\sigma,i} > \lambda_{\sigma,j}\}$. The algebraic stack $\mathcal{X}_{K,d}^{\mathrm{crys},\underline{\lambda},\tau} \times_{\mathrm{Spf}\,\mathcal{O}} \operatorname{Spec} \mathbf{F}$ is of finite type over $\mathbf{F}$, and we need to show that it is equidimensional of dimension $d_{\underline{\lambda}}$. Let $x\colon \operatorname{Spec} \mathbf{F}' \to \mathcal{X}_{K,d}^{\mathrm{crys},\underline{\lambda},\tau}$ be a finite type point corresponding to a Galois representation $\overline{\rho}$, with corresponding versal morphism $\operatorname{Spf} R_{\overline{\rho}}^{\mathrm{crys},\underline{\lambda},\mathcal{O}'}/\varpi \to \overline{\mathcal{X}}_{K,d}^{\mathrm{crys},\underline{\lambda},\tau}$. We have the fiber product

$$\operatorname{Spf} R_{\overline{\rho}}^{\mathrm{crys},\underline{\lambda},\mathcal{O}'}/\varpi \times_{\widehat{(\overline{\mathcal{X}}_{K,d}^{\mathrm{crys},\underline{\lambda},\tau})}_x} \operatorname{Spf} R_{\overline{\rho}}^{\mathrm{crys},\underline{\lambda},\mathcal{O}'}/\varpi \xrightarrow{\sim} \widehat{\mathrm{GL}}_{d, R_{\overline{\rho}}^{\mathrm{crys},\underline{\lambda},\mathcal{O}'}/\varpi, 1},$$

where $\widehat{\mathrm{GL}}_{d, R_{\overline{\rho}}^{\mathrm{crys},\underline{\lambda},\mathcal{O}'}/\varpi, 1}$ denotes the completion of $(\mathrm{GL}_d)_{R_{\overline{\rho}}^{\mathrm{crys},\underline{\lambda},\mathcal{O}'}/\varpi}$ along the identity element in its special fiber. By [EG19, Lem. 2.40], it is therefore enough to recall that since $R_{\overline{\rho}}^{\mathrm{crys},\underline{\lambda},\mathcal{O}'}$ is equidimensional of dimension $d_{\underline{\lambda}}+1$, $R_{\overline{\rho}}^{\mathrm{crys},\underline{\lambda},\mathcal{O}'}/\varpi$ is equidimensional of dimension $d_{\underline{\lambda}}$ (see [BM14, Lem. 2.1]). □

Chapter Five

Families of extensions

In this chapter we extend the theory of the Herr complex to the context of (φ,Γ)-modules with coefficients, and use it to develop a theory of families of extensions of (φ,Γ)-modules. We then inductively construct families which cover the underlying reduced stack $\mathcal{X}_{K,d,\mathrm{red}}$, and employ obstruction theory to deduce that $\mathcal{X}_{K,d}$ is a Noetherian formal algebraic stack.

5.1 THE HERR COMPLEX

In this chapter we consider the Herr complex; our approach is informed by the papers [Her98, Her01, CC98, Liu08a, Pot13, KPX14], and in particular we follow [Pot13, KPX14] in considering it as an object of the derived category. Our main technical result is to show that it is a perfect complex, which we will do by showing that it satisfies the following well-known criterion.

5.1.1 Lemma. *Let A be a Noetherian commutative ring, and let $C^\bullet$ be an object of $D(A)$. Then the following two conditions are equivalent:*

1. *There is a quasi-isomorphism $F^\bullet \to C^\bullet$ where $F^\bullet$ is a complex of flat A-modules, concentrated in a finite number of degrees; and the cohomology groups of $C^\bullet$ are all finite A-modules.*
2. *$C^\bullet$ is perfect.*

Proof. This follows from [Sta, Tag 0658], [Sta, Tag 0654] and [Sta, Tag 066E]. □

Let A be a p-adically complete $\mathcal{O}$-algebra, and let M be an A-module with commuting A-linear endomorphisms φ, Γ (in our main applications, M will be a projective étale (φ,Γ)-module with A-coefficients, but it will be convenient in our arguments to be able to consider more general possibilities, such as subquotients of base changes of such modules). Then the *Herr complex* of M is by definition the complex of A-modules $\mathcal{C}^\bullet(M)$ in degrees $0, 1, 2$ given by

$$M \xrightarrow{(\varphi-1,\ \gamma-1)} M \oplus M \xrightarrow{(\gamma-1)\oplus(1-\varphi)} M.$$

We have the following useful interpretation of the cohomology groups of the Herr complex.

5.1.2 Lemma. *Let M_1, M_2 be projective étale (φ, Γ)-modules with A-coefficients. Then for $i=0,1$ there are natural isomorphisms*

$$H^i(\mathcal{C}^\bullet(M_1^\vee \otimes M_2)) \cong \mathrm{Ext}^i_{\mathbf{A}_{K,A},\varphi,\Gamma}(M_1, M_2).$$

Proof. This is straightforward; in particular the case $i=0$ is immediate. When $i=1$, any extension M of M_1 by M_2 splits on the level of the underlying projective $\mathbf{A}_{K,A}$-modules, and such a splitting is unique up to an element of $\mathrm{Hom}_{\mathbf{A}_{K,A}}(M_1, M_2) = M_1^\vee \otimes M_2$. Given such a splitting of M, we obtain two elements f, g of $M_1^\vee \otimes M_2$ by writing

$$\varphi_M = \begin{pmatrix} \varphi_{M_2} & \varphi_{M_2} \circ f \\ 0 & \varphi_{M_1} \end{pmatrix}, \ \gamma_M = \begin{pmatrix} \gamma_{M_2} & \gamma_{M_2} \circ g \\ 0 & \gamma_{M_1} \end{pmatrix}.$$

The condition that φ_M, γ_M commute shows that (f, g) determines a class in $H^1(\mathcal{C}^\bullet(M_1^\vee \otimes M_2))$ (this class is easily seen to be well-defined, by our above remark about the ambiguity in the choice of splitting). We leave the verification that this gives the claimed bijection to the sufficiently enthusiastic reader. □

In order to show that the Herr complex is a perfect complex, we will need to develop a little of the theory of the ψ-operator on (φ, Γ)-modules. We will only need this in the case that A is an $\mathbf{F}$-algebra, and we mostly work in this context from now on. Note that for any K, if A is an $\mathbf{F}$-algebra, then $\mathbf{A}^+_{K,A}$ is (φ, Γ)-stable, and $(\mathbf{A}'_{K,A})^+$ is $(\varphi, \widetilde{\Gamma})$-stable; indeed, we have $\varphi(T) = T^p$ and $\gamma(T) - T \in T^2 \mathbf{A}^+_{K,A}$ by Lemma 3.2.18, and the stability under the action of $\widetilde{\Gamma}$ follows by an identical argument to the proof of Lemma 3.2.18 (that is, it follows from the continuity of the action of $\widetilde{\Gamma}$).

We claim that for any K, $1, \varepsilon, \ldots, \varepsilon^{p-1}$ is a basis for $\mathbf{E}'_K$ as a $\varphi(\mathbf{E}'_K)$-vector space. To see this, note that since the extension $\mathbf{E}'_K/\varphi(\mathbf{E}'_K)$ is inseparable of degree p, while $\mathbf{E}'_K/\mathbf{E}'_{\mathbf{Q}_p}$ is a separable extension, it is enough to show the result for $\mathbf{E}'_{\mathbf{Q}_p}$. In this case we have $\mathbf{E}'_{\mathbf{Q}_p} = \mathbf{F}_p((\varepsilon - 1))$, so $\varphi(\mathbf{E}'_{\mathbf{Q}_p}) = \mathbf{F}_p((\varepsilon^p - 1))$ and the claim is clear. Then for any $x \in \mathbf{E}'_K$ we can write $x = \sum_{i=0}^{p-1} \varepsilon^i \varphi(x_i)$, and we define $\psi\colon \mathbf{E}'_K \to \mathbf{E}'_K$ by $\psi(x) = x_0$. By definition ψ is continuous, $\mathbf{F}_p$-linear, commutes with $\widetilde{\Gamma}$, and satisfies $\psi \circ \varphi = \mathrm{id}$. Since ψ commutes with Δ, we have an induced map $\psi\colon \mathbf{E}_K \to \mathbf{E}_K$, which again is continuous, $\mathbf{F}_p$-linear, commutes with Γ, and satisfies $\psi \circ \varphi = \mathrm{id}$.

If A is an $\mathbf{F}_p$-algebra, then since ψ is continuous we can extend scalars from $\mathbf{F}_p$ to A and complete to obtain continuous A-linear maps $\psi\colon \mathbf{A}'_{K,A} \to \mathbf{A}'_{K,A}$ and $\psi\colon \mathbf{A}_{K,A} \to \mathbf{A}_{K,A}$. Again, these are continuous and commute with $\widetilde{\Gamma}$

and Γ, and satisfy $\psi\circ\varphi=\mathrm{id}$. We have

$$\mathbf{A}'_{K,A}=\oplus_{i=0}^{p-1}\varepsilon^i\varphi(\mathbf{A}'_{K,A}),$$

and if $x\in\mathbf{A}'_{K,A}$ and we write $x=\sum_{i=0}^{p-1}\varepsilon^i\varphi(x_i)$, then $\psi(x)=x_0$.

5.1.3 Proposition. *Let A be an $\mathbf{F}$-algebra, and let M be a projective étale (φ,Γ)-module with A-coefficients. Then there is a continuous and open A-linear surjection $\psi\colon M\to M$ such that ψ commutes with Γ, and we have*

$$\psi(\varphi(a)m)=a\psi(m),$$

$$\psi(a\varphi(m))=\psi(a)m$$

for any $a\in\mathbf{A}_{K,A}$, $m\in M$.

Proof. We define $\psi\colon M\to M$ to be the composite

$$M\stackrel{\Phi_M^{-1}}{\to}\mathbf{A}_{K,A}\otimes_{\mathbf{A}_{K,A},\varphi}M\stackrel{\psi\otimes 1}{\to}\mathbf{A}_{K,A}\otimes_{\mathbf{A}_{K,A}}M=M.$$

The relations $\psi(\varphi(a)m)=a\psi(m)$, $\psi(a\varphi(m))=\psi(a)m$ are immediate from the definitions. That ψ is continuous follows from the continuity of Φ_M^{-1} and the continuity of ψ on $\mathbf{A}_{K,A}$, while the surjectivity of ψ is immediate from the relation $\psi(\varphi(m))=m$. That ψ is open follows from Lemma 5.1.4 below. □

5.1.4 Lemma. *Let A be an $\mathbf{F}$-algebra, let M be a projective étale (φ,Γ)-module with A-coefficients, and let $\mathfrak{M}$ be a φ-stable lattice in M. Then there is an integer $h\geq 0$ such that for any integer n, we have*

$$\psi(T^{h+np}\mathfrak{M})\subseteq T^n\mathfrak{M}\subseteq\psi(T^{np}\mathfrak{M}).$$

Proof. Since $\psi\circ\varphi=\mathrm{id}$, for any n we have

$$\psi(T^{np}\varphi(\mathfrak{M}))=T^n\mathfrak{M}.$$

This implies that

$$T^n\mathfrak{M}=\psi(T^{np}\varphi(\mathfrak{M}))\subseteq\psi(T^{np}\mathfrak{M}).$$

For the other direction, choose h such that

$$T^h\mathfrak{M}\subseteq\Phi_M(\varphi^*\mathfrak{M})\subseteq\mathfrak{M}.$$

Then we have

$$\psi(T^h\mathfrak{M})\subseteq\psi(\Phi_M(\varphi^*\mathfrak{M}))=\mathfrak{M}.$$

It follows that

$$\psi(T^{h+np}\mathfrak{M}) = T^n\psi(T^h\mathfrak{M}) \subseteq T^n\mathfrak{M},$$

as required. $\square$

5.1.5 Lemma. *Let A be a Noetherian $\mathbf{F}$-algebra, and let M be a projective étale (φ,Γ)-module with A-coefficients. Then M contains a (φ,Γ)-stable lattice.*

Proof. By [EG21, Lem. 5.2.15], M contains a φ-stable lattice $\mathfrak{M}$. Let $\Gamma\mathfrak{M}$ be the $\mathbf{A}^+_{K,A}$-submodule of M generated by the elements γm with $\gamma \in \Gamma$, $m \in \mathfrak{M}$. We claim that $\Gamma\mathfrak{M}$ is a lattice; note that since Γ and φ commute, $\Gamma\mathfrak{M}$ is φ-stable, and it is Γ-stable because $\mathbf{A}^+_{K,A}$ is Γ-stable. To see that it is a lattice, it is enough (since $\mathbf{A}^+_{K,A}$ is Noetherian) to show that it is contained in a lattice (as it certainly spans M). In particular, it is enough to show that $\Gamma\mathfrak{M} \subseteq T^{-n}\mathfrak{M}$ for some $n \geq 0$. This follows easily from the compactness of Γ; for example, if $e_1, \dots, e_m$ are generators of the finitely generated $\mathbf{A}^+_{K,A}$-module $\mathfrak{M}$, then we have a continuous map $j\colon \Gamma \to M^m$, $\gamma \mapsto (\gamma e_1, \dots, \gamma e_d)$, and $\Gamma = \cup_n j^{-1}(T^{-n}\mathfrak{M}^m)$ is an open cover of Γ. Since this has a finite subcover we are done. $\square$

5.1.6 Lemma. *Let A be an $\mathbf{F}$-algebra. Then for any $i \in \mathbf{Z}$ we have $\gamma(T^i) \in T^i(\mathbf{A}^+_{K,A})^\times$.*

Proof. This follows from Lemma 3.2.18. Indeed, we can write $\gamma(T) = T + \lambda$ with $\lambda \in T^2\mathbf{A}^+_{K,A}$, so for any $i \geq 1$, we have $\gamma(T^i) = (T+\lambda)^i$, and $\gamma(T^i) - T^i \in T^{i+1}\mathbf{A}^+_{K,A}$, whence $\gamma(T^i) \in T^i(\mathbf{A}^+_{K,A})^\times$. Since $\gamma(T^i)\gamma(T^{-i}) = 1$, the result then also holds for $i \leq 0$. $\square$

5.1.7 Corollary. *Let A be a Noetherian $\mathbf{F}$-algebra, and let M be a projective étale (φ,Γ)-module with A-coefficients. Then M contains a lattice $\mathfrak{M}'$ such that for all $m \geq 0$, $T^{-m}\mathfrak{M}'$ is (ψ,Γ)-stable.*

Proof. By Lemma 5.1.5 there is a (φ,Γ)-stable lattice $\mathfrak{M}$, and by Lemma 5.1.4, the lattice $T^{-n}\mathfrak{M}$ is ψ-stable for all $n \gg 0$. It is also γ-stable for all $n \geq 0$ by Lemma 5.1.6, so we may take $\mathfrak{M}' = T^{-n}\mathfrak{M}$ for any sufficiently large value of n. $\square$

If M is an étale (φ,Γ)-module with A-coefficients, then we write

$$M' := \mathbf{A}'_{K,A} \otimes_{\mathbf{A}_{K,A}} M,$$

an étale $(\varphi,\widetilde{\Gamma})$-module with A-coefficients.

5.1.8 Lemma. *Let A be an $\mathbf{F}$-algebra, and let M be a projective étale (φ,Γ)-module with A-coefficients. Then we have a decomposition of $\varphi(\mathbf{A}'_{K,A})$-modules $M' = \ker(\psi) \oplus \varphi(M')$, and we may write $\ker(\psi) = \oplus_{i=1}^{p-1} \varepsilon^i \varphi(M')$.*

Proof. Since $\psi \circ \phi$ is the identity on M', the endomorphism $\phi \circ \psi$ of M' is idempotent, and so M' decomposes as the direct sum of its kernel and image. Since ϕ is injective, while ψ is surjective, we may express this decomposition as $M' = \ker(\psi) \oplus \varphi(M')$, as claimed.

As M' is étale, we have $\mathbf{A}'_{K,A} \otimes_{\varphi(\mathbf{A}'_{K,A})} \varphi(M') \xrightarrow{\sim} M'$. Since $\mathbf{A}'_{K,A} = \oplus_{i=0}^{p-1} \varepsilon^i \varphi(\mathbf{A}'_{K,A})$, we find that $M' = \oplus_{i=0}^{p-1} \varepsilon^i \varphi(\mathbf{A}'_{K,A}) \varphi(M') = \oplus_{i=0}^{p-1} \varepsilon^i \varphi(M')$. Now, for $1 \le i \le p-1$ we have $\psi(\varepsilon^i \varphi(M')) = \psi(\varepsilon^i) M' = 0$, so $\oplus_{i=1}^{p-1} \varepsilon^i \varphi(M') \subseteq \ker(\psi)$. Since we have $\oplus_{i=1}^{p-1} \varepsilon^i \varphi(M') \oplus \varphi(M') = \ker(\psi) \oplus \varphi(M')$, it follows that this inclusion is an equality, as required. ☐

The proof of the following result follows the approach of [CC98], although one difference in our situation is that because we are working in characteristic p, we do not need to reduce to the case that $K/\mathbf{Q}_p$ is unramified.

5.1.9 Proposition. *Let A be a Noetherian $\mathbf{F}$-algebra. If M is a projective étale (φ,Γ)-module with A-coefficients, then $(1-\gamma)$ is bijective on $\ker(\psi)$.*

Proof. We begin with some preliminaries on the action of γ on $\varepsilon \in (\mathbf{A}'_{K,A})^+$. We have $\gamma(\varepsilon) = \varepsilon^{\chi(\gamma)}$, and we write $\chi(\gamma) = 1 + p^N u$ where $N \ge 1$ and $u \in \mathbf{Z}_p^\times$. Then for each $n \ge 1$ we can write $\chi(\gamma^{p^{n-1}}) = 1 + p^{n+N-1} u_n$ with $u_n \in \mathbf{Z}_p^\times$, and if $0 \le r \le n+N-1$ and $i \in \mathbf{Z}$ then we have

$$\gamma^{p^{n-1}}(\varepsilon^i) = \varepsilon^i \varphi^r(\varepsilon^{ip^{n+N-1-r} u_n}). \tag{5.1.10}$$

Note that since u_n is a p-adic unit, $(\varepsilon^{u_n} - 1)/(\varepsilon - 1)$ is a unit in $(\mathbf{A}'_{K,A})^+$, and we can write

$$(\gamma^{p^{n-1}} - 1)(\varepsilon - 1) = \varepsilon(\varepsilon^{u_n} - 1)^{p^{n+N-1}} \in (\varepsilon - 1)^{p^{n+N-1}}((\mathbf{A}'_{K,A})^+)^\times. \tag{5.1.11}$$

It follows that for any $r \in \mathbf{Z}$, we have

$$(\gamma^{p^{n-1}} - 1)((\varepsilon - 1)^r) \in (\varepsilon - 1)^{r + p^{n+N-1} - 1}(\mathbf{A}'_{K,A})^+. \tag{5.1.12}$$

Indeed, if $r \ge 0$ then we may write

$$(\gamma^{p^{n-1}} - 1)((\varepsilon - 1)^r) = (\gamma^{p^{n-1}} - 1)((\varepsilon - 1)) \cdot \sum_{j=0}^{r-1} (\gamma^{p^{n-1}}(\varepsilon - 1)^{r-1-j}) \cdot (\varepsilon - 1)^j,$$

so (5.1.12) follows from (5.1.10) and the fact that $\gamma^{p^{n-1}}(\varepsilon-1)\in(\varepsilon-1)(\mathbf{A}'_{K,A})^+$, which in turn follows from (5.1.11). The case $r\leq 0$ then follows from the result for $-r$ by writing

$$(\gamma^{p^{n-1}}-1)((\varepsilon-1)^r)=-\frac{(\gamma^{p^{n-1}}-1)((\varepsilon-1)^{-r})}{(\gamma^{p^{n-1}}(\varepsilon-1)^{-r})(\varepsilon-1)^{-r}},$$

and noting that $(\gamma^{p^{n-1}}(\varepsilon-1))/(\varepsilon-1)=(\varepsilon^{1+p^{n+N-1}u_n}-1)/(\varepsilon-1)$ is a unit.

Since both ψ and γ commute with the action of Δ, in order to prove the proposition it suffices to show that $(1-\gamma)$ is bijective on the kernel of $\psi\colon M'\to M'$. By Lemma 5.1.8, this kernel equals $\ker(\psi)=\oplus_{i=1}^{p-1}\varepsilon^i\varphi(M')$. It follows from (5.1.10) that γ preserves $\varepsilon^i\varphi(M')$ for any integer i, so it is enough to prove that if $(i,p)=1$, then $(1-\gamma)$ acts invertibly on $\varepsilon^i\varphi(M')$.

In fact, we will prove the stronger statement that for each $n\geq 1$, each $0\leq t\leq N-1$, and each $(i,p)=1$, the operator $(1-\gamma^{p^{n-1}})$ acts invertibly on $\varepsilon^i\varphi^{n+t}(M')$ (note that $(1-\gamma^{p^{n-1}})$ acts on this space by (5.1.10)).

Writing

$$\varepsilon^i\varphi^{n+t-1}(M')=\varepsilon^i\varphi^{n+t-1}(\oplus_{j=0}^{p-1}\varepsilon^j\varphi(M'))=\oplus_{j=0}^{p-1}\varepsilon^{i+p^{n+t-1}j}\varphi^{n+t}(M'),\quad (5.1.13)$$

we see that if $t>0$ then the statement for some pair (n,t) (and all i) implies the statement for $(n,t-1)$. Writing $(1-\gamma^{p^{n-1}})=(1-\gamma^{p^{n-2}})(1+\gamma^{p^{n-2}}+\cdots+\gamma^{p^{n-2}(p-1)})$, and again using (5.1.13), we see also that the statement for (n,t) implies the statement for $(n-1,t)$. We can therefore assume that n is arbitrarily large and that $t=N-1$.

Let $\mathfrak{M}$ be a (ψ,Γ)-stable lattice in M (which exists by Corollary 5.1.7), let $\mathfrak{M}'=(\mathbf{A}'_{K,A})^+\otimes_{\mathbf{A}_{K,A}}\mathfrak{M}$, and choose n large enough that $p^{n+N-1}\geq 3$ and

$$(\gamma^{p^{n-1}}-1)(\mathfrak{M}')\subseteq(\varepsilon-1)^2\mathfrak{M}'.\quad (5.1.14)$$

It follows that

$$(\gamma^{p^{n-1}}-1)((\varepsilon-1)^r\mathfrak{M}')\subseteq(\varepsilon-1)^{r+2}\mathfrak{M}'\quad (5.1.15)$$

for all $r\in\mathbf{Z}$, because if $m\in\mathfrak{M}'$ then

$$\begin{aligned}(\gamma^{p^{n-1}}-1)((\varepsilon-1)^r m)&=(\gamma^{p^{n-1}}-1)((\varepsilon-1)^r)\cdot\gamma^{p^{n-1}}(m)\\&+(\varepsilon-1)^r(\gamma^{p^{n-1}}-1)(m),\end{aligned}$$

so that (5.1.15) follows from (5.1.12) and (5.1.14), together with our assumption that $p^{n+N-1}\geq 3$.

By (5.1.10), we have

$$(\gamma^{p^{n-1}}-1)(\varepsilon^i\varphi^{n+N-1}(x))=\varepsilon^i\varphi^{n+N-1}(\varepsilon^{iu_n}\gamma^{p^{n-1}}(x)-x),$$

so it is enough to check that the map $f\colon M' \to M'$ given by

$$f(x) = \varepsilon^{iu_n}\gamma^{p^{n-1}}(x) - x$$

is invertible.

Let $\alpha = \varepsilon^{iu_n} - 1$, so that $\alpha/(\varepsilon - 1)$ is a unit in $(\mathbf{A}'_{K,A})^+$. Then for any $r \in \mathbf{Z}$ and $x \in (\varepsilon - 1)^r\mathfrak{M}'$, it follows from (5.1.15) that

$$\left(\frac{1}{\alpha}f - 1\right)(x) = \frac{\varepsilon^{iu_n}}{\alpha}(\gamma^{p^{n-1}} - 1)(x) \in (\varepsilon - 1)^{r+1}\mathfrak{M}'.$$

In particular the sum

$$g(x) := \sum_{j=0}^{\infty}\left(1 - \frac{1}{\alpha}f\right)^j(x)$$

converges. Since f, g are additive by definition, we have

$$\left(1 - \frac{1}{\alpha}f\right) \circ g(x) = g \circ \left(1 - \frac{1}{\alpha}f\right)(x) = g(x) - x,$$

so that the function $g\colon M \to M$ is a left and right inverse to $\frac{1}{\alpha}f$. Thus f is invertible, as required. □

Suppose that A is an $\mathbf{F}$-algebra. Using the ψ operator, we can give an alternative description of the Herr complex, which will be important in establishing that it is a perfect complex. Let $\mathcal{C}^\bullet_\psi(M)$ be the complex in degrees $0, 1, 2$ given by

$$M \xrightarrow{(\psi-1,\gamma-1)} M \oplus M \xrightarrow{(\gamma-1)\oplus(1-\psi)} M.$$

We have a morphism of complexes $\mathcal{C}^\bullet(M) \to \mathcal{C}^\bullet_\psi(M)$ given by

$$\begin{array}{ccccc} M & \xrightarrow{(\varphi-1,\gamma-1)} & M \oplus M & \xrightarrow{(\gamma-1)\oplus(1-\varphi)} & M \\ \downarrow{\scriptstyle 1} & & \downarrow{\scriptstyle (-\psi,1)} & & \downarrow{\scriptstyle -\psi} \\ M & \xrightarrow{(\psi-1,\gamma-1)} & M \oplus M & \xrightarrow{(\gamma-1)\oplus(1-\psi)} & M \end{array} \tag{5.1.16}$$

(That this is a morphism of complexes follows from the facts that $\psi \circ \varphi = \mathrm{id}$, and that ψ commutes with γ.)

5.1.17 Proposition. *Let A be a Noetherian $\mathbf{F}$-algebra, and let M be a projective étale (φ, Γ)-module with A-coefficients. Then the morphism of complexes $\mathcal{C}^\bullet(M) \to \mathcal{C}^\bullet_\psi(M)$ defined by (5.1.16) is a quasi-isomorphism.*

Proof. The cokernel of (5.1.16) is the zero complex, while its kernel is the complex

$$0 \longrightarrow \ker(\psi) \xrightarrow{(\gamma-1)} \ker(\psi),$$

which is acyclic by Proposition 5.1.9. □

In fact, we require a slightly stronger statement than the quasi-isomorphism of the preceding lemma. Namely, we need a statement that takes into account topologies, which will make use of the following lemma.

5.1.18 Lemma. *If $f: X \to Y$ is a continuous open map of topological spaces, and if $Y' \subseteq Y$ is the inclusion of a subspace into Y (i.e., Y' is a subset of Y endowed with the induced topology), then the base-changed map $X' := f^{-1}(Y') \to Y'$ is again open when X' is endowed with the topology induced by that of X.*

Proof. Let U' be an open subset of X'; then we may write $U' = U \cap X'$, for some open subset U of X. Thus

$$f(U') = f(U \cap X') = f\big(U \cap f^{-1}(Y')\big) = f(U) \cap Y'.$$

Since f is open, we conclude that $f(U')$ is an open subset of Y', as required. □

We now have the following strengthening of Proposition 5.1.17, which takes into account the topologies on the complexes.

5.1.19 Proposition. *Let A be a countable Noetherian $\mathbf{F}$-algebra and let M be a projective étale (φ, Γ)-module with A-coefficients. Then the morphism of complexes $\mathcal{C}^\bullet(M) \to \mathcal{C}^\bullet_\psi(M)$ defined by* (5.1.16) *induces topological isomorphisms on each of the associated cohomology modules.*

Proof. Proposition 5.1.17 shows that (5.1.16) is a quasi-isomorphism. By definition, the induced map on H^0 is the identity map, and is therefore a homeomorphism. Since ψ is continuous and open, we see that the maps $\mathcal{C}^i(M) \to \mathcal{C}^i_\psi(M)$ are continuous and open for each i; in particular, the isomorphism on H^2 is induced from the continuous open map $-\psi: \mathcal{C}^2(M) \to \mathcal{C}^2_\psi(M)$, and is therefore a homeomorphism. (Here and below we use the standard fact that a quotient morphism of topological groups is necessarily open.)

The case of H^1 requires a little more work. Since the maps $\mathcal{C}^0(M) \to \mathcal{C}^0_\psi(M)$ and $H^1(\mathcal{C}^\bullet(M)) \to H^1(\mathcal{C}^\bullet_\psi(M))$ are isomorphisms, we see that the continuous morphism $\mathcal{C}^1(M) \to \mathcal{C}^1_\psi(M)$ induces a continuous isomorphism of the modules of cocycles $Z^1(M) \to Z^1_\psi(M)$, and it suffices to show that this map is open (since this will imply that the induced bijection on H^1 is both continuous and open, and thus is an isomorphism of topological groups).

If we let $\widetilde{Z}^1(M)$ denote the preimage of $Z^1_\psi(M)$ in $\mathcal{C}^1(M)$, and if we let $K \subseteq \mathcal{C}^1(M) = M \oplus M$ denote $\ker(\psi) \oplus 0$, then we have inclusions $Z^1(M), K \subseteq \widetilde{Z}^1(M)$, and hence a continuous morphism

$$Z^1(M) \oplus K \to \widetilde{Z}^1(M), \tag{5.1.20}$$

which is in fact a bijection. Each of $Z^1(M)$, K, and $\widetilde{Z}^1(M)$ is a closed subgroup of $\mathcal{C}^1(M) = M \oplus M$. Lemma D.3 shows that this latter topological group is Polish, and thus so is any of its closed subgroups. Corollary C.3 then shows that (5.1.20) is in fact a homeomorphism, while Lemma 5.1.18 shows that the morphism $\widetilde{Z}^1(M) \to Z^1_\psi(M)$ is open. Consequently, we find that the morphism $Z^1(M) \to Z^1_\psi(M)$ is open, as required. □

5.1.21 Lemma. *Let A be a Noetherian $\mathbf{F}$-algebra, and let X be an A-module subquotient of a finitely generated projective $A[[T]]$-module, endowed with its natural subquotient topology. If the topology on X is discrete, then X is finitely generated as an A-module.*

Proof. Write $X = Y/Z$, where Y is an A-submodule of a finitely generated projective $A[[T]]$-module $\mathfrak{M}$. Since the topology on X is discrete, Z is open in Y, and thus $Z \supseteq U \cap Y$, for some open neighborhood U of zero in $\mathfrak{M}$. Such a neighborhood U contains $T^n\mathfrak{M}$ for some $n \gg 0$, and so we find that X is a subquotient of the finitely generated A-module $\mathfrak{M}/T^n\mathfrak{M}$, as required. □

5.1.22 Theorem. *Let A be a Noetherian $\mathcal{O}/\varpi^a$-algebra such that A/ϖ is countable, and let M be a projective étale (φ, Γ)-module with A-coefficients.*

1. *The Herr complex $\mathcal{C}^\bullet(M)$ is a perfect complex of A-modules, concentrated in degrees $[0,2]$.*
2. *If either (i) B is a finite A-algebra; or (ii) B is a finite type A-algebra and A itself is of finite type over $\mathcal{O}/\varpi^a$; then there is a natural isomorphism in the derived category*

$$\mathcal{C}^\bullet(M) \otimes^{\mathbf{L}}_A B \xrightarrow{\sim} \mathcal{C}^\bullet(M \otimes_{\mathbf{A}_{K,A}} \mathbf{A}_{K,B}).$$

In particular, there is a natural isomorphism

$$H^2(\mathcal{C}^\bullet(M)) \otimes_A B \xrightarrow{\sim} H^2(\mathcal{C}^\bullet(M \otimes_{\mathbf{A}_{K,A}} \mathbf{A}_{K,B})).$$

Proof. Since A is Noetherian, $\mathbf{A}_{K,A}$ is a flat A-algebra (being a localization of the power series ring $\mathbf{A}^+_{K,A}$), so $\mathcal{C}^\bullet(M)$ is a complex of flat A-modules. By [Sta, Tag 07LU] (for part (1), taking $R = \mathcal{O}/\varpi^a$) and [Pil20, Prop. 2.2.2] (for part (2), again taking $R = \mathcal{O}/\varpi^a$), it suffices to prove the result in the case that A is an $\mathbf{F}$-algebra, which we assume from now on. We begin with (1). By Lemma 5.1.1, we need only check that the cohomology groups of $\mathcal{C}^\bullet(M)$ are finitely generated A-modules. In order to do this, we will make two truncation arguments.

Firstly, by Lemma 5.1.5, we can choose a (φ,Γ)-stable lattice $\mathfrak{M} \subseteq M$. We claim that for every $n \geq 1$ the Herr complex $\mathcal{C}^\bullet(T^n\mathfrak{M})$ is acyclic; consequently, the natural morphism of complexes $\mathcal{C}^\bullet(M) \to \mathcal{C}^\bullet(M/T^n\mathfrak{M})$ is a quasi-isomorphism.

To see the claim, it suffices to show that $(1-\varphi)\colon T^n\mathfrak{M} \to T^n\mathfrak{M}$ is an isomorphism; indeed, the exactness of $\mathcal{C}^\bullet(T^n\mathfrak{M})$ is a formal consequence of this. We begin by checking injectivity. If $m \in T^n\mathfrak{M}$ and $(1-\varphi)(m)=0$, then we have $m=\varphi(m) \in T^{pn}\mathfrak{M}$ so that in particular we have $m \in T^{n+1}\mathfrak{M}$. By induction, we see that $m \in T^n\mathfrak{M}$ for all n, so that $m=0$, as required.

We now prove surjectivity. If $m \in T^n\mathfrak{M}$, then we have seen in the previous paragraph that $\varphi(m) \in T^{n+1}\mathfrak{M}$, $\varphi^2(m) \in T^{n+2}\mathfrak{M}$, and so on. Since $\mathfrak{M}$ is T-adically complete, we can set $x=\sum_{i\geq 0}\varphi^i(m) \in T^n\mathfrak{M}$; then $(1-\varphi)(x)=m$, as required.

We now turn to our other truncation argument. Let $\mathfrak{M}'$ be as in Lemma 5.1.7, so that for each $n' \leq 0$, $T^{n'}\mathfrak{M}'$ is a (ψ,Γ)-stable lattice in M. In particular, for each $n' \leq 0$, $\mathcal{C}_\psi^\bullet(T^{n'}\mathfrak{M})$ is a subcomplex of $\mathcal{C}_\psi^\bullet(M)$. We claim that if $n' \leq -2$ then $\mathcal{C}_\psi^\bullet(M/T^{n'}\mathfrak{M}')$ is acyclic, so that the natural morphism of complexes $\mathcal{C}_\psi^\bullet(T^{n'}\mathfrak{M}') \to \mathcal{C}_\psi^\bullet(M)$ is a quasi-isomorphism.

As above, it is enough to show that $(1-\psi)$ is bijective on $M/T^{n'}\mathfrak{M}'$. Note firstly that if $r \leq n'$ then $\psi(T^r\mathfrak{M}') \subseteq T^{r+1}\mathfrak{M}'$. Indeed, for each integer $r \leq 0$ we have

$$\psi(T^r\mathfrak{M}') \subseteq \psi(T^{p\lfloor r/p\rfloor}\mathfrak{M}') = T^{\lfloor r/p\rfloor}\psi(\mathfrak{M}') \subseteq T^{\lfloor r/p\rfloor}\mathfrak{M}',$$

and the claim follows since if $r \leq -2$ then $\lfloor r/p\rfloor \geq r+1$.

We now show that $(1-\psi)$ is injective on $M/T^{-n}\mathfrak{M}'$. Suppose that $m \in M$ with $(1-\psi)(m) \in T^{n'}\mathfrak{M}'$. Then $m \in T^r\mathfrak{M}'$ for some $r \ll 0$. If $r \geq n'$ then we are done, and if not then since $\psi(m) \in T^{r+1}\mathfrak{M}'$ and $(1-\psi)(m) \in T^{n'}\mathfrak{M} \subseteq T^{r+1}\mathfrak{M}$, we have $m \in T^{r+1}\mathfrak{M}$. It follows by induction that $m \in T^{n'}\mathfrak{M}'$, as required.

For surjectivity, take $m \in M$, and choose $r \ll 0$ such that $m \in T^r\mathfrak{M}'$. If $r < n'$ then $\psi(m) \in T^{r+1}\mathfrak{M}'$, so by induction we see that there is some $s \geq 0$ such that $\psi^s(m) \in T^{n'}\mathfrak{M}'$. If we set $x=\sum_{i=0}^{s-1}\psi^i(m)$, then $(1-\psi)(x)=m-\psi^s(m) \in m+T^{n'}\mathfrak{M}$, as required.

We now consider the quasi-isomorphisms

$$\mathcal{C}_\psi^\bullet(T^{n'}\mathfrak{M}') \to \mathcal{C}_\psi^\bullet(M) \leftarrow \mathcal{C}^\bullet(M) \to \mathcal{C}^\bullet(M/T^n\mathfrak{M}).$$

Proposition 5.1.19 shows that the middle quasi-isomorphism induces a topological isomorphism on cohomology modules, and thus altogether we obtain morphisms

$$H^i(\mathcal{C}_\psi^\bullet(T^{n'}\mathfrak{M}')) \to H^i(\mathcal{C}^\bullet(M/T^n\mathfrak{M}))$$

which are isomorphisms of A-modules, and continuous with respect to the natural topologies on the source and target. Since the target is endowed with the discrete topology, we find that the source is also endowed with the discrete topology. By Lemma 5.1.21, it follows that these cohomology modules are indeed finitely generated over A, and hence so are the cohomology modules of $\mathcal{C}^\bullet(M)$, as required.

We now turn to (2), where we follow the proof of [KPX14, Thm. 4.4.3]. Since both (i) and (ii) in particular require B to be a finite type A-algebra, we see that B is Noetherian, and B is countable, so that in particular $\mathcal{C}^\bullet(M \otimes_{\mathbf{A}_{K,A}} \mathbf{A}_{K,B})$ is a perfect complex of B-modules by part (1) (replacing A by B, and M by $M \otimes_{\mathbf{A}_{K,A}} \mathbf{A}_{K,B}$). Since $\mathcal{C}^\bullet(M)$ is a perfect complex of A-modules, $\mathcal{C}^\bullet(M) \otimes_A^{\mathbf{L}} B$ is also a perfect complex of B-modules, and the natural map $\mathbf{A}_{K,A} \otimes_A B \to \mathbf{A}_{K,B}$ induces a morphism

$$\mathcal{C}^\bullet(M) \otimes_A^{\mathbf{L}} B \to \mathcal{C}^\bullet(M \otimes_{\mathbf{A}_{K,A}} \mathbf{A}_{K,B}). \tag{5.1.23}$$

In case (i), when B is in fact finite as an A-module, the natural map $\mathbf{A}_{K,A} \otimes_A B \to \mathbf{A}_{K,B}$ is an isomorphism (indeed, we have $A[[T]] \otimes_A B = B[[T]]$, because $A[[T]] \otimes_A B$ is finitely generated over $A[[T]]$ and thus T-adically complete and separated by the Artin–Rees lemma), so (5.1.23) is certainly an isomorphism in this case.

We now turn to case (ii), when B is only assumed to be a finite type A-algebra, but A, and hence also B, is furthermore of finite type over $\mathcal{O}/\varpi^a$. This implies in particular that A is Jacobson, so that if $\mathfrak{m}$ is any maximal ideal of B, then $B/\mathfrak{m}$ is finite as an A-module. From this, and the already-proved case (i), we conclude that (5.1.23) is an isomorphism if we replace B by $B/\mathfrak{m}$.

It follows that we have a chain of quasi-isomorphisms

$$\begin{aligned}(\mathcal{C}^\bullet(M) \otimes_A^{\mathbf{L}} B) \otimes_B^{\mathbf{L}} B/\mathfrak{m} &\xrightarrow{\sim} \mathcal{C}^\bullet(M) \otimes_A^{\mathbf{L}} B/\mathfrak{m} \xrightarrow{\sim} \mathcal{C}^\bullet(M \otimes_A B/\mathfrak{m}) \\ &\xrightarrow{\sim} \mathcal{C}^\bullet(M \otimes_{\mathbf{A}_{K,A}} \mathbf{A}_{K,B}) \otimes_B^{\mathbf{L}} B/\mathfrak{m},\end{aligned}$$

where the final quasi-isomorphism comes from replacing A by B, B by $B/\mathfrak{m}$, and M by $M \otimes_{\mathbf{A}_{K,A}} \mathbf{A}_{K,B}$ in (5.1.23). (That this is indeed a quasi-isomorphism follows by case (i), since $B/\mathfrak{m}$ is a finite B-module, while we have $(M \otimes_{\mathbf{A}_{K,A}} \mathbf{A}_{K,B}) \otimes_{\mathbf{A}_{K,B}} \mathbf{A}_{K,B/\mathfrak{m}} = M \otimes_{\mathbf{A}_{K,A}} \mathbf{A}_{K,B/\mathfrak{m}} = M \otimes_{\mathbf{A}_{K,A}} (\mathbf{A}_{K,A} \otimes_A B/\mathfrak{m}) = M \otimes_A B/\mathfrak{m}$).

Thus (5.1.23) becomes a quasi-isomorphism after applying $\otimes_B^{\mathbf{L}} B/\mathfrak{m}$, for any maximal ideal $\mathfrak{m}$ of B. It follows from [KPX14, Lem. 4.1.5] that (5.1.23) is a quasi-isomorphism, as required. Finally, the compatibility of the formation of H^2 with base change is an immediate consequence of this isomorphism in the derived category, together with the vanishing of all of the higher degree cohomology groups. □

5.1.24 Corollary. *Let A be a p-adically complete Noetherian $\mathcal{O}$-algebra such that A/ϖ is countable, and let M be a projective étale (φ,Γ)-module. Then the Herr complex $\mathcal{C}^\bullet(M)$ is a perfect complex concentrated in degrees $[0,2]$.*

Proof. This follows from [Sta, Tag 0CQG] and Theorem 5.1.22. (We apply the base change statement of part (2) of the theorem to the surjective (and hence finite) morphisms $A/\varpi^{a+1} \to A/\varpi^a$.) □

We also note the following rather technical corollary Theorem 5.1.22, which we will use in our discussion of families of extensions below.

5.1.25 Corollary. *If A is of finite type over $\mathcal{O}/\varpi^a$ for some $a \geq 1$, and M is a projective étale (φ,Γ)-module, then the Herr complex $\mathcal{C}^\bullet(M)$ can be represented by a complex $C^0 \to C^1 \to C^2$ of finite projective A-modules in degrees $[0,2]$.*

Proof. Theorem 5.1.22 shows that $\mathcal{C}^\bullet(M)$ is perfect, and so by [Sta, Tag 0658], the present corollary will follow provided that we show that $\mathcal{C}^\bullet(M)$ has tor-amplitude in [0,2]. Since any A-module is a filtered direct limit of finite A-modules, it suffices to show that $\mathcal{C}^\bullet(M)\otimes^{\mathbf{L}}{}_A M$ has cohomological amplitude lying in $[0,2]$ for any finite A-module M. If we let $B = A \oplus M$, thought of as an A-algebra by declaring M to be a square zero ideal, then it is equivalent to show that $\mathcal{C}^\bullet(M)\otimes^{\mathbf{L}}{}_A B$ has cohomological amplitude lying in $[0,2]$. But Theorem 5.1.22 shows that this latter complex is isomorphic (in the derived category of B-modules) to $\mathcal{C}^\bullet(M \otimes_{\mathbf{A}_{K,A}} \mathbf{A}_{K,B})$, whose cohomological amplitude lies in [0,2] by construction. □

We now explain the comparison between the Herr complex and Galois cohomology, and the relationship to Tate local duality. These results are essentially due to Herr and are proved in [Her98, Her01] but we follow [KPX14] and formulate them as statements in the derived category.

Exactly as in [KPX14, Defn. 2.3.10], if M_1, M_2 are projective étale (φ,Γ)-modules with A-coefficients, then there is a cup product

$$\mathcal{C}^\bullet(M_1) \otimes_A \mathcal{C}^\bullet(M_2) \to \mathcal{C}^\bullet(M_1 \otimes_{\mathbf{A}_{K,A}} M_2).$$

More precisely, the cup product arises from the following generalities. If we have two complexes of A-modules $C^\bullet$ and $D^\bullet$, then the tensor product $C^\bullet \otimes D^\bullet$ has differential given by

$$d(x \otimes y) = dx \otimes y + (-1)^i x \otimes y$$

if $x \in C^i$, $y \in D^j$. If $f^\bullet : C^\bullet \to C^\bullet$, then we write $\mathrm{Fib}(f|C^\bullet) := \mathrm{Cone}(f)[-1]$, which by definition has $\mathrm{Fib}(f)^i = C^i \oplus C^{i-1}$ and $d^i((x,y)) = (d^i(x), -d^{i-1}(y) - f^i(x))$.

As a special case of [KPX14, Lem. 2.3.9], if $f_1 \colon C_1^\bullet \to C_1^\bullet$, $f_2 \colon C_2^\bullet \to C_2^\bullet$, then we have a natural morphism

$$\mathrm{Fib}(1-f_1|C_1^\bullet) \otimes \mathrm{Fib}(1-f_2|C_2^\bullet) \to \mathrm{Fib}(1-f_1 \otimes f_2|C_1^\bullet \otimes C_2^\bullet). \tag{5.1.26}$$

Then by definition we have $\mathcal{C}^\bullet(M) = \mathrm{Fib}\big(1-\gamma|\,\mathrm{Fib}(1-\varphi|M)\big)$, so that by multiple applications of (5.1.26) we have morphisms

$$\begin{aligned}\mathcal{C}^\bullet(M_1) \otimes_A \mathcal{C}^\bullet(M_2) &= \mathrm{Fib}\big(1-\gamma|\,\mathrm{Fib}(1-\varphi|M_1)\big) \otimes_A \mathrm{Fib}\big(1-\gamma|\,\mathrm{Fib}(1-\varphi|M_2)\big)\\ &\to \mathrm{Fib}\Big(1-\gamma|\big(\mathrm{Fib}(1-\varphi|M_1) \otimes_A \mathrm{Fib}(1-\varphi|M_2)\big)\Big)\\ &\to \mathrm{Fib}\Big(1-\gamma|\,\mathrm{Fib}\big(1-\varphi|(M_1 \otimes_A M_2)\big)\Big)\\ &\twoheadrightarrow \mathrm{Fib}\Big(1-\gamma|\,\mathrm{Fib}\big(1-\varphi|(M_1 \otimes_{\mathbf{A}_{K,A}} M_2)\big)\Big)\\ &= \mathcal{C}^\bullet(M_1 \otimes M_2)\end{aligned}$$

whose composite defines the cup product.

5.1.27 Lemma. *If A is a finite type $\mathcal{O}/\varpi^a$-algebra, then there is an isomorphism*

$$H^2(\mathcal{C}^\bullet(\mathbf{A}_{K,A}(1))) \cong A,$$

compatible with base change.

Proof. Since by Theorem 5.1.22 (2) the formation of $H^2(\mathcal{C}^\bullet(M))$ is compatible with base change, the result follows from the case $A = \mathcal{O}/\varpi^a$, which is immediate from Theorem 5.1.29 below (that is, from the natural isomorphism $H^2\big(G_K, (\mathcal{O}/\varpi^a)(1)\big) \cong \mathcal{O}/\varpi^a$). □

If M is a projective étale (φ,Γ)-module over a finite type $\mathcal{O}/\varpi^a$-algebra, then we define the Tate duality pairing between the Herr complexes of M and of its Cartier dual M^* as the following composite of the cup product, truncation, and the isomorphism of Lemma 5.1.27:

$$\mathcal{C}^\bullet(M) \times \mathcal{C}^\bullet(M^*) \to \mathcal{C}^\bullet(\mathbf{A}_{K,A}(1))) \to H^2(\mathcal{C}^\bullet(\mathbf{A}_{K,A}(1)))[-2] \cong A[-2].$$

5.1.28 Proposition. *Let A be a finite type $\mathcal{O}/\varpi^a$-algebra, and let M be a projective étale (φ,Γ)-module.*

1. *The Tate duality pairing induces a quasi-isomorphism*

$$\mathcal{C}^\bullet(M) \xrightarrow{\sim} \mathrm{RHom}_A(\mathcal{C}^\bullet(M^*), A))[-2].$$

2. *If A is a finite extension of $\mathbf{F}$, then the Euler characteristic $\chi_A(\mathcal{C}^\bullet(M))$ is equal to $-[K:\mathbf{Q}_p]d$.*

Proof. For part (1), by [KPX14, Lem. 4.1.5], it is enough to treat the case that A is a field, in which case A is a finite extension of $\mathbf{F}$. Then both parts follow from Theorem 5.1.29 below (that is, from the corresponding statements for Galois representations). □

Finally we recall the relationship between the Herr complex and Galois cohomology. As in [Pot13, §2] it is possible to upgrade the following theorem to an isomorphism in the derived category, but as we do not need this we do not give the details here. If A is a complete local Noetherian $\mathcal{O}$-algebra with finite residue field, and M is a formal projective étale (φ,Γ)-module with A-coefficients, then we can define the Herr complex $\mathcal{C}^\bullet(M)$ exactly as for étale (φ,Γ)-modules, so that by definition we have $\mathcal{C}^\bullet(M) = \varprojlim_n \mathcal{C}^\bullet(M_{A/\mathfrak{m}_A^n})$. By [Sta, Tag 0CQG], $\mathcal{C}^\bullet(M)$ is a perfect complex, concentrated in degrees $[0,2]$.

5.1.29 Theorem. *If A is a complete local Noetherian $\mathcal{O}$-algebra with finite residue field, and T is a finitely generated projective A-module with a continuous action of G_K, then there are isomorphisms of A-modules*

$$H^i(G_K,T) \xrightarrow{\sim} H^i(\mathcal{C}^\bullet(\mathbf{D}_A(T)))$$

which are functorial in T and compatible with cup products and duality.

Proof. This is [Dee01, Prop. 3.1.1], which is deduced from the results of [Her01, Her98] by passage to the limit. □

5.1.30 Remark. In accordance with our general convention of writing ρ_T for a family $T \to \mathcal{X}_d$, if T is an affine scheme of finite type over $\mathcal{O}/\varpi^a$, then we write $H^2(G_K,\rho_T)$ for the pull-back to T of the cohomology group $H^2(\mathcal{C}^\bullet(M))$ on $\mathcal{X}_d$, where M is the étale (φ,Γ)-module corresponding to the morphism $T \to \mathcal{X}_d$. By Theorem 5.1.22, $H^2(G_K,\rho_T)$ is a coherent sheaf, whose formation is compatible with arbitrary finite type base change, and so in particular by Theorem 5.1.29 its specializations at $\overline{\mathbf{F}}_p$-points coincide with the usual Galois cohomology groups. This compatibility with base change also allows us to extend the definition of $H^2(G_K,\rho_T)$ to arbitrary (not necessarily affine) schemes T of finite type over $\mathcal{O}/\varpi^a$.

5.1.31 Lemma. *Let T be a scheme of finite type over $\mathcal{O}/\varpi^a$, and let ρ_T, ρ'_T be families of Galois representations over T. Then sections of $H^2(G_K,\rho_T \otimes (\rho'_T)^\vee(1))$ are in natural bijection with homomorphisms $\rho_T \to \rho'_T$.*

5.1.32 Remark. Here, by a homomorphism $\rho_T \to \rho'_T$ we mean a homomorphism of the corresponding (φ, Γ)-modules.

Proof of Lemma 5.1.31. This follows from Proposition 5.1.28 and Lemma 5.1.2. □

5.1.33 Obstruction theory

We now show that the Herr complex provides $\mathcal{X}_d$ with a nice obstruction theory in the sense of Definition A.34.

Let A be a p-adically complete $\mathcal{O}$-algebra, and let $x\colon \operatorname{Spec} A \to \mathcal{X}_d$ be a morphism, corresponding to a projective (φ, Γ)-module M. We wish to consider the problem of deforming M to a square zero thickening of A. Specifically, if

$$0 \to I \to A' \to A \to 0$$

is a square zero extension, then we let $\operatorname{Lift}(x, A')$ be the set of isomorphism classes of projective (φ, Γ)-modules M' with A'-coefficients which have the property that $M' \otimes_{A'} A \cong M$.

For any such thickening A', we define a corresponding obstruction class as follows. The underlying $\mathbf{A}_{K,A}$-module of M has a unique (up to isomorphism) lifting to a projective $\mathbf{A}_{K,A'}$-module $\widetilde{M}$, and we may lift φ, γ to semi-linear endomorphisms $\widetilde{\varphi}, \widetilde{\gamma}$ of $\widetilde{M}$. (Indeed, $\mathbf{A}_{K,A'}$ is a square zero thickening of $\mathbf{A}_{K,A}$, and a finite projective module P over any ring R deforms uniquely through any square zero extension $R' \to R$, as its deformations are controlled by

$$H^1(\operatorname{Spec} R, \ker(R' \to R) \otimes \operatorname{End}_R(P)),$$

which vanishes (as $\operatorname{Spec} R$ is affine). To see that φ, γ lift, think of them as $\mathbf{A}_{K,A}$-linear maps $\varphi^* M \to M$, $\gamma^* M \to M$.)

However, there is no guarantee that we can find lifts $\widetilde{\varphi}, \widetilde{\gamma}$ which commute. To measure the obstruction to the existence of such lifts, let $\operatorname{ad} M = \operatorname{Hom}_{\mathbf{A}_{K,A}}(M, M)$ be the adjoint of M. This naturally has the structure of a (φ, Γ)-module; indeed, we have a natural identification $\varphi^* \operatorname{ad} M = \operatorname{Hom}_{\mathbf{A}_{K,A}}(\varphi^* M, \varphi^* M)$, and we define $\Phi_{\operatorname{ad} M}\colon \varphi^* \operatorname{ad} M \to \operatorname{ad} M$ by

$$(\Phi_{\operatorname{ad} M}(f))(x) = \Phi_M(f(\Phi_M^{-1}(x))).$$

We define the action of Γ in the analogous way. Then we let $o_x(A')$ be the image in $H^2(\mathcal{C}^\bullet(\operatorname{ad} M)) \otimes_A I = H^2(\mathcal{C}^\bullet(\operatorname{ad} M \otimes_A I))$ of

$$\widetilde{\varphi}\widetilde{\gamma}\widetilde{\varphi}^{-1}\widetilde{\gamma}^{-1} - 1 \in \operatorname{ad} M \otimes_A I = \mathcal{C}^2(\operatorname{ad} M \otimes_A I).$$

5.1.34 Lemma. *The cohomology class $o_x(A')$ is well-defined independently of the choice of $\widetilde{\varphi}, \widetilde{\gamma}$, and vanishes if and only if* $\operatorname{Lift}(x, A') \neq 0$.

Proof. Any other liftings $\widetilde{\varphi}', \widetilde{\gamma}'$ are obtained from our given liftings $\widetilde{\varphi}, \widetilde{\gamma}$ by setting $\widetilde{\varphi}' = (1+X)\widetilde{\varphi}$, $\widetilde{\gamma}' = (1-Y)\widetilde{\gamma}$, for some $X, Y \in \operatorname{ad} M \otimes_A I$. A simple computation shows that

$$\widetilde{\varphi}'\widetilde{\gamma}'(\widetilde{\varphi}')^{-1}(\widetilde{\gamma}')^{-1} - \widetilde{\varphi}\widetilde{\gamma}\widetilde{\varphi}^{-1}\widetilde{\gamma}^{-1} = (1-\gamma)X + (1-\varphi)Y,$$

which shows that the cohomology class $o_x(A')$ is well-defined, and that it vanishes if and only if we can choose $\widetilde{\varphi}', \widetilde{\gamma}'$ so that $\widetilde{\varphi}'\widetilde{\gamma}'(\widetilde{\varphi}')^{-1}(\widetilde{\gamma}')^{-1} = 1$, which is in turn equivalent to $\operatorname{Lift}(x, A') \neq 0$, as required. □

If F is an A-module, we let $A[F] := A \oplus F$ be the A-algebra with multiplication given by

$$(a, m)(a', m') := (aa', am' + a'm).$$

This is a square zero thickening of A, and $\operatorname{Lift}(x, A[F]) \neq 0$, because we have the trivial lifting given by $M \otimes_A A[F]$.

5.1.35 Lemma. *Suppose that F is a finitely generated A-module. Then there is a natural isomorphism of A-modules* $\operatorname{Lift}(x, A[F]) \xrightarrow{\sim} H^1(\mathcal{C}^\bullet(\operatorname{ad} M) \otimes_A^{\mathbf{L}} F)$.

Proof. We begin by constructing the isomorphism on the level of sets. Note that $\mathcal{C}^\bullet(\operatorname{ad} M) \otimes^{\mathbf{L}} F$ is computed by $\mathcal{C}^\bullet(\operatorname{ad} M \otimes_A F)$. Liftings of M to $A[F]$ are determined by the corresponding liftings $\widetilde{\varphi}, \widetilde{\gamma}$ of φ, γ, and we obtain a class in $H^1(\mathcal{C}^\bullet(\operatorname{ad} M \otimes_A F))$ by taking the image of

$$(\widetilde{\varphi}\varphi^{-1} - 1, \widetilde{\gamma}\gamma^{-1} - 1) \in \operatorname{ad} M \otimes_A F \oplus \operatorname{ad} M \otimes_A F = \mathcal{C}^1(\operatorname{ad} M \otimes_A F).$$

An elementary calculation shows that $(\widetilde{\varphi}\varphi^{-1} - 1, \widetilde{\gamma}\gamma^{-1} - 1)$ is in the kernel of $\mathcal{C}^1(\operatorname{ad} M \otimes_A F) \to \mathcal{C}^2(\operatorname{ad} M \otimes_A F)$ if and only if $\widetilde{\varphi}\widetilde{\gamma} = \widetilde{\gamma}\widetilde{\varphi}$, so it only remains to check that the lifting given by $\widetilde{\varphi}, \widetilde{\gamma}$ is trivial if and only if the corresponding cohomology class vanishes.

To see this, note that the endomorphisms of the trivial lifting are of the form $1 + X$, for some $X \in \operatorname{ad} M \otimes F$. The corresponding $\widetilde{\varphi}, \widetilde{\gamma}$ are given by $\widetilde{\varphi} = (1+X)\varphi(1+X)^{-1}$, $\widetilde{\gamma} = (1+X)\gamma(1+X)^{-1}$, which is equivalent to $(\widetilde{\varphi}\varphi^{-1} - 1, \widetilde{\gamma}\gamma^{-1} - 1) = ((1-\varphi)X, (1-\gamma)X)$, which by definition is equivalent to $(\widetilde{\varphi}\varphi^{-1} - 1, \widetilde{\gamma}\gamma^{-1} - 1)$ being a coboundary, as required.

To compare the A-module structures, recall that by definition the A-module structure on $\operatorname{Lift}(x, A[F])$ is defined as follows (see [Sta, Tag 07Y9]). If $r \in A$, then we have a homomorphism $f_r \colon A[F] \to A[F]$ given by $f_r(a, f) = (a, rf)$, and given a lifting $\widetilde{M}$ of M, we let $r\widetilde{M}$ be the base change of $\widetilde{M}$ via f_r.

Explicitly, this means that we replace $(\widetilde{\varphi}-\varphi)$ and $(\widetilde{\gamma}-\gamma)$ by $r(\widetilde{\varphi}-\varphi)$ and $r(\widetilde{\gamma}-\gamma)$.

Similarly, the addition map $\mathrm{Lift}(x,A[F])\times\mathrm{Lift}(x,A[F])\to\mathrm{Lift}(x,A[F])$ comes from the obvious identification $\mathrm{Lift}(x,A[F])\times\mathrm{Lift}(x,A[F])=\mathrm{Lift}(x,A[F\times F])$ together with base change via the homomorphism $A[F\times F]\to A[F]$ given by $(a,f_1,f_2)\mapsto(a,f_1+f_2)$. With obvious notation, this amounts to setting $\widetilde{\varphi}_1\boxplus\widetilde{\varphi}_2:=\widetilde{\varphi}_1+\widetilde{\varphi}_2-\varphi$, $\widetilde{\gamma}_1\boxplus\widetilde{\gamma}_2:=\widetilde{\gamma}_1+\widetilde{\gamma}_2-\gamma$.

On the other hand, by definition the A-module structure on $H^1(\mathcal{C}^\bullet(\mathrm{ad}\,M)\otimes^{\mathbf{L}}F)$ is given by the obvious A-module structure on the pairs $(\widetilde{\varphi}\varphi^{-1}-1, \widetilde{\gamma}\gamma^{-1}-1)$. This is obviously the same as the A-module structure that we have just explicated on $\mathrm{Lift}(x,A[F])$. □

5.1.36 Proposition. *$\mathcal{X}_d$ admits a nice obstruction theory in the sense of Definition* A.34.

Proof. Since $\mathcal{X}_d$ is limit preserving by Lemma 3.2.19, it follows from Theorem 5.1.22 and Lemmas 5.1.34 and 5.1.35 that the Herr complex $\mathcal{C}^\bullet(\mathrm{ad}\,M)$ provides the required obstruction theory. □

5.2 RESIDUAL GERBES AND ISOTRIVIAL FAMILIES

In this section we briefly discuss the notion of isotrivial families of (φ,Γ)-modules over reduced $\mathbf{F}$-schemes; i.e., of families which are pointwise constant. The language of residual gerbes (see Appendix E) provides a convenient framework for doing this. In the following discussion we allow $\mathbf{F}'$ to denote any algebraic extension of $\mathbf{F}$.

We have seen that $\mathcal{X}_d$ is a quasi-separated Ind-algebraic stack, which by Proposition 3.4.12 may be written as the inductive limit of algebraic stacks with closed immersions (and so, in particular, monomorphisms) as transition morphisms. Thus the discussion at the end of Appendix E applies, and in particular, for each point $x\in|\mathcal{X}_d|$, the residual gerbe $\mathcal{Z}_x$ at x in $\mathcal{X}_d$ exists. Furthermore, if x is a finite type point, then the canonical monomorphism $\mathcal{Z}_x\hookrightarrow\mathcal{X}_d$ is an immersion.

If we consider an $\mathbf{F}'$-valued point $x\colon\operatorname{Spec}\mathbf{F}'\to\mathcal{X}_d$ (by abuse of notation we use x to denote both this point, and its image $x\in|\mathcal{X}_d|$, which is a finite type point of $\mathcal{X}_d$), then we may base-change the residual gerbe to $\mathcal{X}_d$ at x (which is a gerbe over $\mathbf{F}$) over $\mathbf{F}'$ via the composite $\operatorname{Spec}\mathbf{F}'\xrightarrow{x}\mathcal{X}_d\to\operatorname{Spec}\mathbf{F}$; equivalently, we may regard x as a point of the base change $(\mathcal{X}_d)_{\mathbf{F}'}$, and then consider the residual gerbe of $(\mathcal{X}_d)_{\mathbf{F}'}$ at this point. This residual gerbe is then of the form $[\operatorname{Spec}\mathbf{F}'/G]$, for a finite type affine group scheme G over $\mathbf{F}'$. (By [Sta, Tag 06QG] the group G may be described as the fiber product $x\times_{\mathcal{X}_d}x$, and hence its claimed properties follow from the fact that by Proposition 3.4.12, the diagonal of $\mathcal{X}_d$

is affine and of finite type.) In fact, we have the following more precise result regarding G.

5.2.1 Lemma. *If x is an $\mathbf{F}'$-valued point of $(\mathcal{X}_d)_{\mathbf{F}'}$, then* $\mathrm{Aut}(x)$ *is an irreducible smooth closed algebraic subgroup of* $\mathrm{GL}_{d/\mathbf{F}'}$.

Proof. As already noted, it follows from Proposition 3.4.12 that $\mathrm{Aut}(x)$ is a finite type affine group scheme over $\mathbf{F}'$. Let D denote the étale (φ,Γ)-module over $\mathbf{F}'$ corresponding to x, and let $R := \mathrm{End}_{\mathbf{A}_{K,\mathbf{F}'},\varphi,\Gamma}(D)$. Then, if $\overline{\rho}\colon G_K \to \mathrm{GL}_d(\mathbf{F}')$ denotes the Galois representation corresponding to x, we see that also $R = \mathrm{End}_{G_K}(\overline{\rho})$, and hence that R is an $\mathbf{F}'$-subalgebra of $M_d(\mathbf{F}')$. Furthermore, if A is any finite type $\mathbf{F}'$-algebra, and if D_A denotes the base change of D over A, then the natural morphism $R \otimes_{\mathbf{F}'} A \to \mathrm{End}_{\mathbf{A}_{K,\mathbf{F}'},\varphi,\Gamma}(D_A)$ is an isomorphism by Theorem 5.1.22 (2) and Lemma 5.1.2 (note that A is automatically flat over $\mathbf{F}'$). Thus

$$\mathrm{Aut}(x)(A) := \mathrm{Aut}_{\mathbf{A}_{K,A},\varphi,\Gamma}(D_A) = (R \otimes_{\mathbf{F}'} A)^{\times} = (R \otimes_{\mathbf{F}'} A) \bigcap \mathrm{GL}_d(A),$$

so that $\mathrm{Aut}(x)$, as a scheme over $\mathbf{F}'$, is precisely the open subscheme $R \cap \mathrm{GL}_d$, where we think of R as an affine subspace of the affine space M_d. In particular, we see that $\mathrm{Aut}(x)$ is an open subscheme of an affine space, and thus smooth and irreducible. □

Suppose that S is a reduced $\mathbf{F}'$-scheme of finite type, and that $S \to \mathcal{X}_d$ is a morphism such that every closed point of S maps to some fixed $\mathbf{F}'$-valued point $x \in |\mathcal{X}_d|$. We can think of this morphism as classifying a family of Galois representations $\overline{\rho}_S$ over S whose fiber at each closed point is isomorphic to the fixed representation $\overline{\rho}\colon G_K \to \mathrm{GL}_d(\mathbf{F}')$ classified by x. If $G := \mathrm{Aut}_{G_K}(\overline{\rho})$, then Lemma E.5 shows that the morphism $S \to \mathcal{X}$ factors through the residual gerbe $[\operatorname{Spec} \mathbf{F}'/G]$, and thus corresponds to an *étale* locally trivial G-bundle E over S. (A priori, the G-bundle E over S is *fppf* locally trivial. However, since G is smooth, by Lemma 5.2.1, we see that E is in fact smooth locally trivial. By taking an étale slice of a smooth cover over which E trivializes, we see that E is in fact étale locally trivial.) We can then describe the family $\overline{\rho}_S$ concretely as a twist by E of the constant family $S \times \overline{\rho}$, i.e., $\overline{\rho}_S = E \times_G \overline{\rho}$.

5.3 TWISTING FAMILIES

We now begin our study of the dimensions of certain families of (φ,Γ)-representations, and the behaviors of these dimensions under the operations of twisting by families of one-dimensional representations, and forming extensions of families.

We will need in particular to be able to twist families of representations by unramified characters. Given an $\mathbf{F}_p$-algebra A and an element $a \in A^\times$, we have a (φ, Γ)-module M_a whose underlying $\mathbf{A}_{K,A}$-module is free of rank 1, generated by some $v \in M_a$ for which $\varphi(v) = av$ and $\gamma(v) = v$. If $A = \mathbf{F}$ then the corresponding representation of G_K is the unramified character ur_a taking a geometric Frobenius to a. The universal instance of this construction comes by taking $a = x \in A = \mathbf{F}[x, x^{-1}]$; the corresponding (φ, Γ)-module M_x is then classified by a morphism which we denote $\mathrm{ur}_x : \mathbf{G}_m := \operatorname{Spec} \mathbf{F}[x, x^{\pm 1}] \to \mathcal{X}_1$, which evidently factors through $\mathcal{X}_{1,\mathrm{red}}$.

Given any morphism $T \to \mathcal{X}_{d,\mathrm{red}}$, with T a reduced finite type $\mathbf{F}$-scheme, corresponding to a family $\overline{\rho}_T$ of G_K-representations over T, we may consider the family $\overline{\rho}_T \boxtimes \mathrm{ur}_x$ over $T \times \mathbf{G}_m$, as in Section 3.8. We refer to this operation on (φ, Γ)-modules as *unramified twisting*.

5.3.1 Definition. We say that $\overline{\rho}_T$ is *twistable* if whenever $\overline{\rho}_t \cong \overline{\rho}_{t'} \otimes \mathrm{ur}_a$ where $t, t' \in T(\overline{\mathbf{F}}_p)$ and $a \in \overline{\mathbf{F}}_p^\times$, then $a = 1$. We say that it is *essentially twistable* if for each $t \in T(\overline{\mathbf{F}}_p)$, the set of $a \in \overline{\mathbf{F}}_p^\times$ for which there exists $t' \in T(\overline{\mathbf{F}}_p)$ with $\overline{\rho}_t \cong \overline{\rho}_{t'} \otimes \mathrm{ur}_a$ is finite.

5.3.2 Lemma. *If the dimension of the scheme-theoretic image of T in $\mathcal{X}_{d,\mathrm{red}}$ is e, then the dimension of the scheme-theoretic image of $T \times \mathbf{G}_m$ in $\mathcal{X}_{d,\mathrm{red}}$ is at most $e+1$. If T contains a dense open subscheme U such that $\overline{\rho}_U$ is essentially twistable, then equality holds.*

5.3.3 Remark. Since a twistable representation is essentially twistable, we see that if T contains a dense open subscheme U such that $\overline{\rho}_U$ is twistable, then equality holds in Lemma 5.3.2.

Proof of Lemma 5.3.2. We may assume that T is irreducible. Write f for the morphism $T \to \mathcal{X}_{d,\mathrm{red}}$ and g for the morphism $T \times \mathbf{G}_m \to \mathcal{X}_{d,\mathrm{red}}$. By [Sta, Tag 0DS4], we may, after possibly replacing T by a non-empty open subscheme, assume that for each $\overline{\mathbf{F}}_p$-point t of T we have $\dim T_{f(t)} = \dim T - e$.

Let $v = (t, \lambda)$ be an $\overline{\mathbf{F}}_p$-point of $T \times \mathbf{G}_m$. Then $(T \times \mathbf{G}_m)_{g(v)}$ contains $T_{f(t)} \times \{\lambda\}$, so that $\dim (T \times \mathbf{G}_m)_{g(v)} \geq \dim T_{f(t)} = \dim T - e = \dim (T \times \mathbf{G}_m) - (e+1)$, so the first claim follows from another application of [Sta, Tag 0DS4].

For the last part, we may replace T by U, and then the hypothesis that $\overline{\rho}_T$ is essentially twistable means that we can identify $(T \times \mathbf{G}_m)_{g(v)}$ with a finite union of fibres $T_{f(t')}$ (indexed by the finitely many a for which there is a t' and an isomorphism $\overline{\rho}_t \cong \overline{\rho}_{t'} \otimes \mathrm{ur}_a$), so that equality holds in the above inequality, and the result again follows from [Sta, Tag 0DS4]. □

Recall that for each embedding $\overline{\sigma} : k \hookrightarrow \mathbf{F}$, we have a corresponding fundamental character $\omega_{\overline{\sigma}} : I_K \to \mathbf{F}^\times$, which corresponds via local class field theory (normalized to take uniformizers to geometric Frobeneii) to the composite of $\overline{\sigma}$ and the natural map $\mathcal{O}_K^\times \to k^\times$. It is well known, and easy to check, that $\prod_{\overline{\sigma}} \omega_{\overline{\sigma}}^{e(K/\mathbf{Q}_p)} = \overline{\epsilon}|_{I_K}$. If $\mathbf{F}'$ is an algebraic extension of $\mathbf{F}$, then we can and do identify the embeddings $k \hookrightarrow \mathbf{F}$ and $k \hookrightarrow \mathbf{F}'$.

Let $\underline{n} = (n_{\overline{\sigma}})_{\overline{\sigma}:\, k \hookrightarrow \mathbf{F}}$ be a tuple of integers $0 \le n_{\overline{\sigma}} \le p-1$. The characters $\omega_{\overline{\sigma}}$ can all be extended to G_K, and the restriction to I_K of any character $G_K \to \overline{\mathbf{F}}_p^\times$ is equal to $\prod_{\overline{\sigma}} \omega_{\overline{\sigma}}^{-n_{\overline{\sigma}}}$ for some $\underline{n}$ (which is unique unless the character is unramified). We write $\psi_{\underline{n}}\colon G_K \to \mathbf{F}^\times$ for a fixed choice of an extension of $\prod_{\overline{\sigma}} \omega_{\overline{\sigma}}^{-n_{\overline{\sigma}}}$ to G_K. If $n_{\overline{\sigma}} = 0$ for all $\overline{\sigma}$, or $n_{\overline{\sigma}} = p-1$ for all $\overline{\sigma}$ (these are exactly the cases corresponding to unramified characters), then we fix the choice $\psi_{\underline{n}} = 1$. We can and do make our choice so that if $\underline{n}$, $\underline{n}'$ are such that $(\psi_{\underline{n}}\psi_{\underline{n}'}^{-1})|_{I_K} = \overline{\epsilon}|_{I_K}$, then in fact $\psi_{\underline{n}}\psi_{\underline{n}'}^{-1} = \overline{\epsilon}$.

Somewhat abusively, we will also write $\psi_{\underline{n}}$ for the constant family of (φ, Γ)-modules $\mathbf{D}(\psi_{\underline{n}})$ over any $\overline{\mathbf{F}}_p$-scheme T.

5.3.4 Remark. The irreducible representations $\overline{\alpha}\colon G_K \to \mathrm{GL}_d(\overline{\mathbf{F}}_p)$ are easily classified; indeed, since the wild inertia subgroup must act trivially, they are tamely ramified, so that the restriction of $\overline{\alpha}$ to I_K is diagonalizable. Considering the action of Frobenius on tame inertia, it is then easy to see that $\overline{\alpha}$ is induced from a character of the unramified extension of K of degree d. It follows that each irreducible $\overline{\alpha}$ is absolutely irreducible.

As we have just seen, there are only finitely many such characters up to unramified twist, so that in particular there are only finitely many such $\overline{\alpha}$ up to unramified twist. It follows that there are up to unramified twist only finitely many irreducible (φ, Γ)-modules over $\overline{\mathbf{F}}_p$ of any fixed rank.

It is convenient to extend the notation ρ_T to algebraic stacks. To this end, if $\mathcal{T}$ is a reduced algebraic stack of finite type over $\overline{\mathbf{F}}_p$, we will denote a morphism $\mathcal{T} \to (\mathcal{X}_{d,\mathrm{red}})_{\overline{\mathbf{F}}_p}$ by $\overline{\rho}_{\mathcal{T}}$, and will sometimes abusively refer to $\rho_{\mathcal{T}}$ as a family of G_K-representations parameterized by $\mathcal{T}$.

5.3.5 Definition. We say that a representation $\overline{\rho}\colon G_K \to \mathrm{GL}_d(\overline{\mathbf{F}}_p)$ is *maximally nonsplit of niveau* 1 if it has a unique filtration by G_K-stable $\overline{\mathbf{F}}_p$-subspaces such that all of the graded pieces are one-dimensional representations of G_K. We say that the family $\overline{\rho}_{\mathcal{T}}$ is *maximally nonsplit of niveau* 1 if $\overline{\rho}_t$ is maximally nonsplit of niveau 1 for all $t \in T(\overline{\mathbf{F}}_p)$, and that $\overline{\rho}_T$ is *generically maximally nonsplit of niveau* 1 if there is a dense open substack $\mathcal{U}$ of $\mathcal{T}$ such that $\overline{\rho}_{\mathcal{U}}$ is maximally nonsplit of niveau 1.

In particular, we say that an algebraic substack $\mathcal{T}$ of $(\mathcal{X}_{d,\mathrm{red}})_{\overline{\mathbf{F}}_p}$ is *maximally nonsplit of niveau* 1 if $\overline{\rho}_t$ is maximally nonsplit of niveau 1 for all $t \in \mathcal{T}(\overline{\mathbf{F}}_p)$, and that $\mathcal{T}$ is *generically maximally nonsplit of niveau* 1 if there is a dense open substack $\mathcal{U}$ of $\mathcal{T}$ such that $\mathcal{U}$ is maximally nonsplit of niveau 1.

5.3.6 Remark. We will eventually see in Theorem 6.5.1 below that $(\mathcal{X}_{d,\mathrm{red}})_{\overline{\mathbf{F}}_p}$ itself is an algebraic stack of finite type over $\overline{\mathbf{F}}_p$, and is generically maximally nonsplit of niveau 1.

Our next goal is to prove a structure theorem (Proposition 5.3.8) for families which are generically maximally nonsplit of niveau 1. We begin with the following technical lemma.

5.3.7 Lemma. *Let T be a reduced scheme of finite type over an algebraically closed field k, let M be a coherent sheaf on $T \times_k \mathbf{G}_m$, and for each $t \in T(k)$, let M_t denote the restriction of M to $\mathbf{G}_m \xrightarrow{\sim} t \times_k \mathbf{G}_m \hookrightarrow T \times_k \mathbf{G}_m$. Suppose that for each point $t \in T(k)$, M_t is a length 1 skyscraper sheaf supported at a single point of $\mathbf{G}_m$. Then there is a dense open subscheme U of T such that, over U, the composite $\mathrm{Supp}(M) \hookrightarrow T \times_k \mathbf{G}_m \to T$ pulls back to an isomorphism, while the pull-back of M over $U \times_k \mathbf{G}_m$ is locally free of rank 1 over its support.*

Proof. Let $\mathrm{Fitt}\,(M) \subseteq \mathcal{O}_{T\times\mathbf{G}_m}$ denote the Fitting ideal sheaf of M, and write $Z := \mathrm{Spec}\,\mathcal{O}_{T\times\mathbf{G}_m}/\mathrm{Fitt}\,(M)$; recall that $\mathrm{Supp}(M)$ is a closed subscheme of Z, supported on the same underlying closed subset of $T \times_k \mathbf{G}_m$. For each $t \in T(k)$, we find that the fiber Z_t of Z over t is equal to $\mathrm{Spec}\,\mathcal{O}_{\mathbf{G}_m}/\mathrm{Fitt}\,(M_t)$ (recall that the formation of Fitting ideals is compatible with base change), which, by assumption, is a single reduced point of $\mathbf{G}_m$. Thus the composite $Z \to T \times_k \mathbf{G}_m \to T$ is a morphism with reduced singleton fibres, and hence we may find a dense open subscheme U of T such that this morphism restricts to an isomorphism over U. In particular, the pull-back of Z over U is reduced (since U is), and thus coincides with the pull-back of $\mathrm{Supp}(M)$ over U. This shows that the morphism $\mathrm{Supp}(M) \to T$ pulls back to an isomorphism over U.

If we use the inverse of the isomorphism just constructed to identify U with an open subscheme of $\mathrm{Supp}(M)$, then our assumption on the nature of M_t for $t \in T(k)$ implies that the fiber of M over each k-point of U is one-dimensional. Since U is reduced, we find that the restriction of M to U is furthermore locally free of rank 1, as claimed. $\square$

5.3.8 Proposition. *If $\mathcal{T}$ is a reduced finite type algebraic stack over $\overline{\mathbf{F}}_p$, and $\overline{\rho}_{\mathcal{T}}$ is generically maximally nonsplit of niveau 1, then there exist:*

- *a dense open substack $\mathcal{U}$ of $\mathcal{T}$;*
- *tuples $\underline{n}_i$, $1 \le i \le d$;*
- *morphisms $\lambda_i : \mathcal{U} \to \mathbf{G}_m$, $1 \le i \le d$;*
- *morphisms $x_i : \mathcal{U} \to (\mathcal{X}_{i,\mathrm{red}})_{\overline{\mathbf{F}}_p}$, $0 \le i \le d$, corresponding to families of i-dimensional representations ρ_i over $\mathcal{U}$;*

such that x_d is the restriction of $\overline{\rho}_{\mathcal{T}}$ to $\mathcal{U}$, and for each $0 \le i \le d-1$, we have short exact sequences of families over $\mathcal{U}$

$$0 \to \rho_i \to \rho_{i+1} \to \mathrm{ur}_{\lambda_{i+1}} \otimes \psi_{\underline{n}_{i+1}} \to 0.$$

In particular, for each $t \in \mathcal{U}$, we have

$$\overline{\rho}_t \cong \begin{pmatrix} \mathrm{ur}_{\lambda_1(t)} \otimes \psi_{\underline{n}_1} & * & \cdots & * \\ 0 & \mathrm{ur}_{\lambda_2(t)} \otimes \psi_{\underline{n}_2} & \cdots & * \\ \vdots & & \ddots & \vdots \\ 0 & \cdots & 0 & \mathrm{ur}_{\lambda_d(t)} \otimes \psi_{\underline{n}_d} \end{pmatrix}. \tag{5.3.9}$$

5.3.10 Remark. We cannot necessarily attain (5.3.9) over all of $\mathcal{T}$. For example, there are families of two-dimensional representations of G_K which are generically maximally nonsplit of niveau 1, but which also specialize to irreducible representations. (The existence of such families follows from Theorem 6.5.1 below.)

Proof of Proposition 5.3.8. We begin by assuming that $\mathcal{T}$ is a scheme, say $\mathcal{T} = T$. Without loss of generality, we can replace by T by a dense open subscheme, and thus assume that $\overline{\rho}_T$ is maximally nonsplit of niveau 1, i.e., that $\overline{\rho}_t$ is maximally nonsplit of niveau 1 for each $t \in T(\overline{\mathbf{F}}_p)$, and also assume that T is a disjoint union of finitely many integral open and closed subschemes. Replacing T by each of these subschemes in turn, we may assume that T itself is integral.

For each $\underline{n}$, we may form the family

$$\rho_T \boxtimes (\mathrm{ur}_x^{-1} \otimes \psi_{\underline{n}}^{-1})(1)$$

over $T \times_{\overline{\mathbf{F}}_p} \mathbf{G}_m$ (where as above x denotes the variable on $\mathbf{G}_m$, so that $\mathbf{G}_m := \operatorname{Spec} \overline{\mathbf{F}}_p[x, x^{-1}]$), and then consider (in the notation of Remark 5.1.30)

$$H_{\underline{n}} := H^2\big(G_K, \rho_T \boxtimes (\mathrm{ur}_x^{-1} \otimes \psi_{\underline{n}}^{-1})(1)\big),$$

a coherent sheaf on $T \times_{\overline{\mathbf{F}}_p} \mathbf{G}_m$. Since the formation of $H^2(G_K, -)$ is compatible with arbitrary finite type base change, we see that for any point $t \in T(\overline{\mathbf{F}}_p)$, the pull-back of $H_{\underline{n}}$ over $t \times_{\overline{\mathbf{F}}_p} \mathbf{G}_m \cong \mathbf{G}_m$ admits the description

$$(H_{\underline{n}})_t \cong H^2\big(G_K, \rho_t \otimes (\mathrm{ur}_x^{-1} \otimes \psi_{\underline{n}}^{-1})(1)\big),$$

and that the fiber $(H_{\underline{n}})_{(t,x_0)}$ of this pull-back at a point $x_0 \in \mathbf{G}_m(\overline{\mathbf{F}}_p)$ admits the description

$$(H_{\underline{n}})_{(t,x_0)} \cong H^2\big(G_K, \rho_t \otimes (\mathrm{ur}_{x_0}^{-1} \otimes \psi_{\underline{n}}^{-1})(1)\big) \cong \operatorname{Hom}_{G_K}(\rho_t, \mathrm{ur}_{x_0} \otimes \psi_{\underline{n}})$$

(the second isomorphism following from Tate local duality). The assumption that ρ_t is maximally nonsplit shows that this fiber is nonzero for exactly one choice of $\underline{n}$ and one choice of x_0, and is then precisely one-dimensional. In particular, we see, for this distinguished choice of $\underline{n}$, that $(H_{\underline{n}})_t$ is set-theoretically supported at x_0. Now let $\xi_0\colon \operatorname{Spec}\overline{\mathbf{F}}_p[\epsilon]/(\epsilon^2) \hookrightarrow \mathbf{G}_m$ be the (unique up to scalar) non-trivial tangent vector to x_0, and let $(H_{\underline{n}})_{(t,\xi_0)}$ denote the pull-back of $(H_{\underline{n}})_t$ over ξ_0. A similar computation using Tate local duality and the maximal nonsplitness of ρ_t shows that $(H_{\underline{n}})_{(t,\xi_0)}$ is also one-dimensional; thus $(H_{\underline{n}})_t$ is in fact a skyscraper sheaf of length 1 supported at x_0. For all other choices of $\underline{n}$, we see that $(H_{\underline{n}})_t$ vanishes (since all its fibres do).

Applying Lemma 5.3.7 to $M := \bigoplus_{\underline{n}} H_{\underline{n}}$, and replacing T by an appropriately chosen dense open subscheme, we find that $\operatorname{Supp}(M)$ maps isomorphically to T, and that M is locally free of rank 1 over its support. Taking into account the definition of M as a direct sum, and the irreducibility of T, we find that in fact $M = H_{\underline{n}}$ for exactly one choice $\underline{n}_d$ of $\underline{n}$, and that $H_{\underline{n}} = 0$ for all other possible choices. Let $\lambda\colon T \to \mathbf{G}_m$ denote the composite of the inverse isomorphism $T \xrightarrow{\sim} \operatorname{Supp}(M)$ with the projection $\operatorname{Supp}(M) \to \mathbf{G}_m$. Again using the fact that the formation of H^2 commutes with base change, we find that the pull-back M_T of M to $T \cong \operatorname{Supp}(M)$ may be identified with

$$H^2\big(\rho_T \boxtimes (\mathrm{ur}_\lambda^{-1} \otimes \psi_{\underline{n}_d}^{-1})(1)\big).$$

We have already remarked that M_T is locally free of rank 1; replacing T by a non-empty open subscheme once more, we may in fact assume that M_T is free of rank 1, and so choose a nowhere zero section of M_T, which by Lemma 5.1.31 we may and do regard as a surjection

$$\rho_T \twoheadrightarrow \mathrm{ur}_\lambda \otimes \psi_{\underline{n}_d}.$$

The kernel of this surjection is a rank $(d-1)$ family of projective étale (φ,Γ)-modules, so the result for T follows by induction on d.

We now return to the general case that $\mathcal{T}$ is a reduced finite type algebraic stack over $\overline{\mathbf{F}}_p$. Since $\mathcal{X}_d$ is Ind-algebraic by Proposition 3.4.12, we can form the scheme-theoretic image $\mathcal{T}'$ of the morphism $\mathcal{T} \to (\mathcal{X}_{d,\mathrm{red}})_{\overline{\mathbf{F}}_p}$, which is again a reduced finite type algebraic stack over $\overline{\mathbf{F}}_p$. The family $\overline{\rho}_{\mathcal{T}'}$ is again generically maximally nonsplit of niveau 1 (since by the stacky version of Chevalley's theorem (see [Ryd11, App. D]), the image of $\mathcal{T}$ in $\mathcal{T}'$ is constructible, and so contains a dense open subset of $\mathcal{T}'$), and since the morphism $\mathcal{T} \to \mathcal{T}'$ is scheme-theoretically dominant, any dense open substack of $\mathcal{T}'$ pulls back to a dense open substack of $\mathcal{T}$. It therefore suffices to prove the result for $\mathcal{T}'$. Replacing $\mathcal{T}$ by $\mathcal{T}'$, then, we can and do assume that $\mathcal{T}$ is a closed substack of $(\mathcal{X}_{d,\mathrm{red}})_{\overline{\mathbf{F}}_p}$.

Let $T \to \mathcal{T}$ be a smooth cover by a (necessarily) reduced and finite type $\overline{\mathbf{F}}_p$-scheme (we can ensure that T is finite type over $\overline{\mathbf{F}}_p$, since $\mathcal{T}$ is so), and let $\overline{\rho}_T$

be the induced family. We wish to deduce the result for $\mathcal{T}$ from the result for T, which we have already proved. We thus replace T by a dense open subscheme (and correspondingly replace $\mathcal{T}$ by a dense open substack, namely the image of this open subscheme of T), so that the family $\overline{\rho}_T$ is maximally nonsplit, and so that the morphisms $\lambda_i\colon T \to \mathbf{G}_m$ and $x_i\colon T \to (\mathcal{X}_{i,\mathrm{red}})_{\overline{\mathbf{F}}_p}$ exist. We will then show that these morphisms factor through the smooth surjection $T \to \mathcal{T}$. This amounts to showing that we can factor the morphism of groupoids

$$T \times_{\mathcal{T}} T \to T \times T \tag{5.3.11}$$

through a morphism of groupoids $T \times_{\mathcal{T}} T \to T \times_{\mathbf{G}_m} T$ or $T \times_{\mathcal{T}} T \to T \times_{(\mathcal{X}_{i,\mathrm{red}})_{\overline{\mathbf{F}}_p}} T$ respectively.

Since $\mathcal{T}$ is a substack of $\mathcal{X}$, the natural morphism induces an isomorphism $T \times_{\mathcal{T}} T \xrightarrow{\sim} T \times_{\mathcal{X}} T$. Thus for any test scheme T', a T'-valued point of $T \times_{\mathcal{T}} T$ consists of a pair of morphisms $f_0, f_1\colon T' \rightrightarrows T$, and an isomorphism of families

$$f_0^*\rho_T \xrightarrow{\sim} f_1^*\rho_T \tag{5.3.12}$$

(where $f_i^*\rho_T$ denotes the pull-back of ρ_T to T' via f_i).

In the first case, in which we consider one of the morphisms λ_i, we need to show (employing the notation just introduced) that the morphism $f_0 \times f_1\colon T' \times T' \to T \times T$ necessarily factors through $T \times_{\mathbf{G}_m} T$ (this latter fiber product formed using the morphisms λ_i). More concretely, this amounts to checking that $\lambda_i \circ f_0 = \lambda_i \circ f_1$. Since the source of (5.3.11) is reduced and of finite type over $\overline{\mathbf{F}}_p$, it suffices to check this for reduced test schemes T' of finite type over $\overline{\mathbf{F}}_p$, and thus to check that these morphisms coincide at the $\overline{\mathbf{F}}_p$-valued points t' of T'. This follows from the assumed isomorphism (5.3.12), and the uniqueness of the filtration at each $\overline{\mathbf{F}}_p$-valued point of the family.

In the second case, in which we consider one of the morphisms x_i, the morphism $T \times_{(\mathcal{X}_{i,\mathrm{red}})_{\overline{\mathbf{F}}_p}} T \to T \times T$ is no longer a monomorphism, and so we have to actually construct a morphism of groupoids $T \times_{\mathcal{T}} T \to T \times_{(\mathcal{X}_{i,\mathrm{red}})_{\overline{\mathbf{F}}_p}} T$ (rather than simply verify a factorization). So, given $f_0, f_1\colon T' \to T$, and an isomorphism (5.3.12), we have to construct a corresponding isomorphism

$$f_0^*(\rho_T)_i \xrightarrow{\sim} f_1^*(\rho_T)_i, \tag{5.3.13}$$

where $(\rho_T)_i$ denotes the family of i-dimensional subrepresentations of ρ_T corresponding to the morphism x_i. We will define the desired isomorphism (5.3.13) simply to be the restriction of the isomorphism (5.3.12). For this definition to make sense, we have to show that (5.3.12) does in fact restrict to an isomorphism (5.3.13). In fact, it suffices to check that (5.3.12) induces an embedding $f_0^*(\rho_T)_i \hookrightarrow f_1^*(\rho_T)_i$; reversing the roles of f_0 and f_1 then allows us to promote this to an isomorphism.

Again, it suffices to check this on reduced test schemes T' of finite type over $\overline{\mathbf{F}}_p$, and then it suffices to show that the composite

$$f_0^*(\rho_T)_i \hookrightarrow f_0^*\rho_T \overset{(5.3.12)}{\longrightarrow} f_1^*\rho_T \to f_1^*\rho_T/f_1^*(\rho_T)_i$$

vanishes. This can be checked on $\overline{\mathbf{F}}_p$-valued points, where it again follows from the uniqueness of the filtrations that define the $(\rho_T)_i$. □

5.4 DIMENSIONS OF FAMILIES OF EXTENSIONS

Suppose that we have a morphism $T \to (\mathcal{X}_{d,\mathrm{red}})_{\overline{\mathbf{F}}_p}$, with T a reduced finite type $\overline{\mathbf{F}}_p$-scheme, corresponding to a family $\overline{\rho}_T$ of G_K-representations over T. Fix a representation $\overline{\alpha}\colon G_K \to \mathrm{GL}_a(\overline{\mathbf{F}}_p)$.

5.4.1 Lemma. *If* $\mathrm{Ext}^2_{G_K}(\overline{\alpha},\overline{\rho}_t)$ *is of constant rank* r *for all* $t \in T(\overline{\mathbf{F}}_p)$, *then* $H^2(G_K,\overline{\rho}_T \otimes \overline{\alpha}^\vee)$ *is locally free of rank* r *as an* $\mathcal{O}_T$*-module.*

Proof. Since the formation of H^2 is compatible with arbitrary finite type base-change (by Theorem 5.1.22 (2)), we see that $H^2(G_K,\overline{\rho}_T \otimes \overline{\alpha}^\vee)$ is a coherent sheaf on T of constant fiber rank r. Since T is reduced, the lemma follows. □

By Corollary 5.1.25, if T is affine, then we can and do choose a good complex (that is, a bounded complex of finite rank locally free $\mathcal{O}_T$-modules)

$$C_T^0 \to C_T^1 \to C_T^2$$

computing $H^\bullet(G_K,\overline{\rho}_T \otimes \overline{\alpha}^\vee)$. Suppose that we are in the context of the preceding lemma, i.e., that $\mathrm{Ext}^2_{G_K}(\overline{\alpha},\overline{\rho}_T)$ is locally free of some rank r. It follows that the truncated complex

$$C_T^0 \to Z_T^1$$

is again good (here $Z_T^1 := \ker(C_T^1 \to C_T^2)$). As in Remark 5.1.30, we write

$$H^1(G_K,\overline{\rho}_T \otimes \overline{\alpha}^\vee)$$

for the cohomology group of this complex in degree 1; its formation is compatible with arbitrary finite type base change, and its specializations to finite type points of T agree with the usual Galois cohomology.

In particular, if we choose another integer n, we may use the surjection $Z_T^1 \to H^1(G_K,\overline{\rho}_T \otimes \overline{\alpha}^\vee)$ (together with Lemma 5.1.2) to construct a universal family of extensions

$$0 \to \overline{\rho}_T \to \mathcal{E} \to \overline{\alpha}^{\oplus n} \to 0 \tag{5.4.2}$$

parameterized by the vector bundle V corresponding to the finite rank locally free sheaf $(Z^1_T)^{\oplus n}$, giving rise to a morphism

$$V \to (\mathcal{X}_{d+an,\mathrm{red}})_{\overline{\mathbf{F}}_p}. \tag{5.4.3}$$

(See [CEGS19, §4.2] for a similar construction.)

Write $G_{\overline{\alpha}} := \mathrm{Aut}_{G_K}(\overline{\alpha})$, thought of as an affine algebraic group over $\overline{\mathbf{F}}_p$.

5.4.4 Proposition. *Maintain the notation and assumptions above; so we assume in particular that* $\mathrm{Ext}^2_{G_K}(\overline{\alpha}, \overline{\rho}_t)$ *is of constant rank* r *for all* $t \in T(\overline{\mathbf{F}}_p)$.

1. *If* e *denotes the dimension of the scheme-theoretic image of* T *in* $(\mathcal{X}_{d,\mathrm{red}})_{\overline{\mathbf{F}}_p}$, *then the dimension of the scheme-theoretic image, with respect to the morphism* (5.4.3), *of* V *in* $(\mathcal{X}_{d+an,\mathrm{red}})_{\overline{\mathbf{F}}_p}$ *is bounded above by*

$$e + n([K:\mathbf{Q}_p]ad + r) - n^2 \dim G_{\overline{\alpha}}.$$

2. *Suppose that:*
   (a) $n = 1$;
   (b) $\overline{\alpha}$ *is one-dimensional;*
   (c) $\overline{\rho}_T$ *is generically maximally nonsplit of niveau* 1;
   (d) *if* $K = \mathbf{Q}_p$, *then after replacing* T *by a dense open subscheme, so that the tuples* $\underline{n}_i$ *exist, and so that the morphisms* λ_i *and* x_i *as in the statement of Proposition* 5.3.8 *are defined on* T, *at least one of the following conditions holds at each* $t \in T(\overline{\mathbf{F}}_p)$:
      i. $\mathrm{ur}_{\lambda_{d-1}(t)} \otimes \psi_{\underline{n}_{d-1}} \neq \overline{\alpha}(1)$.
      ii. $\mathrm{ur}_{\lambda_d(t)} \otimes \psi_{\underline{n}_d} = \overline{\alpha}$.
      iii. $\mathrm{ur}_{\lambda_d(t)} \otimes \psi_{\underline{n}_d} = \overline{\alpha}(1)$.

   Then equality holds in the inequality of (1). Furthermore the family of extensions $\mathcal{E}_V$ *corresponding to* V *is generically maximally nonsplit of niveau* 1.
3. *Suppose that we are in the setting of (2), and that after replacing* T *by a dense open subscheme, the following conditions hold:*
   (a) *For each* $t \in T(\overline{\mathbf{F}}_p)$ *the representation* $\overline{\rho}_t$ *is maximally nonsplit, and furthermore the unique quotient character of* $\overline{\rho}_t$ *(which exists by the assumption that* $\overline{\rho}_T$ *is generically maximally nonsplit of niveau* 1) *is equal to* $\overline{\alpha}(1)$.
   (b) *Furthermore, if we write*

$$0 \to \overline{r}_t \to \overline{\rho}_t \to \overline{\alpha}(1) \to 0$$

   for the corresponding filtration of $\overline{\rho}_t$, *and write* γ_t *for the unique quotient character of* $\overline{r}_t$ *then for each* $t \in T(\overline{\mathbf{F}}_p)$ *either i. we have* $\gamma_t \neq \overline{\alpha}(1)$, *or*

ii. the cyclotomic character $\overline{\epsilon}$ *is trivial, we have* $\gamma_t = \overline{\alpha}(1) = \overline{\alpha}$, *and the extension of* $\overline{\alpha}$ *by* $\gamma_t = \overline{\alpha}(1)$ *induced by* $\overline{\rho}_t$ *is très ramifiée.*

Then, after replacing V *by a dense open subscheme, we have that for all* $t \in V(\overline{\mathbf{F}}_p)$ *the extension of* $\overline{\alpha}$ *by* $\overline{\alpha}(1)$ *induced by the extension of* $\overline{\alpha}$ *by* $\overline{\rho}_t$ *is très ramifiée.*

5.4.5 Remark. See Remark 5.4.10 below for a further discussion of condition (2d).

Proof of Proposition 5.4.4. Writing T as the union of its irreducible components, we obtain a corresponding decomposition of V into the union of its irreducible components, and we can prove the theorem one component at a time. Thus we may assume that T (and hence V) is irreducible. Replacing T by a non-empty affine open subscheme, we may furthermore assume that it is affine, so that we may find a good complex $C_T^\bullet$ supported in degrees $[0,2]$ that is isomorphic in the derived category to the Herr complex, as above.

Let $f\colon T \to (\mathcal{X}_{d,\mathrm{red}})_{\overline{\mathbf{F}}_p}$ be the given morphism, and let $g\colon V \to (\mathcal{X}_{d+an,\mathrm{red}})_{\overline{\mathbf{F}}_p}$ denote the morphism (5.4.3). In order to relate the scheme-theoretic image of g to the scheme-theoretic image of f, we will also consider (in a slightly indirect manner) the scheme-theoretic image of the morphism $V \to (\mathcal{X}_{d,\mathrm{red}})_{\overline{\mathbf{F}}_p} \times (\mathcal{X}_{d+an,\mathrm{red}})_{\overline{\mathbf{F}}_p}$ (the first factor being the composite of f and the projection from V to T, and the second factor being g). If $(\mathcal{X}_{d,\mathrm{red}})_{\overline{\mathbf{F}}_p}$ were a scheme or an algebraic space, then the dimension of this latter scheme-theoretic image would be greater than or equal to the dimension of the former; but since $(\mathcal{X}_{d,\mathrm{red}})_{\overline{\mathbf{F}}_p}$ is a stack, we have to be slightly careful in comparing the dimensions of these two images.

Let $v\colon \operatorname{Spec} \overline{\mathbf{F}}_p \to V$ be an $\overline{\mathbf{F}}_p$-point, and let t denote the composite $\operatorname{Spec} \overline{\mathbf{F}}_p \to V \to T$, so that t is an $\overline{\mathbf{F}}_p$-point of T. We also write $f(t)$ for the composite $f \circ t$, and $g(v)$ for the composite $g \circ v$. We write $T_{f(t)}$ for the fiber product $T \times_{(\mathcal{X}_{d,\mathrm{red}})_{\overline{\mathbf{F}}_p}} \operatorname{Spec} \overline{\mathbf{F}}_p$ (with respect to the morphisms $f\colon T \to (\mathcal{X}_{d,\mathrm{red}})_{\overline{\mathbf{F}}_p}$ and $f(t)\colon \operatorname{Spec} \overline{\mathbf{F}}_p \to (\mathcal{X}_{d,\mathrm{red}})_{\overline{\mathbf{F}}_p}$), and write $V_{g(v)}$ and $V_{(f(t),g(v))}$ for the evident analogous fiber products.

We begin with (1). In order to prove the required bound on the dimension of the scheme-theoretic image of g, it suffices (e.g., by [Sta, Tag 0DS4]) to show that

$$\dim V_{g(v)} \overset{?}{\geq} \dim V - e - n([K:\mathbf{Q}_p]ad + r) + n^2 \dim G_{\overline{\alpha}}$$

for v lying in some non-empty open subscheme of V. As already alluded to above, what we will actually be able to do is to estimate the dimension of the fiber $V_{(f(t),g(v))}$, and so our first job is to compare the dimension of this fiber with that of $V_{g(v)}$.

Let $\overline{\rho}_{f(t)}$ denote the Galois representation corresponding to $f(t)$, and let $G_t := \operatorname{Aut}(\overline{\rho}_{f(t)})$. Then, by the discussion of Section 5.2, the morphism $t\colon$ Spec

$\overline{\mathbf{F}}_p \to (\mathcal{X}_{d,\mathrm{red}})_{\overline{\mathbf{F}}_p}$ induces an immersion

$$[\operatorname{Spec} \overline{\mathbf{F}}_p/G_t] \hookrightarrow (\mathcal{X}_{d,\mathrm{red}})_{\overline{\mathbf{F}}_p},$$

which in turn induces a monomorphism

$$[\operatorname{Spec} \overline{\mathbf{F}}_p/G_t] \times_{(\mathcal{X}_{d,\mathrm{red}})_{\overline{\mathbf{F}}_p}} V_{g(v)} \hookrightarrow V_{g(v)}. \tag{5.4.6}$$

Since $\operatorname{Spec} \overline{\mathbf{F}}_p$ is a G_t-torsor over $[\operatorname{Spec} \overline{\mathbf{F}}_p/G_t]$, we see that $V_{(f(t),g(v))}$ is a G_t-torsor over $[\operatorname{Spec} \overline{\mathbf{F}}_p/G_t] \times_{(\mathcal{X}_{d,\mathrm{red}})_{\overline{\mathbf{F}}_p}} V_{g(v)}$, and hence that

$$\dim V_{(f(t),g(v))} \leq \dim G_t + \dim V_{g(v)}. \tag{5.4.7}$$

Thus it suffices to prove the inequality

$$\dim V_{(f(t),g(v))} \overset{?}{\geq} \dim V - e - n([K:\mathbf{Q}_p]ad + r) + n^2 \dim G_{\overline{\alpha}} + \dim G_t.$$

For t lying in some non-empty open subscheme of T, we have that $e = \dim T - \dim T_{f(t)}$, and so, replacing T by this non-empty open subscheme and V by its preimage, we have to show that

$$\begin{aligned}\dim V_{(f(t),g(v))} \overset{?}{\geq} \dim V - \dim T + \dim T_{f(t)} - n([K:\mathbf{Q}_p]ad + r) \\ + n^2 \dim G_{\overline{\alpha}} + \dim G_t\end{aligned}$$

for v lying in some non-empty open subscheme of V. In fact we will show that this inequality holds for every $\overline{\mathbf{F}}_p$-point v of V.

The right-hand side of our putative inequality can be rewritten as

$$n \operatorname{rk} Z^1_T + \dim T_{f(t)} - n([K:\mathbf{Q}_p]ad + r) + n^2 \dim G_{\overline{\alpha}} + \dim G_t,$$

while the local Euler characteristic formula shows that

$$H^0(G_K, \overline{\rho}_{f(t)} \otimes \overline{\alpha}^\vee) - H^1(G_K, \overline{\rho}_{f(t)} \otimes \overline{\alpha}^\vee) + r = -[K:\mathbf{Q}_p]ad.$$

Thus we may rewrite our desired inequality as

$$\begin{aligned}\dim V_{(f(t),g(v))} \overset{?}{\geq} n\big(\operatorname{rk} Z^1_T - \dim H^1(G_K, \overline{\rho}_{f(t)} \otimes \overline{\alpha}^\vee)\big) \\ + n \dim H^0(G_K, \overline{\rho}_{f(t)} \otimes \overline{\alpha}^\vee) + \dim T_{f(t)} + n^2 \dim G_{\overline{\alpha}} + \dim G_t.\end{aligned}$$

There is a canonical isomorphism $V_{(f(t),g(v))} \cong T_{f(t)} \times_T V_{g(v)}$, and so, writing $S := (T_{f(t)})_{\mathrm{red}}$, we find that $\dim V_{(f(t),g(v))} = \dim S \times_T V_{g(v)}$. If we let $\overline{\rho}_S$ denote the pull-back of the family $\overline{\rho}_T$ over S, then the Galois representations attached to all the closed points of S are isomorphic to $\overline{\rho}_{f(t)}$. The discussion of Section 5.2 shows we may find an étale cover $S' \to S$ such that the pulled-back family $\overline{\rho}_{S'}$ is constant. We may replace S by S' without changing the dimension we are trying to estimate, and thus we may furthermore assume that $\overline{\rho}_S$ is the trivial family with fibres $\overline{\rho}_{f(t)}$.

If we let $C_S^0 \to Z_S^1$ denote the pull-back of the good complex $C_T^0 \to Z_T^1$ to S, then this complex is also the pull-back to S of the good complex $C_t^0 \to Z_t^1$ corresponding to the representation $\overline{\rho}_t$. If we let W denote the vector space $H^1(G_K, \overline{\rho}_t \otimes \overline{\alpha}^\vee)^{\oplus n}$, thought of as an affine space over $\overline{\mathbf{F}}_p$, then there is a projection $S \times_T V \to S \times_{\overline{\mathbf{F}}_p} W$, whose kernel is a (trivial) vector bundle which we denote by V'.

The affine space W parameterizes a universal family of extensions $0 \to \overline{\rho}_{f(t)} \to \overline{\rho}_W \to \overline{\alpha}^{\oplus n} \to 0$, giving rise to a morphism $h\colon W \to (\mathcal{X}_{d+an,\mathrm{red}})_{\overline{\mathbf{F}}_p}$, and the composite morphism $S \times_T V \to V \xrightarrow{g} (\mathcal{X}_{d+an,\mathrm{red}})_{\overline{\mathbf{F}}_p}$ admits an alternative factorization as

$$S \times_T V \to S \times_{\overline{\mathbf{F}}_p} W \to W \xrightarrow{h} (\mathcal{X}_{d+an,\mathrm{red}})_{\overline{\mathbf{F}}_p}.$$

Thus

$$S \times_T V_{g(v)} = S \times_T V \times_W W_{h(w)}$$

(where w denotes the image of v in W under the projection), and hence

$$\begin{aligned}\dim S \times_T V_{g(v)} &= \operatorname{rk} V' + \dim S + \dim W_{h(w)} \\ &= n\big(\operatorname{rk} Z_T^1 - \dim H^1(G_K, \overline{\rho}_{f(t)} \otimes \overline{\alpha}^\vee)\big) + \dim S + \dim W_{h(w)},\end{aligned}$$

so that our desired inequality becomes equivalent to the inequality

$$\dim W_{h(w)} \overset{?}{\geq} n \dim H^0(G_K, \overline{\rho}_{f(t)} \otimes \overline{\alpha}^\vee) + n^2 \dim G_{\overline{\alpha}} + \dim G_t. \tag{5.4.8}$$

There is an action of $G_t \times \operatorname{Aut}_{G_K}(\overline{\alpha}^{\oplus n})$ on

$$W = H^1(G_K, \overline{\rho}_{f(t)} \otimes \overline{\alpha}^\vee)^{\oplus n} \cong \operatorname{Ext}^1_{G_K}(\overline{\alpha}^{\oplus n}, \overline{\rho}_t),$$

which lifts to an action on the family $\overline{\rho}_W$. There is furthermore a (unipotent) action of $H^0(G_K, \overline{\rho}_t \otimes \overline{\alpha}^\vee)^{\oplus n}$ on $\overline{\rho}_W$ (lying over the trivial action on W). Altogether, we find that the morphism h factors through $[W/\big(H^0(G_K, \overline{\rho}_t \otimes \overline{\alpha}^\vee)^{\oplus n} \rtimes (G_t \times \operatorname{Aut}_{G_K}(\overline{\alpha}^{\oplus n}))\big)]$, implying that

$$\dim W_{h(w)} \geq n \dim H^0(G_K, \overline{\rho}_t \otimes \overline{\alpha}^\vee) + n^2 \dim G_{\overline{\alpha}} + \dim G_t,$$

as required.

We now turn to (2), so that we assume that $n=1$, that $\overline{\alpha}$ is one-dimensional, and that $\overline{\rho}_T$ is generically maximally nonsplit of niveau 1. By Proposition 5.3.8, after possibly replacing T with a non-empty open subscheme, we can and do assume that $\overline{\rho}_T$ is maximally nonsplit of niveau 1, that we can write $\overline{\rho}_T$ as an extension of families

$$0 \to \overline{r}_T \to \overline{\rho}_T \to \overline{\beta}_T \to 0 \tag{5.4.9}$$

where the family $\overline{\beta}_T$ arises from twisting a character $\overline{\beta}\colon G_K \to \overline{\mathbf{F}}_p^\times$ via a morphism $T \to \mathbf{G}_m$, and that if $K = \mathbf{Q}_p$, then (2) holds for all finite type points t of T. (Indeed, $\overline{\beta}_T$ is the family corresponding to the character $\mathrm{ur}_{\lambda_d} \otimes \psi_{\underline{n}_d}$ in the notation of Proposition 5.3.8.)

We want to show that $\mathcal{E}_V$ is generically maximally nonsplit of niveau 1. Assume that this is the case; we now show that we then have equality in the bound of part (1). After replacing V with a non-empty open subscheme, we can and do suppose that $\mathcal{E}_V$ is maximally nonsplit of niveau 1, so that in particular, for every finite type point v of V, the extension

$$0 \to \overline{\rho}_{f(t)} \to \mathcal{E}_{g(v)} \to \overline{\alpha} \to 0$$

is maximally nonsplit. From the proof of part (1), we see that it is enough to show that (after shrinking V as we have) equality holds in both (5.4.7) and (5.4.8).

We begin by considering (5.4.7). For equality to hold here, it is enough to check that the immersion (5.4.6) induces an isomorphism on underlying reduced substacks

$$([\operatorname{Spec} \overline{\mathbf{F}}_p / G_t] \times_{(\mathcal{X}_{d,\mathrm{red}})_{\overline{\mathbf{F}}_p}} V_{g(v)})_{\mathrm{red}} \xrightarrow{\sim} (V_{g(v)})_{\mathrm{red}}.$$

We can check this on the level of $\overline{\mathbf{F}}_p$-valued points, for which it suffices to show that the representation $\mathcal{E}_{g(v)}$ has a unique d-dimensional subrepresentation, namely $\overline{\rho}_{f(t)}$; but this is immediate from $\mathcal{E}_{g(v)}$ being maximally nonsplit of niveau 1.

Similarly, to show that equality holds in (5.4.8), it is enough to show that the morphism

$$[W/\big(H^0(G_K, \overline{\rho}_{f(t)} \otimes \overline{\alpha}^\vee) \rtimes \big(G_t \times \operatorname{Aut}_{G_K}(\overline{\alpha})\big)\big)] \to (\mathcal{X}_{d+a,\mathrm{red}})_{\overline{\mathbf{F}}_p}$$

is a monomorphism. To see this, note that it follows from the definition of "maximally nonsplit" that any automorphism of $\mathcal{E}_{g(v)}$ induces automorphisms of $\overline{\rho}_{f(t)}$ and $\overline{\alpha}$, so that we have

$$W \times_{(\mathcal{X}_{d+a,\mathrm{red}})_{\overline{\mathbf{F}}_p}} W = W \times_{\overline{\mathbf{F}}_p} \Big(H^0(G_K, \overline{\rho}_t \otimes \overline{\alpha}^\vee) \rtimes \big(G_t \times \operatorname{Aut}_{G_K}(\overline{\alpha})\big)\Big).$$

Hence it is enough to show that the morphism $[W/W \times_{(\mathcal{X}_{d+a,\mathrm{red}})_{\overline{\mathbf{F}}_p}} W] \to (\mathcal{X}_{d+a,\mathrm{red}})_{\overline{\mathbf{F}}_p}$ is a monomorphism; this follows from Lemma A.33.

In order to prove (2), it remains to prove that $\mathcal{E}_V$ is generically maximally nonsplit of niveau 1. We need to show that after possibly shrinking V, for each $v \in V$ the image in $\mathrm{Ext}^1_{G_K}(\overline{\alpha}, \overline{\beta}_{f(t)})$ of the element of $\mathrm{Ext}^1_{G_K}(\overline{\alpha}, \overline{\rho}_{f(t)})$ corresponding to $\mathcal{E}_{g(v)}$ is nonzero. To this end, note that, after possibly shrinking T (and remembering that T is reduced), we may assume that $H^2(G_K, \overline{\beta}_T \otimes \overline{\alpha}^\vee)$ is locally free of some constant rank. We can therefore repeat the construction that we carried out after Lemma 5.4.1, and find a good complex

$$C^0_T(\overline{\beta}) \to Z^1_T(\overline{\beta})$$

whose cohomology in degree 1, which we denote by $H^1_T(\overline{\beta})$, is compatible with base change and computes $H^1(G_K, \overline{\beta}_t \otimes \overline{\alpha}^\vee) = \mathrm{Ext}^1_{G_K}(\overline{\alpha}, \overline{\beta}_t)$ at finite type points t of T. By [Sta, Tag 064E] we have a morphism of complexes

$$(C^0_T \to Z^1_T) \to (C^0_T(\overline{\beta}) \to Z^1_T(\overline{\beta})),$$

compatible with the natural maps on the cohomology groups induced by the corresponding morphism of Herr complexes.

The kernel of $Z^1_T \to H^1_T(\overline{\beta})$ is a coherent sheaf, so after possibly shrinking the reduced scheme T, we can suppose that it is a vector subbundle of Z^1_T. By definition, we see that if we delete from V the corresponding subbundle then the required condition holds. It therefore suffices to show that we are deleting a *proper* subbundle of V. If this were not the case, then (considering the fiber of V and the subbundle under consideration over some closed point t) we would have an exact sequence

$$0 \to \mathrm{Ext}^1_{G_K}(\overline{\alpha}, \overline{\beta}_t) \to \mathrm{Ext}^2_{G_K}(\overline{\alpha}, \overline{r}_t) \to \mathrm{Ext}^2_{G_K}(\overline{\alpha}, \overline{\rho}_t) \to \mathrm{Ext}^2_{G_K}(\overline{\alpha}, \overline{\beta}_t) \to 0;$$

equivalently, by Tate local duality we would have an exact sequence

$$0 \to \mathrm{Hom}_{G_K}(\overline{\beta}_t, \overline{\alpha}(1)) \to \mathrm{Hom}_{G_K}(\overline{\rho}_t, \overline{\alpha}(1)) \to \mathrm{Hom}_{G_K}(\overline{r}_t, \overline{\alpha}(1)) \to$$

$$\mathrm{Ext}^1_{G_K}(\overline{\beta}_t, \overline{\alpha}(1)) \to 0.$$

Assume for the sake of contradiction that this is the case. Since $\overline{\rho}_t$ is maximally nonsplit, the map $\mathrm{Hom}_{G_K}(\overline{\beta}_t, \overline{\alpha}(1)) \to \mathrm{Hom}_{G_K}(\overline{\rho}_t, \overline{\alpha}(1))$ is an isomorphism; so it suffices to show that we cannot have an isomorphism $\mathrm{Hom}_{G_K}(\overline{r}_t, \overline{\alpha}(1)) \to \mathrm{Ext}^1_{G_K}(\overline{\beta}_t, \overline{\alpha}(1))$.

Since $\overline{r}_t$ is maximally nonsplit, $\mathrm{Hom}_{G_K}(\overline{r}_t, \overline{\alpha}(1))$ is at most one-dimensional with equality if and only if $\mathrm{ur}_{\lambda_{d-1}(t)} \otimes \psi_{\underline{n}_{d-1}} = \overline{\alpha}(1)$; while by the local Euler characteristic formula, $\mathrm{Ext}^1_{G_K}(\overline{\beta}_t, \overline{\alpha}(1))$ has dimension at least $[K : \mathbf{Q}_p]$, with equality if and only if $\overline{\beta}_t \neq \overline{\alpha}$ and $\overline{\beta}_t \neq \overline{\alpha}(1)$. It follows that we must have $K = \mathbf{Q}_p$, $\mathrm{ur}_{\lambda_{d-1}(t)} \otimes \psi_{\underline{n}_{d-1}} = \overline{\alpha}(1)$, $\overline{\beta}_t \neq \overline{\alpha}$ and $\overline{\beta}_t \neq \overline{\alpha}(1)$; but this case was excluded by assumption (2d), and we have our a contradiction.

Finally, suppose that we are in the situation of (3). We wish to argue in the same way as in part (2), but rather than deleting the split locus, we need to delete the peu ramifiée locus. To be more precise, note that the surjection $\overline{\rho}_T \to \overline{\alpha}(1)_T$ (where $\overline{\alpha}(1)_T$ denotes the constant rank 1 family given by spreading out $\overline{\alpha}(1)$ over T) induces a composite morphism $Z^1_T \to H^1(G_K, \overline{\rho}_T \otimes \overline{\alpha}^\vee) \to H^1(G_K, \overline{\epsilon}_T)$ (where $\overline{\epsilon}_T$ denotes the spreading out of the mod p cyclotomic character $\overline{\epsilon}$ over T), which in turn induces a morphism $V \to Y$ of total spaces of vector bundles over T, with Y denoting the total space of the trivial bundle over T having fiber $H^1(G_K, \overline{\epsilon})$. Now Y contains a trivial subbundle $Y_{\text{p.r.}}$ of codimension 1, parameterizing the peu ramifiée classes. The preimage of $Y_{\text{p.r.}}$ is a closed subscheme $V_{\text{p.r.}}$ of V, and we wish to show that $V_{\text{p.r.}}$ is a proper closed subscheme of V.

For this, it is enough to show that for each t (perhaps after replacing T by a non-empty open subset), there is some extension of $\overline{\alpha}$ by $\overline{\rho}_{f(t)}$ with the property that the induced extension of $\overline{\alpha}$ by $\overline{\beta}_t = \overline{\alpha}(1)$ is très ramifiée. We have an exact sequence

$$\operatorname{Ext}^1_{G_K}(\overline{\alpha}, \overline{\rho}_t) \to \operatorname{Ext}^1_{G_K}(\overline{\alpha}, \overline{\beta}_t) \to \operatorname{Ext}^2_{G_K}(\overline{\alpha}, \overline{r}_t),$$

so we are done if $\operatorname{Ext}^2_{G_K}(\overline{\alpha}, \overline{r}_t) = 0$. Since $\overline{r}_t$ is maximally nonsplit, this means that we are done unless $\overline{r}_t$ admits $\overline{\alpha}(1)$ as its unique quotient character.

By hypothesis, this implies that $\overline{\epsilon}$ is trivial, and we can assume that the extension of $\overline{\beta}_t = \overline{\alpha}(1) = \overline{\alpha}$ by $\overline{\alpha}(1) = \overline{\alpha}$ induced by $\overline{\rho}_t$ is très ramifiée. Write c_t for the corresponding class in $H^1(G_K, \overline{\epsilon})$. As in Remark 5.4.10 below, it is enough to show that there is a très ramifiée extension of $\overline{\alpha}$ by $\overline{\alpha}(1)$, corresponding to a class $d \in H^1(G_K, \overline{\epsilon})$ such that the cup product of c_t and d vanishes. If this is not the case, then c_t must be an unramified class in $H^1(G_K, 1) = H^1(G_K, \overline{\epsilon})$; but the extension of K cut out by c_t considered as a class in $H^1(G_K, 1)$ is an extension given by adjoining a pth root of a uniformizer, so is in particular a ramified extension, and we are done. □

5.4.10 Remark. The curious looking condition Proposition 5.4.4 (2) is not merely an artefact of our arguments. Indeed, if we fix characters $\overline{\alpha}, \overline{\beta} \colon G_{\mathbf{Q}_p} \to \overline{\mathbf{F}}_p^\times$, with $\overline{\beta} \neq \overline{\alpha}$, $\overline{\beta} \neq \overline{\alpha}(1)$, and let

$$\overline{r} = \begin{pmatrix} \overline{\alpha}(1) & * \\ 0 & \overline{\beta} \end{pmatrix}$$

be a nonsplit extension, then any extension of $\overline{\alpha}$ by $\overline{r}$ induces the trivial extension of $\overline{\alpha}$ by $\overline{\beta}$. To see this, note that if we fix an extension of $\overline{\alpha}$ by $\overline{\beta}$, then the condition that we can form an extension of $\overline{\alpha}$ by $\overline{r}$ realising this extension of $\overline{\alpha}$ by $\overline{\beta}$ is that the cup product of the corresponding classes in $H^1(G_{\mathbf{Q}_p}, \overline{\alpha}\overline{\beta}^{-1}(1))$ and $H^1(G_{\mathbf{Q}_p}, \overline{\beta}\overline{\alpha}^{-1})$ vanishes. But by Tate local duality, this cup product is a perfect pairing of one-dimensional vector spaces, so one of the two classes must vanish, as required.

5.5 $\mathcal{X}_d$ IS A FORMAL ALGEBRAIC STACK

We now use the theory of families of extensions to show that $\mathcal{X}_d$ is a formal algebraic stack. The key ingredient is Theorem 5.5.12, which we prove by induction on d. We begin by setting up some useful terminology, which is motivated by the generalizations of the weight part of Serre's conjecture formulated in [GHS18]. Recall that the notion of a *Serre weight* was defined in Section 1.12. Since $\mathbf{F}$ is assumed to contain k, we can and do identify embeddings $k \hookrightarrow \overline{\mathbf{F}}_p$ and $k \hookrightarrow \mathbf{F}$.

5.5.1 Definition. If $\underline{k}$ is a Serre weight, and $\overline{\rho}\colon G_K \to \mathrm{GL}_d(\overline{\mathbf{F}}_p)$ is maximally nonsplit of niveau 1, then we say that $\overline{\rho}$ is of weight $\underline{k}$ if we can write

$$\overline{\rho} \cong \begin{pmatrix} \chi_1 & * & \dots & * \\ 0 & \chi_2 & \dots & * \\ \vdots & & \ddots & \vdots \\ 0 & \dots & 0 & \chi_d \end{pmatrix}$$

where

- $\chi_i|_{I_K} = \overline{\epsilon}^{1-i} \prod_{\overline{\sigma}\colon k \hookrightarrow \mathbf{F}} \omega_{\overline{\sigma}}^{-k_{\overline{\sigma},d+1-i}}$, and
- if $(\chi_{i+1}\chi_i^{-1})|_{I_K} = \overline{\epsilon}^{-1}$, then $k_{\overline{\sigma},d-i} - k_{\overline{\sigma},d+1-i} = p-1$ for all $\overline{\sigma}$ if and only if $\chi_{i+1}\chi_i^{-1} = \overline{\epsilon}^{-1}$ and the element of $\mathrm{Ext}^1_{G_K}(\chi_i, \chi_{i+1}) = H^1(G_K, \overline{\epsilon})$ determined by $\overline{\rho}$ is très ramifiée (and otherwise $k_{\overline{\sigma},d-i} - k_{\overline{\sigma},d+1-i} = 0$ for all $\overline{\sigma}$).

Note that each maximally nonsplit representation of niveau 1 is of weight $\underline{k}$ for a unique $\underline{k}$. For each $\underline{k}$, there exists a $\overline{\rho}$ which is maximally nonsplit of niveau 1 and weight $\underline{k}$; the existence of such a $\overline{\rho}$ follows by an easy induction, and is in any case an immediate consequence of Theorem 5.5.12 below.

5.5.2 Definition. We say that a weight $\underline{k}'$ is a *shift* of a weight $\underline{k}$ if $k'_{\underline{\sigma},d} = k_{\underline{\sigma},d}$ for all $\overline{\sigma}$, and if for each $1 \le i \le d-1$, we either have $k'_{\overline{\sigma},i} - k'_{\overline{\sigma},i+1} = k_{\overline{\sigma},i} - k_{\overline{\sigma},i+1}$ for all $\overline{\sigma}$, or we have $k'_{\overline{\sigma},i} - k'_{\overline{\sigma},i+1} = p-1$ and $k_{\overline{\sigma},i} - k_{\overline{\sigma},i+1} = 0$ for all $\overline{\sigma}$. (Note in particular that any weight is a shift of itself.)

5.5.3 Definition. We say that a representation $\rho\colon G_K \to \mathrm{GL}_d(\mathcal{O})$ or $\rho\colon G_K \to \mathrm{GL}_d(\overline{\mathbf{Z}}_p)$ is *crystalline*, resp. *crystalline of weight* λ, if $\rho[1/p]$ is crystalline, resp. crystalline of weight λ. We say that ρ is a *crystalline lift* of $\overline{\rho}\colon G_K \to \mathrm{GL}_d(\mathbf{F})$ (resp. of $\overline{\rho}\colon G_K \to \mathrm{GL}_d(\overline{\mathbf{F}}_p)$).

The motivation for Definitions 5.5.1 and 5.5.2 is the following lemma.

5.5.4 Lemma. *If $\overline{\rho}$ is maximally nonsplit of niveau* 1 *and weight $\underline{k}$, if $\underline{k}'$ is a shift of $\underline{k}$, and if $\underline{\lambda}$ is a lift of $\underline{k}'$, then $\overline{\rho}$ has a crystalline lift (defined over $\overline{\mathbf{Z}}_p$) of weight $\underline{\lambda}$. Furthermore, this lift can be chosen to be ordinary in the sense of Definition* 6.4.1 *below.*

Proof. Write $\overline{\rho}$ as in Definition 5.5.1, and assume (by replacing E by a finite unramified extension if necessary) that both $\overline{\rho}$ and the filtration in Definition 5.5.1 are defined over $\mathbf{F}$. Note firstly that by [GHS18, Lem. 5.1.6] (which uses the opposite conventions for the sign of the Hodge–Tate weights to those of this book), together with the observation made above that $\prod_{\overline{\sigma}} \omega_{\overline{\sigma}}^{e(K/\mathbf{Q}_p)} = \overline{\epsilon}|_{I_K}$, we can find crystalline characters $\psi_i : G_K \to \mathcal{O}^\times$ such that $\overline{\psi}_i = \chi_i$ and ψ_i has Hodge–Tate weights given by $\lambda_{\sigma,d+1-i}$.

It is enough to prove that $\overline{\rho}$ has a crystalline lift which is a successive extension of characters ψ_i of the form

$$\begin{pmatrix} \psi'_1 & * & \dots & * \\ 0 & \psi'_2 & \dots & * \\ \vdots & & \ddots & \vdots \\ 0 & \dots & 0 & \psi'_d \end{pmatrix},$$

where each ψ'_i is an unramified twist of ψ_i; note that any such representation is by definition ordinary in the sense of Definition 6.4.1. We prove this by induction on d, the case $d=1$ being trivial. In the general case, write $\overline{\rho}$ as an extension

$$0 \to \overline{r} \to \overline{\rho} \to \overline{\chi}_d \to 0,$$

and suppose inductively that $\overline{r}$ has a lift of the required form. Write $r : G_K \to \mathrm{GL}_{d-1}(\mathcal{O})$ for this lift; then we inductively seek a lift of $\overline{\rho}$ of the form

$$0 \to r \to \rho \to \psi'_d \to 0, \tag{5.5.5}$$

where ψ'_d is an unramified twist of ψ_d. We insist also that

$$\psi'_{d-1}(\psi'_d)^{-1} \neq \epsilon; \tag{5.5.6}$$

note that this could only fail if $k'_{\overline{\sigma},d} = k'_{\overline{\sigma},d-1}$ for all $\overline{\sigma}$, and in this case we are only ruling out a single unramified twist. Then since $\overline{\rho}$ is maximally nonsplit, we have $\mathrm{Hom}_{G_K}(r, \psi_d \epsilon) = 0$, and it follows as in the proof of Lemma 6.3.1 that such a ρ, if it exists, is automatically crystalline.

It is therefore enough to show that there is a lift of the form (5.5.5) satisfying (5.5.6). As in the proof of [GHLS17, Thm. 2.1.8], we have an exact sequence

$$H^1(G_K, (\psi'_d)^{-1} \otimes r) \to H^1(G_K, \chi_d^{-1} \otimes \overline{r}) \xrightarrow{\delta} H^2(G_K, (\psi'_d)^{-1} \otimes r). \tag{5.5.7}$$

Thus, if $c \in H^1(G_K, \chi_d^{-1} \otimes \overline{r})$ is the class determined by $\overline{\rho}$, it is enough to show that we can choose the unramified twist ψ'_d so that $\delta(c) = 0$. Now, $H^2(G_K, (\psi'_d)^{-1} \otimes r)$ is dual to $H^0(G_K, r^\vee \otimes (\psi'_d \epsilon) \otimes E/\mathcal{O})$, and since $\overline{r}$ is maximally nonsplit, we see that this latter group is equal to $H^0(G_K, (\psi'_{d-1})^{-1} \psi'_d \epsilon \otimes E/\mathcal{O})$. In other words, to show that $\delta(c)=0$ in (5.5.7), it is enough to show the corresponding statement where r is replaced by ψ'_{d-1}; so we need only show that we can choose ψ'_d so that the extension of χ_d by χ_{d-1} lifts to a crystalline extension of ψ'_d by ψ'_{d-1}. This is an immediate consequence of the weight part of Serre's conjecture for GL_2, as proved in [GLS15]; see in particular [GLS15, Thm. 6.1.18] and the references therein (and note that these results also hold for $p=2$, by [Wan17]). □

5.5.8 Remark. We emphasize that the preceding lemma applies only to $\overline{\rho}$ that are maximally split of niveau 1. In Theorem 6.4.4 below we prove an analogous result for arbitrary $\overline{\rho}$.

Suppose that $\overline{\rho}_{\mathcal{T}}$ is generically maximally nonsplit of niveau 1, where $\mathcal{T}$ is some integral finite type $\overline{\mathbf{F}}_p$-stack. We say that $\overline{\rho}_{\mathcal{T}}$ is generically of weight $\underline{k}$ if there is a dense open substack $\mathcal{U}$ of $\mathcal{T}$ such that every $\overline{\mathbf{F}}_p$-point of $\overline{\rho}_{\mathcal{U}}$ is maximally nonsplit of niveau 1, and is of weight $\underline{k}$.

5.5.9 Lemma. *If $\overline{\rho}_{\mathcal{T}}$ is generically maximally nonsplit of niveau 1, then $\overline{\rho}_{\mathcal{T}}$ is generically of weight $\underline{k}$ for a unique Serre weight $\underline{k}$.*

Proof. Consider a dense open substack $\mathcal{U}\subseteq\mathcal{T}$, tuples $\underline{n}_i$ and morphisms $\lambda_i: U\to\mathbf{G}_m$ as in Proposition 5.3.8. We inductively determine the $k_{\overline{\sigma},i}$ as follows. We let $k_{\overline{\sigma},d}$ be the unique choice with $p-1\geq k_{\overline{\sigma},d}\geq 0$ not all equal to $p-1$ and satisfying

$$\psi_{\underline{n}_1}|_{I_K}=\prod_{\overline{\sigma}:\,k\hookrightarrow\overline{\mathbf{F}}_p}\omega_{\overline{\sigma}}^{-k_{\overline{\sigma},d}}.$$

Inductively, if $k_{\overline{\sigma},i+1}$ has been determined, then we demand that $k_{\overline{\sigma},i}$ satisfies $p-1\geq k_{\overline{\sigma},i}-k_{\overline{\sigma},i+1}\geq 0$ and

$$(\psi_{\underline{n}_{d-i}}\psi^{-1}_{\underline{n}_{d+1-i}})|_{I_K}=\overline{\epsilon}\prod_{\overline{\sigma}:\,k\hookrightarrow\overline{\mathbf{F}}_p}\omega_{\overline{\sigma}}^{k_{\overline{\sigma},i}-k_{\overline{\sigma},i+1}}.$$

This determines $k_{\overline{\sigma},i}$, except in the case that $(\psi_{\underline{n}_{d-i}}\psi^{-1}_{\underline{n}_{d+1-i}})|_{I_K}=\overline{\epsilon}$. In this case, we consider the character $\lambda_{d+1-i}\lambda^{-1}_{d-i}: U\to\mathbf{G}_m$. Since $\mathcal{U}$ is integral, this character is either constant or has open image. After possibly shrinking $\mathcal{U}$, we may therefore assume that either $\lambda_{d+1-i}(t)\neq\lambda_{d-i}(t)$ for all $t\in\mathcal{U}$, in which case we set $k_{\overline{\sigma},d+1-i}-k_{\overline{\sigma},d-i}=0$ for all $\overline{\sigma}$, or that the character is trivial.

If the character is trivial, we may consider the locus in $\mathcal{U}$ over which the extension between χ_{i+1} and χ_i is peu ramifiée. The argument used in the proof of part (3) of Proposition 5.4.4 shows that this is a constructible subset of $\mathcal{U}$; since $\mathcal{U}$ is integral, after shrinking $\mathcal{U}$ we can thus suppose that it is either equal to $\mathcal{U}$, or that it is empty. In the former case we set $k_{\overline{\sigma},d+1-i}-k_{\overline{\sigma},d-i}=0$ for all $\overline{\sigma}$, and in the latter we take $k_{\overline{\sigma},d+1-i}-k_{\overline{\sigma},d-i}=p-1$ for all $\overline{\sigma}$. It follows from the construction that every $\overline{\mathbf{F}}_p$-point of $\overline{\rho}_{\mathcal{U}}$ is maximally nonsplit of niveau 1 and weight $\underline{k}$, as required. □

It will be convenient to use the following variant of the notation established in Proposition 5.3.8. If $\overline{\rho}_{\mathcal{T}}$ is generically maximally nonsplit of niveau 1 and weight $\underline{k}$, then we set $\nu_i := \lambda_{d+1-i}$, and write $\omega_{\underline{k},i}$ for the character $\overline{\epsilon}^{d-i}\psi_{\underline{n}_{d+1-i}}$, so that

$$\omega_{\underline{k},i}|I_K = \prod_{\overline{\sigma}:\, k \hookrightarrow \overline{\mathbf{F}}_p} \omega_{\overline{\sigma}}^{-k_{\overline{\sigma},i}},$$

and (5.3.9) can be rewritten as

$$\overline{\rho}_t \cong \begin{pmatrix} \mathrm{ur}_{\nu_d(t)} \otimes \omega_{\underline{k},d} & * & \cdots & * \\ 0 & \mathrm{ur}_{\nu_{d-1}(t)} \otimes \epsilon^{-1}\omega_{\underline{k},d-1} & \cdots & * \\ \vdots & & \ddots & \vdots \\ 0 & \cdots & 0 & \mathrm{ur}_{\nu_1(t)} \otimes \epsilon^{1-d}\omega_{\underline{k},1} \end{pmatrix}. \tag{5.5.10}$$

We have the *eigenvalue morphism* $\underline{\nu}: \mathcal{U} \to (\mathbf{G}_m)^d$ given by $(\nu_1, \ldots, \nu_d)$.

5.5.11 Definition. If $\underline{k}$ is a Serre weight, we let $(\mathbf{G}_m)^d_{\underline{k}}$ denote the closed subgroup scheme of $(\mathbf{G}_m)^d$ parameterizing tuples $(x_1, \ldots, x_d)$ for which $x_i = x_{i+1}$ whenever $k_{\overline{\sigma},i} - k_{\overline{\sigma},i+1} = p-1$ for all $\overline{\sigma}$.

Note that $(\mathbf{G}_m)^d_{\underline{k}}$ is closed under the simultaneous multiplication action of $\mathbf{G}_m$ on $(\mathbf{G}_m)^d_{\underline{k}}$ (i.e., the action $a \cdot (x_1, \ldots, x_d) := (ax_1, \ldots, ax_d)$).

It follows from the definitions that the eigenvalue morphism $\underline{\nu}$ is valued in $(\mathbf{G}_m)^d_{\underline{k}}$.

We have seen in Proposition 3.4.12 that $\mathcal{X}_d$ is an Ind-algebraic stack, which is in fact the 2-colimit of algebraic stacks with respect to transition morphisms that are closed immersions. We let $\mathcal{X}_{d,\mathrm{red}}$ be the underlying reduced substack of $\mathcal{X}_d$; it is then a closed substack of $\mathcal{X}_d$ with the same structure: namely, it is the 2-colimit of closed algebraic stacks (which can even be assumed to be reduced) with respect to transition morphisms that are closed immersions.

5.5.12 Theorem

1. The Ind-algebraic stack $\mathcal{X}_{d,\mathrm{red}}$ is an algebraic stack, of finite presentation over $\mathbf{F}$.
2. We can write $(\mathcal{X}_{d,\mathrm{red}})_{\overline{\mathbf{F}}_p}$ as a union of closed algebraic substacks of finite presentation over $\overline{\mathbf{F}}_p$,

$$(\mathcal{X}_{d,\mathrm{red}})_{\overline{\mathbf{F}}_p} = \mathcal{X}^{\mathrm{small}}_{d,\mathrm{red},\overline{\mathbf{F}}_p} \cup \bigcup_{\underline{k}} \mathcal{X}^{\underline{k}}_{d,\mathrm{red},\overline{\mathbf{F}}_p},$$

where:

- $\mathcal{X}^{\text{small}}_{d,\text{red},\overline{\mathbf{F}}_p}$ is empty if $d=1$, and otherwise is non-empty of dimension strictly less than $[K:\mathbf{Q}_p]d(d-1)/2$.
- each $\mathcal{X}^{\underline{k}}_{d,\text{red},\overline{\mathbf{F}}_p}$ is a closed irreducible substack of dimension $[K:\mathbf{Q}_p]d(d-1)/2$, and is generically maximally nonsplit of niveau 1 and weight $\underline{k}$. The corresponding eigenvalue morphism is dominant (i.e., has dense image in $(\mathbf{G}_m)^d_{\underline{k}}$).

3. If we fix an irreducible representation $\overline{\alpha}\colon G_K \to \mathrm{GL}_a(\overline{\mathbf{F}}_p)$ (for some $a \geq 1$), then the locus of $\overline{\rho}$ in $\mathcal{X}_{d,\text{red}}(\overline{\mathbf{F}}_p)$ for which $\dim \mathrm{Hom}_{G_K}(\overline{\rho},\overline{\alpha}) \geq r$ (for any $r \geq 1$) is (either empty, or) of dimension at most

$$[K:\mathbf{Q}_p]d(d-1)/2 - \lceil r\big((a^2+1)r-a\big)/2\rceil.$$

Furthermore, the locus of $\overline{\rho}$ in $\mathcal{X}^{\text{small}}_{d,\text{red},\overline{\mathbf{F}}_p}(\overline{\mathbf{F}}_p)$ for which $\dim \mathrm{Hom}_{G_K}(\overline{\rho},\overline{\alpha}) \geq r$ is of dimension strictly less than this.

4. If we fix an irreducible representation $\overline{\alpha}\colon G_K \to \mathrm{GL}_a(\overline{\mathbf{F}}_p)$ (for some $a \geq 1$), then the locus of $\overline{\rho}$ in $\mathcal{X}_{d,\text{red}}(\overline{\mathbf{F}}_p)$ for which $\dim \mathrm{Ext}^2_{G_K}(\overline{\alpha},\overline{\rho}) \geq r$ is of dimension at most

$$[K:\mathbf{Q}_p]d(d-1)/2 - r.$$

5.5.13 Remark. Note that in parts (3) and (4), the locus of points in question corresponds to a closed substack of $(\mathcal{X}_{d,\text{red}})_{\overline{\mathbf{F}}_p}$, by upper-semicontinuity of fiber dimension.

5.5.14 Remark. Since the $\mathcal{X}^{\underline{k}}_{d,\text{red},\overline{\mathbf{F}}_p}$ are irreducible, have dimension equal to that of $(\mathcal{X}_{d,\text{red}})_{\overline{\mathbf{F}}_p}$, and have pairwise disjoint open substacks (corresponding to maximally nonsplit representations of niveau 1 and weight $\underline{k}$), they are in fact distinct irreducible components of $(\mathcal{X}_{d,\text{red}})_{\overline{\mathbf{F}}_p}$.

5.5.15 Remark. We can, and will, be much more precise about the structure of $\mathcal{X}_{d,\text{red}}$. Namely, in Chapter 6 we combine Theorem 5.5.12 with additional arguments to show that $\mathcal{X}_{d,\text{red}}$ is equidimensional of dimension $[K:\mathbf{Q}_p]d(d-1)/2$; accordingly, the irreducible components of $(\mathcal{X}_{d,\text{red}})_{\overline{\mathbf{F}}_p}$ are precisely the $\mathcal{X}^{\underline{k}}_{d,\text{red},\overline{\mathbf{F}}_p}$, and in particular are in bijection with the Serre weights $\underline{k}$. We also show that these irreducible components are all defined over $\mathbf{F}$.

5.5.16 Remark. Note that since the eigenvalue morphism has dense image, it follows that $\mathcal{X}^{\underline{k}}_{d,\text{red}}$ contains a dense open substack with the property that $\nu_i(t) \neq \nu_{i+1}(t)$ unless $k_{\overline{\sigma},i} - k_{\overline{\sigma},i+1} = p-1$ for all $\overline{\sigma}$.

5.5.17 Remark. As will be evident from the proof, the upper bound of Theorem 5.5.12 (3) is quite crude when $[K:\mathbf{Q}_p] > 1$, although it is reasonably sharp in the case $K = \mathbf{Q}_p$. However, it suffices for our purposes, and indeed we will only use the (even cruder) consequence Theorem 5.5.12 (4) in the rest of the book.

Proof of Theorem 5.5.12. Recall that a closed immersion of reduced algebraic stacks that are locally of finite type over $\mathbf{F}_p$ which is surjective on finite type points is necessarily an isomorphism. As recalled above, $\mathcal{X}_{d,\mathrm{red}}$ is an inductive limit of such stacks (indeed, by Lemma A.9, we have $\mathcal{X}_{d,\mathrm{red}} = \varinjlim \mathcal{X}^a_{d,h,s,\mathrm{red}}$, where the $\mathcal{X}^a_{d,h,s}$ are as in Section 3.4), and so if we produce closed algebraic substacks $\mathcal{X}^{\mathrm{small}}_{d,\mathrm{red},\overline{\mathbf{F}}_p}$ and $\mathcal{X}^{\underline{k}}_{d,\mathrm{red},\overline{\mathbf{F}}_p}$ of $\mathcal{X}_d$, the union of whose $\overline{\mathbf{F}}_p$-points exhausts those of $\mathcal{X}_{d,\mathrm{red}}$, then $\mathcal{X}_{d,\mathrm{red},\overline{\mathbf{F}}_p}$ will in fact be an algebraic stack which is the union of its closed substacks $\mathcal{X}^{\mathrm{small}}_{d,\mathrm{red},\overline{\mathbf{F}}_p}$ and $\mathcal{X}^{\underline{k}}_{d,\mathrm{red},\overline{\mathbf{F}}_p}$. Thus (1) is an immediate consequence of (2) (where the "union" statement in (2) is now to be understood on the level of $\overline{\mathbf{F}}_p$-points).

Claim (4) follows from (3) (with $\overline{\alpha}$ replaced by $\overline{\alpha} \otimes \overline{\epsilon}$) by Tate local duality and the easily verified inequality

$$\lceil r\big((a^2+1)r - a\big)/2 \rceil \geq r.$$

Thus it is enough to prove (2) and (3), which we do simultaneously by induction on d.

As recalled in Remark 5.3.4, there are up to twist by unramified characters only finitely many irreducible $\overline{\mathbf{F}}_p$-representations of G_K of any fixed dimension. Accordingly, we let $\{\overline{\alpha}_i\}$ be a finite set of irreducible continuous representations $\overline{\alpha}_i \colon G_K \to \mathrm{GL}_{d_i}(\overline{\mathbf{F}}_p)$, such that any irreducible continuous representation of G_K over $\overline{\mathbf{F}}_p$ of dimension at most d arises as an unramified twist of exactly one of the $\overline{\alpha}_i$. We let the one-dimensional representations in this set be the characters $\psi_{\underline{n}}$ defined in Section 5.3.

Each $\overline{\alpha}_i$ corresponds to a finite type point of $\mathcal{X}_{d_i,\mathrm{red}}$, whose associated residual gerbe is a substack of $\mathcal{X}_{d_i,\mathrm{red}}$ of dimension -1: the morphism $\operatorname{Spec} \overline{\mathbf{F}}_p \to \mathcal{X}_{d_i,\mathrm{red}}$ corresponding to $\overline{\alpha}_i$ factors through a monomorphism $[\operatorname{Spec} \overline{\mathbf{F}}_p/\mathbf{G}_m] \to \mathcal{X}_{d_i,\mathrm{red}}$. It follows from Lemma 5.3.2 that for each $\overline{\alpha}_i$ there is an irreducible closed zero-dimensional algebraic substack of $\mathcal{X}_{d_i,\mathrm{red}}$ of finite presentation over $\overline{\mathbf{F}}_p$ whose $\overline{\mathbf{F}}_p$-points are exactly the unramified twists of $\overline{\alpha}_i$.

In particular, if $d=1$, then we let $\mathcal{X}^{\underline{k}}_{1,\mathrm{red}}$ be the zero-dimensional stack constructed in the previous paragraph, whose $\overline{\mathbf{F}}_p$-points are the unramified twists of $\psi_{\underline{k}}$; this satisfies the required properties by definition, so (2) holds when $d=1$. For (3), note that if $r>0$ then we must have $a=1$, and then the locus where $\mathrm{Hom}_{G_K}(\overline{\rho}, \overline{\alpha})$ is nonzero (equivalently, one-dimensional) is exactly the closed substack of dimension -1 corresponding to $\overline{\alpha}$, so the required bound holds.

We now begin the inductive proof of (2) and (3) for $d>1$. In fact, it will be helpful to simultaneously prove another statement (2'), which is as follows: for each $\underline{k}$, there is a closed irreducible algebraic substack $\mathcal{X}^{\underline{k},\mathrm{fixed}}_{d,\mathrm{red},\overline{\mathbf{F}}_p}$ of $(\mathcal{X}_{d,\mathrm{red}})_{\overline{\mathbf{F}}_p}$ of finite presentation over $\overline{\mathbf{F}}_p$ and dimension $[K:\mathbf{Q}_p]d(d-1)/2-1$, which is generically maximally nonsplit of niveau 1 and weight $\underline{k}$, and furthermore has the property that the corresponding character ν_1 is trivial. The discussion of the previous paragraph also proves (2') when $d=1$. The point of hypothesis (2') is that we can use Proposition 5.4.4 (2) to construct $\mathcal{X}^{\underline{k},\mathrm{fixed}}_{d,\mathrm{red},\overline{\mathbf{F}}_p}$ by using (2) and (2') in dimension $(d-1)$, and then construct $\mathcal{X}^{\underline{k}}_{d,\mathrm{red},\overline{\mathbf{F}}_p}$ from it using Lemma 5.3.2.

We now prove the inductive step, so we assume that (2), (2') and (3) hold in dimension less than d. We begin by constructing the closed substacks $\mathcal{X}^{\underline{k},\mathrm{fixed}}_{d,\mathrm{red}}$ and $\mathcal{X}^{\underline{k}}_{d,\mathrm{red},\overline{\mathbf{F}}_p}$. Let $\underline{k}_{d-1}$ be the Serre weight in dimension $(d-1)$ obtained by deleting the first entry in $\underline{k}$. We set $\overline{\alpha}:=\overline{\epsilon}^{1-d}\omega_{\underline{k},1}$. If $k_{\overline{\sigma},1}-k_{\overline{\sigma},2}=p-1$ for all $\overline{\sigma}$ then we say that we are in the *très ramifiée case*, and we let $\mathcal{T}$ denote $\mathcal{X}^{\underline{k}_{d-1},\mathrm{fixed}}_{d-1,\mathrm{red},\overline{\mathbf{F}}_p}$; otherwise, we let $\mathcal{T}$ denote $\mathcal{X}^{\underline{k}_{d-1}}_{d-1,\mathrm{red},\overline{\mathbf{F}}_p}$. In the latter case, it follows from our inductive hypothesis (more precisely, from the assumption that (2) and (3) hold in dimension $(d-1)$), together with Tate local duality, that after replacing $\mathcal{T}$ with an open substack we can and do assume that for each $\overline{\mathbf{F}}_p$-point t of $\mathcal{T}$, we have $\mathrm{Ext}^2_{G_K}(\overline{\alpha},\overline{r}_t)=0$, where $\overline{r}_t$ is the $(d-1)$-dimensional representation corresponding to t. If we are in the très ramifiée case then it follows similarly that after replacing $\mathcal{T}$ with an open substack we can and do assume that for each $\overline{\mathbf{F}}_p$-point t of $\mathcal{T}$, $\mathrm{Ext}^2_{G_K}(\overline{\alpha},\overline{r}_t)$ is one-dimensional. In either case we can in addition assume that $\overline{r}_t$ is maximally nonsplit of niveau 1 and weight $\underline{k}_{d-1}$.

We let T be an irreducible scheme which smoothly covers $\mathcal{T}$, and we let $\mathcal{X}^{\underline{k},\mathrm{fixed}}_{d,\mathrm{red},\overline{\mathbf{F}}_p}$ be the irreducible closed substack of $(\mathcal{X}_{d,\mathrm{red}})_{\overline{\mathbf{F}}_p}$ constructed as the scheme-theoretic image of V in the notation of Proposition 5.4.4. Part (2) of that proposition, together with the inductive hypothesis, implies that $\mathcal{X}^{\underline{k},\mathrm{fixed}}_{d,\mathrm{red},\overline{\mathbf{F}}_p}$ has the claimed dimension. (Note in particular that if $K=\mathbf{Q}_p$, then condition (2d) of Proposition 5.4.4 holds. Indeed, if we are in the très ramifiée case then condition (2d)(iii) holds, and otherwise the inductive hypothesis that the image of the eigenvalue morphism is dense in $(\mathbf{G}_m)^{d-1}_{\underline{k}_{d-1}}$ implies that condition (2d)(i) holds after shrinking T.) We then let $\mathcal{X}^{\underline{k}}_{d,\mathrm{red},\overline{\mathbf{F}}_p}$ be the substack obtained from $\mathcal{X}^{\underline{k},\mathrm{fixed}}_{d,\mathrm{red}}$ by twisting by unramified characters, which has the claimed dimension by Lemma 5.3.2 (and the fact that by Proposition 5.4.4 (2), there is a dense open substack of $\mathcal{X}^{\underline{k},\mathrm{fixed}}_{d,\mathrm{red}}$ whose $\overline{\mathbf{F}}_p$-points correspond to representations which are of niveau 1 and are maximally nonsplit, so that in particular the corresponding family is twistable).

By construction, we see that the stacks $\mathcal{X}^{\underline{k},\mathrm{fixed}}_{d,\mathrm{red},\overline{\mathbf{F}}_p}$ and $\mathcal{X}^{\underline{k}}_{d,\mathrm{red},\overline{\mathbf{F}}_p}$ satisfy the properties required of them, except possibly that in the très ramifiée case, we need to check that $\mathcal{X}^{\underline{k},\mathrm{fixed}}_{d,\mathrm{red},\overline{\mathbf{F}}_p}$ (and consequently $\mathcal{X}^{\underline{k}}_{d,\mathrm{red},\overline{\mathbf{F}}_p}$) is generically maximally nonsplit of niveau 1 and weight $\underline{k}$. More precisely, in this case we need to check the condition that the final extension is not just nonsplit, but is (generically) très ramifiée. This follows from Proposition 5.4.4 (3). Indeed, by our inductive assumption that $\mathcal{X}^{\underline{k}_{d-1}}_{d-1,\mathrm{red},\overline{\mathbf{F}}_p}$ is generically maximally nonsplit of niveau 1 and weight $\underline{k}_{d-1}$, and the image of the eigenvalue morphism is dense, we see that we are in case (3)(b)(i) unless $\overline{\epsilon}$ is trivial and furthermore we have $k_{\overline{\sigma}_2} - k_{\overline{\sigma}_3} = p-1$ for all $\overline{\sigma}$, in which case we satisfy (3)(b)(ii).

To complete the proof of (2), we need only construct $\mathcal{X}^{\mathrm{small}}_{d,\mathrm{red},\overline{\mathbf{F}}_p}$. By Proposition 5.4.4, Tate local duality, upper semi-continuity of the fiber dimension, and the inductive hypothesis, we see that for each $1 \le i \le d$, each $\overline{\alpha}_i$ (of dimension a_i, say), and each $s \ge 0$ there is a finitely presented closed algebraic substack $\mathcal{X}_{s,\overline{\alpha}_i,\overline{\mathbf{F}}_p}$ of $(\mathcal{X}_{d,\mathrm{red}})_{\overline{\mathbf{F}}_p}$, whose $\overline{\mathbf{F}}_p$-points contain all the representations of the form $0 \to \overline{\rho}_{d-a_i} \to \overline{\rho} \to \overline{\alpha}_i \to 0$ for which $\dim_{\overline{\mathbf{F}}_p} \mathrm{Ext}^2_{G_K}(\overline{\alpha}_i, \overline{\rho}_{d-a_i}) = s$, and whose dimension is at most

$$\begin{aligned}&[K:\mathbf{Q}_p](d-a_i)(d-a_i-1)/2 - \lceil s((a_i{}^2+1)s - a_i)/2 \rceil \\ &\quad + [K:\mathbf{Q}_p]a_i(d-a_i) + s - 1;\end{aligned}$$

furthermore, the locus where $\overline{\rho}_{d-a_i}$ is an $\overline{\mathbf{F}}_p$-point of $\mathcal{X}^{\mathrm{small}}_{d-a_i,\mathrm{red},\overline{\mathbf{F}}_p}$ is of dimension strictly less than this. These stacks are only nonzero for finitely many values of s. For fixed a_i, we see that as a function of s, this quantity is maximized by $s=0$, as well as by $s=1$ when $a_i=1$. (To see this, we have to maximize the quantity $s - \lceil s((a_i{}^2+1)s - a_i)/2 \rceil$. Suppose firstly that $a_i > 1$. Then if $s=0$ we have 0, while if $s>0$ we have $s - \lceil s((a_i{}^2+1)s - a_i)/2 \rceil \le s - s((a_i{}^2+1)s - a_i)/2 \le s - s(a_i{}^2 + 1 - a_i)/2 \le s - 3s/2 < 0$. Meanwhile if $a_i = 1$, then for $s=0$ we have 0, for $s=1$ we have $1 - \lceil 1/2 \rceil = 1$, while for $s>1$ we have $s - \lceil s(2s-1)/2 \rceil \le s - s(2s-1)/2 \le s - 3s/2 < 0$.) It follows that as a function of a_i the bound is maximized at $a_i = 1$ and $s=0$ or 1, when it is equal to $[K:\mathbf{Q}_p]d(d-1)/2 - 1$, and it is otherwise strictly smaller.

By Lemma 5.3.2, it follows that the locus in $(\mathcal{X}_{d,\mathrm{red}})_{\overline{\mathbf{F}}_p}$ of representations of the form $0 \to \overline{\rho}_{d-a_i} \to \overline{\rho} \to \overline{\alpha}' \to 0$, with $\overline{\alpha}'$ an unramified twist of $\overline{\alpha}_i$ for which $\dim_{\overline{\mathbf{F}}_p} \mathrm{Ext}^2_{G_K}(\overline{\alpha}', \overline{\rho}_{d-a_i}) = s$, is of dimension at most $[K:\mathbf{Q}_p]d(d-1)/2$, with equality holding only if $a_i = 1$ and $s = 0$ or 1. Furthermore, the locus of those representations for which $\overline{\rho}_{d-a_i}$ is an $\overline{\mathbf{F}}_p$-point of $\mathcal{X}^{\mathrm{small}}_{d-a_i,\mathrm{red},\overline{\mathbf{F}}_p}$ is of dimension strictly less than $[K:\mathbf{Q}_p]d(d-1)/2$. Putting this together, we see that (2) holds in dimension d if we take $\mathcal{X}^{\mathrm{small}}_{d,\mathrm{red},\overline{\mathbf{F}}_p}$ to be the union of the twists by unramified characters of the substacks $\mathcal{X}_{s,\overline{\alpha}_i}$ for which $\dim \overline{\alpha}_i > 1$ or $s > 1$, together with the union of the twists by unramified characters of the substacks of the $\mathcal{X}_{s,\overline{\alpha}_i,\overline{\mathbf{F}}_p}$ for which $\dim \overline{\alpha}_i = 1$, $s = 0$ or 1, and $\overline{\rho}_{d-1}$ is an $\overline{\mathbf{F}}_p$-point of $\mathcal{X}^{\mathrm{small}}_{d-1,\mathrm{red},\overline{\mathbf{F}}_p}$.

Finally we prove (3) in dimension d. In the case $r=0$, there is nothing to prove. If $\dim \mathrm{Hom}_{G_K}(\overline{\rho}, \overline{\alpha}) \ge r \ge 1$, then we may place $\overline{\rho}$ in a short exact sequence

$$0 \to \overline{\theta} \to \overline{\rho} \to \overline{\alpha}^{\oplus r} \to 0,$$

where $\overline{\theta}$ is of dimension $d-ra<d$. We may apply part (2) so as to find that $\mathcal{X}_{d-ar,\mathrm{red},\overline{\mathbf{F}}_p}$ has dimension at most $[K:\mathbf{Q}_p](d-ar)(d-ar-1)/2$. Let $\mathcal{U}_s$ be the locally closed substack of $\mathcal{X}_{d-ar,\mathrm{red},\overline{\mathbf{F}}_p}$ over which $\dim H^2(G_K,\overline{\theta}\otimes\overline{\alpha}^\vee)=s$; by the inductive hypothesis, this locus has dimension at most $[K:\mathbf{Q}_p](d-ar)(d-ar-1)/2-s\big((a^2+1)s-a\big)/2$, and over this locus we may construct a universal family of extensions

$$0\to\overline{\theta}\to\overline{\rho}_{\mathcal{U}_s}\to\overline{\alpha}^{\oplus r}\to 0.$$

The locus of $\overline{\rho}$ we are interested in is contained in the scheme-theoretic image of this family in $(\mathcal{X}_{d,\mathrm{red}})_{\overline{\mathbf{F}}_p}$, and Proposition 5.4.4 shows that this scheme-theoretic image has dimension bounded above by

$$\begin{aligned}
&[K:\mathbf{Q}_p](d-ar)(d-ar-1)/2-s((a^2+1)s-a)/2\\
&\quad+r([K:\mathbf{Q}_p]a(d-ar)+s)-r^2\\
&\quad=[K:\mathbf{Q}_p]d(d-1)/2-(r(a^2+1)r-a)/2-(r-s)^2/2-(ar(ar-1))/2\\
&\quad\leq[K:\mathbf{Q}_p]d(d-1)/2-(r(a^2+1)r-a)/2.
\end{aligned}$$

Since this conclusion holds for each of the finitely many values of s (and since the dimension is an integer, allowing us to take the floor of this upper bound), we are done. □

5.5.18 Corollary. *$\mathcal{X}_d$ is a Noetherian formal algebraic stack.*

Proof. Theorem 5.5.12 shows that $\mathcal{X}_{d,\mathrm{red}}$ is an algebraic stack that is of finite presentation over $\mathbf{F}$ (and hence quasi-compact and quasi-separated). Together with Proposition 3.4.12, which shows that $\mathcal{X}_d$ is isomorphic to the inductive limit of a sequence of finitely presented algebraic stacks with respect to transition morphisms that are closed immersions, this implies that $\mathcal{X}_d$ is indeed a (locally countably indexed) formal algebraic stack, by Proposition A.12. By Remark A.20, $\mathcal{X}_d$ is Ind-locally of finite type over $\mathcal{O}$, and it then follows from Proposition 5.1.36 and Theorem A.35 that $\mathcal{X}_d$ is locally Noetherian, and hence Noetherian (as we have already seen that it is quasi-compact and quasi-separated). □

Chapter Six

Crystalline lifts and the finer structure of $\mathcal{X}_{d,\mathrm{red}}$

Up to this point, we have shown that $\mathcal{X}_d$ is a Noetherian formal algebraic stack. In this chapter, we will additionally show that $(\mathcal{X}_d)_{\mathrm{red}}$ is equidimensional of dimension $[K:\mathbf{Q}_p]d(d-1)/2$, and enumerate its irreducible components. In the course of doing this, we will also prove a result on the existence of crystalline lifts which is of independent interest. We furthermore determine the closed points of $(\mathcal{X}_d)_{\mathrm{red}}$, and describe the maximal substack of $\mathcal{X}_d$ over which the universal (φ,Γ)-module gives rise to a continuous G_K-representation.

6.1 THE FIBER DIMENSION OF H^2 ON CRYSTALLINE DEFORMATION RINGS

Let $\overline{\rho}$: $G_K \to \mathrm{GL}_d(\mathbf{F}')$ be a continuous representation, for some finite extension $\mathbf{F}'$ of $\mathbf{F}$, and let $\mathcal{O}'$ be the ring of integers in a finite extension E'/E, with residue field $\mathbf{F}'$. Let $\underline{\lambda}$ be a Hodge type, and consider the lifting ring $R:=R_{\overline{\rho}}^{\mathrm{crys},\underline{\lambda},\mathcal{O}'}$. Let M denote the universal étale (φ,Γ)-module over R (note that there is a universal étale (φ,Γ)-module over R, and not merely a universal *formal* étale (φ,Γ)-module, as a consequence of Corollary 4.8.13). Let $\alpha^\circ: G_K \to \mathrm{GL}_a(\mathcal{O}')$ be a representation with $\overline{\alpha}: G_K \to \mathrm{GL}_a(\mathbf{F}')$ being absolutely irreducible, and let N be the base change to R of the étale (φ,Γ)-module over $\mathcal{O}'$ which corresponds to α°.

Let $\mathcal{C}^\bullet(N^\vee \otimes M)$ denote the Herr complex associated to $N^\vee \otimes M$, as in Section 5.1; this is a perfect complex of R-modules, with cohomology concentrated in degrees $[0,2]$, as a consequence of [Sta, Tag 0CQG] and Theorem 5.1.22 (applied to the quotients $R/\mathfrak{m}_R^a$). Write $\mathrm{Ext}^2 := H^2\big(\mathcal{C}^\bullet(N^\vee \otimes M)\big)$. For each $r \geq 1$, write

$$X_r := \{x \in \operatorname{Spec} R \mid \dim \kappa(x) \otimes_R \mathrm{Ext}^2 \geq r\}.$$

The theorem on upper semi-continuity of fiber dimension for coherent sheaves shows that X_r is a closed subset of $\operatorname{Spec} R$.

6.1.1 Theorem. *If* Ext^2 *is ϖ-power torsion, then for each $r \geq 1$, X_r is of codimension $\geq r+1$ in* $\operatorname{Spec} R$.

Proof. The assumption that Ext^2 is ϖ-power torsion implies that Ext^2 is (set-theoretically) supported on the closed subscheme $\mathrm{Spec}(R/\varpi)_{\mathrm{red}}$ of $\mathrm{Spec}\, R$. Write $\overline{M}$ and $\overline{N}$ to denote the pull-backs of M and N respectively, over $(R/\varpi)_{\mathrm{red}}$, write $\overline{\mathrm{Ext}}^2 := \mathrm{Ext}^2\big(\mathcal{C}^\bullet(\overline{N}^\vee \otimes \overline{M})\big)$, and define

$$\overline{X}_r := \{x \in \mathrm{Spec}(R/\varpi)_{\mathrm{red}} \mid \dim \kappa(x) \otimes_{(R/\varpi)_{\mathrm{red}}} \overline{\mathrm{Ext}}^2 \geq r\}.$$

Since Ext^2 is compatible with base change, and is supported on $\mathrm{Spec}(R/\varpi)_{\mathrm{red}}$, while R is $\mathcal{O}$-flat, it suffices to show that the codimension of $\overline{X}_r$ in $\mathrm{Spec}(R/\varpi)_{\mathrm{red}}$ is at least r. Since $\mathrm{Spec}(R/\varpi)_{\mathrm{red}}$ is equidimensional of dimension $[K:\mathbf{Q}_p]d(d-1)/2+d^2$, it suffices in turn to show that $\overline{X}_r$ is of dimension at most $([K:\mathbf{Q}_p]d(d-1)/2)+d^2-r$.

By Proposition 3.6.3, we have a versal morphism $\mathrm{Spf}\, R_{\overline{\rho}}^{\square,\mathcal{O}'} \to \mathcal{X}_d$. Consider the fiber product $\mathrm{Spf}\, R_{\overline{\rho}}^{\square,\mathcal{O}'} \times_{\mathcal{X}_d} \mathcal{X}_{d,\mathrm{red}}$. This is a closed formal algebraic subspace of $\mathrm{Spf}\, R_{\overline{\rho}}^{\square,\mathcal{O}'}$, and so by Lemma A.3 is of the form $\mathrm{Spf}\, S$ for some quotient S of $R_{\overline{\rho}}^{\square,\mathcal{O}'}$. Its description as a fiber product shows that the morphism $\mathrm{Spf}\, S \to \mathcal{X}_{d,\mathrm{red}}$ is versal. Since $\mathcal{X}_{d,\mathrm{red}}$ is an algebraic stack, this morphism is effective, i.e., arises from a morphism

$$\mathrm{Spec}\, S \to \mathcal{X}_{d,\mathrm{red}} \tag{6.1.2}$$

([Sta, Tag 0DR1]), which is furthermore flat, by [Sta, Tag 0DR2]. Let $x_0 = \mathrm{Spec}\, \mathbf{F}'$, thought of as the closed point of $\mathrm{Spec}\, S$, and endowed with a morphism $x_0 \to \mathcal{X}_{d,\mathrm{red}}$ corresponding to the representation $\overline{\rho}$. Since $\mathrm{Spf}\, S \to \mathcal{X}_{d,\mathrm{red}}$ is versal, it follows from the Artin Approximation Theorem that we may find a finite type $\mathbf{F}$-scheme U, equipped with a point u_0 with residue field $\mathbf{F}'$, such that the complete local ring of U at u_0 is isomorphic to S, and such that (6.1.2) may be promoted to a smooth morphism

$$U \to \mathcal{X}_{d,\mathrm{red}}, \tag{6.1.3}$$

in the sense that on $u_0 (= \mathrm{Spec}\, \mathbf{F}')$, the morphism (6.1.3) induces the given morphism $x_0 (= \mathrm{Spec}\, \mathbf{F}') \to \mathcal{X}_{d,\mathrm{red}}$, and on $\mathrm{Spec}\, \widehat{\mathcal{O}}_{U,u_0} \cong \mathrm{Spec}\, S$, the morphism (6.1.3) induces the morphism (6.1.2). (See [Sta, Tag 0DR0].)

It follows from Proposition 3.6.3, by pulling back over $\mathcal{X}_{d,\mathrm{red}}$, that $\mathrm{Spf}\, S \times_{\mathcal{X}_{d,\mathrm{red}}} \mathrm{Spf}\, S$ is isomorphic to $\widehat{\mathrm{GL}}_{d,S}$, a certain completion of $(\mathrm{GL}_d)_S$, and thus that $\mathrm{Spf}\, S \times_{\mathcal{X}_{d,\mathrm{red}}} x_0$ is isomorphic to $\widehat{\mathrm{GL}}_{d,\mathbf{F}'}$, a certain completion of $(\mathrm{GL}_d)_{\mathbf{F}'}$. This latter fiber product may be identified with the completion of $\mathrm{Spec}\, S \times_{\mathcal{X}_{d,\mathrm{red}}} x_0$ along its closed subspace $x_0 \times_{\mathcal{X}_{d,\mathrm{red}}} x_0$, and so we deduce that the morphism (6.1.2) has relative dimension d^2 at the point x_0 in its domain. Thus the morphism (6.1.3) has relative dimension d^2 at the point u_0 in its domain. Since this latter morphism is also smooth, its relative dimension is locally constant on its domain [Sta, Tag 0DRQ], and thus (since S is a local ring, so that $\operatorname{Spec} S$ is connected), we see that (6.1.2) in fact has relative dimension d^2.

Let $\widetilde{M}$ denote the universal étale (φ,Γ)-module over S, and $\widetilde{N}$ denote the base change of N to S. Write $\widetilde{\operatorname{Ext}}^2 := H^2(\mathcal{C}^\bullet(\widetilde{N}^\vee \otimes \widetilde{M}))$, and write

$$Y_r := \{x \in \operatorname{Spec} S \mid \dim \kappa(x) \otimes_S \widetilde{\operatorname{Ext}}^2 \geq r\}.$$

We claim that Y_r has dimension at most $([K:\mathbf{Q}_p]d(d-1)/2) + d^2 - r$. In order to see this, we note that $(Y_r)_{\overline{\mathbf{F}}_p}$ is the pull-back to $\operatorname{Spec} S$ via the flat morphism (6.1.2) of the locus considered in Theorem 5.5.12 (4) which has dimension at most $[K:\mathbf{Q}_p]d(d-1)/2 - r$. The required bound on the dimension of Y_r follows, for example, from [Sta, Tag 02RE].

The composite $\operatorname{Spf} R/\varpi \to \operatorname{Spf} R_{\overline{\rho}}^{\square,\mathcal{O}'} \to \mathcal{X}_d$ is effective by Corollary 4.8.13, i.e., arises from a morphism $\operatorname{Spec} R/\varpi \to \mathcal{X}_d$. Thus the composite $\operatorname{Spf}(R/\varpi)_{\mathrm{red}} \to \operatorname{Spf} R/\varpi \to \mathcal{X}_d$ arises from a morphism $\operatorname{Spec}(R/\varpi)_{\mathrm{red}} \to \mathcal{X}_d$, and hence factors through $\mathcal{X}_{d,\mathrm{red}}$. Consequently, we find that the surjection $R_{\overline{\rho}}^{\square,\mathcal{O}'} \to (R/\varpi)_{\mathrm{red}}$ factors through S. The closed immersion $\operatorname{Spec}(R/\varpi)_{\mathrm{red}} \hookrightarrow \operatorname{Spec} S$ then induces a closed immersion $\overline{X}_r \hookrightarrow Y_r$ for each $r \geq 0$. The upper bound on the dimension of Y_r computed in the preceding paragraph then provides the desired upper bound on the dimension of $\overline{X}_r$, completing the proof of the theorem. $\square$

6.2 TWO GEOMETRIC LEMMAS

In this section we establish two lemmas which will be used in the proof of Theorem 6.3.2 below. We begin with the following simple lemma.

6.2.1 Lemma. *Suppose that $0 \to \mathcal{F} \to \mathcal{G} \to \mathcal{H} \to 0$ is a short exact sequence of coherent sheaves on a reduced Noetherian scheme X, with $\mathcal{F}$ and $\mathcal{G}$ being locally free, and $\mathcal{H}$ being torsion, in the sense that its support does not contain any irreducible component of X. Then there is an effective Cartier divisor D which contains the scheme-theoretic support of $\mathcal{H}$, and whose set-theoretic support coincides with the set-theoretic support of $\mathcal{H}$, with the property that if $f: T \to X$ is any morphism which meets D properly, in the sense that the pull-back $f^*\mathcal{O}_X \to f^*\mathcal{O}_X(D)$ of the canonical morphism $\mathcal{O}_X \to \mathcal{O}_X(D)$ is injective, then $\mathbf{L}_i f^*\mathcal{H} = 0$ if $i > 0$.*

Proof. Replacing X by each of its finitely many connected components in turn, we see that it is no loss of generality to assume that X is connected, and we do so; this ensures that each locally free sheaf on X is of constant rank. If η is any generic point of X, then $\mathcal{H}_\eta = 0$ by assumption, and so $\mathcal{F}_\eta \to \mathcal{G}_\eta$ is an isomorphism. In particular, $\mathcal{F}$ and $\mathcal{G}$ are of the same rank, say r, and the induced morphism

$$\wedge^r \mathcal{F} \to \wedge^r \mathcal{G} \tag{6.2.2}$$

is an injection of invertible sheaves. (Indeed, it follows from [Bou98, III, §7, Prop. 3] that a map of locally free sheaves of rank r is injective if and only if the map on $\wedge^r$ is injective.) If we let D denote the support of the cokernel of (6.2.2), then D is an effective Cartier divisor, and twisting by $\wedge^r \mathcal{F}^\vee$ identifies (6.2.2) with the canonical section $\mathcal{O}_X \hookrightarrow \mathcal{O}_X(D)$. The scheme-theoretic support of $\mathcal{H}$ is contained in D, and the set-theoretic supports of $\mathcal{H}$ and of D coincide (this is a special case of the general fact that the Fitting ideal is contained in the annihilator, and has the same radical as it).

Now suppose that $f\colon T \to X$ meets D properly in the sense described in the statement of the theorem. Then we see that

$$\wedge^r f^*\mathcal{F} \to \wedge^r f^*\mathcal{G}$$

is injective, and thus that $f^*\mathcal{F} \to f^*\mathcal{G}$ is also injective. Consequently

$$0 \to f^*\mathcal{F} \to f^*\mathcal{G} \to f^*\mathcal{H} \to 0$$

is short exact, and so $\mathbf{L}_i f^*\mathcal{H} = 0$ for $i > 0$, as claimed. □

We now introduce some notation related to the second of the lemmas (Lemma 6.2.7 below).

6.2.3 Hypothesis. Let X be a Noetherian scheme, and suppose that we have a short exact sequence of coherent sheaves

$$0 \to \mathcal{F} \to \mathcal{G} \to \mathcal{H} \to 0, \tag{6.2.4}$$

such that

- $\mathcal{G}$ is locally free;
- $\mathcal{H}$ has the property that, for each $r \geq 1$, the locus

  $$X_r := \{x \in X \mid \dim \kappa(x) \otimes_{\mathcal{O}_X} \mathcal{H} \geq r\}$$

  is of codimension $\geq r+1$ in X.

In the setting of Hypothesis 6.2.3, if $\pi\colon \widetilde{X} \to X$ is a morphism of Noetherian schemes, then we write $\widetilde{\mathcal{G}} := \pi^*\mathcal{G}$ and $\widetilde{\mathcal{H}} := \pi^*\mathcal{H}$, and we let $\widetilde{\mathcal{F}}$ denote the image of the pulled-back morphism $\pi^*\mathcal{F} \to \pi^*\mathcal{G} = \widetilde{\mathcal{G}}$, so that we again have a short exact sequence

$$0 \to \widetilde{\mathcal{F}} \to \widetilde{\mathcal{G}} \to \widetilde{\mathcal{H}} \to 0. \tag{6.2.5}$$

6.2.6 Remark. If $\mathcal{E}$ is a locally free coherent sheaf on a Noetherian scheme X, and $s\colon \mathcal{O}_X \to \mathcal{E}$ is a morphism, then we can think of $\mathcal{E}$ as being the sheaf of sections of a vector bundle over X, and think of s as being a section of this vector bundle. We can then speak of the zero locus of s; it is the closed subscheme of X which locally, if we choose an isomorphism $\mathcal{E} \cong \mathcal{O}_X^r$ and write $s = (a_1, \ldots, a_r)$, is cut out by the ideal sheaf $(a_1, \ldots, a_r)$; more globally, it is cut out by the ideal sheaf which is the image of the composite

$$\mathcal{E}^\vee = \mathcal{O}_X \otimes_{\mathcal{O}_X} \mathcal{E}^\vee \xrightarrow{s \otimes \mathrm{id}} \mathcal{E} \otimes_{\mathcal{O}_X} \mathcal{E}^\vee \to \mathcal{O}_X,$$

where the second arrow is the canonical pairing.

We note that s is nowhere-vanishing, i.e., the zero locus of s is empty, if and only if s is a split injection (so that s gives rise to a morphism of vector bundles, rather than merely of sheaves). We also note that if $\mathcal{E}$ has rank r, then the Hauptidealsatz shows that the zero locus of a section of $\mathcal{E}$ has codimension at most r around each of its points.

6.2.7 Lemma. *Suppose that we are in the setting of Hypothesis* 6.2.3, *that the* $\pi\colon \widetilde{X} \to X$ *is a surjective morphism whose domain is a Noetherian scheme, and that* $\widetilde{\mathcal{F}}$ *is locally free. Suppose further that* $\mathcal{O}_X \to \mathcal{F}$ *is a morphism with the property that the pulled-back morphism* $\mathcal{O}_{\widetilde{X}} \to \widetilde{\mathcal{F}}$ *is a nowhere-vanishing section. Then the composite* $\mathcal{O}_X \to \mathcal{F} \to \mathcal{G}$ *is nowhere-vanishing.*

Proof. The statement may be checked by working locally at each of the points of X; more precisely, we may replace X by $\operatorname{Spec} \mathcal{O}_{X,x}$ and $\widetilde{X}$ by its pull-back over $\operatorname{Spec} \mathcal{O}_{X,x}$. We have the stratification of X by the closed subsets X_r (where in the case $r = 0$, we declare $X_0 := X$), and if $x \in X$, then we set $r(x) := \dim \kappa(x) \otimes_{\mathcal{O}_X} \mathcal{H}$, or equivalently, the maximal value of r for which $x \in X_r$; we then prove the theorem by induction on $r(x)$. In fact, we assume that the statement is true after localizing at points x for which $r(x) < r$ for *all* exact sequences (6.2.4) which satisfy the hypotheses of the lemma, and prove it for our given exact sequence after localizing at an x for which $r(x) = r$.

In the case $r = 0$, the sheaf $\mathcal{H}$ is zero, so that $\mathcal{F} \xrightarrow{\sim} \mathcal{G}$. Thus the induced morphism $\mathcal{O}_{\widetilde{X}} \to \widetilde{\mathcal{G}}$ gives a nowhere-vanishing section. Since π is surjective, this ensures that the induced morphism $\mathcal{O}_X \to \mathcal{G}$ is also nowhere-vanishing, as required.

We now consider the case $r \geq 1$. Since we have replaced X by $\operatorname{Spec} \mathcal{O}_{X,x}$, we may assume that $\mathcal{G}$ is free. By Nakayama's lemma and our assumption that $r(x) = r$, we may choose a surjection

$$\mathcal{O}_X^r \to \mathcal{H}, \tag{6.2.8}$$

which we may then lift to a morphism $\mathcal{O}_X^r \to \mathcal{G}$. If we let $\mathcal{K}$ denote the kernel of (6.2.8), then we obtain a morphism of short exact sequences

$$\begin{array}{ccccccccc} 0 & \longrightarrow & \mathcal{K} & \longrightarrow & \mathcal{O}_X^r & \longrightarrow & \mathcal{H} & \longrightarrow & 0 \\ & & \downarrow & & \downarrow & & \| & & \\ 0 & \longrightarrow & \mathcal{F} & \longrightarrow & \mathcal{G} & \longrightarrow & \mathcal{H} & \longrightarrow & 0 \end{array} \tag{6.2.9}$$

Pulling back this diagram via π, and letting $\widetilde{\mathcal{K}}$ denote the image of $\pi^*\mathcal{K}$ in $\mathcal{O}_{\widetilde{X}}^r$, we obtain a corresponding morphism of short exact sequences

$$\begin{array}{ccccccccc} 0 & \longrightarrow & \widetilde{\mathcal{K}} & \longrightarrow & \mathcal{O}_{\widetilde{X}}^r & \longrightarrow & \widetilde{\mathcal{H}} & \longrightarrow & 0 \\ & & \downarrow & & \downarrow & & \| & & \\ 0 & \longrightarrow & \widetilde{\mathcal{F}} & \longrightarrow & \widetilde{\mathcal{G}} & \longrightarrow & \widetilde{\mathcal{H}} & \longrightarrow & 0 \end{array} \tag{6.2.10}$$

The morphism of short exact sequences (6.2.9) induces a short exact sequence

$$0 \to \mathcal{K} \to \mathcal{F} \oplus \mathcal{O}_X^r \to \mathcal{G} \to 0, \tag{6.2.11}$$

which is furthermore split (since $\mathcal{G}$ is free and we are over an affine scheme), so that we may write

$$\mathcal{F} \oplus \mathcal{O}_X^r \cong \mathcal{K} \oplus \mathcal{G}. \tag{6.2.12}$$

Correspondingly, the short exact sequence (6.2.10) induces a short exact sequence

$$0 \to \widetilde{\mathcal{K}} \to \widetilde{\mathcal{F}} \oplus \mathcal{O}_{\widetilde{X}}^r \to \widetilde{\mathcal{G}} \to 0,$$

which can be thought of as being obtained from (6.2.11) by pulling back via π and then taking the quotient by the (naturally identified) copies of $\mathbf{L}_1\pi^*\mathcal{H}$ sitting inside $\pi^*\mathcal{K}$ and $\pi^*\mathcal{F}$. Since (6.2.11) is split, so is this latter short exact sequence, and so we also obtain an isomorphism

$$\widetilde{\mathcal{F}} \oplus \mathcal{O}_{\widetilde{X}}^r \cong \widetilde{\mathcal{K}} \oplus \widetilde{\mathcal{G}} \tag{6.2.13}$$

which can be thought of as being obtained by pulling back the isomorphism (6.2.12) along π, and then taking the quotient of each side by the appropriate copy of $\mathbf{L}_1\pi^*\mathcal{H}$. Since all the other summands appearing in the isomorphism (6.2.13) are locally free, we see that the same is true of $\widetilde{\mathcal{K}}$.

Suppose, by way of obtaining a contradiction, that the composite $\mathcal{O}_X \to \mathcal{F} \to \mathcal{G}$ is not nowhere-vanishing; then we see that its image lies in $\mathfrak{m}_x\mathcal{G}$. Thus the pulled-back morphism $\mathcal{O}_{\widetilde{X}} \to \widetilde{\mathcal{G}}$ vanishes at each point $\widetilde{x}$ lying over x (and there is at least one such point, since π is surjective by assumption). Since the pulled-back morphism $\mathcal{O}_{\widetilde{X}} \to \widetilde{\mathcal{F}}$ *is* nowhere-vanishing, so is the induced morphism $\mathcal{O}_{\widetilde{X}} \to \widetilde{\mathcal{F}} \oplus \mathcal{O}_{\widetilde{X}}^r$. A consideration of (6.2.13) then shows that the induced morphism $\mathcal{O}_{\widetilde{X}} \to \widetilde{\mathcal{K}}$, which is pulled back from the induced morphism $\mathcal{O}_X \to \mathcal{K}$, must be nowhere-vanishing.

Thus, if we consider the exact sequence

$$0 \to \mathcal{K} \to \mathcal{O}_X^r \to \mathcal{H} \to 0,$$

it satisfies Hypothesis 6.2.3, and we are given a morphism $\mathcal{O}_X \to \mathcal{K}$ whose pull-back to $\widetilde{X}$ is a nowhere-vanishing section of $\widetilde{\mathcal{K}}$. Applying the inductive hypothesis to this situation, we find that the composite $\mathcal{O}_X \to \mathcal{O}_X^r$ is a section of a rank r locally free sheaf whose zero locus is contained in X_r (because if we localize at a point not in X_r, then the section is nowhere vanishing by the inductive hypothesis), and is therefore of codimension at least $r+1$ by hypothesis. This zero locus contains x (since this section factors through $\mathcal{K}$, and $\mathcal{H} \cong \mathcal{O}_X^r/\mathcal{K}$ has fiber dimension exactly r at x, so the map $\mathcal{O}_X^r \otimes_{\mathcal{O}_X} \kappa(x) \to \mathcal{H} \otimes_{\mathcal{O}_X} \kappa(x)$ is an isomorphism), and hence is non-empty. On the other hand, as we noted above, the zero locus of a section of a rank r locally free sheaf, if it is non-empty, has codimension at most r. This contradiction completes the proof of the lemma. □

6.3 CRYSTALLINE LIFTS

Given a d-tuple of labeled Hodge–Tate weights $\underline{\lambda}$, and a d'-tuple of labeled Hodge–Tate weights $\underline{\lambda}'$, we say that $\underline{\lambda}'$ is *slightly greater than* $\underline{\lambda}$ (and that $\underline{\lambda}$ is *slightly less than* $\underline{\lambda}'$) if for each $\sigma\colon K \hookrightarrow \overline{\mathbf{Q}}_p$ we have $\lambda'_{\sigma,d} \geq \lambda_{\sigma,1}+1$, and the inequality is strict for at least one σ.

This is not standard terminology, but it will be convenient for us; it is motivated by the following well-known result. Here, as throughout the book, we slightly abuse terminology and refer to lattices in crystalline representations as themselves being crystalline.

6.3.1 Lemma. *Let $0 \to \rho_1^\circ \to \rho^\circ \to \rho_2^\circ \to 0$ be an extension of $\overline{\mathbf{Z}}_p$-valued representations of G_K, with ρ_1° and ρ_2° being crystalline of Hodge–Tate weights $\underline{\lambda}_1$, $\underline{\lambda}_2$ respectively. If $\underline{\lambda}_1$ is slightly less than $\underline{\lambda}_2$, then ρ° is crystalline.*

Proof. This follows easily from the formulae in [Nek93, Prop. 1.24]. Indeed, if we let ρ_1, ρ_2 be the $\overline{\mathbf{Q}}_p$-representations corresponding to ρ_1°, ρ_2°, and we set $V := \rho_1 \otimes \rho_2^\vee$, then we need to show that $h^1_f(V) = h^1(V)$. Since the Hodge–Tate weights of V are all negative, and for at least one embedding $\sigma\colon K \hookrightarrow \overline{\mathbf{Q}}_p$ the σ-labeled Hodge–Tate weights are all less than -1, we have $h^1(V) - h^1_f(V) = h^2(V) = h^0(V^\vee(1)) = 0$, as required. □

Arguing inductively, the following theorem will allow us to construct crystalline lifts of any given $\overline{\rho}$. In particular it implies Theorem 1.2.2 from the introduction, in the more refined form of Theorem 6.4.4 below.

6.3.2 Theorem. *Suppose we are given a representation $\overline{\rho}_d\colon G_K \to \mathrm{GL}_d(\overline{\mathbf{F}}_p)$ that admits a lift $\rho_d^\circ\colon G_K \to \mathrm{GL}_d(\overline{\mathbf{Z}}_p)$ which is crystalline with labeled Hodge–Tate weights $\underline{\lambda}$. Let $0 \to \overline{\rho}_d \to \overline{\rho}_{d+a} \to \overline{\alpha} \to 0$ be any extension of G_K-representations over $\overline{\mathbf{F}}_p$, with $\overline{\alpha}\colon G_K \to \mathrm{GL}_a(\overline{\mathbf{F}}_p)$ irreducible, and let $\alpha^\circ\colon G_K \to \mathrm{GL}_a(\overline{\mathbf{Z}}_p)$ be any crystalline lifting of $\overline{\alpha}$ with labeled Hodge–Tate weights $\underline{\lambda}'$, which we assume to be slightly greater than $\underline{\lambda}$.*

Then we may find a lifting of the given extension to an extension

$$0 \to \theta_d^\circ \to \theta_{d+a}^\circ \to \alpha^\circ \to 0$$

of G_K-representations over $\overline{\mathbf{Z}}_p$, where $\theta_d^\circ\colon G_K \to \mathrm{GL}_d(\overline{\mathbf{Z}}_p)$ again has the property that the associated p-adic representation $\theta_d\colon G_K \to \mathrm{GL}_d(\overline{\mathbf{Q}}_p)$ is crystalline with labeled Hodge–Tate weights $\underline{\lambda}$. Furthermore, θ_{d+a}° is crystalline, and we may choose θ_d° to lie on the same irreducible component of $\mathrm{Spec}\, R_{\overline{\rho}_d}^{\mathrm{crys},\underline{\lambda}}$ *that ρ_d° does.*

Proof of Theorem 6.3.2. We may and do choose our field of coefficients E to be large enough that the various Galois representations are defined over $\mathcal{O}$ or $\mathbf{F}$ as the case may be. We write $R := R_{\overline{\rho}}^{\mathrm{crys},\underline{\lambda}}$, and we let X denote the component of the crystalline lifting scheme $\mathrm{Spec}\, R_{\overline{\rho}}^{\mathrm{crys},\underline{\lambda}}$ on which ρ_d° lies. As recalled in Section 1.12, X is reduced. Over X we have a good complex $C^\bullet$ supported in degrees $[0,2]$ computing $\mathrm{Ext}^\bullet_{G_K}(\alpha^\circ, \rho^\circ)$ for the universal deformation ρ°; indeed, by a straightforward variant of Corollary 5.1.25, we can choose a good complex supported in degrees $[0,2]$ and quasi-isomorphic to the Herr complex for $\rho^\circ \otimes (\alpha^\circ)^\vee$. Our assumption on the Hodge–Tate weights implies that Ext^2 is supported (set-theoretically) on the special fiber $\overline{X}$ of X. Indeed, the formation of Ext^2 is compatible with base change, and at any closed point x of the generic fiber of R, we have $\mathrm{Ext}^2_{G_K}(\alpha^\circ, \rho_x^\circ) = \mathrm{Hom}_{G_K}(\rho_x^\circ, \alpha^\circ(1))$ by Tate local duality, and this space vanishes by the assumption that $\underline{\lambda}'$ is slightly greater than $\underline{\lambda}$.

As usual, we let B^2 denote the image of C^1 in C^2; it is a subsheaf of the locally free sheaf C^2, and so is torsion free.[1] By [Sta, Tag 0815], we may find a blow up $\pi\colon \widetilde{X} \to X$, whose center lies (set-theoretically) in the special fiber of X, such that the torsion-free quotient of $\pi^* B^2$ becomes locally free (recall that Ext^2 is supported on the special fiber of X); in other words, if we let $\widetilde{C}^\bullet$ denote the pull-back by π of $C^\bullet$, the corresponding $\mathcal{O}_{\widetilde{X}}$-module of 2-coboundaries $\widetilde{B}^2$ (that is, the image of $\widetilde{C}^1$ in $\widetilde{C}^2$) is locally free. Thus, if we let $\widetilde{Z}^1$ denote the $\mathcal{O}_{\widetilde{X}}$-module of 1-cocycles for $\widetilde{C}^\bullet$, then $\widetilde{Z}^1$ is also locally free; so, in particular, the complex $\widetilde{C}^0 \to \widetilde{Z}^1$ is a good complex.

The given extension $\overline{\rho}_{d+a}$ is classified by a class in $\mathrm{Ext}^1_{G_K}(\overline{\alpha}, \overline{\rho}_d)$, which is to say, a class in H^1 of the complex $\kappa \otimes C^\bullet$. (Here we write κ to denote $\mathbf{F}$ thought of as the residue field of R.) Lifting this class to an element of $\kappa \otimes C^1$,

[1] Since X is reduced, *torsion free* is a reasonable notion for a finitely generated module, equivalent to the associated primes lying among the minimal primes.

and then to an element of C^1, we find ourselves in the following situation: we have a morphism $c\colon R \to C^1$, whose image under the coboundary lies in $\mathfrak{m}_R C^2$ (reflecting the fact that we have a cochain which becomes a cocycle at the closed point).

Consider the composite $b\colon R \xrightarrow{c} C^1 \to B^2$, which pulls back to a section $\widetilde{b}\colon \mathcal{O}_{\widetilde{X}} \to \widetilde{B}^2$. It follows from Lemma 6.2.7 above (applied with $\mathcal{F} = B^2$, $\mathcal{G} = C^2$, and $\mathcal{H} = \mathrm{Ext}^2$), together with Theorem 6.1.1, that the lifted section $\widetilde{b}$ must have non-empty zero locus, which (being closed) must contain a point $\widetilde{x}$ lying over the closed point $x \in X$. The section c itself pulls back to a section $\widetilde{c}\colon \mathcal{O}_{\widetilde{X}} \to \widetilde{C}^1$, whose value at the point $\widetilde{x}$ lies in the fiber of $\widetilde{Z}^1$. In other words, the fiber of $\widetilde{c}$ at $\widetilde{x}$ is a one-cocycle in the complex $\kappa(\widetilde{x}) \otimes [\widetilde{C}^0 \to \widetilde{Z}^1]$, giving rise to a class $\overline{e} \in H^1\big(\kappa(\widetilde{x}) \otimes [\widetilde{C}^0 \to \widetilde{Z}^1]\big)$ lifting the original class in $\mathrm{Ext}^1_{G_K}(\overline{\alpha}, \overline{\rho}_d)$ that classifies $\overline{\rho}_{d+a}$.

Since $\widetilde{X}$ is flat over $\mathbf{Z}_p$ (because X is), we may find a morphism $\widetilde{f}\colon \operatorname{Spec} \overline{\mathbf{Z}}_p \to \widetilde{X}$ passing through the point $\widetilde{x} \in \widetilde{X}(\overline{\mathbf{F}}_p)$ (in the sense that the closed point of $\operatorname{Spec} \overline{\mathbf{Z}}_p$ maps to $\widetilde{x}$). The morphism $\widetilde{f}$ determines (and, by the valuative criterion of properness, is determined by) a morphism $f\colon \operatorname{Spec} \overline{\mathbf{Z}}_p \to X$, lifting the closed point $x \in X$, which in turn corresponds to a representation $\theta_d^\circ\colon G_K \to \mathrm{GL}_d(\overline{\mathbf{Z}}_p)$, as in the statement of the theorem. Now $H^2(\widetilde{C}^\bullet)$ is the cokernel of an inclusion of locally free sheaves (the inclusion $\widetilde{B}^2 \hookrightarrow \widetilde{C}^2$), and it is torsion (because it is pulled back from $H^2(C^\bullet)$, which is set-theoretically supported in the special fiber), and so Lemma 6.2.1 above shows that there is an effective Cartier divisor D contained (set-theoretically) in the special fiber of $\widetilde{X}$ with the property that for any morphism to $\widetilde{X}$ that meets D properly, the higher derived pull-backs of $H^2(\widetilde{C}^\bullet)$ under this morphism vanish. Since the domain of $\widetilde{f}$ is p-torsion free, it meets the special fiber of $\widetilde{X}$ properly, and in particular meets D properly; thus we infer that

$$\mathbf{L}_i \widetilde{f}^* H^2(\widetilde{C}^\bullet) = 0$$

for $i > 0$. From this vanishing we deduce that

$$\begin{aligned}\mathrm{Ext}^1_{G_K}(\alpha^\circ, \theta_d^\circ) = H^1(f^* C^\bullet) = H^1(\widetilde{f}^* \widetilde{C}^\bullet) = \widetilde{f}^* H^1(\widetilde{C}^\bullet) \\ = \widetilde{f}^* H^1\big([\widetilde{C}^0 \to \widetilde{Z}^1]\big) = H^1\big(\widetilde{f}^* [\widetilde{C}^0 \to \widetilde{Z}^1]\big).\end{aligned}$$

(The first of these identifications is the general base change property of the perfect complex $C^\bullet$, the second follows from the fact that $f = \pi \circ \widetilde{f}$, the third follows from the vanishing of $\mathbf{L}_i \widetilde{f}^* H^2(\widetilde{C}^\bullet)$ for $i > 0$ and the base change spectral sequence

$$E_2^{p,q} := L_{-p} \widetilde{f}^* H^q(\widetilde{C}^\bullet) \implies H^{p+q}(\widetilde{f}^* \widetilde{C}^\bullet)$$

(see, e.g., [Sta, Tag 0662]), the fourth holds by definition of H^1, and the fifth follows from right-exactness of $\widetilde{f}^*$.)

Since $\widetilde{f}$ identifies the closed point of $\operatorname{Spec}\overline{\mathbf{Z}}_p$ with the point $\widetilde{x}$, we may find a class $e \in H^1(\widetilde{f}^*[\widetilde{C}^0 \to \widetilde{Z}^1])$ lifting the class $\overline{e}$, which is then identified with an element of $\operatorname{Ext}^1_{G_K}(\alpha^\circ, \theta^\circ_d)$. This element classifies an extension $0 \to \theta^\circ_d \to \theta^\circ_{d+a} \to \alpha^\circ \to 0$ which by construction lifts the given extension $0 \to \overline{\rho}^\circ_d \to \overline{\rho}^\circ_{d+a} \to \overline{\alpha}^\circ \to 0$, and so by Lemma 6.3.1 satisfies the conclusions of the theorem. □

6.4 POTENTIALLY DIAGONALIZABLE CRYSTALLINE LIFTS

We now use Theorem 6.3.2 to prove Theorem 1.2.2. There are at least two differences between the proof of Theorem 1.2.2 and previous work on the problem (in particular the results of [Mul13, GHLS17]). One is that, rather than working only with lifts which are extensions of inductions of characters, we utilize extensions of more general potentially diagonalizable representations. The other difference is that our argument exploits our control of the support of H^2. In previous work, the only tool used to deal with classes in H^2 was twisting the various irreducible representations by unramified characters; at points where H^2 has dimension greater than 1, this seems to be insufficient.

In fact, we prove various refinements of this result, allowing us to produce potentially diagonalizable lifts; such lifts are important for automorphy lifting theorems. We can also control the Hodge–Tate weights of these potentially diagonalizable lifts; for example, we can insist that the gaps between the weights are arbitrarily large, which is useful in applications to automorphy lifting. Alternatively, we can produce lifts whose Hodge–Tate weights lift some Serre weight, proving a conjecture which is important for the formulation of general Serre weight conjectures (see the discussion after [GHS18, Rem. 5.1.8]).

We begin by recalling some definitions and some basic lemmas about extensions of crystalline representations. Let $\underline{\lambda}$ be a regular d-tuple of Hodge–Tate weights.

6.4.1 Definition. A crystalline representation of weight $\underline{\lambda}$ is *ordinary* if it has a G_K-invariant decreasing filtration whose associated graded pieces are all one-dimensional, such that the σ-labeled Hodge–Tate weight of the ith graded piece is $\lambda_{\sigma,i}$. In other words, we can write the representation in the form

$$\begin{pmatrix} \chi_1 & * & \dots & * \\ 0 & \chi_2 & \dots & * \\ \vdots & & \ddots & \vdots \\ 0 & \dots & 0 & \chi_d \end{pmatrix}$$

where χ_i is a crystalline character whose σ-labeled Hodge–Tate weight is $\lambda_{\sigma,d+1-i}$.

If $\rho^\circ_1, \rho^\circ_2 \colon G_K \to \mathrm{GL}_d(\overline{\mathbf{Z}}_p)$ are two crystalline representations, then we write $\rho^\circ_1 \sim \rho^\circ_2$, and say that ρ°_1 *connects to* ρ°_2, if and only if the following conditions hold: ρ_1° and ρ_2° have the same labeled Hodge–Tate weights, have isomorphic reductions modulo $\mathfrak{m}_{\overline{\mathbf{Z}}_p}$, and determine points on the same irreducible component of the corresponding crystalline lifting rings. The following definition was originally made in [BLGGT14, §1.4].

6.4.2 Definition. A representation $\rho^\circ\colon G_K \to \mathrm{GL}_d(\overline{\mathbf{Z}}_p)$ is *potentially diagonalizable* if there is a finite extension K'/K and crystalline characters $\chi_1, \dots, \chi_d\colon G_{K'} \to \overline{\mathbf{Z}}_p^\times$ such that $\rho^\circ|_{G_{K'}} \sim \chi_1 \oplus \cdots \oplus \chi_d$.

6.4.3 Lemma. *Let $0 \to \rho_1^\circ \to \rho^\circ \to \rho_2^\circ \to 0$ be an extension of $\overline{\mathbf{Z}}_p$-valued representations of G_K. If ρ° is crystalline, and ρ_1° and ρ_2° are potentially diagonalizable, then ρ° is also potentially diagonalizable.*

Proof. We may choose K'/K such that $\rho_1^\circ|_{G_{K'}}$ and $\rho_2^\circ|_{G_{K'}}$ both connect to direct sums of crystalline characters, and such that $\overline{\rho}|_{G_{K'}}$ is trivial. It then follows from points (5) and (7) of the list before [BLGGT14, Lem. 1.4.1] that $\rho^\circ|_{G_{K'}}$ connects to the direct sum of the union of the sets of crystalline characters for $\rho_1^\circ|_{G_{K'}}$ and $\rho_2^\circ|_{G_{K'}}$, as required. □

Note that it follows in particular from Lemma 6.4.3 that ordinary representations are potentially diagonalizable.

6.4.4 Theorem. *Let $K/\mathbf{Q}_p$ be a finite extension, and let $\overline{\rho}\colon G_K \to \mathrm{GL}_d(\overline{\mathbf{F}}_p)$ be a continuous representation. Then $\overline{\rho}$ admits a lift to a crystalline representation $\rho^\circ\colon G_K \to \mathrm{GL}_d(\overline{\mathbf{Z}}_p)$ of some regular labeled Hodge–Tate weights $\underline{\lambda}$. Furthermore:*

1. *ρ° can be taken to be potentially diagonalizable.*
2. *If every Jordan–Hölder factor of $\overline{\rho}$ is one-dimensional, then ρ° can be taken to be ordinary.*
3. *ρ° can be taken to be potentially diagonalizable, and $\underline{\lambda}$ can be taken to be a lift of a Serre weight.*
4. *ρ° can be taken to be potentially diagonalizable, and $\underline{\lambda}$ can be taken to have arbitrarily spread-out Hodge–Tate weights: that is, for any $C > 0$, we can choose ρ° such that for each $\sigma\colon K \hookrightarrow \overline{\mathbf{Q}}_p$, we have $\lambda_{\sigma,i} - \lambda_{\sigma,i+1} \geq C$ for each $1 \leq i \leq n-1$.*

Proof. We prove all of these results by induction on d, using Theorem 6.3.2. We begin by proving (1), and then explain how to refine the proof to give each of (2)–(4).

Suppose firstly that $\overline{\rho}$ is irreducible (this is the base case of the induction). Then $\overline{\rho}$ is of the form $\mathrm{Ind}_{G_{K'}}^{G_K} \overline{\psi}$ for some character $\overline{\psi}\colon G_{K'} \to \overline{\mathbf{F}}_p^\times$, where K'/K is unramified of degree d. We can choose (for example by [GHS18, Lem. 7.1.1]) a crystalline lift $\psi\colon G_{K'} \to \overline{\mathbf{Z}}_p^\times$ of $\overline{\psi}$ such that $\rho^\circ := \mathrm{Ind}_{G_{K'}}^{G_K} \psi$ has regular Hodge–Tate weights. Then ρ° is crystalline (because K'/K is unramified), and it is potentially diagonalizable, because $\rho^\circ|_{G'_K}$ is a direct sum of crystalline characters.

For the inductive step, we may therefore suppose that we can write $\overline{\rho}$ as an extension $0 \to \overline{\rho}_1 \to \overline{\rho} \to \overline{\rho}_2 \to 0$, with $\overline{\rho}_2$ irreducible, and we may assume that we have regular potentially diagonalizable lifts ρ_1°, ρ_2° of $\overline{\rho}_1$, $\overline{\rho}_2$ respectively. Twisting ρ_1° by a crystalline character with sufficiently large Hodge–Tate weights and trivial reduction modulo p (which exists by [GHS18, Lem. 7.1.1]), we may suppose that the Hodge–Tate weights of ρ_1° are slightly less than those of ρ_2°. By Theorem 6.3.2, there is a lift θ_1° of $\overline{\rho}_1$ with $\theta_1^\circ \sim \rho_1^\circ$, and a lift ρ° of $\overline{\rho}$ which is an extension

$$0 \to \theta_1^\circ \to \rho^\circ \to \rho_2^\circ \to 0.$$

Since ρ_1° is potentially diagonalizable, so is θ_1°, and it follows from Lemmas 6.3.1 and 6.4.3 that ρ° is crystalline and potentially diagonalizable. This completes the proof of (1).

We claim that if every Jordan–Hölder factor of $\overline{\rho}$ is one-dimensional, then the lift that we produced in proving (1) is actually automatically ordinary. Examining the proof, we see that it is enough to show that if ρ_1° is ordinary and $\theta_1^\circ \sim \rho_1^\circ$ then θ_1° is also ordinary; this is immediate from [Ger19, Lem. 3.3.3(2)].

To show (3) and (4) we have to check that we can choose the Hodge–Tate weights of the lifts of the irreducible pieces of $\overline{\rho}$ appropriately. This requires substantially more effort in the case of (3), but that effort has already been made in the proof of [GHS18, Thm. B.1.1] (which proves the case that $\overline{\rho}$ is semisimple). Indeed, examining that proof, we see that it produces lifts satisfying all of the conditions we need; we need only note that the "slightly less" condition is automatic, except in the case that for each σ we have (in the notation of the proof of [GHS18, Thm. B.1.1]) $h_\sigma + x_\sigma = H_\sigma + 1$, in which case we can take $h_\sigma + x_\sigma = H_\sigma + p$ for each σ instead.

Finally, to prove (4) it is enough to check the case that $\overline{\rho}$ is irreducible (because we can choose the Hodge–Tate weights of the character that we twisted ρ_1° by to be arbitrarily large). For this, note that in our application of [GHS18, Lem. 7.1.1], we can change any labeled Hodge–Tate by any multiple of $(p^d - 1)$, so we can certainly arrange that the gaps between labeled Hodge–Tate weights are as large as we please. □

6.4.5 Remark. To complete the proof of Theorem 1.2.2, it is enough to note that by definition the lifts produced in Theorem 6.4.4 (3) have all their Hodge–Tate weights in the interval $[0, dp-1]$.

6.4.6 Remark. The proof of Theorem 6.4.4 (3) actually proves the slightly stronger statement that there is a Serre weight such that if $\underline{\lambda}$ is any lift of that Serre weight, then $\overline{\rho}$ admits a potentially diagonalizable crystalline lift of weight $\underline{\lambda}$. (Note that in the case that $K/\mathbf{Q}_p$ is ramified, there may be many such choices of a lift of a given Serre weight, because by definition, the choice of a lift of $\underline{\lambda}$ depends on a choice of a lift of each embedding $k \hookrightarrow \overline{\mathbf{F}}_p$ to an embedding $K \hookrightarrow \overline{\mathbf{Q}}_p$. However, it is noted in the first paragraph of the proof of [GHS18, Thm. B.1.1] that in the case that $\overline{\rho}$ is semisimple, there is a crystalline lift of weight $\underline{\lambda}$ for any choice of lift $\underline{\lambda}$, so the same is true in our construction.)

As a consequence of Theorem 6.4.4, we can remove a hypothesis made in [EG14, App. A], proving in particular the following result on the existence of globalizations of local Galois representations.

6.4.7 Corollary. *Suppose that $p \nmid 2d$. Let $K/\mathbf{Q}_p$ be a finite extension, and let $\overline{\rho}\colon G_K \to \mathrm{GL}_n(\overline{\mathbf{F}}_p)$ be a continuous representation. Then there is an imaginary CM field F and a continuous irreducible representation $\overline{r}\colon G_F \to \mathrm{GL}_n(\overline{\mathbf{F}}_p)$ such that*

- *each place $v|p$ of F satisfies $F_v \cong K$,*
- *for each place $v|p$ of F, either $\overline{r}|_{G_{F_v}} \cong \overline{\rho}$ or $\overline{r}|_{G_{F_{v^c}}} \cong \overline{\rho}$, and*
- *$\overline{r}$ is automorphic, in the sense that it may be lifted to a representation $r\colon G_F \to \mathrm{GL}_n(\overline{\mathbf{Q}}_p)$ coming from a regular algebraic conjugate self dual cuspidal automorphic representation of GL_n/F.*

Proof. This is immediate from [EG14, Cor. A.7] (since [EG14, Conj. A.3] is a special case of Theorem 6.4.4). □

6.5 THE IRREDUCIBLE COMPONENTS OF $\mathcal{X}_{d,\mathrm{red}}$

We now complete the analysis of the irreducible components of $\mathcal{X}_{d,\mathrm{red}}$ that we began in Section 5.5. Recall that Theorem 5.5.12 shows that $\mathcal{X}_{d,\mathrm{red}}$ is an algebraic stack of finite presentation over $\mathbf{F}$, and has dimension $[K:\mathbf{Q}_p]d(d-1)/2$. Furthermore, for each Serre weight $\underline{k}$, there is a corresponding irreducible component $\mathcal{X}^{\underline{k}}_{d,\mathrm{red},\overline{\mathbf{F}}_p}$ of $(\mathcal{X}_{d,\mathrm{red}})_{\overline{\mathbf{F}}_p}$, and the components for different weights $\underline{k}$ are distinct (see Remark 5.5.14).

We are now finally in a position to prove the following result.

6.5.1 Theorem. *$\mathcal{X}_{d,\mathrm{red}}$ is equidimensional of dimension $[K:\mathbf{Q}_p]d(d-1)/2$, and the irreducible components of $(\mathcal{X}_{d,\mathrm{red}})_{\overline{\mathbf{F}}_p}$ are precisely the various closed substacks $\mathcal{X}^{\underline{k}}_{d,\mathrm{red},\overline{\mathbf{F}}_p}$ of Theorem 5.5.12; in particular, $(\mathcal{X}_{d,\mathrm{red}})_{\overline{\mathbf{F}}_p}$ is maximally nonsplit of niveau 1. Furthermore each $\mathcal{X}^{\underline{k}}_{d,\mathrm{red},\overline{\mathbf{F}}_p}$ can be defined over $\mathbf{F}$, i.e., is the base change of an irreducible component $\mathcal{X}^{\underline{k}}_{d,\mathrm{red}}$ of $\mathcal{X}_{d,\mathrm{red}}$.*

Proof. By Theorem 5.5.12, it is enough to prove that each irreducible component of $(\mathcal{X}_{d,\mathrm{red}})_{\overline{\mathbf{F}}_p}$ is of dimension of at least $[K:\mathbf{Q}_p]d(d-1)/2$. Indeed, this shows that the irreducible components of $(\mathcal{X}_{d,\mathrm{red}})_{\overline{\mathbf{F}}_p}$ are precisely the $\mathcal{X}^{\underline{k}}_{d,\mathrm{red},\overline{\mathbf{F}}_p}$; to see that these may all be defined over $\mathbf{F}$, we need to show that the action of $\mathrm{Gal}(\overline{\mathbf{F}}_p/\mathbf{F})$ on the irreducible components of $(\mathcal{X}_{d,\mathrm{red}})_{\overline{\mathbf{F}}_p}$ is trivial. This follows immediately by considering its action on the maximally nonsplit representations of niveau 1 (since the action of $\mathrm{Gal}(\overline{\mathbf{F}}_p/\mathbf{F})$ preserves the property of being maximally nonsplit of niveau 1 and weight $\underline{k}$).

In order to see that each irreducible component of $(\mathcal{X}_{d,\mathrm{red}})_{\overline{\mathbf{F}}_p}$ is of dimension of at least $[K:\mathbf{Q}_p]d(d-1)/2$, it suffices to show that every finite type point x of $(\mathcal{X}_{d,\mathrm{red}})_{\overline{\mathbf{F}}_p}$ is contained in an irreducible substack of $(\mathcal{X}_{d,\mathrm{red}})_{\overline{\mathbf{F}}_p}$ of dimension at least $[K:\mathbf{Q}_p]d(d-1)/2$; so it suffices in turn to show that each x is contained in an equidimensional substack of dimension $[K:\mathbf{Q}_p]d(d-1)/2$. But by Theorem 6.4.4, there is a regular Hodge type $\underline{\lambda}$ such that x is contained in $\mathcal{X}_d^{\mathrm{crys},\underline{\lambda}} \times_{\mathrm{Spf}\,\mathcal{O}} \mathrm{Spec}\,\mathbf{F}$ (for $\mathcal{O}$ sufficiently large), and this stack is equidimensional of dimension $[K:\mathbf{Q}_p]d(d-1)/2$ by Theorem 4.8.14, as required. □

We end this section with the following result, showing that in contrast to its substacks $\mathcal{X}_d^{\mathrm{crys},\underline{\lambda},\tau}$ and $\mathcal{X}_d^{\mathrm{ss},\underline{\lambda},\tau}$, the formal algebraic stack $\mathcal{X}_d$ is not a p-adic formal algebraic stack.

6.5.2 Proposition. *$\mathcal{X}_d$ is not a p-adic formal algebraic stack.*

Proof. Assume that $\mathcal{X}_d$ is a p-adic formal algebraic stack, so that its special fiber $\overline{\mathcal{X}}_d := \mathcal{X}_d \times_{\mathcal{O}} \mathbf{F}$ is an algebraic stack, which is furthermore of finite type over $\mathbf{F}$ (since $\mathcal{X}_d$ is a Noetherian formal algebraic stack, by Corollary 5.5.18, and $\mathcal{X}_{d,\mathrm{red}}$ is of finite type over $\mathbf{F}$, by Theorem 5.5.12). Since the underlying reduced substack of $\overline{\mathcal{X}}_d$ is $\mathcal{X}_{d,\mathrm{red}}$, which is equidimensional of dimension $[K:\mathbf{Q}_p]d(d-1)/2$, we see that $\overline{\mathcal{X}}_d$ also has dimension $[K:\mathbf{Q}_p]d(d-1)/2$.

Consider a finite type point $x\colon \mathrm{Spec}\,\mathbf{F}' \to \mathcal{X}_{d,\mathrm{red}}$, corresponding to a representation $\overline{\rho}\colon G_K \to \mathrm{GL}_d(\mathbf{F}')$. By Proposition 3.6.3 there is a corresponding versal morphism $\mathrm{Spf}\,R^{\square}_{\overline{\rho}}/\varpi \to \overline{\mathcal{X}}_d$, and applying [EG19, Lem. 2.40] as in the proof of Theorem 4.8.14, we conclude that $R^{\square}_{\overline{\rho}}/\varpi$ must have dimension $d^2 + [K:\mathbf{Q}_p]d(d-1)/2$. However, it is known that there are representations $\overline{\rho}$ for which $R^{\square}_{\overline{\rho}}/\varpi$ is formally smooth of dimension $d^2 + [K:\mathbf{Q}_p]d^2$ (see for example [All19, Lem. 3.3.1]; indeed, a Galois cohomology calculation shows that the dimension is always at least $d^2 + [K:\mathbf{Q}_p]d^2$, which is all that we need). Thus we must have $d^2 = d(d-1)/2$, a contradiction. □

6.5.3 Remark. A similar argument shows that the versal morphisms $\mathrm{Spf}\,R^{\square,\mathcal{O}'}_{\overline{\rho}} \to \mathcal{X}_d$ and $\mathrm{Spf}\,R^{\square,\mathcal{O}'}_{\overline{\rho}}/\varpi \to \mathcal{X}_d \otimes_{\mathcal{O}} \mathbf{F}$ of Proposition 3.6.3 are *not* effective.

6.6 CLOSED POINTS

Recall that if $\mathcal{Y}$ is an algebraic stack, then $|\mathcal{Y}|$ denotes its underlying topological space. The points of $|\mathcal{Y}|$, which we also refer to as the points of $\mathcal{Y}$, are the equivalence classes of morphisms $\operatorname{Spec} k \to \mathcal{Y}$, with k a field; two morphisms $\operatorname{Spec} k \to \mathcal{Y}$ and $\operatorname{Spec} l \to \mathcal{Y}$ are deemed to be equivalent if we may find a field Ω and embeddings $k, l \hookrightarrow \Omega$ such that the pull-backs to $\operatorname{Spec} \Omega$ of the two given morphisms coincide. A point of $\mathcal{Y}$ is called *finite type* if it is representable by a morphism $\operatorname{Spec} k \to \mathcal{Y}$ which is locally of finite type. If $\mathcal{Y}$ is locally of finite type over a base-scheme S, then a point of $\mathcal{Y}$ is of finite type if and only if its image in S is a finite type point of S in the usual sense [Sta, Tag 01T9].

We say that a point of $\mathcal{Y}$ is *closed* if the corresponding point of $|\mathcal{Y}|$ is closed in the topology on $|\mathcal{Y}|$. Closed points are necessarily of finite type. However, if $\mathcal{X}$ is not quasi-DM, then finite type points of $\mathcal{X}$ need not be closed, even if $\mathcal{X}$ is Jacobson (in contrast to the situation for schemes [Sta, Tag 01TB]).

If $\mathcal{Y}$ is locally of finite type over an algebraically closed field k, then the map $\mathcal{Y}(k) \to |\mathcal{Y}|$ (taking a k-valued point of $\mathcal{Y}$ to the point it represents) identifies $\mathcal{Y}(k)$ with the set of finite type points of $\mathcal{Y}$. (This is a standard consequence of the Nullstellensatz, applied to a scheme chart of $\mathcal{Y}$.)

Throughout this section, we fix a value of $d \geq 1$, and write $\mathcal{X}$ rather than $\mathcal{X}_d$. In order to describe the finite type and closed points of $(\mathcal{X}_{\mathrm{red}})_{\overline{\mathbf{F}}_p}$, we introduce the following terminology.

6.6.1 Definition. Let $\overline{\rho}, \overline{\theta}\colon G_K \to \mathrm{GL}_d(\overline{\mathbf{F}}_p)$ be continuous representations. Then we say that $\overline{\theta}$ is a *partial semi-simplification* of $\overline{\rho}$ if there are short exact sequences

$$0 \to \overline{r}_i \to \overline{\rho}_i \to \overline{r}'_i \to 0$$

for $i = 1, \ldots$ such that $\overline{\rho}_1 = \overline{\rho}$, $\overline{\rho}_{i+1} = \overline{r}_i \oplus \overline{r}'_i$, and $\overline{\rho}_n = \overline{\theta}$ for n sufficiently large.

Note in particular that the semi-simplification of $\overline{\rho}$ is a partial semi-simplification of $\overline{\rho}$. We also consider the following related notion.

6.6.2 Definition. Let $\overline{\rho}, \overline{\theta}\colon G_K \to \mathrm{GL}_d(\overline{\mathbf{F}}_p)$ be continuous representations. Then we say that $\overline{\theta}$ is a *virtual partial semi-simplification* of $\overline{\rho}$ if there is a short exact sequence of the form

$$0 \to \overline{r} \to \overline{r} \oplus \overline{\rho} \to \overline{\theta} \to 0.$$

It follows from an evident induction on the integer n in Definition 6.6.1 that if $\overline{\theta}$ is a partial semi-simplification of $\overline{\rho}$, then it is in particular a virtual partial semisimplification; however, the converse does not hold in general. (See the introduction to [Zwa00].)

Our promised description of the finite type and closed points of $(\mathcal{X}_{\mathrm{red}})_{\overline{\mathbf{F}}_p}$ is given by the following theorem.

6.6.3 Theorem

1. *The morphism* $\mathcal{X}(\overline{\mathbf{F}}_p) = \mathcal{X}_{\mathrm{red}}(\overline{\mathbf{F}}_p) \to |(\mathcal{X}_{\mathrm{red}})_{\overline{\mathbf{F}}_p}|$ *that sends each morphism* $\operatorname{Spec} \overline{\mathbf{F}}_p \to \mathcal{X}$ *to the point of* $(\mathcal{X}_{\mathrm{red}})_{\overline{\mathbf{F}}_p}$ *that it represents is injective, and its image consists precisely of the finite type points. Thus the finite type points of* $(\mathcal{X}_{\mathrm{red}})_{\overline{\mathbf{F}}_p}$ *are in natural bijection with the isomorphism classes of continuous representations* $\overline{\rho}\colon G_K \to \mathrm{GL}_d(\overline{\mathbf{F}}_p)$.
2. *A finite type point of* $|(\mathcal{X}_{\mathrm{red}})_{\overline{\mathbf{F}}_p}|$ *is closed if and only if the associated Galois representation* $\overline{\rho}$ *is semi-simple.*
3. *If* $x \in |(\mathcal{X}_{\mathrm{red}})_{\overline{\mathbf{F}}_p}|$ *is a finite type point, corresponding to the Galois representation* $\overline{\rho}$, *then the closure* $\overline{\{x\}}$ *contains a unique closed point, whose corresponding Galois representation is the semi-simplification* $\overline{\rho}^{\mathrm{ss}}$ *of* $\overline{\rho}$. *More generally, if* x *and* y *are two finite type points of* $|(\mathcal{X}_{\mathrm{red}})_{\overline{\mathbf{F}}_p}|$, *corresponding to Galois representations* $\overline{\rho}$ *and* $\overline{\theta}$ *respectively, then* y *lies in* $\overline{\{x\}}$ *if and only if* $\overline{\theta}$ *is a virtual partial semi-simplification of* $\overline{\rho}$.

Proof. Since $\overline{\mathbf{F}}_p$-valued points of $\mathcal{X}$ coincide with $\overline{\mathbf{F}}_p$-valued points of $\mathcal{X}_{\mathrm{red}}$, and so also with $\overline{\mathbf{F}}_p$-valued points of $(\mathcal{X}_{\mathrm{red}})_{\overline{\mathbf{F}}_p}$, the claim of (1) is a particular case of the more general consequence of the Nullstellensatz that we recalled above.

Suppose now that we have a short exact sequence

$$0 \to \overline{\tau} \to \overline{\tau} \oplus \overline{\rho} \to \overline{\theta} \to 0. \tag{6.6.4}$$

We now follow the proof of [Rie86, Prop. 3.4] to show that the point y corresponding to $\overline{\theta}$ lies in the closure of the point x corresponding to $\overline{\rho}$. Denote the morphism $\overline{\tau} \to \overline{\tau} \oplus \overline{\rho}$ in (6.6.4) by (f, g). Then for t in a sufficiently small open neighborhood U of $0 \in \mathbf{A}^1$, the morphism $(f + t\mathbf{1}_{\overline{\tau}}, g)\colon \overline{\tau} \to \overline{\tau} \oplus \overline{\rho}$ is an injection with projective cokernel (to see this one can for example consider a splitting of(6.6.4) on the level of vector spaces); so we have a morphism (where we as usual abuse notation by writing families of (φ, Γ)-modules as families of Galois representations) $U \to \mathcal{X}$ given by $t \mapsto \overline{\rho}_t := (\overline{\tau} \oplus \overline{\rho}) / \operatorname{im}(f + t\mathbf{1}_{\overline{\tau}}, g)$. After possibly shrinking U further we may assume that $f + t\mathbf{1}_{\overline{\tau}}$ is an automorphism of $\overline{\tau}$ for all $\overline{\mathbf{F}}_p$-points $t \neq 0$ of U, so that $\overline{\rho}_t \cong \overline{\rho}$; while for $t = 0$ we have $\overline{\rho}_t \cong \overline{\theta}$. Thus $\overline{\theta}$ is indeed in the closure of $\overline{\rho}$, as claimed. This proves the "if" direction of (3), and the "only if" direction of (2).

The "if" direction of (2) follows from the "only if" direction of (3), and so it remains to prove this latter statement. To this end, we fix $\overline{\rho}, \overline{\theta}\colon G_K \to \mathrm{GL}_d(\overline{\mathbf{F}}_p)$, and assume that the point y corresponding to $\overline{\theta}$ lies in the closure of the point x corresponding to $\overline{\rho}$. In order to avoid discussing deformation theory over $\overline{\mathbf{F}}_p$, we extend our coefficients $\mathcal{O}$ if necessary so that $\overline{\rho}$ and $\overline{\theta}$ are both defined over $\mathbf{F}$. We may and do also assume that x and y are distinct; i.e., that $\overline{\rho}$ and $\overline{\theta}$ are not isomorphic (as representations defined over $\overline{\mathbf{F}}_p$, or equivalently, as representations defined over $\mathbf{F}$). Let x_0 and y_0 denote the images of x and

y respectively in $\mathcal{X}_{\text{red}}$; then x_0 and y_0 are again distinct. Let $\mathcal{Z}$ denote the closure of $\{x_0\}$, thought of as a reduced closed substack of $\mathcal{X}_{\text{red}}$; then y_0 is a point of $\mathcal{Z}$.

Let R be the framed deformation ring of $\overline{\theta}$ over $\mathcal{O}$; then R is a versal ring to $\mathcal{X}$ at y. Let $\operatorname{Spf} S = \operatorname{Spf} R \times_{\mathcal{X}} \mathcal{Z}$; then S is a versal ring to $\mathcal{Z}$ at y_0. Since $\mathcal{Z}$ is algebraic, the versal ring S is effective [Sta, Tag 07X8], and the corresponding morphism $\operatorname{Spec} S \to \mathcal{Z}$ is furthermore flat [Sta, Tag 0DR2]. The morphism $\operatorname{Spec} \mathbf{F} \to \mathcal{Z}$ corresponding to x_0 is quasi-compact and scheme-theoretically dominant; thus the base-changed morphism

$$\operatorname{Spec} S \times_{\mathcal{Z}, x_0} \operatorname{Spec} \mathbf{F} \to \operatorname{Spec} S \tag{6.6.5}$$

is again scheme-theoretically dominant.

Let η be a point in the image of (6.6.5), and let $\operatorname{Spec} T$ be the Zariski closure of η in $\operatorname{Spec} S$. Then T is a quotient of R which is an integral domain. If $\mathcal{K}$ denotes the fraction field of T, then the morphism $\operatorname{Spec} \mathcal{K} \to \mathcal{X}_{\text{red}}$ factors through the morphism $\operatorname{Spec} \mathbf{F} \to \mathcal{X}_{\text{red}}$ corresponding to x_0, and so the Galois representation $G_K \to \mathrm{GL}_d(\mathcal{K})$ classified by the point η of $\operatorname{Spec} T$ is isomorphic to $\overline{\rho}$ (see Section 3.6.5 for the definition of this Galois representation). Thus, if L denotes the splitting field of $\overline{\rho}$, then we see that the deformations of $\overline{\theta}$ classified by T all factor through the finite group $\mathrm{Gal}(L/K)$. It follows immediately from [Zwa00, Thm. 1] that $\overline{\theta}$ is a virtual partial semi-simplification of $\overline{\rho}$, as required. $\square$

6.7 THE SUBSTACK OF G_K-REPRESENTATIONS

We now briefly explain the relationship between our stacks and the stacks of G_K-representations constructed by Wang-Erickson in [WE18]. Write $\mathcal{X}_d^{\mathrm{Gal}}$ for the formal algebraic stack characterized by the following property: if A is an $\mathcal{O}$-algebra in which p is nilpotent, then $\mathcal{X}_d^{\mathrm{Gal}}(A)$ is the groupoid of continuous morphisms

$$\rho \colon G_K \to \mathrm{GL}_d(A)$$

(where A has the discrete topology, and G_K its natural profinite topology). That this *is* a formal algebraic stack follows from [WE18, Thm. 3.8, Rem. 3.9]. Equivalently, we can think of $\mathcal{X}_d^{\mathrm{Gal}}(A)$ as the groupoid of rank d projective A-modules T_A with a continuous action of G_K (where each T_A has the discrete topology).

For any finite type $\mathcal{O}/\varpi^a$-algebra A, and any $T_A \in \mathcal{X}_d^{\mathrm{Gal}}(A)$, we set

$$\mathbf{D}_A(T_A) := W(\mathbf{C}^\flat)_A \otimes_A T_A,$$

which naturally has the structure of a rank d projective (φ, G_K)-module with A-coefficients (with φ acting on the first factor in the tensor product, and G_K

acting diagonally). (In the case that A is actually a finite $\mathcal{O}/\varpi^a$-algebra, this agrees with the construction given in Section 3.6.4.)

By Proposition 2.7.8, for each finite type $\mathcal{O}/\varpi^a$-algebra A the assignment $T_A \mapsto \mathbf{D}_A(T_A)$ gives a functor $\mathcal{X}_d^{\mathrm{Gal}}(A) \to \mathcal{X}_d(A)$. Since both $\mathcal{X}_d^{\mathrm{Gal}}$ and $\mathcal{X}_d$ are limit preserving (for $\mathcal{X}_d^{\mathrm{Gal}}$ this follows easily from the definition, since any ρ as above factors through a finite quotient of G_K), this defines a morphism of stacks

$$\mathcal{X}_d^{\mathrm{Gal}} \to \mathcal{X}_d. \tag{6.7.1}$$

We now show that this morphism is in fact a monomorphism.

6.7.2 Theorem. *The morphism* (6.7.1) *is a monomorphism, and is furthermore versal at finite type points in the sense of Definition* 7.1.1 *below.*

Proof. The versality statement follows from the fact that for any finite Artinian $\mathbf{Z}_p$-algebra A, the A-valued points of $\mathcal{X}_d^{\mathrm{Gal}}$ and $\mathcal{X}_d$ coincide. (Over such a ring A, étale (φ, Γ)-modules do arise from Galois representations.)

To prove the monomorphism claim, we need to show that for any finite type $\mathcal{O}/\varpi^a$-algebra A, the functor $\mathcal{X}_d^{\mathrm{Gal}}(A) \to \mathcal{X}_d(A)$ is fully faithful; that is, we need to show that given $T_1, T_2 \in \mathcal{X}_d^{\mathrm{Gal}}(A)$, we have

$$\mathrm{Hom}_{G_K,A}(T_1,T_2) \stackrel{?}{=} \mathrm{Hom}_{\varphi,G_K}(\mathbf{D}_A(T_1), \mathbf{D}_A(T_2)).$$

Writing $T := T_1^\vee \otimes_A T_2$, it suffices to show that

$$T^{G_K} \stackrel{?}{=} \mathbf{D}_A(T)^{\varphi=1,G_K},$$

or even that

$$T \stackrel{?}{=} \mathbf{D}_A(T)^{\varphi=1}.$$

Since T is a finite projective A-module, we have

$$\mathbf{D}_A(T)^{\varphi=1} := (W(\mathbf{C}^\flat)_A \otimes_A T)^{\varphi=1} = (W(\mathbf{C}^\flat)_A)^{\varphi=1} \otimes_A T,$$

and the result follows from Lemma 2.2.19. □

6.7.3 Remark. In [WE18, §4], Wang-Erickson explains how to associate étale φ-modules to G_{K_∞}-representations with open kernel. A similar argument would allow us to associate projective étale (φ, Γ)-modules to objects of $\mathcal{X}_d^{\mathrm{Gal}}(A)$. However, if we directly followed the strategy of [WE18], the resulting étale (φ, Γ)-modules would have as coefficients the rings $\mathbf{A}_K \otimes_{\mathbf{Z}_p} A$, rather than the rings $\mathbf{A}_{K,A}$ that we use throughout this book. Thus this approach would not obviously yield Theorem 6.7.2, which is why we've preferred to make a more direct argument in terms of étale (φ, G_K)-modules.

6.7.4 Remark. It follows immediately from Theorem 6.7.2 that $\mathcal{X}_d^{\mathrm{Gal}}$ is the largest substack of $\mathcal{X}_d$ over which the universal (φ, Γ)-module can be realized as a G_K-representation.

In fact, it should be possible to strengthen Theorem 6.7.2. For example, [WE18, Thm. 3.8] shows that we can write

$$\mathcal{X}_d^{\mathrm{Gal}} = \coprod_D \mathcal{X}_{d,D}^{\mathrm{Gal}}, \tag{6.7.5}$$

where D runs over the isomorphism classes of d-dimensional semisimple $\overline{\mathbf{F}}_p$-representations of G_K, and $\mathcal{X}_{d,D}^{\mathrm{Gal}}(A)$ is the groupoid of those ρ the semisimplification of whose reductions modulo ϖ is D. In view of Theorem 6.6.3, we expect that the monomorphism $(\mathcal{X}_{d,D}^{\mathrm{Gal}})_{\mathrm{red}} \to \mathcal{X}_{d,\mathrm{red}}$ (induced by the monomorphism of Theorem 6.7.2) will actually be a closed immersion, and hence (given the versality statement of Theorem 6.7.2) that $\mathcal{X}_{d,D}^{\mathrm{Gal}}$ will be identified with the formal completion of $\mathcal{X}_d$ along the image of this closed immersion. In fact, this image will contain a unique closed point (corresponding to the semisimple Galois representation D), and we imagine that $\mathcal{X}_{d,D}^{\mathrm{Gal}}$ could also be identified with the *coherent completion* (in the sense [AHR20, Defn. 2.1]) of $\mathcal{X}_d$ at this closed point. However, we don't pursue these ideas further here.

6.7.6 Remark. As well as considering G_K-representations, it is also possible to consider a larger substack of $\mathcal{X}_d$ over which our (φ, Γ)-modules can be realized as "Weil–Deligne"-representations. (Here the notion of Weil–Deligne representations is not quite the usual one, but rather is given by representations of certain discretizations of G_K, as described in [EG20, §1.2] and the references therein.)

In the case $K = \mathbf{Q}_p$ and $d = 2$, this locus is discussed in [EG20, §1]; roughly speaking, the difference between it and $\mathcal{X}_d^{\mathrm{Gal}}$ is that it contains families of semisimple representations given by direct sums of unramified twists of fixed irreducible representations (the key point being that the universal unramified character does not correspond to a G_K-representation, but does correspond to a representation of the Weil group W_K; see Remark 7.2.19 below).

Chapter Seven

The rank 1 case

In this chapter we describe some of our key constructions explicitly in the case where $d=1$. More precisely, following the notation of Chapter 3, we give explicit descriptions of the stacks $\mathcal{R}_d$, $\mathcal{R}_d^{\Gamma_{\mathrm{disc}}}$, and $\mathcal{X}_d$, all in the case when $d=1$ (and K is arbitrary).

7.1 PRELIMINARIES

While it may be possible to find an explicit description of the stacks we are interested in directly from their definitions, this is not how we proceed. Rather, we use the relationship between étale φ-modules (resp. étale (φ,Γ)-modules) and Galois representations to construct certain affine formal algebraic spaces U, V, and W, along with morphisms $U\to\mathcal{R}_1$, $V\to\mathcal{R}_1^{\Gamma_{\mathrm{disc}}}$, and $W\to\mathcal{X}_1$, each satisfying the conditions of Lemma 7.1.8 below. An application of this lemma then yields a description of each of $\mathcal{R}_1$, $\mathcal{R}_1^{\Gamma_{\mathrm{disc}}}$, and $\mathcal{X}_1$.

7.1.1 Definition. Let S be a locally Noetherian scheme, let $\mathcal{X}$ and $\mathcal{Y}$ be stacks over S and let $f\colon \mathcal{X}\to\mathcal{Y}$ be a morphism. We say that f is *versal at finite type points* if for each morphism $x\colon \operatorname{Spec} k\to\mathcal{X}$, with k a finite type $\mathcal{O}_S$-field, the morphism f is versal at x, in the sense of Definition A.26.

7.1.2 Remark. As the discussion of versality in Appendix A should make clear, this definition is related to the various properties considered in [EG21, Def. 2.4.4] and the surrounding discussion. As noted in [EG21, Rem. 2.4.6], it is somewhat complicated to define the notion of versality at literal points of a stack, since the points of a stack are (by definition) equivalence classes of morphisms from the spectrum of a field to the stack, and it is not immediately clear in general whether or not the versality condition would be independent of the choice of equivalence class representative. In the preceding definition we obviate this point, by directly requiring versality at all (finite type) representatives of all (finite type) points.

7.1.3 Lemma. *If $\mathcal{X}\to\mathcal{Y}$ is a morphism of stacks over a locally Noetherian base S which is versal at finite type points, and if $\mathcal{Z}\to\mathcal{Y}$ is a morphism of stacks, then the induced morphism $\mathcal{X}\times_{\mathcal{Y}}\mathcal{Z}\to\mathcal{Z}$ is again versal at finite type points.*

Proof. Let $\widetilde{x}\colon \operatorname{Spec} k \to \widetilde{\mathcal{X}} := \mathcal{X} \times_{\mathcal{Y}} \mathcal{Z}$ with $\operatorname{Spec} k$ of finite type over S; from the definition of the 2-fiber product $\widetilde{\mathcal{X}}$, we see that giving the morphism $\widetilde{x}$ amounts to giving the pair of morphisms $x\colon \operatorname{Spec} k \to \mathcal{X}$ and $z\colon \operatorname{Spec} k \to \mathcal{Z}$ (the result of composing $\widetilde{x}$ with each of the projections), as well as an isomorphism $y \xrightarrow{\sim} y'$, where y and y' are the k-valued points of $\mathcal{X}$ obtained respectively by composing x with the morphism $\mathcal{X} \to \mathcal{Y}$ and by composing z with the morphism $\mathcal{Z} \to \mathcal{Y}$. It is then a straightforward diagram chase, working from the various definitions, to deduce the versal property of the morphism $\widehat{(\widetilde{\mathcal{X}})}_{\widetilde{x}} \to \widehat{\mathcal{Z}}_z$ (where we use the notation of Definition A.25) from the versal property of the morphism $\widehat{\mathcal{X}}_x \to \widehat{\mathcal{Y}}_y$. □

The following result is standard, but we indicate the proof for the sake of completeness.

7.1.4 Lemma. *If $f\colon \mathcal{X} \to \mathcal{Y}$ is a morphism of algebraic stacks, each of finite type over a locally Noetherian scheme S, then f is versal at finite type points if and only if f is smooth.*

Proof. The "if" direction follows from an application of the infinitesimal lifting property of smooth morphisms; see, e.g., the implication "(4) $\Longrightarrow$ (1)" of [EG21, Lem. 2.4.7 (4)]. For the converse, choose a smooth surjective morphism $U \to \mathcal{X}$ with U a scheme (necessarily locally of finite type over S); then by the direction already proved, this morphism is also versal at finite type points, and hence so is the composite morphism $U \to \mathcal{X} \to \mathcal{Y}$. To show that f is smooth, it suffices to show the same for this composite. It follows from the implication "(1) $\Longrightarrow$ (4)" of [EG21, Lem. 2.4.7 (4)] that f is smooth in a neighborhood of each finite type point of U; but since these points are dense in U, the desired result follows. (We have cited [EG21] in this argument purely for our own convenience; of course the present lemma is just a version of Grothendieck's result relating smoothness and formal smoothness.) □

We now strengthen the previous result to cover the case where $\mathcal{X}$ and $\mathcal{Y}$ are not necessarily assumed to be algebraic, but merely the map between them is assumed to be representable by algebraic stacks.

7.1.5 Lemma. *If $f\colon \mathcal{X} \to \mathcal{Y}$ is a morphism of limit preserving stacks over a locally Noetherian base S, and f is representable by algebraic stacks, then f is versal at finite type points if and only if f is smooth.*

Proof. If f is smooth, then by Lemma A.2 (2) it satisfies the infinitesimal lifting property, which immediately implies that it is versal at finite type points. Conversely, suppose now that f is versal at finite type points.

It follows from [EG21, Cor. 2.1.8] that f is limit preserving on objects, and since it is also representable by algebraic stacks, it is locally of finite presentation, by Lemma A.2 (1). We now have to verify that for any test morphism $Z \to \mathcal{Y}$ whose source is a scheme, the induced morphism

$$\mathcal{X} \times_{\mathcal{Y}} Z \to Z, \tag{7.1.6}$$

which is again locally of finite presentation, is in fact smooth. Of course it suffices to do this for affine schemes, and then, since $\mathcal{Y}$ is limit preserving, for affine schemes that are of finite type over S. Since f is representable by algebraic stacks, we find that the source of (7.1.6) is an algebraic stack. Furthermore, since this morphism is locally of finite presentation, its source is again locally of finite type over S. Finally, Lemma 7.1.3 shows that (7.1.6) is versal at finite type points. The claimed result then follows from Lemma 7.1.4. □

We may apply the previous lemma to obtain a criterion for representing a stack as the quotient of a sheaf by a smooth groupoid. We first make another definition.

7.1.7 Definition. We say that a morphism $\mathcal{X} \to \mathcal{Y}$ of stacks over a locally Noetherian scheme S is *surjective on finite type points* if for each morphism $x\colon \operatorname{Spec} k \to \mathcal{Y}$, with k a finite type $\mathcal{O}_S$-field, we may find a field extension l of k and a morphism $\operatorname{Spec} l \to \mathcal{X}$ which makes the diagram

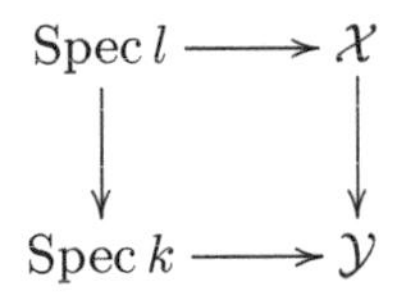

commutative.

7.1.8 Lemma. *Suppose that $\mathcal{Z}$ is a stack over a locally Noetherian scheme S, and that $f\colon U \to \mathcal{Z}$ is a morphism from a sheaf to $\mathcal{Z}$ which is representable by algebraic spaces*[1] *and surjective on finite type points (in the sense of Definition 7.1.7). Suppose furthermore that both U and $\mathcal{Z}$ are limit preserving, and that f is versal at finite type points. Then the morphism f is in fact smooth and surjective, and, if we let $R := U \times_{\mathcal{Z}} U$, then the induced morphism $[U/R] \to \mathcal{Z}$ is an isomorphism.*

Proof. It follows from Lemma 7.1.5 that the morphism f is smooth. We claim that it is furthermore surjective; by definition, this means that if $T \to \mathcal{Z}$ is any

[1] Since the source of the morphism is a sheaf, this is equivalent to being representable by algebraic stacks.

test morphism from an affine scheme, we must show that the morphism of algebraic spaces $g: T \times_{\mathcal{Z}} U \to T$ induced by f is surjective. Since $\mathcal{Z}$ is limit preserving, we may assume that T is of finite type over S; also, since g is a base-change of the smooth morphism f, it is itself smooth, and so in particular has open image. Our assumption that f is surjective on finite type points implies furthermore that the image of g contains all the finite type points of T; thus g is indeed surjective.

Since smooth morphisms are in particular flat, the second assertion of the lemma follows from Lemma A.33. $\square$

7.1.9 Remark. If we omit the assumption in Lemma 7.1.8 that the morphism $U \to \mathcal{Z}$ is representable by algebraic spaces, then the morphism $[U/R] \to \mathcal{Z}$ need not be an isomorphism.

An example illustrating this is given by taking $\mathcal{Z}$ to be a scheme Z, locally of finite type over a locally Noetherian base scheme S, choosing a closed subscheme Y of Z which is *not* also open, and defining U to be the formal scheme obtained as the disjoint union $U := (Z \setminus Y) \coprod \widehat{Z}$, where $\widehat{Z}$ denotes the completion of Z along Y. We let $U \to Z$ be the obvious morphism; this is then a surjective monomorphism, which is not an isomorphism, although it is versal at every finite type point, and (since it is a monomorphism) we have that $R := U \times_Z U$ coincides with the diagonal copy of U inside $U \times_S U$ (so that $[U/R] = U$).

We will find the following lemmas useful in verifying the hypotheses of Lemma 7.1.8 in our application.

7.1.10 Lemma. *Let $f: \mathcal{X} \to \mathcal{Y}$ and $g: \mathcal{Y} \to \mathcal{Z}$ be morphisms of stacks over a locally Noetherian scheme S. Assume that f is representable by algebraic stacks and locally of finite type, and that both g and the morphism $\mathcal{Z} \to S$ are limit preserving on objects. Then f is an isomorphism if and only if the induced morphism*

$$\operatorname{Spec} A \times_{\mathcal{Z}} \mathcal{X} \to \operatorname{Spec} A \times_{\mathcal{Z}} \mathcal{Y} \tag{7.1.11}$$

is an isomorphism for every morphism $\operatorname{Spec} A \to \mathcal{Z}$ *with A a locally of finite type Artinian local $\mathcal{O}_S$-algebra.*

Proof. The "only if" direction is evident, and so we focus on the "if" direction. We begin by reducing to the case $\mathcal{Z} = S$. We first recall a basic fact about fiber products: if $T \to \mathcal{Y}$ is any morphism from a scheme to the stack $\mathcal{Y}$, then the base change $T \times_{\mathcal{Y}} \mathcal{X}$ may be described as an iterated fiber product $T \times_{(T \times_{\mathcal{Z}} \mathcal{Y})} (T \times_{\mathcal{Z}} \mathcal{X})$ (where we regard T as lying over $\mathcal{Z}$ via the composite $T \to \mathcal{Y} \to \mathcal{Z}$, and as lying over $T \times_{\mathcal{Z}} \mathcal{Y}$ via the graph of the given morphism $T \to \mathcal{Y}$).

In order to show that f is an isomorphism, we have to show that for any morphism $T \to \mathcal{Y}$, the base-changed morphism $T \times_{\mathcal{Y}} \mathcal{X} \to T$ is an isomorphism.

By the fact about fiber products just recalled (and remembering that the base-change of an isomorphism is an isomorphism), it suffices to show that for any morphism $T \to \mathcal{Z}$ whose source is a scheme, the base-changed morphism

$$T \times_{\mathcal{Z}} \mathcal{X} \to T \times_{\mathcal{Z}} \mathcal{Y} \tag{7.1.12}$$

is an isomorphism. Since $\mathcal{Z} \to S$ is limit preserving on objects, we may and do assume that T is locally of finite type over S. We will then replace both $\mathcal{Z}$ and S by T (mapping identically to itself), and the morphism f by (7.1.12). Note that the projection $T \times_{\mathcal{Z}} \mathcal{Y} \to T$ is again limit preserving on objects (being a base change of g, which has this property). We now have to reinterpret our original hypothesis on A-valued points of $\mathcal{Z}$ (for locally of finite type Artinian local $\mathcal{O}_S$-algebras A) in this new context. To this end, note that if $\operatorname{Spec} A \to T$ is locally of finite type with A being Artinian local, then $\operatorname{Spec} A$ is also locally of finite type over S. The fiber product of (7.1.12) with $\operatorname{Spec} A$ over T may then be identified with the morphism (7.1.11), and is thus an isomorphism (by assumption). Putting all this together, we see that we have reduced to the case when $\mathcal{Z} = S$; so we return to our original notation, but assume in addition that $\mathcal{Z} = S$ (mapping to itself via the identity).

What we have to show is that if $T \to \mathcal{Y}$ is any morphism, then

$$T \times_{\mathcal{Y}} \mathcal{X} \to T \tag{7.1.13}$$

is an isomorphism. Since $\mathcal{Y} \to S$ is limit preserving, we may assume that T is locally of finite type over S. Any finite type Artinian local T-algebra is then also finite type over S, and so, replacing both $\mathcal{Y}$ and S by T (and again applying the fiber product fact recalled above, with T replaced by $\operatorname{Spec} A$) we find that we may make another reduction, to the case when $\mathcal{Y} = S$. The representability assumption on f then implies that $\mathcal{X}$ is an algebraic stack. In this case the lemma is standard, but we recall a proof for completeness.

Our hypothesis that f induces an isomorphism on A-valued points implies that the morphism f is versal at finite type points, and so by Lemma 7.1.4, it is smooth. It also implies that f contains all finite type points in its image; since the image of f is constructible, we see that f is in fact surjective. This same argument implies that the diagonal morphism $\mathcal{X} \to \mathcal{X} \times_S \mathcal{X}$ is surjective, and thus that f is universally injective. Since f is smooth, it is flat and locally of finite type. Since it is also universally injective, it is an open immersion. Since it is surjective, it is an isomorphism, as claimed. □

7.1.14 Lemma. *Let $\mathcal{Y}$ be a stack over a locally Noetherian scheme S, let $\mathcal{Z}$ be a closed substack of $\mathcal{Y}$, and let $f\colon \mathcal{X} \to \mathcal{Y}$ be a morphism of stacks. Suppose that both f and the morphism $\mathcal{Y} \to S$ are limit preserving on objects; then f factors through $\mathcal{Z}$ if and only if and only if for any finite local Artinian $\mathcal{O}_S$-algebra A,* *and any morphism* $\operatorname{Spec} A \to \mathcal{Y}$ *over* S*, the morphism* $\operatorname{Spec} A \times_{\mathcal{Y}} \mathcal{X} \to \mathcal{Y}$ *factors through* $\mathcal{Z}$*.*

Proof. The "only if" direction is trivial. For the converse, consider the base-change via f of the closed immersion $\mathcal{Z} \hookrightarrow \mathcal{Y}$; this is a closed immersion $g\colon \mathcal{X} \times_{\mathcal{Y}} \mathcal{Z} \hookrightarrow \mathcal{X}$ of stacks over $\mathcal{Y}$, which we must show is an isomorphism. The hypothesis implies that for any morphism $\operatorname{Spec} A \to \mathcal{Y}$ whose source is a finite type local Artinian $\mathcal{O}_S$-algebra, the pull-back of g

$$\operatorname{Spec} A \times_{\mathcal{Y}} (\mathcal{X} \times_{\mathcal{Y}} \mathcal{Z}) \to \operatorname{Spec} A \times_{\mathcal{Y}} \mathcal{X}$$

is an isomorphism; it then follows from Lemma 7.1.10 that g is an isomorphism. □

The next lemma lets us compute fiber products over stacks in certain situations. We recall that if $\mathcal{Y}$ is a stack (over some base scheme S), then the inertia stack $\mathcal{I}_{\mathcal{Y}}$ is the stack over S whose T-valued points (for any S-scheme T) consist of pairs (y, α), where y is a T-valued point of $\mathcal{Y}$, and α is an automorphism of y. It is a group object in the category of stacks lying over $\mathcal{Y}$ via morphisms that are representable by sheaves. We also remind the reader that there is a canonical isomorphism

$$\mathcal{I}_{\mathcal{Y}} \xrightarrow{\sim} \mathcal{Y} \times_{\Delta_{\mathcal{Y}}, \mathcal{Y} \times_S \mathcal{Y}, \Delta_{\mathcal{Y}}} \mathcal{Y},$$

where the subscripts $\Delta_{\mathcal{Y}}$ indicate that both copies of $\mathcal{Y}$ are regarded as lying over the product $\mathcal{Y} \times_S \mathcal{Y}$ via the diagonal morphism $\Delta_{\mathcal{Y}}\colon \mathcal{Y} \to \mathcal{Y} \times_S \mathcal{Y}$, and where, under this identification, the forgetful morphism $\mathcal{I}_{\mathcal{Y}} \to \mathcal{Y}$ may be identified with projection onto the first copy of $\mathcal{Y}$.

7.1.15 Lemma. *Let* $X \to \mathcal{Y}$ *be a morphism from a sheaf to a stack (both over some base-scheme* S*), and suppose that the natural morphism* $X \times_{\mathcal{Y}} X \to X \times_S X$ *factors through the diagonal copy of* X *lying in the target (so that the groupoid* $X \times_{\mathcal{Y}} X$ *is in fact a group object in the category of sheaves over* X*). Then there is a canonical isomorphism* $X \times_{\mathcal{Y}} X \xrightarrow{\sim} X \times_{\mathcal{Y}} \mathcal{I}_{\mathcal{Y}}$ *of group objects over* X*.*

Proof. By assumption, we have a commutative diagram

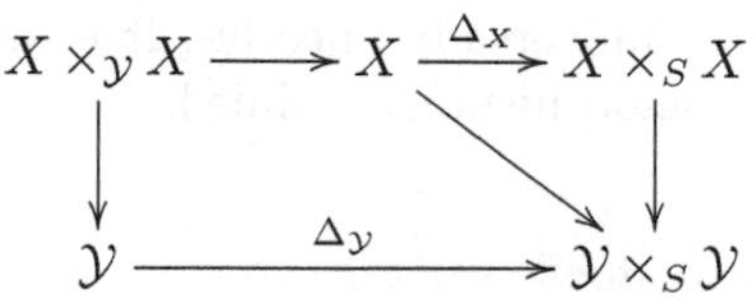

$$\begin{array}{ccccc} X \times_{\mathcal{Y}} X & \longrightarrow & X & \xrightarrow{\Delta_X} & X \times_S X \\ \downarrow & & & \searrow & \downarrow \\ \mathcal{Y} & & \xrightarrow{\Delta_{\mathcal{Y}}} & & \mathcal{Y} \times_S \mathcal{Y} \end{array}$$

in which the outer rectangle is 2-Cartesian. Since X is a sheaf, the diagonal $\Delta_X\colon X \to X \times_S X$ is a monomorphism, and so one immediately checks that the left-hand trapezoid is also 2-Cartesian. Thus we obtain the required isomorphism

$$X \times_{\mathcal{Y}} X \xrightarrow{\sim} X \times_{\mathcal{Y} \times_S \mathcal{Y}} \mathcal{Y} \xrightarrow{\sim} X \times_{\mathcal{Y}} (\mathcal{Y} \times_{\mathcal{Y} \times_S \mathcal{Y}} \mathcal{Y}) \xrightarrow{\sim} X \times_{\mathcal{Y}} \mathcal{I}_{\mathcal{Y}}. \qquad \square$$

7.2 MODULI STACKS IN THE RANK 1 CASE

We now give concrete descriptions of our various stacks in the rank 1 case. Before doing this, we prove a result that describes the stack structure on families of rank 1 objects. (It encodes the fact that the automorphisms of a rank 1 object are simply the scalars.)

7.2.1 Lemma. *If $\mathcal{Z}$ denotes either of the stacks $\mathcal{R}_1$ or $\mathcal{X}_1$, then there is a canonical isomorphism*

$$\mathcal{Z} \times_{\mathcal{O}} \widehat{\mathbf{G}}_m \xrightarrow{\sim} \mathcal{I}_{\mathcal{Z}} \tag{7.2.2}$$

of group objects over $\mathcal{Z}$ (where $\widehat{\mathbf{G}}_m$ is the ϖ-adic completion of $\mathbf{G}_m$ over $\mathcal{O}$).

Proof. For definiteness we give the proof in the case of $\mathcal{R}_1$; the proof in the $\mathcal{X}_1$ case is identical. We begin by defining the morphism (7.2.2): if A is any ϖ-adically complete $\mathcal{O}$-algebra, and M is an étale φ-module over $\mathbf{A}_{K,A}$, then (7.2.2) is defined on A-valued points of the source lying over M by mapping an element $a \in A^\times$ to the automorphism of M given by multiplication by a.

Since $\mathcal{R}_1$ has affine diagonal which is of finite presentation, the morphism (7.2.2) is a morphism between finite type group objects over $\mathcal{R}_1$, each of whose structure morphisms is representable by algebraic spaces, indeed affine, and of finite presentation. Furthermore, $\mathcal{R}_1$ itself is limit preserving over $\mathcal{O}$. Thus to show that (7.2.2) is an isomorphism, it suffices (by Lemma 7.1.10) to show that it induces an isomorphism on A-valued points for A an Artinian local $\mathcal{O}$-algebra of finite type. Returning to the notation of the preceding paragraph (but now assuming that A is Artinian local), we have to show that any automorphism of M is given by multiplication by an element of $A^\times$. Since M is of rank 1, we find that

$$\mathrm{Hom}_{\mathbf{A}_{K,A},\varphi}(M,M) \cong (M^\vee \otimes_{\mathbf{A}_{K,A}} M)^{\varphi=1} \cong (\mathbf{A}_{K,A})^{\varphi=1} \cong A$$

(the final isomorphism following from Lemma 2.2.19, since $\mathbf{A}_{K,A} \subseteq W(\mathbf{C}^\flat)_A$ by Proposition 2.2.12). The lemma follows. $\square$

7.2.3 Local Galois theory

We briefly recall the local Galois theory that is relevant to our computation. If L/K is an algebraic extension, we write I_L^{ab} for the image of I_L in G_L^{ab} (note that this is not the abelianization of I_L). We have short exact sequences

$$1 \to I_K \to G_K \to \mathrm{Frob}_K^{\widehat{\mathbf{Z}}} \to 1 \tag{7.2.4}$$

and

$$1 \to I_{K_{\mathrm{cyc}}} \to G_{K_{\mathrm{cyc}}} \to \mathrm{Frob}_K^{f\widehat{\mathbf{Z}}} \to 1, \tag{7.2.5}$$

where $f := [k_\infty : k]$.

These induce corresponding short exact sequences

$$1 \to I_K^{\mathrm{ab}} \to G_K^{\mathrm{ab}} \to \mathrm{Frob}_K^{\widehat{\mathbf{Z}}} \to 1 \tag{7.2.6}$$

and

$$1 \to I_{K_{\mathrm{cyc}}}^{\mathrm{ab}} \to G_{K_{\mathrm{cyc}}}^{\mathrm{ab}} \to \mathrm{Frob}_K^{f\widehat{\mathbf{Z}}} \to 1. \tag{7.2.7}$$

If G is any of the various profinite Galois groups appearing in the preceding discussion, we let $\mathcal{O}[[G]]$ denote the corresponding completed group ring over $\mathcal{O}$. This is a pro-Artinian ring, which gives rise to the affine formal algebraic space $\mathrm{Spf}\,\mathcal{O}[[G]]$. We endow this space with the trivial action of $\widehat{\mathbf{G}}_m$ (the ϖ-adic completion of $\mathbf{G}_m$ over $\mathcal{O}$).

7.2.8 Galois lifting rings for characters

If $\mathbf{F}'/\mathbf{F}$ is a finite extension, and $\overline{\rho}\colon G_{K_{\mathrm{cyc}}} \to (\mathbf{F}')^\times$ is a continuous character, then we can write $\mathcal{O}' = \mathcal{O} \otimes_{W(\mathbf{F})} W(\mathbf{F}')$, and consider the universal lifting $\mathcal{O}$-algebra $R_{\overline{\rho}}^{\square,\mathcal{O}'}$. If $x\colon \mathrm{Spf}\,\mathbf{F}' \to \mathcal{R}_1$ is the corresponding finite type point, then we have a versal morphism $\mathrm{Spf}\,R_{\overline{\rho}}^{\square,\mathcal{O}'} \to \mathcal{R}_1$; indeed it follows exactly as in the proof of Proposition 3.6.3 that we have an isomorphism

$$\mathrm{Spf}\,R_{\overline{\rho}}^{\square,\mathcal{O}'} \times_{\mathcal{R}_1} \mathrm{Spf}\,R_{\overline{\rho}}^{\square,\mathcal{O}'} \xrightarrow{\sim} (\widehat{\mathbf{G}}_m)_{R_{\overline{\rho}}^{\square,\mathcal{O}'}}, \tag{7.2.9}$$

where $(\widehat{\mathbf{G}}_m)_{R_{\overline{\rho}}^{\square,\mathcal{O}'}}$ denotes the completion of $(\mathbf{G}_m)_{R_{\overline{\rho}}^{\square,\mathcal{O}'}}$ along $(\mathbf{G}_m)_{\mathbf{F}'}$.

7.2.10 Descriptions of the rank 1 stacks

We can now establish our explicit description of $\mathcal{R}_1$.

7.2.11 Proposition. *There is an isomorphism*

$$\left[\left(\mathrm{Spf}\,\mathcal{O}[[I_{K_{\mathrm{cyc}}}^{\mathrm{ab}}]] \times \widehat{\mathbf{G}}_m\right)/\widehat{\mathbf{G}}_m\right] \xrightarrow{\sim} \mathcal{R}_1$$

(*where, in the formation of the quotient stack, the* $\widehat{\mathbf{G}}_m$*-action is taken to be trivial*).

Proof. We begin by constructing a morphism

$$\operatorname{Spf} \mathcal{O}[[I^{\mathrm{ab}}_{K_{\mathrm{cyc}}}]] \times \widehat{\mathbf{G}}_m \to \mathcal{R}_1. \tag{7.2.12}$$

For this, we choose a lift of $\sigma_{\mathrm{cyc}} \in G_{K_{\mathrm{cyc}}}$ of Frob_K^f, splitting the short exact sequences (7.2.5) and (7.2.7). If A is any discrete Artinian quotient of $\mathcal{O}[[I^{\mathrm{ab}}_{K_{\mathrm{cyc}}}]]$, then we extend the continuous morphism $\mathcal{O}[[I^{\mathrm{ab}}_{K_{\mathrm{cyc}}}]] \to A$ to a continuous morphism $\mathcal{O}[[G^{\mathrm{ab}}_{K_{\mathrm{cyc}}}]] \to A$ by mapping σ_{cyc} to $1 \in A$. We may view this latter morphism as a rank 1 representation of $G_{K_{\mathrm{cyc}}}$ with coefficients in A; it thus gives rise to a rank 1 projective étale φ-module M_A over $\mathbf{A}_{K,A}$, and therefore to a morphism $\operatorname{Spf} \mathcal{O}[[I^{\mathrm{ab}}_{K_{\mathrm{cyc}}}]] \to \mathcal{R}_1$. We extend this to the morphism (7.2.12) by unramified twisting; more precisely, we have an induced morphism $\operatorname{Spf} \mathcal{O}[[I^{\mathrm{ab}}_{K_{\mathrm{cyc}}}]] \times \widehat{\mathbf{G}}_m \to \mathcal{R}_1 \times \widehat{\mathbf{G}}_m$, and we compose with the morphism $\mathcal{R}_1 \times \widehat{\mathbf{G}}_m \to \mathcal{R}_1$ given by taking the tensor product with the universal unramified rank 1 étale φ-module.

By construction (and the usual explicit description of the universal Galois deformation ring of a character), (7.2.12) is versal at finite type points, as well as surjective on finite type points. We claim that the canonical morphism

$$\begin{aligned}(\operatorname{Spf} \mathcal{O}[[I^{\mathrm{ab}}_{K_{\mathrm{cyc}}}]] \times \widehat{\mathbf{G}}_m) &\times_{\mathcal{R}_1} (\operatorname{Spf} \mathcal{O}[[I^{\mathrm{ab}}_{K_{\mathrm{cyc}}}]] \times \widehat{\mathbf{G}}_m) \\ &\to (\operatorname{Spf} \mathcal{O}[[I^{\mathrm{ab}}_{K_{\mathrm{cyc}}}]] \times \widehat{\mathbf{G}}_m) \times_{\mathcal{O}} (\operatorname{Spf} \mathcal{O}[[I^{\mathrm{ab}}_{K_{\mathrm{cyc}}}]] \times \widehat{\mathbf{G}}_m)\end{aligned} \tag{7.2.13}$$

factors through the diagonal copy of $\operatorname{Spf} \mathcal{O}[[I^{\mathrm{ab}}_{K_{\mathrm{cyc}}}]] \times \widehat{\mathbf{G}}_m$ in the target; given this, it follows from Lemmas 7.1.15 and 7.2.1 that there is an isomorphism of groupoids

$$(\operatorname{Spf} \mathcal{O}[[I^{\mathrm{ab}}_{K_{\mathrm{cyc}}}]] \times \widehat{\mathbf{G}}_m) \times \widehat{\mathbf{G}}_m \xrightarrow{\sim} (\operatorname{Spf} \mathcal{O}[[I^{\mathrm{ab}}_{K_{\mathrm{cyc}}}]] \times \widehat{\mathbf{G}}_m) \times_{\mathcal{R}_1} (\operatorname{Spf} \mathcal{O}[[I^{\mathrm{ab}}_{K_{\mathrm{cyc}}}]] \times \widehat{\mathbf{G}}_m)$$

with $\widehat{\mathbf{G}}_m$ acting trivially on $\operatorname{Spf} \mathcal{O}[[I^{\mathrm{ab}}_{K_{\mathrm{cyc}}}]] \times \widehat{\mathbf{G}}_m$. The proposition will then follow from Lemma 7.1.8 provided we show that (7.2.12) is representable by algebraic spaces (or, equivalently, by algebraic stacks).

It remains to verify the claimed factorization of (7.2.13), as well as the claimed representability by algebraic stacks of (7.2.12). We establish each of these claims in turn.

To show that (7.2.13) factors through the diagonal diagonal copy of Spf $\mathcal{O}[[I^{\mathrm{ab}}_{K_{\mathrm{cyc}}}]] \times \widehat{\mathbf{G}}_m$ in the target, it suffices, by Lemma 7.1.14, to show that if

$$\operatorname{Spec} A \to (\operatorname{Spf} \mathcal{O}[[I^{\mathrm{ab}}_{K_{\mathrm{cyc}}}]] \times \widehat{\mathbf{G}}_m) \times (\operatorname{Spf} \mathcal{O}[[I^{\mathrm{ab}}_{K_{\mathrm{cyc}}}]] \times \widehat{\mathbf{G}}_m)$$

is a morphism whose source is a finite type Artinian local $\mathcal{O}$-algebra, then the pull back of (7.2.13) to $\operatorname{Spec} A$ factors through the pull-back to $\operatorname{Spec} A$ of the diagonal morphism

$$\operatorname{Spf}\mathcal{O}[[I_{K_{\mathrm{cyc}}}^{\mathrm{ab}}]]\times\widehat{\mathbf{G}}_m\to\big(\operatorname{Spf}\mathcal{O}[[I_{K_{\mathrm{cyc}}}^{\mathrm{ab}}]]\times\widehat{\mathbf{G}}_m\big)\times_{\mathcal{O}}\big(\operatorname{Spf}\mathcal{O}[[I_{K_{\mathrm{cyc}}}^{\mathrm{ab}}]]\times\widehat{\mathbf{G}}_m\big).$$

Since morphisms $\operatorname{Spec} A\to\operatorname{Spf}\mathcal{O}[[I_{K_{\mathrm{cyc}}}^{\mathrm{ab}}]]\times\widehat{\mathbf{G}}_m$ correspond to characters $\chi\colon G_{K_{\mathrm{cyc}}}^{\mathrm{ab}}\to A^\times$, this amounts to the evident fact that if χ,χ' are two such characters, then the locus in $\operatorname{Spec} A$ over which χ and χ' become isomorphic coincides with the locus over which they coincide.

We now turn to proving that (7.2.12) is representable by algebraic stacks. To see this, it suffices to study the corresponding question modulo some power ϖ^a of the uniformizer in $\mathcal{O}$, so we work with the stack $\mathcal{R}_{K,1}^a$ from now on.

We next recall some of the constructions we made in Section 3.2. We have the subfield $K^{\mathrm{basic}}\subseteq K$ of Definition 3.2.3, and the natural morphism $\mathcal{R}_{K,1}^a\to\mathcal{R}_{K^{\mathrm{basic}},[K:K^{\mathrm{basic}}]}^a$ given by the forgetful map which regards a rank 1 $\mathbf{A}_{K,A}$-module as an $\mathbf{A}_{K^{\mathrm{basic}},A}$-module of rank $[K:K^{\mathrm{basic}}]$. By Lemma 3.2.11, we can write $\mathcal{R}_{K,1}^a=\varinjlim_h\mathcal{R}_{K,1,K^{\mathrm{basic}},h}^a$, where the algebraic stack $\mathcal{R}_{K,1,K^{\mathrm{basic}},h}^a$ is the scheme-theoretic image of the base-changed morphism

$$\mathcal{C}_{[K:K^{\mathrm{basic}}],h}^a\times_{\mathcal{R}_{K^{\mathrm{basic}},[K:K^{\mathrm{basic}}]}}\mathcal{R}_{K,1}\to\mathcal{R}_{K,1},$$

where $\mathcal{C}_{[K:K^{\mathrm{basic}}],h}^a$ is the algebraic stack of φ-modules over $\mathbf{A}_{K^{\mathrm{basic}},A}^+$ of rank $[K:K^{\mathrm{basic}}]$ and T-height at most h. Accordingly, it suffices to show that each base-change

$$\mathcal{Y}_h^a:=\Big(\operatorname{Spf}\mathcal{O}[[I_{K_{\mathrm{cyc}}}^{\mathrm{ab}}]]\times\widehat{\mathbf{G}}_m\Big)\times_{\mathcal{R}_{K,1}}\mathcal{R}_{K,1,K^{\mathrm{basic}},h}^a$$

is an algebraic stack.

By construction, $\mathcal{Y}_h^a$ is a closed subsheaf of $\operatorname{Spf}(\mathcal{O}/\varpi^a)[[I_{K_{\mathrm{cyc}}}^{\mathrm{ab}}]]\times(\mathbf{G}_m)_{\mathcal{O}/\varpi^a}$. We claim that it is in fact a closed subsheaf of $\operatorname{Spec}(\mathcal{O}/\varpi^a)[I_{K_{\mathrm{cyc}}}^{\mathrm{ab}}/U]\times(\mathbf{G}_m)_{\mathcal{O}/\varpi^a}$, for some open subgroup U of $I_{K_{\mathrm{cyc}}}^{\mathrm{ab}}$.

Since $\mathcal{Y}_h^a$ and $\operatorname{Spec}(\mathcal{O}/\varpi^a)[I_{K_{\mathrm{cyc}}}^{\mathrm{ab}}/U]\times(\mathbf{G}_m)_{\mathcal{O}/\varpi^a}$ are both closed subsheaves of $\operatorname{Spf}(\mathcal{O}/\varpi^a)[[I_{K_{\mathrm{cyc}}}^{\mathrm{ab}}]]\times(\mathbf{G}_m)_{\mathcal{O}/\varpi^a}$, it follows from Lemma 7.1.14 that it is enough to show that if A is a finite local Artinian $\mathcal{O}/\varpi^a$-algebra, then given any morphism $\operatorname{Spec} A\to\mathcal{Y}_h^a$, the composite $\operatorname{Spec} A\to\mathcal{Y}_h^a\to\operatorname{Spf}(\mathcal{O}/\varpi^a)[[I_{K_{\mathrm{cyc}}}^{\mathrm{ab}}]]\times(\mathbf{G}_m)_{\mathcal{O}/\varpi^a}$ factors through $\operatorname{Spec}(\mathcal{O}/\varpi^a)[I_{K_{\mathrm{cyc}}}^{\mathrm{ab}}/U]\times(\mathbf{G}_m)_{\mathcal{O}/\varpi^a}$ for the open subgroup $U:=I^{a,h}$ of Proposition 7.3.17 below. If A is in fact a field, then by [EG21, Lem. 3.2.14], the étale φ-module M_A corresponding to the morphism $\operatorname{Spec} A\to\mathcal{R}_{K,1}$ attains (K^{basic},T)-height at most h in the sense of Definition 7.2.14 below after passing to a finite extension of A, and the required factorization is immediate from Proposition 7.3.17.

Having established the result in the case that A is a finite type field, in order to prove the general local Artinian case, it suffices to prove a factorization on the level of versal rings. More precisely, let $\mathbf{F}'/\mathbf{F}$ be a finite extension, and let $x\colon \operatorname{Spec}\mathbf{F}' \to \mathcal{Y}_h^a$ be a finite type point of $\mathcal{Y}_h^a$, corresponding to a representation $\overline{\rho}\colon G_{K_{\mathrm{cyc}}} \to (\mathbf{F}')^\times$. As above, we write $R_{\overline{\rho}}^{\square,\mathcal{O}'}$ for the corresponding universal lifting $\mathcal{O}'$-algebra. Write $\operatorname{Spf} R_h^a$ for the scheme-theoretic image of the morphism

$$\mathcal{C}^a_{[K:K^{\mathrm{basic}}],h} \times_{\mathcal{R}_{K^{\mathrm{basic}},[K:K^{\mathrm{basic}}]}} \mathcal{R}_{K,1} \times_{\mathcal{R}_{K,1}} \operatorname{Spf} R_{\overline{\rho}}^{\square,\mathcal{O}'} \to \operatorname{Spf} R_{\overline{\rho}}^{\square,\mathcal{O}'}.$$

By Lemma A.30, the induced morphism $\operatorname{Spf} R_h^a \to \mathcal{R}^a_{K,1,K^{\mathrm{basic}},h}$ is a versal morphism at x. Write $R_{\overline{\rho}}^{\square,\mathcal{O}',U}$ for the quotient of $R_{\overline{\rho}}^{\square,\mathcal{O}'}$ corresponding to liftings which are trivial on U; then it suffices to show that $\operatorname{Spf} R_h^a$ is a closed formal subscheme of $\operatorname{Spf} R_{\overline{\rho}}^{\square,\mathcal{O}',U}$.

We now employ Lemma A.32. Exactly as in the proof of Lemma 3.4.8, it is enough (after possibly increasing $\mathbf{F}'$) to show that if A is a finite type local Artinian $R_{\overline{\rho}}^{\square,\mathcal{O}'}$-algebra with residue field $\mathbf{F}'$ for which the induced morphism $\mathcal{C}^a_{[K:K^{\mathrm{basic}}],h} \times_{\mathcal{R}_{K^{\mathrm{basic}},[K:K^{\mathrm{basic}}]}} \mathcal{R}_{K,1} \times_{\mathcal{R}_{K,1}} \operatorname{Spec} A \to \operatorname{Spec} A$ admits a section, then the morphism $\operatorname{Spec} A \to \operatorname{Spf} R_{\overline{\rho}}^{\square,\mathcal{O}'}$ factors through $\operatorname{Spf} R_{\overline{\rho}}^{\square,\mathcal{O}',U}$. Since the existence of the section implies (by definition) that the étale φ-module M_A has (T, K^{basic})-height at most h (in the sense of Definition 7.2.14), we are done by Proposition 7.3.17. $\square$

In the previous argument, we used the following definition, which will play an important role in the technicalities of Section 7.3 below. (See in particular the statements of Propositions 7.3.17 and 7.3.18.)

7.2.14 Definition. Let A be a finite local Artinian $\mathcal{O}/\varpi^a$-algebra for some $a \geq 1$, and let M be a rank 1 projective étale φ-module over $\mathbf{A}_{K,A}$. We regard M as a rank $[K:K^{\mathrm{basic}}]$ étale φ-module M' over $\mathbf{A}_{K^{\mathrm{basic}},A}$, and we say that M *has* (K^{basic}, T)*-height at most* h if there is a projective φ-module $\mathfrak{M}$ over $\mathbf{A}^+_{K^{\mathrm{basic}},A}$ of T-height at most h such that $M' = \mathfrak{M}[1/T]$.

We next turn to describing $\mathcal{R}_1^{\Gamma_{\mathrm{disc}}}$ explicitly. We first note that since $I_K \cap G_{K_{\mathrm{cyc}}} = I_{K_{\mathrm{cyc}}}$, we have an exact sequence $1 \to I_{K_{\mathrm{cyc}}} \to G_K \to \mathrm{Frob}_K^{\widehat{\mathbf{Z}}} \times \Gamma$. We let $H \subseteq \mathrm{Frob}_K^{\widehat{\mathbf{Z}}} \times \Gamma$ denote the image of G_K; it is an open subgroup of $\mathrm{Frob}_K^{\widehat{\mathbf{Z}}} \times \Gamma$. We write $H_{\mathrm{disc}} := H \cap (\mathrm{Frob}_K^{\mathbf{Z}} \times \Gamma_{\mathrm{disc}})$; then H_{disc} is dense in H, and is isomorphic to $\mathbf{Z} \times \mathbf{Z}$. Write $\mathcal{O}[H_{\mathrm{disc}}]$ for the group ring of H_{disc} over $\mathcal{O}$, and $\widehat{\mathcal{O}[H_{\mathrm{disc}}]}$ for its ϖ-adic completion. Then $\operatorname{Spec} \mathcal{O}[H_{\mathrm{disc}}] \cong \mathbf{G}_m \times_{\mathcal{O}} \mathbf{G}_m$, and $\operatorname{Spf} \widehat{\mathcal{O}[H_{\mathrm{disc}}]} \cong \widehat{\mathbf{G}}_m \times_{\mathcal{O}} \widehat{\mathbf{G}}_m$. Let I' denote the image of $I_{K_{\mathrm{cyc}}}$ in G_K^{ab}; equivalently, I' is the quotient of $I_{K_{\mathrm{cyc}}}^{\mathrm{ab}}$ by the closure of its subgroup of commutators $[I_{K_{\mathrm{cyc}}}^{\mathrm{ab}}, \Gamma]$.

7.2.15 Proposition. *There is an isomorphism*

$$\Big[\Big(\operatorname{Spf}\mathcal{O}[[I']]\times\operatorname{Spf}\widehat{\mathcal{O}[H_{\mathrm{disc}}]}\Big)/\widehat{\mathbf{G}}_m\Big]\xrightarrow{\sim}\mathcal{R}_1^{\Gamma_{\mathrm{disc}}}$$

(where, in the formation of the quotient stack, the $\widehat{\mathbf{G}}_m$-action is taken to be trivial).

Proof. We leave the construction of this isomorphism to the reader, by combining the isomorphism of Proposition 7.2.11 with the definition of $\mathcal{R}_1^{\Gamma_{\mathrm{disc}}}$. □

7.2.16 Remark. Note that the case $d=1$ is not representative of the general case insofar as the structure of $\mathcal{R}$, and so also of $\mathcal{R}^{\Gamma_{\mathrm{disc}}}$, is concerned. Namely, when $d=1$, the stacks $\mathcal{R}$ and $\mathcal{R}^{\Gamma_{\mathrm{disc}}}$ are formal algebraic stacks. This will not be the case when $d>1$. (They will instead be Ind-algebraic stacks whose underlying reduced substacks are Ind-algebraic but not algebraic.) The reason for this is that the one-dimensional mod p representations of $G_{K_{\mathrm{cyc}}}$ may be described as Frobenius twists of a finite number of characters when $d=1$, so that in this case $\mathcal{R}_{\mathrm{red}}$ is indeed an algebraic stack, while the spaces $\operatorname{Ext}^1_{G_{K_{\mathrm{cyc}}}}(\chi,\psi)$ are infinite dimensional, for any two mod p characters χ and ψ of $G_{K_{\mathrm{cyc}}}$, so that already in the case $d=2$, the stack $\mathcal{R}_{\mathrm{red}}$ is merely Ind-algebraic.

We now give an explicit description of the stack $\mathcal{X}_1$.

7.2.17 Proposition. *There is an isomorphism*

$$\Big[\Big(\operatorname{Spf}(\mathcal{O}[[I_K^{\mathrm{ab}}]])\times\widehat{\mathbf{G}}_m\Big)/\widehat{\mathbf{G}}_m\Big]\xrightarrow{\sim}\mathcal{X}_1$$

(where, in the formation of the quotient stack, the $\widehat{\mathbf{G}}_m$-action is taken to be trivial).

Proof. We prove this in the same way as Proposition 7.2.11; we leave the details to the reader, indicating only the key differences. We can construct a morphism $\operatorname{Spf}(\mathcal{O}[[I_K^{\mathrm{ab}}]])\times\widehat{\mathbf{G}}_m\to\mathcal{X}_1$ in exactly the same way as in the proof of Proposition 7.2.11 (by choosing a lift $\sigma_K\in G_K$ of Frob_K, splitting the short exact sequence (7.2.4)), and we need to prove that this morphism is representable by algebraic spaces. This can be done by arguing exactly as in the proof of Proposition 7.2.11, with the isomorphism $\mathcal{X}^a_{K,1}=\varinjlim_{h,s}\mathcal{X}^a_{K,1,K^{\mathrm{basic}},h,s}$ of Lemma 3.5.5 replacing the isomorphism $\mathcal{R}^a_{K,1}=\varinjlim_h\mathcal{R}^a_{K,1,K^{\mathrm{basic}},h}$.

Bearing in mind the definition of $\mathcal{X}^a_{K,1,K^{\mathrm{basic}},h,s}$, we find that we have to show that there is an open subgroup $I_K^{h,s,a}$ of I_K^{ab} such that if A is a finite local Artinian $\mathcal{O}/\varpi^a$-algebra, and M is a rank 1 étale φ-module over $\mathbf{A}_{K,A}$ with the property that there is a rank $[K:K^{\mathrm{basic}}]$ φ-module $\mathfrak{M}$ over $\mathbf{A}^+_{K^{\mathrm{basic}},A}$, such that

- $\mathfrak{M}[1/T]=M$ (where we are regarding M as an étale φ-module over $\mathbf{A}_{K^{\text{basic}},A}$),
- $\mathfrak{M}$ is of T-height at most h, and
- the action of $\Gamma_{K^{\text{basic}},\text{disc}}=\Gamma_{K,\text{disc}}$ on $\mathfrak{M}[1/T]$ extends the canonical action of $\Gamma_{K_s^{\text{basic}},\text{disc}}$ given by Corollary 3.5.2,

then the action of $I_K^{h,s,a}$ on $T_A(M)$ is trivial. This is immediate from Proposition 7.3.18. □

7.2.18 Remark. Note that Propositions 7.2.15 and 7.2.17 describe $\mathcal{X}$ as a certain formal completion of $\mathcal{R}^{\Gamma_{\text{disc}}}$ (in the case $d=1$). In particular, the monomorphism $\mathcal{X}\hookrightarrow\mathcal{R}^{\Gamma_{\text{disc}}}$ is not a closed immersion (in the case $d=1$, and presumably not in the general case either). This helps to explain why somewhat elaborate arguments were required in Chapter 3 to deduce properties of $\mathcal{X}$ from the corresponding properties of $\mathcal{R}^{\Gamma_{\text{disc}}}$.

7.2.19 Remark. It follows easily from Proposition 7.2.17 that the stack $\mathcal{X}_1$ may be described as a moduli stack of one-dimensional continuous representations of the Weil group W_K; indeed, we see that if A is a p-adically complete $\mathcal{O}$-algebra, then $\mathcal{X}_1(\operatorname{Spf} A)$ is the groupoid of continuous characters $I_K^{\text{ab}}\times\mathbf{Z}\to A^\times$ (if we identify $\widehat{\mathbf{G}}_m$ with the p-adically completed group ring of $\mathbf{Z}$). Mapping the generator $1\in\mathbf{Z}$ to σ_K (a lift of Frobenius, as in the proof of Proposition 7.2.17) yields an isomorphism $I_K^{\text{ab}}\times\mathbf{Z}\xrightarrow{\sim}W_K^{\text{ab}}$.

This phenomenon doesn't persist in the general case. Indeed, as noted in Section 1.4, in the case $d=2$, there seems to be no description of $\mathcal{X}_2$ as the moduli space of representations of a group that is compatible with the description of its closed points in terms of representations of G_K.

One can attempt to adapt the proof of Proposition 7.2.17 to the case $d>1$, by constructing a morphism from the moduli stack of continuous d-dimensional Weil–Deligne representations (suitably understood)[2] to the stack $\mathcal{X}_d$. However, when $d>1$, this morphism of stacks will not be representable by algebraic spaces. The precise point where the proof that we've given in the rank 1 case fails to generalize is that once we are not passing to abelianized Galois groups, the upper numbered ramification groups are not open in the inertia group.

7.3 A RAMIFICATION BOUND

We end this chapter by proving the bounds on the ramification of a character valued in a finite Artinian $\mathcal{O}$-algebra that were used in the proofs of Propositions

[2]Since we are considering representations in rings that are p-power torsion, we cannot use the usual formulation of the Weil–Deligne group. Rather, we choose a finitely generated dense subgroup $\mathbf{Z}\ltimes\mathbf{Z}[1/p]$ of the tame Galois group of G_K (by choosing a lift of Frobenius and a lift of a topological generator of tame inertia), and form the topological group WD_K by taking its preimage in G_K (and equipping $\mathbf{Z}\ltimes\mathbf{Z}[1/p]$ with its discrete topology).

7.2.11 and 7.2.17. There are well-established techniques for proving such a bound, going back to [Fon85]. We find it convenient to follow the arguments of [CL11], and indeed we will follow some of its arguments very closely. The main differences between our setting and that of [CL11] are that we are working in the cyclotomic setting, rather than the Kummer setting; that we are considering representations of $G_{K_{\mathrm{cyc}}}$ of finite height, rather than semistable representations of G_K; and that to define a representation of finite height, we need to pass between K and K^{basic}. None of these changes make a fundamental difference to the argument.

In fact, while [CL11] go to some effort to work modulo an arbitrary power of p, and to optimize the bounds that they obtain, we are content to give the simplest proof that we can of the existence of a bound of the kind that we need. In particular, we are able to reduce to the case of mod p representations, which simplifies much of the discussion.

Recall that K_{cyc}/K is a Galois extension with Galois group $\Gamma_K \cong \mathbf{Z}_p$. For each $s \geq 0$, we write $K_{\mathrm{cyc},s}$ for the unique subfield of K_{cyc} which is cyclic over K of degree p^s. We write $e = e(K/K_0)$.

Suppose that K is basic, and that $\mathfrak{M}$ is a free φ-module over $\mathbf{A}_K^+/p$ of T-height at most h. Then $M := \mathbf{A}_K \otimes_{\mathbf{A}_K^+} \mathfrak{M}$ is an étale φ-module over $\mathbf{A}_K/p$, and we have a corresponding $G_{K_{\mathrm{cyc}}}$-representation $T(M)$ as in Section 3.6. (Note that while elsewhere in the book we have worked with $\mathcal{O}/\varpi^a$-coefficients, it is more convenient to use $\mathbf{Z}/p\mathbf{Z}$-coefficients in most of this section.)

In order to compare directly to the arguments of [CL11], it is more convenient to work with the contragredient $G_{K_{\mathrm{cyc}}}$-representation $T(M)^\vee$, which can also be computed as $T(M^\vee) = (\mathbf{C}^\flat \otimes_{\mathbf{A}_K} M^\vee)^{\varphi=1} = \mathrm{Hom}_{\mathbf{A}_K,\varphi}(M, \mathbf{C}^\flat) = \mathrm{Hom}_{\mathbf{A}_K^+,\varphi}(M, \mathbf{C}^\flat)$. By [Fon90, A.1.2.7, B.1.8.3], this can also be computed via the functor

$$T^\vee(\mathfrak{M}) := \mathrm{Hom}_{\mathbf{A}_K^+,\varphi}(\mathfrak{M}, \mathcal{O}_{\mathbf{C}}^\flat).$$

More precisely, there is a diagram

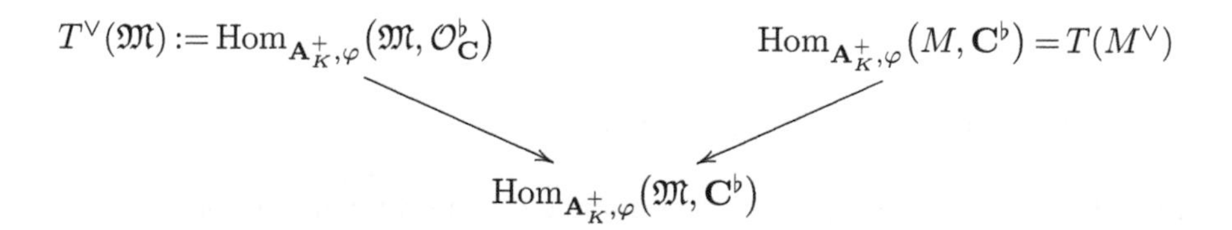

(the first arrow being induced by the inclusion of $\mathcal{O}_{\mathbf{C}}^\flat$ in $\mathbf{C}^\flat$, and the second by restriction from M to $\mathfrak{M}$), with both arrows in fact being isomorphisms (by op. cit.).

We will now follow the arguments of [CL11, §4]; note that the ring denoted R in [CL11] is $\mathcal{O}_{\mathbf{C}}^\flat$. For the time being we will assume that we are in the following situation; the rather complicated hypotheses here will allow us to interpret the canonical actions of Corollary 3.5.2 in terms of the constructions of [CL11, §4].

7.3.1 Hypothesis. Assume that K is basic, and fix some $h \geq 1$. Let $\mathfrak{M}$ be a free φ-module over $\mathbf{A}_K^+/p$ of T-height at most h. Fix the following:

- an integer $b \geq eh/(p-1)$.
- An integer $N > pb/e$ such that $N \geq N(1,h)$, where $N(1,h)$ is as in Corollary 3.5.2.
- An integer s such that $p^{s-1} > b/e$, and $s \geq s(1,h,N)$, where $s(1,h,N)$ is as in Corollary 3.5.2.

Note in particular that under Hypothesis 7.3.1 we have $p^s > ph/(p-1) > h$.

We normalize the valuation v on K so that $v(K^\times) = \mathbf{Z}$, and continue to write v for the unique compatible valuation on $\overline{K}$, and for the induced valuation on $\mathcal{O}_{\mathbf{C}}^\flat = \varprojlim \mathcal{O}_{\overline{K}}/p$. Note that with this convention we have $v(T) = v(p) = e$ (where we are abusively continuing to denote the image of T in $\mathcal{O}_{\mathbf{C}}^\flat$ by T; this follows straightforwardly from the definition of T as the trace of $T' := ([\varepsilon]-1)$, which shows (after a simple computation) that $v(T) = (p-1)v(T')$, along with the formula $v(\zeta_{p^s} - 1) = v(p)/p^{s-1}(p-1)$, which shows that $v(T') = v(p)/(p-1)$). We write $\mathfrak{a}_{\mathcal{O}_{\mathbf{C}}^\flat}^{>c}$ for the set of elements of $\mathcal{O}_{\mathbf{C}}^\flat$ of valuation greater than c, and define $\mathfrak{a}_{\mathcal{O}_{\mathbf{C}}^\flat}^{\geq c}$ in the same way. For any $c \geq 0$, we write

$$T^\vee_{\mathcal{O}_{\mathbf{C}}^\flat, c}(\mathfrak{M}) := \operatorname{Hom}_{\mathbf{A}_K^+, \varphi}(\mathfrak{M}, \mathcal{O}_{\mathbf{C}}^\flat / \mathfrak{a}_{\mathcal{O}_{\mathbf{C}}^\flat}^{>c}),$$

so that there is a natural map $T^\vee(\mathfrak{M}) \to T^\vee_{\mathcal{O}_{\mathbf{C}}^\flat, c}(\mathfrak{M})$. For any $c' \geq c$ we write

$$T^\vee_{\mathcal{O}_{\mathbf{C}}^\flat, c', c}(\mathfrak{M}) = \operatorname{Im}\,(T^\vee_{\mathcal{O}_{\mathbf{C}}^\flat, c'}(\mathfrak{M}) \to T^\vee_{\mathcal{O}_{\mathbf{C}}^\flat, c}(\mathfrak{M})),$$

where the morphism is induced by the natural map $\mathfrak{a}_{\mathcal{O}_{\mathbf{C}}^\flat}^{>c'} \to \mathfrak{a}_{\mathcal{O}_{\mathbf{C}}^\flat}^{>c}$.

7.3.2 Lemma. *Assume that we are in the setting of Hypothesis* 7.3.1. *Then the induced morphism* $T^\vee(\mathfrak{M}) \to T^\vee_{\mathcal{O}_{\mathbf{C}}^\flat, b}(\mathfrak{M})$ *is injective, with image* $T^\vee_{\mathcal{O}_{\mathbf{C}}^\flat, pb, b}(\mathfrak{M})$.

Proof. This can be proved by a standard Frobenius amplification argument exactly as in the proof of [CL11, Prop. 2.3.3]. For example, to see the injectivity, we note that since $\mathfrak{M}$ is finitely generated, any element of the kernel corresponds to a φ-equivariant morphism $\mathfrak{M} \to \mathfrak{a}_{\mathcal{O}_{\mathbf{C}}^\flat}^{\geq c}$ for some $c > b$. Since $\mathfrak{M}$ has T-height at most h, and $v(T^h) = eh$, the φ-equivariance implies that the image of this morphism is contained in $\mathfrak{a}_{\mathcal{O}_{\mathbf{C}}^\flat}^{\geq pc-eh}$. Since $c > b \geq eh/(p-1)$ by hypothesis, we see that the sequence $c, pc-eh, p(pc-eh)-eh, \ldots$ goes to infinity, and the injectivity follows.

We leave surjectivity to the reader; it is a simpler version of the proof of Lemma 7.3.11 below. (See also the proof of Lemma 4.3.2 for an almost identical argument.) □

Suppose we are given $c \geq 0$ satisfying $p^s > c/e$ (e.g., b and pb both satisfy this condition). The Frobenius φ is an automorphism of the perfect k-algebra $\mathcal{O}_{\mathbf{C}}^\flat$ which multiplies valuations by p, and so φ^{-s} induces the first of the following sequence of G_K-equivariant isomorphisms of k-algebras,

$$\mathcal{O}_{\mathbf{C}}^\flat/\mathfrak{a}_{\mathcal{O}_{\mathbf{C}}^\flat}^{\geq c} \xrightarrow{\sim} k \otimes_{k,\varphi^s} \mathcal{O}_{\mathbf{C}}^\flat/\mathfrak{a}_{\mathbf{C}}^{\geq c/p^s} \xrightarrow{\sim} k \otimes_{k,\varphi^s} \mathcal{O}_{\overline{K}}/\mathfrak{a}_{\overline{K}}^{\geq c/p^s}, \tag{7.3.3}$$

where $\mathfrak{a}_{\overline{K}}^{\geq c/p^s}$ has the evident meaning, and the second isomorphism is induced by projection onto the first factor in the projective limit $\mathcal{O}_{\mathbf{C}}^\flat = \varprojlim_\varphi \mathcal{O}_{\mathbf{C}}/p = \varprojlim_\varphi \mathcal{O}_{\overline{K}}/p$. (The upper bound on c ensures that $p \in \mathfrak{a}_{\overline{K}}^{\geq c/p^s}$, so that this projection does indeed induce the indicated morphism; it is then straightforward to verify that it is an isomorphism.) This isomorphism in turn induces an isomorphism

$$T^\vee_{\mathcal{O}_C^\flat,c}(\mathfrak{M}) \xrightarrow{\sim} \mathrm{Hom}_{\mathbf{A}_K^+,\varphi}(\mathfrak{M}, \mathcal{O}_{\overline{K}}/\mathfrak{a}_{\overline{K}}^{\geq c/p^s}). \tag{7.3.4}$$

7.3.5 Lemma. *If* $p^s > c/e \geq 0$, *then the action of* $G_{K_{\mathrm{cyc},s}}$ *on* $\mathcal{O}_{\mathbf{C}}^\flat$ *induces an action of* $G_{K_{\mathrm{cyc},s}}$ *on* $T^\vee_{\mathcal{O}_C^\flat,c}(\mathfrak{M})$.

Proof. The action of φ on $\mathcal{O}_{\mathbf{C}}^\flat$ is G_K-equivariant, and so the G_K-action on $\mathcal{O}_{\mathbf{C}}^\flat$ induces an action on $\mathrm{Hom}_\varphi(\mathfrak{M}, \mathcal{O}_{\mathbf{C}}^\flat/\mathfrak{a}_{\mathcal{O}_{\mathbf{C}}^\flat}^{\geq c})$. The claim of the lemma is that the restriction of this action to $G_{K_{\mathrm{cyc},s}}$ preserves the subobject $T^\vee_{\mathcal{O}_{\mathbf{C}}^\flat,c}(\mathfrak{M}) := \mathrm{Hom}_{\mathbf{A}_K^+,\varphi}(\mathfrak{M}, \mathcal{O}_{\mathbf{C}}^\flat/\mathfrak{a}_{\mathcal{O}_{\mathbf{C}}^\flat}^{\geq c})$ of $\mathrm{Hom}_\varphi(\mathfrak{M}, \mathcal{O}_{\mathbf{C}}^\flat/\mathfrak{a}_{\mathcal{O}_{\mathbf{C}}^\flat}^{\geq c})$.

Rather than proving this directly, we use the isomorphisms (7.3.3) and (7.3.4); taking these into account, it suffices to prove that the action of $G_{K_{\mathrm{cyc},s}}$ on $\mathrm{Hom}_\varphi(\mathfrak{M}, \mathcal{O}_{\overline{K}}/\mathfrak{a}_{\overline{K}}^{\geq c/p^s})$ preserves the subobject $\mathrm{Hom}_{\mathbf{A}_K^+,\varphi}(\mathfrak{M}, \mathcal{O}_{\overline{K}}/\mathfrak{a}_{\overline{K}}^{\geq c/p^s})$. In other words, we have to check that the $G_{K_{\mathrm{cyc},s}}$-action preserves the property of being $\mathbf{A}_K^+$-equivariant. For this, it suffices to check that $G_{K_{\mathrm{cyc},s}}$ acts trivially on the image of T in $\mathcal{O}_{\overline{K}}/\mathfrak{a}_{\overline{K}}^{\geq c/p^s}$ under the isomorphism (7.3.3). Since T can be expressed as a power series in T', it in fact suffices to check this for the image of T'. But this image is equal to $\zeta_{p^{s+1}} - 1 \bmod \mathfrak{a}_{\overline{K}}^{\geq c/p^s}$, which *is* fixed by $G_{K_{\mathrm{cyc},s}}$ (since $\zeta_{p^{s+1}} \in K_{\mathrm{cyc},s}$). □

7.3.6 Corollary. *Assume that we are in the setting of Hypothesis* 7.3.1. *Then the natural action of* $G_{K_{\mathrm{cyc},s}}$ *on* $\mathcal{O}_{\mathbf{C}}^\flat$ *induces an extension of the action of* $G_{K_{\mathrm{cyc}}}$ *on* $T^\vee(\mathfrak{M})$ *to an action of* $G_{K_{\mathrm{cyc},s}}$.

Proof. This follows directly from Lemmas 7.3.2 and 7.3.5, once we note that, by hypothesis, $p^s > pb/e > b/e$. □

By Corollary 3.5.2 (and our hypothesized bound on s), we can give $\mathfrak{M}[1/T]$ the structure of a $(\varphi, \Gamma_{K_{\mathrm{cyc},s}})$-module; this in particular induces an action of $G_{K_{\mathrm{cyc},s}}$ on $T(M)$. The following corollary expresses the compatibility of this action with the action constructed in Corollary 7.3.6.

7.3.7 Corollary. *Assume that we are in the setting of Hypothesis* 7.3.1. *Then the two actions of* $G_{K_{\mathrm{cyc},s}}$ *on* $T(M)^\vee$ *(from Corollary* 7.3.6, *and from Corollary* 3.5.2) *agree.*

Proof. By the proof of Corollary 3.5.2, the semi-linear action of $\Gamma_{K_{\mathrm{cyc},s}}$ on $\mathfrak{M}$ extends to a semi-linear action of $G_{K_{\mathrm{cyc},s}}$ on $\mathcal{O}_{\mathbf{C}}^\flat \otimes_{\mathbf{A}_K^+} \mathfrak{M}$, which is uniquely determined by the properties that it commutes with φ, and satisfies $(g-1)(\mathfrak{M}) \subset T^N(\mathcal{O}_{\mathbf{C}}^\flat \otimes_{\mathbf{A}_K^+} \mathfrak{M})$ for all $g \in G_{K_{\mathrm{cyc},s}}$. As explained in Section 3.6.4, we have $T(M) = (\mathbf{C}^\flat \otimes_{\mathbf{A}_K^+} \mathfrak{M})^{\varphi=1}$, with the action of $G_{K_{\mathrm{cyc},s}}$ on $T(M)$ being that inherited from its action on $\mathcal{O}_{\mathbf{C}}^\flat \otimes_{\mathbf{A}_K^+} \mathfrak{M}$.

As noted above, we also have

$$\begin{aligned} T(M)^\vee &= T(M^\vee) = T^\vee(\mathfrak{M}) \\ &= \operatorname{Hom}_{\mathbf{A}_K^+,\varphi}(\mathfrak{M}, \mathcal{O}_{\mathbf{C}}^\flat) = \operatorname{Hom}_{\mathbf{C}^\flat,\varphi}(\mathbf{C}^\flat \otimes_{\mathbf{A}_K^+} \mathfrak{M}, \mathbf{C}^\flat). \end{aligned}$$

Now evaluation induces a pairing

$$(\mathbf{C}^\flat \otimes_{\mathbf{A}_K^+} \mathfrak{M}) \times \operatorname{Hom}_{\mathbf{C}^\flat}(\mathbf{C}^\flat \otimes_{\mathbf{A}_K^+} \mathfrak{M}, \mathbf{C}^\flat) \to \mathbf{C}^\flat,$$

which restricts to a pairing

$$(\mathbf{C}^\flat \otimes_{\mathbf{A}_K^+} \mathfrak{M})^{\varphi=1} \times \operatorname{Hom}_{\mathbf{C}^\flat,\varphi}(\mathbf{C}^\flat \otimes_{\mathbf{A}_K^+} \mathfrak{M}, \mathbf{C}^\flat) \to (\mathbf{C}^\flat)^{\varphi=1} = \mathbf{Z}/p\mathbf{Z}, \qquad (7.3.8)$$

and this latter pairing *is* the natural pairing of $G_{K_{\mathrm{cyc}}}$-representations

$$T(M) \times T(M)^\vee \to \mathbf{Z}/p\mathbf{Z}.$$

Therefore, to prove the lemma, we have to show that (7.3.8) is furthermore $G_{K_{\mathrm{cyc},s}}$-equivariant (with the action on $T(M)$ being the action recalled above, coming from Corollary 3.5.2, and the action on $T(M)^\vee = T^\vee(\mathfrak{M})$ given by Corollary 7.3.6).

To see this, note firstly that by the same argument that we used in the proof of Lemma 7.3.2, it follows from Hypothesis 7.3.1 that

$$(\mathbf{C}^\flat \otimes_{\mathbf{A}_K^+} \mathfrak{M})^{\varphi=1} \subseteq (\mathfrak{a}^{\geq b})^{-1} \otimes_{\mathbf{A}_K^+} \mathfrak{M}.$$

In addition, we have

$$\begin{aligned}\mathrm{Hom}_{\mathbf{C}^\flat,\varphi}(\mathbf{C}^\flat \otimes_{\mathbf{A}_K^+} \mathfrak{M}, \mathbf{C}^\flat) &= \mathrm{Hom}_{\mathcal{O}_{\mathbf{C}^\flat},\varphi}(\mathcal{O}_{\mathbf{C}^\flat} \otimes_{\mathbf{A}_K^+} \mathfrak{M}, \mathcal{O}_{\mathbf{C}^\flat})\\ &\subseteq \mathrm{Hom}_{\mathcal{O}_{\mathbf{C}^\flat}}(\mathcal{O}_{\mathbf{C}^\flat} \otimes_{\mathbf{A}_K^+} \mathfrak{M}, \mathcal{O}_{\mathbf{C}^\flat}) \xleftarrow{\sim} \mathrm{Hom}_{\mathcal{O}_{\mathbf{C}^\flat}}((\mathfrak{a}^{\geq b})^{-1} \otimes_{\mathbf{A}_K^+} \mathfrak{M}, (\mathfrak{a}^{\geq b})^{-1})\end{aligned}$$

with the last isomorphism being induced by restricting the domain of a homomorphism from $(\mathfrak{a}^{\geq b})^{-1} \otimes_{\mathbf{A}_K^+} \mathfrak{M}$ to $\mathcal{O}_{\mathbf{C}^\flat} \otimes_{\mathbf{A}_K^+} \mathfrak{M}$. Bearing in mind this isomorphism, we obtain an evaluation pairing

$$((\mathfrak{a}^{\geq b})^{-1} \otimes_{\mathbf{A}_K^+} \mathfrak{M}) \times \mathrm{Hom}_{\mathcal{O}_{\mathbf{C}^\flat}}((\mathcal{O}_{\mathbf{C}^\flat} \otimes_{\mathbf{A}_K^+} \mathfrak{M}, \mathcal{O}_{\mathbf{C}^\flat}) \to (\mathfrak{a}^{\geq b})^{-1}$$

(and taking φ-invariants again recovers the pairing (7.3.8)).

Fix some $\alpha \in \mathbf{C}^\flat$ with $v(\alpha) = -b$, so that $(\alpha) = (\mathfrak{a}^{\geq b})^{-1}$. Let x be an element of $(\mathbf{C}^\flat \otimes_{\mathbf{A}_K^+} \mathfrak{M})^{\varphi=1}$, and write $x = \alpha \sum_{i=1}^n \lambda_i \otimes x_i$ where $\lambda_i \in \mathcal{O}_{\mathbf{C}}^\flat$ and $x_i \in \mathfrak{M}$. Let $f \in \mathrm{Hom}_{\mathcal{O}_{\mathbf{C}}^\flat,\varphi}(\mathcal{O}_{\mathbf{C}}^\flat \otimes_{\mathbf{A}_K^+} \mathfrak{M}, \mathcal{O}_{\mathbf{C}}^\flat)$, so that x and f pair to $\alpha f(\sum_{i=1}^n \lambda_i x_i) \in (\mathfrak{a}^{\geq b})^{-1}$; of course, they in fact pair to an element of $\mathbf{F}_p \subset \mathcal{O}_{\mathbf{C}}^\flat \subset (\mathfrak{a}^{\geq b})^{-1}$. Let $g \in G_{K_{\mathrm{cyc}},s}$ be arbitrary. Then gx and gf pair to $g(\alpha)(gf)(\sum_{i=1}^n g(\lambda_i)g(x_i))$. To establish the claimed $G_{K_{\mathrm{cyc}},s}$-equivariance, we need to show that

$$g(\alpha) \cdot (gf)(\sum_{i=1}^n g(\lambda_i)g(x_i)) = g(\alpha f(\sum_{i=1}^n \lambda_i x_i)).$$

This is an equality of elements of $\mathbf{F}_p$, and it therefore suffices to show that it holds modulo $\mathfrak{m}_{\mathcal{O}_{\mathbf{C}}}$ when each side is considered as an element of $\mathcal{O}_{\mathbf{C}}^\flat$. Equivalently, after multiplication by α^{-1} it suffices to show that

$$(gf)(\sum_{i=1}^n g(\lambda_i)g(x_i)) \equiv g(f(\sum_{i=1}^n \lambda_i x_i)) \pmod{\mathfrak{a}^{>b}}.$$

Recall that $(g-1)(\mathfrak{M}) \subset T^N(\mathcal{O}_{\mathbf{C}}^\flat \otimes_{\mathbf{A}_K^+} \mathfrak{M})$. By Hypothesis 7.3.1, we have $v(T^N) = eN > pb$, so that in particular we have $T^N \in \mathfrak{a}^{>b}_{\mathcal{O}_{\mathbf{C}}^\flat}$. Thus $g(x_i) \equiv x_i \pmod{\mathfrak{a}^{>b}}$, so it suffices to show that

$$(gf)(\sum_{i=1}^n g(\lambda_i)x_i) \equiv g(f(\sum_{i=1}^n \lambda_i x_i)) \pmod{\mathfrak{a}^{>b}},$$

or equivalently that

$$\sum_{i=1}^n g(\lambda_i)(gf)(x_i) \equiv \sum_{i=1}^n g(\lambda_i)g(f(x_i)) \pmod{\mathfrak{a}^{>b}},$$

so in turn it is enough to show that if $x \in \mathfrak{M}$ then $(gf)(x) \equiv g(f(x)) \pmod{\mathfrak{a}^{>b}}$. But this is true by the very definition of gf (which was defined via the action of $G_{K_{\mathrm{cyc},s}}$ on $T^{\vee}_{\mathcal{O}^\flat_{\mathbf{C}},b}(\mathfrak{M})$, which in turn is defined via the action of $G_{K_{\mathrm{cyc},s}}$ on $\mathcal{O}^\flat_{\mathbf{C}}/\mathfrak{a}^{>b}$), so we are done. □

Recall that we write k for the residue field of K, and since K is assumed basic we have $k_\infty = k$, so that we can and do regard $\mathfrak{M}$ as a free $\mathbf{A}^+_K/p = k[[T]]$-module. Write T_s for $\mathrm{tr}_{K(\zeta_{p^{s+1}})/K_{\mathrm{cyc},s}}(\zeta_{p^{s+1}} - 1) \in \mathcal{O}_{K_{\mathrm{cyc},s}}$, and note that we have a homomorphism

$$\mathbf{A}^+_K/p = k[[T]] \to \mathcal{O}_{K_{\mathrm{cyc},s}}/p \tag{7.3.9}$$

which sends $T \to T_s$, and whose restriction to k is given by φ^{-s}.

Let L be any algebraic extension of $K_{\mathrm{cyc},s}$ inside $\overline{K}$. We continue to write v for the unique valuation on L with $v(p) = e$. Note that we have $v(T_s^h) = eh/p^s < e = v(p)$. For any $c \geq 0$ we write $\mathfrak{a}_L^{>c}$ for the set of elements of L of valuation greater than c, and define $\mathfrak{a}_L^{\geq c}$ in the same way.

Choose an (ordered) $\mathbf{A}^+_K/p$-basis $e_1, \ldots, e_d$ of $\mathfrak{M}$, and write $\varphi(e_1, \ldots, e_d) = (e_1, \ldots, e_d)A$ for some $A \in M_d(\mathbf{A}^+_K/p)$. By our assumption that $\mathfrak{M}$ has T-height at most h, we can write $AB = T^h \mathrm{Id}_d$ for some $B \in M_d(\mathbf{A}^+_K/p)$. We can and do choose matrices $\widetilde{A}, \widetilde{B} \in M_d(\mathcal{O}_{K_{\mathrm{cyc},s}})$ which respectively lift the images of A, B under the homomorphism $\mathbf{A}^+_K/p \to \mathcal{O}_{K_{\mathrm{cyc},s}}/p$ defined above. Since the image of T_s^h in $\mathcal{O}_{K_{\mathrm{cyc},s}}/p$ is nonzero, we see that after possibly multiplying $\widetilde{B}$ by an invertible matrix which is trivial mod p/T_s^h, we can and do assume that

$$\widetilde{A}\widetilde{B} = T_s^h \mathrm{Id}_d. \tag{7.3.10}$$

We set

$$\widetilde{T}^*_L(\mathfrak{M}) := \{(x_1, \ldots, x_d) \in \mathcal{O}_L^d \mid (x_1^p, \ldots, x_d^p) = (x_1, \ldots, x_d)\widetilde{A}\}.$$

Note that since $x \mapsto x^p$ is not a ring homomorphism on $\mathcal{O}_L$, this is just a set with a $G_{K_{\mathrm{cyc},s}}$-action, rather than an $\mathcal{O}_L$-module.

For any $c \in [0, ep^s)$, we write

$$T^*_{L,c}(\mathfrak{M}) := \mathrm{Hom}_{\mathbf{A}^+_K/p,\varphi}(\mathfrak{M}, (\mathcal{O}_L/p)/\mathfrak{a}_L^{>c/p^s}),$$

where the $\mathbf{A}^+_K$-algebra structure on $\mathcal{O}_L/p$ is that induced by the composite $\mathbf{A}^+_K/p \overset{(7.3.9)}{\longrightarrow} \mathcal{O}_{K_s}/p \to \mathcal{O}_L/p$. (The use of c/p^s rather than c in the definition of $T^*_{L,c}$ comes from the definition of the homomorphism (7.3.9); see also (7.3.3).) For any $c' \geq c$ we write

$$T^*_{L,c',c}(\mathfrak{M}) = \mathrm{Im}\,(T^*_{L,c'}(\mathfrak{M}) \to T^*_{L,c}(\mathfrak{M})),$$

where the morphism is induced by the natural map $\mathfrak{a}_L^{>c'/p^s} \to \mathfrak{a}_L^{>c/p^s}$.

Evaluating on the members of the basis $e_1, \ldots, e_d$ of $\mathfrak{M}$ yields an identification

$$T^\vee_{L,c}(\mathfrak{M}) \xrightarrow{\sim} \{(x_1, \ldots, x_d) \in ((\mathcal{O}_L/p)/\mathfrak{a}_L^{>c/p^s})^d \mid (x_1^p, \ldots, x_d^p) = (x_1, \ldots, x_d)\overline{A}\}$$

(where $\overline{A}$ denotes the image of A in $M_d((\mathcal{O}_L/p)/\mathfrak{a}_L^{>c/p^s})$), so that there is a natural map $\widetilde{T}^\vee_L(\mathfrak{M}) \to T^\vee_{L,c}(\mathfrak{M})$.

The following lemma is the analogue of [CL11, Lem. 4.1.4] in our setting, and the proof is essentially identical (and in fact simpler, since we are only working modulo p).

7.3.11 Lemma. *Assume that we are in the setting of Hypothesis* 7.3.1. *Then the morphism* $\widetilde{T}^\vee_L(\mathfrak{M}) \to T^\vee_{L,b}(\mathfrak{M})$ *is injective, and has image* $T^\vee_{L,pb,b}(\mathfrak{M})$.

Proof. We begin with injectivity. Suppose that we have two distinct tuples $(x_1, \ldots, x_d)$ and $(y_1, \ldots, y_d) \in \widetilde{T}^*_L(\mathfrak{M})$ whose images in $T^\vee_{L,b}(\mathfrak{M})$ coincide; then if we write $z_i := y_i - x_i$, we have $v(z_i) > b/p^s$ for each i. By (7.3.10), we have

$$((x_1 + z_1)^p - x_1^p, \ldots, (x_d + z_d)^p - x_d^p)\widetilde{B} = (z_1, \ldots, z_d)T_s^h.$$

Writing $(x_i + z_i)^p - x_i^p = z_i^p + pz_i(x_i^{p-1} + \ldots)$, we see that if $z := \min_i v(z_i)$, then we have $eh/p^s + z \geq \min(pz, e + z)$. Since $p^s > h$, we see that necessarily $e + z > eh/p^s + z$. Hence it must be that $eh/p^s + z \geq pz$, and thus that $eh/p^s(p-1) \geq z$. (This final deduction is where we use our assumption that the tuples $(x_1, \ldots, x_d)$ and $(y_1, \ldots, y_d)$ are distinct; this assumption ensures that $z < \infty$, so that it is legitimate to subtract if from the two sides of an inequality.) However this contradicts our assumption that $z > b/p^s$.

We now turn to determining the image. By definition, the morphism $\widetilde{T}^\vee_L(\mathfrak{M}) \to T^\vee_{L,b}(\mathfrak{M})$ factors through $T^\vee_{L,pb,b}(\mathfrak{M})$. Choose some $(\overline{x}_1, \ldots, \overline{x}_d) \in T^\vee_{L,pb}(\mathfrak{M})$. It suffices to construct $(\widetilde{x}_1, \ldots, \widetilde{x}_d) \in \widetilde{T}^\vee_L(\mathfrak{M})$ with $\widetilde{x}_i$ lifting the $\overline{x}_i$ modulo $\mathfrak{a}^{>b/p^s}$ by successive approximation.

Begin by choosing arbitrary lifts x_i of $\overline{x}_i$ to $\mathcal{O}_L$.

We have $T_s^{-h}\mathfrak{a}_L^{>b/p^{s-1}} = \mathfrak{a}_L^{>(pb-eh)/p^s} \subseteq \mathfrak{a}_L^{>b/p^s}$, so by our assumption that $(\overline{x}_1, \ldots, \overline{x}_d) \in T^\vee_{L,pb}(\mathfrak{M})$, we have

$$T_s^{-h}(x_1^p, \ldots, x_d^p)\widetilde{B} - (x_1, \ldots, x_d) \in (T_s^{-h}\mathfrak{a}_L^{>b/p^{s-1}})^d \subseteq (\mathfrak{a}_L^{>b/p^s})^d,$$

so in fact we have

$$T_s^{-h}(x_1^p, \ldots, x_d^p)\widetilde{B} - (x_1, \ldots, x_d) \in (\mathfrak{a}_L^{\geq b'/p^s})^d$$

for some $b' > b$. We will construct Cauchy sequences $(z_i^{(j)}) \in \mathfrak{a}_L^{\geq b'/p^s}$ whose limits z_i are such that taking $\widetilde{x}_i := x_i + z_i$ gives the required lift.

To this end, we set $z_i^{(0)}=0$, and define

$$(z_1^{(j+1)},\ldots,z_d^{(j+1)}):=T_s^{-h}((x_1+z_1^{(j)})^p,\ldots,(x_d+z_d^{(j)})^p)\widetilde{B}-(x_1,\ldots,x_d).$$

To see that $z_i^{(j+1)}\in\mathfrak{a}_L^{\geq b'/p^s}$, we write

$$\begin{aligned}(z_1^{(j+1)},\ldots,z_d^{(j+1)}):=T_s^{-h}((x_1+z_1^{(j)})^p-x_1^p,\ldots,(x_d+z_d^{(j)})^p-x_d^p)\widetilde{B}\\ +(T_s^{-h}(x_1^p,\ldots,x_d^p)\widetilde{B}-(x_1,\ldots,x_d)),\end{aligned}$$

so we need only check that each $(x_i+z_i^{(j)})^p-x_i^p\in T_s^h\mathfrak{a}_L^{\geq b'/p^s}$. Since $(x_i+z_i^{(j)})^p-x_i^p\in(pz_i^{(j)},(z_i^{(j)})^p)$, and $z_i^{(j)}\in\mathfrak{a}_L^{\geq b'/p^s}$, this holds.

It remains to show that each $(z_i^{(j)})$ is Cauchy. Set $\epsilon=\min(e-eh/p^s,((p-1)b'-eh)/p^s)$, so that in particular $\epsilon>0$. It is enough to show that $\min_i v(z_i^{(j+1)}-z_i^{(j)})\geq\min_i v(z_i^{(j+1)}-z_i^{(j)})+\epsilon$. To this end, we have

$$\begin{aligned}&(z_1^{(j+1)}-z_1^{(j)},\ldots,z_d^{(j+1)}-z_d^{(j)})\\ &\quad=T_s^{-h}\big((x_1+z_1^{(j)})^p-(x_1+z_1^{(j-1)})^p,\ldots,(x_d+z_d^{(j)})^p-(x_d+z_d^{(j-1)})^p\big)\widetilde{B},\end{aligned}$$

so that it is enough to show that for each i, we have $(x_i+z_i^{(j)})^p-(x_i+z_i^{(j-1)})^p\in(z_i^{(j)}-z_i^{(j-1)})T_s^h\mathfrak{a}_L^{\geq\epsilon}=(z_i^{(j)}-z_i^{(j-1)})\mathfrak{a}_L^{\geq eh/p^s+\epsilon}$.

For each i, we write

$$(x_i+z_i^{(j)})^p-(x_i+z_i^{(j-1)})^p=\sum_{k=1}^{p}\binom{p}{k}x_i^{p-k}((z_i^{(j)})^k-(z_i^{(j-1)})^k),$$

so it is enough to note that for $1\leq k\leq p-1$ we have $v(\binom{p}{k})=e\geq eh/p^s+\epsilon$, while for $k=p$, we write

$$(z_i^{(j)})^p-(z_i^{(j-1)})^p=(z_i^{(j)}-z_i^{(j-1)})((z_i^{(j)})^{p-1}+\cdots+(z_i^{(j-1)})^{p-1}),$$

and note that since $z_i^{(j)},z_i^{(j-1)}\in\mathfrak{a}_L^{\geq b'/p^s}$, the second term here is contained in $\mathfrak{a}_L^{\geq(p-1)b'/p^s}$, and we have $(p-1)b'/p^s\geq eh/p^s+\epsilon$. □

The following corollary is the analogue of [CL11, Thm. 4.1.1] in our setting.

7.3.12 Corollary. *Assume that we are in the setting of Hypothesis 7.3.1, and let L be an algebraic extension of $K_{\mathrm{cyc},s}$ inside $\overline{K}$, so that G_L acts naturally on $T^\vee(\mathfrak{M})$ by Corollary 7.3.6. Then the natural injection $T^\vee_{L,pb,b}(\mathfrak{M})\subseteq T^\vee_{\overline{K},pb,b}(\mathfrak{M})$ is an isomorphism if and only if the action of G_L on $T^\vee(\mathfrak{M})$ is trivial.*

Proof. By Lemma 7.3.11 (applied to each of L and $\overline{K}$), the inclusion $T^\vee_{L,pb,b}(\mathfrak{M}) \subseteq T^\vee_{\overline{K},pb,b}(\mathfrak{M})$ is an isomorphism if and only if the same is true of the inclusion $\widetilde{T}^\vee_L(\mathfrak{M}) \subseteq \widetilde{T}^\vee_{\overline{K}}(\mathfrak{M})$. Directly from the definition of these latter sets, we see that $\widetilde{T}^\vee_L(\mathfrak{M}) = (\widetilde{T}^\vee_{\overline{K}}(\mathfrak{M}))^{G_L}$, so this containment holds if and only if G_L acts trivially on $\widetilde{T}^\vee_{\overline{K}}(\mathfrak{M})$, or equivalently, if and only if G_L acts trivially on $T^\vee_{\overline{K},pb,b}(\mathfrak{M})$.

Now Lemmas 7.3.2 and 7.3.5 show that $T^\vee(\mathfrak{M}) \xrightarrow{\sim} T^\vee_{\overline{K},pb,b}(\mathfrak{M})$, and by its construction, the $G_{K_{\mathrm{cyc},s}}$-action on $T^\vee(\mathfrak{M})$ arises, via this isomorphism, from the $G_{K_{\mathrm{cyc},s}}$-action on $T^\vee_{\overline{K},pb,b}(\mathfrak{M})$. The lemma follows. □

7.3.13 Corollary. *Assume that we are in the setting of Hypothesis* 7.3.1, *and let ρ denote the representation $T(M)^\vee = T^\vee(\mathfrak{M})$ of $G_{K_{\mathrm{cyc},s}}$ given by Corollary* 7.3.6. *Let L be an algebraic extension of $K_{\mathrm{cyc},s}$ inside $\overline{K}$. If there exists an $\mathcal{O}_{K_{\mathrm{cyc},s}}$-algebra homomorphism $\mathcal{O}_{\overline{K}^{\ker\rho}} \to \mathcal{O}_L/\mathfrak{a}_L^{>b/p^{s-1}}$, then $\overline{K}^{\ker\rho} \subseteq L$.*

Proof. Suppose that we have an $\mathcal{O}_{K_{\mathrm{cyc},s}}$-algebra homomorphism $\eta\colon \mathcal{O}_{\overline{K}^{\ker\rho}} \to \mathcal{O}_L/\mathfrak{a}^{>b/p^{s-1}}$. We claim that η induces an injective homomorphism

$$\mathcal{O}_{\overline{K}^{\ker\rho}}/\mathfrak{a}_{\overline{K}^{\ker\rho}}^{>b/p^{s-1}} \to \mathcal{O}_L/\mathfrak{a}_L^{>b/p^{s-1}}.$$

To begin with, let $F/K_{\mathrm{cyc},s}$ be the maximal unramified subextension of $\overline{K}^{\ker\rho}/K_{\mathrm{cyc},s}$, and let k_F denote its residue field (which is then also the residue field of $\overline{K}^{\ker\rho}$). If k_L denotes the residue field of L, then the morphism η induces an embedding $k_F \hookrightarrow k_L$, which in turn lifts to an embedding $W(k_F) \hookrightarrow \mathcal{O}_L$. This embedding then induces an embedding

$$\mathcal{O}_F = \mathcal{O}_{K_{\mathrm{cyc},s}} \otimes_{W(k)} W(k_L) \hookrightarrow \mathcal{O}_L. \tag{7.3.14}$$

In particular, we find that $\mathcal{O}_L$ contains $\mathcal{O}_F$, or, equivalently, that L contains F, although the embedding (7.3.14) may not be the inclusion, but rather the composite of this inclusion with an element of $\mathrm{Gal}(F/K_{\mathrm{cyc},s})$. If we regard $\mathcal{O}_{\overline{K}^\rho}$ (resp. $\mathcal{O}_L$) as a $\mathcal{O}_F$-algebra via the inclusion (resp. via the embedding (7.3.14)), then we see (from the very construction of (7.3.14)) that η is a morphism of $\mathcal{O}_F$-algebras.

Now, let π be a uniformizer of $\mathcal{O}_{\overline{K}^{\ker\rho}}$, with image $\eta(\pi) \in \mathcal{O}_L/\mathfrak{a}_L^{>b/p^{s-1}}$; we claim that $\eta(\pi)$ is nonzero, so that its valuation is well-defined, and that this valuation is equal to that of π. It follows immediately from this that the kernel of η is generated by $\mathfrak{a}_{\overline{K}^{\ker\rho}}^{>b/p^{s-1}}$, as required. To verify the claim, let E_π be the (Eisenstein) minimal polynomial of π over $\mathcal{O}_F$. Then $\eta(\pi)$ is a root of $\eta(E_\pi)$, and the non-leading coefficients of $\eta(E_\pi)$ all have valuation at least $v(T_s) = e/p^s$, with equality holding for the constant coefficient. Since $e/p^s \le eh/p^s \le (p-1)b/p^s < pb/p^s$, it follows that $v(\eta(\pi)) = v(\pi)$.

We then have a composite of injections

$$T^{\vee}_{\overline{K}^{\ker\rho},pb,b}(\mathfrak{M}) \subseteq T^{\vee}_{L,pb,b}(\mathfrak{M}) \subseteq T^{\vee}_{\overline{K},pb,b}(\mathfrak{M})$$

(the second inclusion being induced by the natural map $\mathcal{O}_L \to \mathcal{O}_{\overline{K}}$). By Corollary 7.3.12 the composite is an isomorphism (being an injection of isomorphic finite sets), so all of these inclusions are isomorphisms. Applying Corollary 7.3.12 again, we conclude that $\overline{K}^{\ker\rho} \subseteq L$, as required. □

Recall that for each finite extension L of $\mathbf{Q}_p$ and each real number v, we have the upper numbered ramification subgroups G_L^v, as defined in [Ser79, Chapter IV]; these are the groups denoted $G_L^{(v+1)}$ in [Fon85] (see [Fon85, Rem. 1.2]). We can now prove our first bound on the ramification of $T(M)$, from which our subsequent bounds will be deduced.

7.3.15 Corollary. *Assume that we are in the setting of Hypothesis* 7.3.1. *If* $v > pb-1$, *then* $G^v_{K_{\mathrm{cyc},s}}$ *acts trivially on* $T(M)$.

Proof. Let ρ denote $T(M)^\vee$. Recall that for a real number m, the extension $K^{\ker\rho}/K_{\mathrm{cyc},s}$ satisfies (P_m) if, for any subfield L of $\overline{K}$ containing $K_{\mathrm{cyc},s}$, the existence of an $\mathcal{O}_{K_{\mathrm{cyc},s}}$-algebra homomorphism $\mathcal{O}_{K^{\ker\rho}} \to \mathcal{O}_L/\mathfrak{a}^{\ge m/p^s}$ implies that $K^{\ker\rho} \subseteq L$. (This property was introduced by Fontaine, and is discussed extensively in [Yos10], to which we refer the reader for further explanations and references. Note that the denominator p^s appears in $\mathfrak{a}^{\ge m/p^s}$ because we always compute valuations with respect to K, and $K_{\mathrm{cyc},s}$ is a totally ramified extension of K of degree p^s.)

Corollary 7.3.13 shows that $\inf\{m \mid (\mathrm{P}_m) \text{ holds for } K^{\ker\rho}/K_{\mathrm{cyc},s}\} \le pb$. The present corollary then follows from this, together with [Yos10, Thm. 1.1]. □

We now return to the setting of a general K (i.e., we do not assume that K is basic). Let A be a finite local Artinian $\mathcal{O}/\varpi^a$-algebra for some $a \ge 1$, and let M be a rank 1 projective étale φ-module over $\mathbf{A}_{K,A}$, which we assume to be of (K^{basic},T)-height at most h in the sense of Definition 7.2.14; that is, we regard M as a rank $[K:K^{\mathrm{basic}}]$ étale φ-module M' over $\mathbf{A}_{K^{\mathrm{basic}},A}$, and we assume there is a projective φ-module $\mathfrak{M}$ over $\mathbf{A}^+_{K^{\mathrm{basic}},A}$ of T-height at most h such that $M' = \mathfrak{M}[1/T]$.

We have the $G_{K^{\mathrm{basic}}_{\mathrm{cyc}}}$-representation $T_A(M')$, which we may compute as

$$T_A(M') = (W(\mathbf{C}^\flat) \otimes_{\mathbf{A}_{K^{\mathrm{basic}}}} M')^{\varphi=1} = (W(\mathbf{C}^\flat) \otimes_{\mathbf{A}_{K^{\mathrm{basic}}}} M)^{\varphi=1}.$$

It follows that there is a $G_{K_{\mathrm{cyc}}}$-equivariant surjection

$$T_A(M') = (W(\mathbf{C}^\flat) \otimes_{\mathbf{A}_{K^{\mathrm{basic}}}} M)^{\varphi=1} \twoheadrightarrow (W(\mathbf{C}^\flat) \otimes_{\mathbf{A}_K} M)^{\varphi=1} = T_A(M). \quad (7.3.16)$$

In fact (2.7.4) shows that $T_A(M')$ can be identified with $\operatorname{Ind}_{G_{K_{\mathrm{cyc}}}}^{G_{K^{\mathrm{basic}}_{\mathrm{cyc}}}} T_A(M)$, and the morphism (7.3.16) then becomes the natural $G_{K_{\mathrm{cyc}}}$-equivariant surjection.

7.3.17 Proposition. *Suppose that A is an Artinian $\mathcal{O}/\varpi^a$-algebra for some $a \geq 1$, and M is a rank 1 projective étale φ-module over $\mathbf{A}_{K,A}$ of (K^{basic}, T)-height at most h.*

Then there is an open subgroup $I^{a,h}$ of $I^{\mathrm{ab}}_{K_{\mathrm{cyc}}}$ depending only on K, a and h (and not on A or M) such that the action of $I^{a,h}$ on $T_A(M)$ is trivial.

Proof. We choose integers b, s, N satisfying the following conditions:

- $b \geq eh/(p-1)$.
- $N > pb/e$ is such that $N \geq N(i,h)$ for each $i \in [1,a]$, where the $N(i,h)$ are as in Corollary 3.5.2, with K there being replaced by K^{basic}.
- s is such that $p^{s-1} > b/e$, and such that $s \geq s(i,h,N)$ for each $i \in [1,a]$, where the $s(i,h,N)$ are as in Corollary 3.5.2, with K there being replaced by K^{basic}.

Note in particular that Hypothesis 7.3.1 holds, with K in the hypothesis replaced by K^{basic}. Note also that our assumptions imply that the hypotheses of Corollary 3.5.2 hold (with the field K there taken to be K^{basic}), so that the action of $G_{K^{\mathrm{basic}}_{\mathrm{cyc}}}$ on $T_A(M')$ has a canonical extension to an action of $G_{K^{\mathrm{basic}}_{\mathrm{cyc},s}}$. This canonical action on $T_A(M')$ induces the canonical action on $\varpi^i T_A(M') = T_A(\varpi^i M')$ for each $i \in [0,a]$, and hence on each of the quotients $\varpi^{i-1} T_A(M')/\varpi^i T_A(M') = T_{A/\varpi}(\varpi^{i-1}M'/\varpi^i M')$, for $i \in [1,a]$.

To ease notation, write $M_i := \varpi^{i-1}M/\varpi^i M$, for $1 \leq i \leq a$, and let M_i' denote M_i regarded as an étale φ-module over $\mathbf{A}_{K^{\mathrm{basic}}, A/\varpi}$. By Corollary 7.3.15, if v is sufficiently large (depending only on K, a and h, and our subsequent choices of b, N and s, and in particular not on A or M), then the canonical action of $G^v_{K^{\mathrm{basic}}_{\mathrm{cyc},s}}$ on each $T_{A/\varpi}(M_i')$ is trivial. It follows that there is some m depending only on a (and the degree $[K : K^{\mathrm{basic}}]$) such that for any $g \in G^v_{K^{\mathrm{basic}}_{\mathrm{cyc},s}}$, the action of g^{p^m} on $T_A(M')$ is trivial.

It follows from the definitions that $G^v_{K^{\mathrm{basic}}_{\mathrm{cyc},s}} \cap G_{K_{\mathrm{cyc},s}} = G_{K_{\mathrm{cyc},s}}^{\psi_{K_{\mathrm{cyc},s}/K^{\mathrm{basic}}_{\mathrm{cyc},s}}(v)}$, where $\psi_{K_{\mathrm{cyc},s}/K^{\mathrm{basic}}_{\mathrm{cyc},s}}$ is as in [Ser79, Chapter IV]), so if we take any $w \geq \psi_{K_{\mathrm{cyc},s}/K^{\mathrm{basic}}_{\mathrm{cyc},s}}(v)$, then $(G^w_{K_{\mathrm{cyc},s}})^{p^m}$ ($:= \{g^{p^m} \mid g \in G^w_{K_{\mathrm{cyc},s}}\}$; note that this is just a subset, not a subgroup, of $G_{K_{\mathrm{cyc},s}}$) acts trivially on $T_A(M')$.

The product $G_{K_{\mathrm{cyc}}}(G^w_{K_{\mathrm{cyc},s}})^{p^m}$ is a priori a subset of G_K, but we claim that it is in fact an open subgroup of G_K, and thus (since it contains $G_{K_{\mathrm{cyc}}}$) is equal to $G_{K_{\mathrm{cyc},s'}}$ for some $s' \geq s$ which depends only on s and m (and so ultimately only on K, a and h).

To prove the claim, we briefly recall the relationship between upper numbered ramification groups and the filtration of the unit group: If L is a finite extension of $\mathbf{Q}_p$, then the Artin map identifies $\mathcal{O}_L^\times$ with I_L^{ab}, and for each integer $w \geq 1$, identifies the subgroup $1+(\mathfrak{m}_L)^v$ of $\mathcal{O}_L^\times$ with the image of G_L^v in G_L^{ab} (see [Ser79, Chapter XV, §2, Thm. 2]). It then identifies the subgroup $(1+(\mathfrak{m}_L)^v)^{p^m}$ with the image of $(G_L^v)^{p^m}$ in G_L^{ab}. This former subgroup is an open subgroup of $\mathcal{O}_L^\times$, and so we find that for each v, the image of $(G_L^v)^{p^m}$ in I_L^{ab} is open.

Now the extension $K_{\mathrm{cyc}}/K_{\mathrm{cyc},s}$ is an infinitely ramified $\Gamma_{K_{\mathrm{cyc},s}} \cong \mathbf{Z}_p$-extension, and so the inertia group $I(K_{\mathrm{cyc}}/K)$ is finite index (equivalently, open) in $\Gamma_{K_{\mathrm{cyc},s}}$. Since $I(K_{\mathrm{cyc}}/K)$ is a quotient of I_K^{ab}, the discussion of the preceding paragraph then shows that the image of $(G_{K_{\mathrm{cyc},s}}^w)^{p^m}$ in $\Gamma_{K_{\mathrm{cyc},s}}$ is also an open subgroup. Since $\Gamma_{K_{\mathrm{cyc},s}}$ is itself open in G_K, this implies the claim.

Since (7.3.16) is $G_{K_{\mathrm{cyc}}}$-equivariant, and since $(G_{K_{\mathrm{cyc},s}}^w)^{p^m}$ acts trivially on $T_A(M')$, we see that the kernel of (7.3.16) is in fact $G_{K_{\mathrm{cyc},s'}}$-invariant, and thus that the $G_{K_{\mathrm{cyc}}}$-action on $T_A(M)$ extends to a $G_{K_{\mathrm{cyc},s'}}$-action, which is abelian (since the $G_{K_{\mathrm{cyc}}}$-action is abelian, as $T_A(M)$ has rank 1 over A, while $(G_{K_{\mathrm{cyc},s}}^w)^{p^m}$ acts trivially).

Now another application of [Ser79, Chapter IV]) shows that $G_{K_{\mathrm{cyc},s}}^w \cap G_{K_{\mathrm{cyc},s'}} = G_{K_{\mathrm{cyc},s'}}^{w'}$ for some w' depending only on w, and our preceding discussion of abelian ramification theory shows that $(G_{K_{\mathrm{cyc},s'}}^{w'})^{p^m}$ has open image in $G_{K_{\mathrm{cyc},s'}}^{\mathrm{ab}}$. If we let I' denote the preimage of this open image (i.e.,

$$I' = \overline{[G_{K_{\mathrm{cyc},s'}}, G_{K_{\mathrm{cyc},s'}}]}(G_{K_{\mathrm{cyc},s'}}^{w'})^{p^m},$$

where the overline denotes closure in $G_{K_{\mathrm{cyc},s'}}$), then I' is an open subgroup of $G_{K_{\mathrm{cyc},s'}}$ which acts trivially on $T_A(M)$. Thus $I'' := I' \cap G_{K_{\mathrm{cyc}}}$ also acts trivially on $T_A(M)$. Finally, we let $I^{a,h}$ denote the image of I'' in I_K^{ab}. (As the notation indicates, $I^{a,h}$ depends only on a, h, and K, but not on A or M, since this is true of each of the groups I' and I'' from which it is constructed.) □

The same argument allows us to prove the following variant of Proposition 7.3.17.

7.3.18 Proposition. *Suppose that A is an Artinian $\mathcal{O}/\varpi^a$-algebra for some $a \geq 1$, and M is a rank 1 projective étale (φ,Γ)-module over $\mathbf{A}_{K,A}$. Assume furthermore that M has (K^{basic},T)-height at most h, and that s is some sufficiently large integer with the property that the action of $\Gamma_{K^{\mathrm{basic}},\mathrm{disc}} = \Gamma_{K,\mathrm{disc}}$ on M extends the canonical action of $\Gamma_{K_s^{\mathrm{basic}},\mathrm{disc}}$ given by Corollary 3.5.2. (In particular, we are assuming that s has been chosen to be large enough that this canonical action is defined.)*

Then there is an open subgroup $I_K^{a,h,s}$ of I_K^{ab} depending only on K, a, h and s (and not on A or M) such that the action of $I_K^{a,h,s}$ on $T_A(M)$ is trivial.

Proof. The proof is very similar to that of Proposition 7.3.17, and we content ourselves with indicating the key modifications. The main difference between the setting of that proposition and the present one is that, in our present setting, the morphism (7.3.16) is a G_K-equivariant morphism from a $G_{K^{\mathrm{basic}}}$-representation to a G_K-representation. Furthermore, by assumption, the $G_{K^{\mathrm{basic}}_{\mathrm{cyc},s}}$-action on its source is the canonical action.

We now follow the argument of Proposition 7.3.17, and find that, after possibly enlarging s in a manner depending only on a, h, and K, some ramification group $G^w_{K_{\mathrm{cyc},s}}$ acts trivially on $T_A(M')$ (the index w also depending only on a, h, and s). In the present argument, there is no need to pass to an auxiliary subgroup $G_{K_{\mathrm{cyc},s'}}$, since the morphism (7.3.16) is in particular already a $G_{K_{\mathrm{cyc},s}}$-equivariant morphism of $G_{K_{\mathrm{cyc},s}}$-representations; and the action of $G_{K_{\mathrm{cyc},s}}$ on $T(M)$ is already abelian. The argument then proceeds in the same manner, to produce an open subgroup I' of the inertia subgroup of $G_{K_{\mathrm{cyc},s}}$ (and thus of I_K) which acts trivially on $T_A(M)$. (And, by construction, the group I' depends only on a, h, and s, but not on A or M.) We then let $I_K^{a,h,s}$ denote the image of I' in I_K^{ab}. □

Chapter Eight

A geometric Breuil–Mézard conjecture

In this chapter we explain a (for the most part conjectural) relationship between the geometry of our potentially semistable and crystalline moduli stacks $\mathcal{X}_d^{\mathrm{crys},\underline{\lambda},\tau}$ and $\mathcal{X}_d^{\mathrm{ss},\underline{\lambda},\tau}$ and the representation theory of $\mathrm{GL}_n(k)$. Throughout the chapter we fix a sufficiently large coefficient field E with ring of integers $\mathcal{O}$ and residue field $\mathbf{F}$, and we largely omit it from our notation.

The starting point for this proposed relationship is the Breuil–Mézard conjecture [BM02], which is a conjectural formula for the Hilbert–Samuel multiplicities of the special fibres of the lifting rings $R_{\overline{\rho}}^{\mathrm{crys},\underline{\lambda},\tau}$ and $R_{\overline{\rho}}^{\mathrm{ss},\underline{\lambda},\tau}$, and which has important applications to proving modularity lifting theorems via the Taylor–Wiles method [Kis09a]. The conjecture was geometrized in [BM14] and [EG14], by refining the conjectural formula for the Hilbert–Samuel multiplicity to a conjectural formula for the underlying cycle of the special fibres of the lifting rings $R_{\overline{\rho}}^{\mathrm{crys},\underline{\lambda},\tau}$ and $R_{\overline{\rho}}^{\mathrm{ss},\underline{\lambda},\tau}$, considered as (equidimensional) closed subschemes of the special fiber of the universal lifting ring $R_{\overline{\rho}}^{\square}$. We refer to this generalization as the "refined Breuil–Mézard conjecture", and to the original conjecture (or rather, its generalizations to GL_n) as the "numerical Breuil–Mézard conjecture".

Our aim in this chapter, then, is to "globalize" the conjectures of [EG14] by formulating versions of them for the stacks $\mathcal{X}_d^{\mathrm{crys},\underline{\lambda},\tau}$ and $\mathcal{X}_d^{\mathrm{crys},\underline{\lambda},\tau}$; the conjectures of [EG14] can be recovered from these conjectures by passing to versal rings at finite type points. (We caution the reader that there is another kind of "globalization" that could be considered, namely realizing local Galois representations as the restrictions to decomposition groups of global representations, as used in the proofs of some of the results of [EG14] and [GK14]. Other than in the proof of Theorem 8.6.1 below, we don't consider this kind of globalization in the present work.) This generalization, which we will call the "geometric Breuil–Mézard conjecture", seems to us to be the natural setting in which to consider the Breuil–Mézard conjecture, and we will use our description of the irreducible components of $\mathcal{X}_{d,\mathrm{red}}$ to deduce new results about both the refined and numerical versions of the Breuil–Mézard conjecture.

8.1 THE QUALITATIVE GEOMETRIC BREUIL–MÉZARD CONJECTURE

If $\underline{\lambda}$ is a regular Hodge type, and τ is any inertial type, then by Theorems 4.8.12 and 4.8.14, the stacks $\mathcal{X}_d^{\mathrm{crys},\underline{\lambda},\tau}$ and $\mathcal{X}_d^{\mathrm{ss},\underline{\lambda},\tau}$ are finite type p-adic formal algebraic stacks over $\mathcal{O}$, which are $\mathcal{O}$-flat and equidimensional of dimension $1+[K:\mathbf{Q}_p]d(d-1)/2$. It follows that their special fibres $\overline{\mathcal{X}}_d^{\mathrm{crys},\underline{\lambda},\tau}$ and $\overline{\mathcal{X}}_d^{\mathrm{ss},\underline{\lambda},\tau}$ are algebraic stacks over $\mathbf{F}$ which are equidimensional of dimension $[K:\mathbf{Q}_p]d(d-1)/2$. Since $\mathcal{X}_d^{\mathrm{crys},\underline{\lambda},\tau}$ and $\mathcal{X}_d^{\mathrm{ss},\underline{\lambda},\tau}$ are closed substacks of $\mathcal{X}_d$, $\overline{\mathcal{X}}_d^{\mathrm{crys},\underline{\lambda},\tau}$ and $\overline{\mathcal{X}}_d^{\mathrm{ss},\underline{\lambda},\tau}$ are closed substacks of the special fiber $\overline{\mathcal{X}}_d$, and their irreducible components (with the induced reduced substack structure) are therefore closed substacks of the algebraic stack $\overline{\mathcal{X}}_{d,\mathrm{red}}$ (see [Sta, Tag 0DR4] for the theory of irreducible components of algebraic stacks and their multiplicities). Since $\overline{\mathcal{X}}_{d,\mathrm{red}}$ is equidimensional of dimension $[K:\mathbf{Q}_p]d(d-1)/2$ by Theorem 6.5.1, it follows that the irreducible components of $\overline{\mathcal{X}}_d^{\mathrm{crys},\lambda,\tau}$ and $\overline{\mathcal{X}}_d^{\mathrm{ss},\lambda,\tau}$ are irreducible components of $\overline{\mathcal{X}}_{d,\mathrm{red}}$, and are therefore of the form $\overline{\mathcal{X}}_{d,\mathrm{red}}^{\underline{k}}$ for some Serre weight $\underline{k}$ (again by Theorem 6.5.1).

For each $\underline{k}$, we write $\mu_{\underline{k}}(\overline{\mathcal{X}}_d^{\mathrm{crys},\underline{\lambda},\tau})$ and $\mu_{\underline{k}}(\overline{\mathcal{X}}_d^{\mathrm{ss},\underline{\lambda},\tau})$ for the multiplicity of $\overline{\mathcal{X}}_{d,\mathrm{red}}^{\underline{k}}$ as a component of $\overline{\mathcal{X}}_d^{\mathrm{crys},\underline{\lambda},\tau}$ and $\overline{\mathcal{X}}_d^{\mathrm{ss},\underline{\lambda},\tau}$. We write $Z_{\mathrm{crys},\underline{\lambda},\tau}=Z(\overline{\mathcal{X}}_d^{\mathrm{crys},\underline{\lambda},\tau})$ and $Z_{\mathrm{ss},\underline{\lambda},\tau}=Z(\overline{\mathcal{X}}_d^{\mathrm{ss},\underline{\lambda},\tau})$ for the corresponding cycles, i.e., for the formal sums

$$Z_{\mathrm{crys},\underline{\lambda},\tau}=\sum_{\underline{k}}\mu_{\underline{k}}(\overline{\mathcal{X}}_d^{\mathrm{crys},\underline{\lambda},\tau})\cdot\overline{\mathcal{X}}_d^{\underline{k}}, \tag{8.1.1}$$

$$Z_{\mathrm{ss},\underline{\lambda},\tau}=\sum_{\underline{k}}\mu_{\underline{k}}(\overline{\mathcal{X}}_d^{\mathrm{ss},\underline{\lambda},\tau})\cdot\overline{\mathcal{X}}_d^{\underline{k}}, \tag{8.1.2}$$

which we regard as elements of the finitely generated free abelian group $\mathbf{Z}[\mathcal{X}_{d,\mathrm{red}}]$ whose generators are the irreducible components $\overline{\mathcal{X}}_d^{\underline{k}}$.

Now fix some representation $\overline{\rho}\colon G_K\to\mathrm{GL}_d(\mathbf{F})$, corresponding to a point $x\colon \operatorname{Spec}\mathbf{F}\to\mathcal{X}_d$. (More generally, we could consider representations valued in $\mathrm{GL}_d(\mathbf{F}')$ for some finite extension $\mathbf{F}'/\mathbf{F}$, but in keeping with our attempt to keep the notation in this chapter as uncluttered as possible by omitting $\mathcal{O}$, it is convenient to suppose that $\mathbf{F}$ has been chosen sufficiently large.) For each regular Hodge type $\underline{\lambda}$ and inertial type τ, we have effective versal morphisms $\operatorname{Spec}R_{\overline{\rho}}^{\mathrm{crys},\underline{\lambda},\tau}/\varpi\to\overline{\mathcal{X}}^{\mathrm{crys},\underline{\lambda},\tau}$ and $\operatorname{Spec}R_{\overline{\rho}}^{\mathrm{ss},\underline{\lambda},\tau}/\varpi\to\overline{\mathcal{X}}^{\mathrm{ss},\underline{\lambda},\tau}$ (see Corollary 4.8.13), as well as a (non-effective; see Remark 6.5.3) versal morphism $\operatorname{Spf}R_{\overline{\rho}}^{\square}\to\mathcal{X}_d$ (see Proposition 3.6.3).

For each $\underline{k}$, we may consider the fiber product $\operatorname{Spf}R_{\overline{\rho}}^{\square}\times_{\mathcal{X}_d}\mathcal{X}_d^{\underline{k}}$. This is a priori a closed formal subscheme of $\operatorname{Spf}R_{\overline{\rho}}^{\square}$, but since $R_{\overline{\rho}}^{\square}$ is a complete local ring, it may equally well be regarded as a closed subscheme of $\operatorname{Spec}R_{\overline{\rho}}^{\square}$ (see Lemma A.3).

8.1.3 Lemma. *The fiber product* $\operatorname{Spf} R^{\square}_{\overline{\rho}} \times_{\mathcal{X}_d} \mathcal{X}^{\underline{k}}_d$, *when we regard it as a closed subscheme of* $\operatorname{Spec} R^{\square}_{\overline{\rho}}$, *is equidimensional of dimension* $d^2 + [K:\mathbf{Q}_p]d(d-1)/2$.

Proof. As in the proof of Theorem 6.5.1, we may find a regular Hodge type $\underline{\lambda}$ such that $\mathcal{X}^{\underline{k}}_d$ is an irreducible component of $(\overline{\mathcal{X}}^{\mathrm{crys},\underline{\lambda}})_{\mathrm{red}}$. Thus $\operatorname{Spf} R^{\square}_{\overline{\rho}} \times_{\mathcal{X}_d} \mathcal{X}^{\underline{k}}_d$ is versal to a union of irreducible components of the spectrum of the versal ring $(R^{\mathrm{crys},\underline{\lambda}}_{\overline{\rho}}/\varpi)_{\mathrm{red}}$ to $(\overline{\mathcal{X}}^{\mathrm{crys},\underline{\lambda}})_{\mathrm{red}}$ (those that correspond to the various formal branches of $\mathcal{X}^{\underline{k}}_d$ passing through x, in the terminology of [Sta, Tag 0DRA]) and hence is of the stated dimension, since $R^{\mathrm{crys},\underline{\lambda}}_{\overline{\rho}}/\varpi$, and so also $(R^{\mathrm{crys},\underline{\lambda}}_{\overline{\rho}}/\varpi)_{\mathrm{red}}$, is equidimensional of this dimension. □

We let $\mathcal{C}_{\underline{k}}(\overline{\rho})$ denote the $d^2 + [K:\mathbf{Q}_p]d(d-1)/2$-dimensional cycle in Spec $R^{\square}_{\overline{\rho}}/\varpi$ underlying the fiber product of Lemma 8.1.3.

The following theorem gives a qualitative version of the refined Breuil–Mézard conjecture [EG14, Conj. 4.2.1]. While its statement is purely local, we do not know how to prove it without making use of the stack $\mathcal{X}_d$.

8.1.4 Theorem. *Let* $\overline{\rho}\colon G_K \to \mathrm{GL}_d(\mathbf{F})$ *be a continuous representation. Then there are finitely many cycles of dimension* $d^2 + [K:\mathbf{Q}_p]d(d-1)/2$ *in* $\operatorname{Spec} R^{\square}_{\overline{\rho}}/\varpi$ *such that for any regular Hodge type* $\underline{\lambda}$ *and any inertial type* τ, *each of the special fibres* $\operatorname{Spec} R^{\mathrm{crys},\underline{\lambda},\tau}_{\overline{\rho}}/\varpi$ *and* $\operatorname{Spec} R^{\mathrm{ss},\underline{\lambda},\tau}_{\overline{\rho}}/\varpi$ *is set-theoretically supported on some union of these cycles.*

Proof. We have $\operatorname{Spf} R^{\mathrm{crys},\underline{\lambda},\tau}_{\overline{\rho}}/\varpi = \operatorname{Spf} R^{\square}_{\overline{\rho}} \times_{\mathcal{X}_d} \overline{\mathcal{X}}^{\mathrm{crys},\underline{\lambda},\tau}$ and $\operatorname{Spf} R^{\mathrm{ss},\underline{\lambda},\tau}_{\overline{\rho}}/\varpi = \operatorname{Spf} R^{\square}_{\overline{\rho}} \times_{\mathcal{X}_d} \overline{\mathcal{X}}^{\mathrm{crys},\underline{\lambda},\tau}$. It follows from (8.1.1) and (8.1.2), together with the definition of $\mathcal{C}_{\underline{k}}(\overline{\rho})$, that we may write the underlying cycles as

$$Z(\operatorname{Spec} R^{\mathrm{crys},\underline{\lambda},\tau}_{\overline{\rho}}/\varpi) = \sum_{\underline{k}} \mu_{\underline{k}}(\overline{\mathcal{X}}^{\mathrm{crys},\underline{\lambda},\tau}_d) \cdot \mathcal{C}_{\underline{k}}(\overline{\rho}), \tag{8.1.5}$$

$$Z(\operatorname{Spec} R^{\mathrm{ss},\underline{\lambda},\tau}_{\overline{\rho}}/\varpi) = \sum_{\underline{k}} \mu_{\underline{k}}(\overline{\mathcal{X}}^{\mathrm{ss},\underline{\lambda},\tau}_d) \cdot \mathcal{C}_{\underline{k}}(\overline{\rho}). \tag{8.1.6}$$

(Note that by [Sta, Tag 0DRD], the multiplicities do not change when passing to versal rings.) The theorem follows immediately (taking our finite set of cycles to be the $\mathcal{C}_{\underline{k}}(\overline{\rho})$). □

We can regard this theorem as isolating the "refined" part of [EG14, Conj. 4.2.1]; that is, we have taken the original numerical Breuil–Mézard conjecture, formulated a geometric refinement of it, and then removed the numerical part of the conjecture. The numerical part of the conjecture (in the optic of this chapter) consists of relating the multiplicities $\mu_{\underline{k}}(\overline{\mathcal{X}}_d^{\mathrm{crys},\underline{\lambda},\tau})$ and $\mu_{\underline{k}}(\overline{\mathcal{X}}_d^{\mathrm{ss},\underline{\lambda},\tau})$ to the representation theory of $\mathrm{GL}_n(k)$, as we recall in the next section.

8.2 SEMISTABLE AND CRYSTALLINE INERTIAL TYPES

We now briefly recall the "inertial local Langlands correspondence" for GL_d. Let rec_p denote the local Langlands correspondence for $\overline{\mathbf{Q}}_p$-representations of $\mathrm{GL}_d(K)$, normalized as in [CEG+16, §1.8]; this is a bijection between the isomorphism classes of irreducible smooth $\overline{\mathbf{Q}}_p$-representations of $\mathrm{GL}_d(K)$ and the isomorphism classes of d-dimensional semisimple Weil–Deligne $\overline{\mathbf{Q}}_p$-representations of the Weil group W_K. We have the following result, which is essentially due to Schneider–Zink [SZ99].

8.2.1 Theorem. *Let $\tau\colon I_K \to \mathrm{GL}_d(\overline{\mathbf{Q}}_p)$ be an inertial type. Then there are finite-dimensional smooth irreducible $\overline{\mathbf{Q}}_p$-representations $\sigma^{\mathrm{crys}}(\tau)$ and $\sigma^{\mathrm{ss}}(\tau)$ of $\mathrm{GL}_d(\mathcal{O}_K)$ with the properties that if π is an irreducible smooth $\overline{\mathbf{Q}}_p$-representation of $\mathrm{GL}_d(K)$, then the $\overline{\mathbf{Q}}_p$-vector space $\mathrm{Hom}_{\mathrm{GL}_d(\mathcal{O}_K)}(\sigma^{\mathrm{crys}}(\tau),\pi)$ (resp. the $\overline{\mathbf{Q}}_p$-vector space $\mathrm{Hom}_{\mathrm{GL}_d(\mathcal{O}_K)}(\sigma^{\mathrm{ss}}(\tau),\pi)$) has dimension at most 1, and is nonzero precisely if $\mathrm{rec}_p(\pi)|_{I_F} \cong \tau$, and $N=0$ on $\mathrm{rec}_p(\pi)$ (resp. if $\mathrm{rec}_p(\pi)|_{I_F} \cong \tau$, and π is generic).*

Proof. See [CEG+16, Thm. 3.7] for $\sigma^{\mathrm{crys}}(\tau)$, and [Sho18, Thm. 3.7] together with [Pyv20, Thm. 2.1, Lem. 2.2] for $\sigma^{\mathrm{ss}}(\tau)$. □

Note that we do not claim that the representations $\sigma^{\mathrm{crys}}(\tau)$ and $\sigma^{\mathrm{ss}}(\tau)$ are unique; the possible non-uniqueness of these representations is of no importance for us.

For each regular Hodge type $\underline{\lambda}$ we let $L_{\underline{\lambda}}$ be the corresponding representation of $\mathrm{GL}_d(\mathcal{O}_K)$, defined as follows: For each $\sigma\colon K \hookrightarrow \overline{\mathbf{Q}}_p$, we write $\xi_{\sigma,i} = \lambda_{\sigma,i} - (d-i)$, so that $\xi_{\sigma,1} \geq \cdots \geq \xi_{\sigma,d}$. We view each $\xi_\sigma := (\xi_{\sigma,1},\dots,\xi_{\sigma,d})$ as a dominant weight of the algebraic group GL_d (with respect to the upper triangular Borel subgroup), and we write M_{ξ_σ} for the algebraic $\mathcal{O}_K$-representation of $\mathrm{GL}_d(\mathcal{O}_K)$ of highest weight ξ_σ. Then we define $L_{\underline{\lambda}} := \otimes_\sigma M_{\xi_\sigma} \otimes_{\mathcal{O}_K,\sigma} \mathcal{O}$.

For each τ we let $\sigma^{\mathrm{crys},\circ}(\tau)$, $\sigma^{\mathrm{ss},\circ}(\tau)$ denote choices of $\mathrm{GL}_d(\mathcal{O}_K)$-stable $\mathcal{O}$-lattices in $\sigma^{\mathrm{crys}}(\tau)$, $\sigma^{\mathrm{ss}}(\tau)$ respectively (the precise choices being unimportant). Then we write $\overline{\sigma}^{\mathrm{crys}}(\lambda,\tau)$, (resp. $\overline{\sigma}^{\mathrm{ss}}(\lambda,\tau)$) for the semisimplification of the $\mathbf{F}$-representation of $\mathrm{GL}_d(k)$ given by $L_{\underline{\lambda}} \otimes_{\mathcal{O}} \sigma^{\mathrm{crys},\circ}(\tau) \otimes_{\mathcal{O}} \mathbf{F}$ (resp. $L_{\underline{\lambda}} \otimes_{\mathcal{O}} \sigma^{\mathrm{ss},\circ}(\tau) \otimes_{\mathcal{O}} \mathbf{F}$). For each Serre weight $\underline{k}$, we write $F_{\underline{k}}$ for the corresponding irreducible $\mathbf{F}$-representation of $\mathrm{GL}_d(k)$ (see for example the appendix to [Her09]). Then there are unique integers $n_{\underline{k}}^{\mathrm{crys}}(\lambda,\tau)$ and $n_{\underline{k}}^{\mathrm{ss}}(\lambda,\tau)$ such that

$$\overline{\sigma}^{\mathrm{crys}}(\lambda,\tau) \cong \oplus_{\underline{k}} F_{\underline{k}}^{\oplus n_{\underline{k}}^{\mathrm{ss}}(\lambda,\tau)},$$

$$\overline{\sigma}^{\mathrm{ss}}(\lambda,\tau) \cong \oplus_{\underline{k}} F_{\underline{k}}^{\oplus n_{\underline{k}}^{\mathrm{ss}}(\lambda,\tau)}.$$

Our geometric Breuil–Mézard conjecture is as follows.

8.2.2 Conjecture. *There are cycles $Z_{\underline{k}}$ with the property that for each regular Hodge type $\underline{\lambda}$ and each inertial type τ, we have $Z_{\mathrm{crys},\underline{\lambda},\tau} = \sum_{\underline{k}} n_{\underline{k}}^{\mathrm{crys}}(\lambda,\tau) \cdot Z_{\underline{k}}$, $Z_{\mathrm{ss},\underline{\lambda},\tau} = \sum_{\underline{k}} n_{\underline{k}}^{\mathrm{ss}}(\lambda,\tau) \cdot Z_{\underline{k}}$.*

For some motivation for the conjecture (coming from the Taylor–Wiles patching method), see for example [EG14, Thm. 5.5.2]. Some evidence for the conjecture is given in the following sections.

The expressions (8.1.1) and (8.1.2) describe $Z_{\mathrm{crys},\underline{\lambda},\tau}$ and $Z_{\mathrm{ss},\underline{\lambda},\tau}$ as linear combinations of (the cycles underlying) the irreducible components $\overline{\mathcal{X}}_d^{\underline{k}'}$, and thus each cycle $Z_{\underline{k}}$ (assuming that such cycles exist) will itself be a linear combination of the various $\overline{\mathcal{X}}_d^{\underline{k}'}$. We expect that the cycles $Z_{\underline{k}}$ are effective, i.e., that each of them is a linear combination of the $\overline{\mathcal{X}}_d^{\underline{k}'}$ with non-negative (integer) coefficients. Note that the finitely many (conjectural) cycles $Z_{\underline{k}}$ are completely determined by the infinitely many equations in Conjecture 8.2.2 (see [EG14, Lem. 4.1.1, Rem. 4.1.7(1)]).

While the original motivation for Conjecture 8.2.2 was to understand potentially semistable deformation rings in terms of the representation theory of GL_d, it can also be thought of as giving a geometric interpretation of the multiplicities $n_{\underline{k}}^{\mathrm{crys}}(\lambda,\tau)$.

8.3 THE RELATIONSHIP BETWEEN THE NUMERICAL, REFINED AND GEOMETRIC BREUIL–MÉZARD CONJECTURES

We now explain the relationship between Conjecture 8.2.2 and the conjectures of [EG14]. Suppose firstly that Conjecture 8.2.2 holds, and fix some $\overline{\rho}\colon G_K \to \mathrm{GL}_d(\mathbf{F})$ corresponding to a point $x\colon \operatorname{Spec} \mathbf{F} \to \mathcal{X}_d$. For each $\underline{k}$, we set

$$Z_{\underline{k}}(\overline{\rho}) := \operatorname{Spf} R_{\overline{\rho}}^{\square} \times_{\mathcal{X}_d} Z_{\underline{k}},$$

which we regard as a $d^2 + [K:\mathbf{Q}_p]d(d-1)/2$-dimensional cycle in $\operatorname{Spec} R_{\overline{\rho}}^{\square}/\varpi$. (As noted above, the $Z_{\underline{k}}$ will be linear combinations of the cycles underlying the various $\overline{\mathcal{X}}_d^{\underline{k}'}$, and so the the $Z_{\underline{k}}(\overline{\rho})$ will be linear combinations of the various cycles $\mathcal{C}_{\underline{k}'}(\overline{\rho})$; thus Lemma 8.1.3 shows that they are indeed $d^2 + [K:\mathbf{Q}_p]d(d-1)/2$-dimensional cycles.)

8.3.1 Remark. Equivalently, we can define $Z_{\underline{k}}(\overline{\rho})$ to be the image of $Z_{\underline{k}}$ under the natural map from $\mathbf{Z}[\mathcal{X}_{d,\mathrm{red}}]$ to the group $\mathbf{Z}_{d^2+[K:\mathbf{Q}_p]d(d-1)/2}(\operatorname{Spec} R_{\overline{\rho}}^{\square}/\varpi)$ of

$d^2+[K:\mathbf{Q}_p]d(d-1)/2$-dimensional cycles in $\operatorname{Spec} R_{\overline{\rho}}^{\square}/\varpi$ which is defined as follows: We let $R_{\overline{\rho}}^{\mathrm{alg}}$ be the quotient of $R_{\overline{\rho}}^{\square}/\varpi$ which is a versal ring to $\mathcal{X}_{d,\mathrm{red}}$ at x, so that by [Sta, Tag 0DRB,Tag 0DRD] we have a multiplicity-preserving surjection from the set of irreducible components of $\operatorname{Spec} R_{\overline{\rho}}^{\mathrm{alg}}$ to the set of irreducible components of $\mathcal{X}_{d,\mathrm{red}}$ containing x; we then send any irreducible component of $\mathcal{X}_{d,\mathrm{red}}$ not containing x to zero, and send each irreducible component containing x to the sum of the corresponding irreducible components of $\operatorname{Spec} R_{\overline{\rho}}^{\mathrm{alg}}$ in its preimage.

Exactly as in the proof of Theorem 8.1.4, it follows that for each regular type $\underline{\lambda}$ and inertial type τ, we have

$$Z(\operatorname{Spec} R_{\overline{\rho}}^{\mathrm{crys},\underline{\lambda},\tau}/\varpi)=\sum_{\underline{k}} n_{\underline{k}}^{\mathrm{crys}}(\lambda,\tau)\cdot Z_{\underline{k}}(\overline{\rho}), \tag{8.3.2}$$

$$Z(\operatorname{Spec} R_{\overline{\rho}}^{\mathrm{ss},\underline{\lambda},\tau}/\varpi)=\sum_{\underline{k}} n_{\underline{k}}^{\mathrm{ss}}(\lambda,\tau)\cdot Z_{\underline{k}}(\overline{\rho}). \tag{8.3.3}$$

The first of these statements is [EG14, Conj. 4.2.1] (with the cycles $\mathcal{C}_a$ there being our cycles $Z_{\underline{k}}(\overline{\rho})$), and the second is the corresponding statement for potentially semistable lifting rings.

We now relate this conjecture to the numerical version of the Breuil–Mézard conjecture. We have a homomorphism $\mathbf{Z}_{d^2+[K:\mathbf{Q}_p]d(d-1)/2}(\operatorname{Spec} R_{\overline{\rho}}^{\square}/\varpi)\to\mathbf{Z}$ defined by sending each cycle to its Hilbert–Samuel multiplicity in the sense of [EG14, §2.1]. Let $\mu_{\underline{k}}(\overline{\rho})$ denote the Hilbert–Samuel multiplicity of the cycle $Z_{\underline{k}}(\overline{\rho})$, and write $e(\operatorname{Spec} R_{\overline{\rho}}^{\mathrm{crys},\underline{\lambda},\tau}/\varpi)$, $e(\operatorname{Spec} R_{\overline{\rho}}^{\mathrm{ss},\underline{\lambda},\tau}/\varpi)$ for the Hilbert–Samuel multiplicities of the indicated rings. Then it follows from (8.3.2) and (8.3.3) that we have

$$e(\operatorname{Spec} R_{\overline{\rho}}^{\mathrm{crys},\underline{\lambda},\tau}/\varpi)=\sum_{\underline{k}} n_{\underline{k}}^{\mathrm{crys}}(\lambda,\tau)\mu_{\underline{k}}(\overline{\rho}), \tag{8.3.4}$$

$$e(\operatorname{Spec} R_{\overline{\rho}}^{\mathrm{ss},\underline{\lambda},\tau}/\varpi)=\sum_{\underline{k}} n_{\underline{k}}^{\mathrm{ss}}(\lambda,\tau)\mu_{\underline{k}}(\overline{\rho}). \tag{8.3.5}$$

Then (8.3.4) is [EG14, Conj. 4.1.6], and (8.3.5) is the corresponding semistable version.

Suppose now that for each $\overline{\rho}\colon G_K\to\mathrm{GL}_d(\mathbf{F})$, all regular Hodge types $\underline{\lambda}$ and all inertial types τ we have (8.3.4) and (8.3.5) for some integers $\mu_{\underline{k}}(\overline{\rho})$, but do not assume Conjecture 8.2.2 (so in particular we do not presuppose any geometric interpretation for the integers $\mu_{\underline{k}}(\overline{\rho})$).

For each $\underline{k}$ we choose a point $x_{\underline{k}}\colon \operatorname{Spec}\mathbf{F}\to\overline{\mathcal{X}}_{d,\mathrm{red}}$ which is contained in $\overline{\mathcal{X}}^{\underline{k}}$ and not in any $\overline{\mathcal{X}}^{\underline{k}'}$ for $\underline{k}'\neq\underline{k}$. We furthermore demand that $x_{\underline{k}}$ is a smooth point of $\overline{\mathcal{X}}_{d,\mathrm{red}}$. (Since $\overline{\mathcal{X}}_{d,\mathrm{red}}$ is reduced and of finite type over $\mathbf{F}$, there is a dense set of points of $\overline{\mathcal{X}}^{\underline{k}}$ satisfying these conditions.) Write $\overline{\rho}_{\underline{k}}: G_K \to \mathrm{GL}_d(\mathbf{F})$ for the representation corresponding to $X_{\underline{k}}$, and set

$$Z_{\underline{k}} := \sum_{\underline{k}'} \mu_{\underline{k}}(\overline{\rho}_{\underline{k}'}) \cdot \overline{\mathcal{X}}^{\underline{k}'}. \tag{8.3.6}$$

Then for each regular Hodge type $\underline{\lambda}$ and inertial type τ, it follows from (8.3.4) that

$$\begin{aligned}\sum_{\underline{k}} n_{\underline{k}}^{\mathrm{crys}}(\lambda,\tau) \cdot Z_{\underline{k}} &= \sum_{\underline{k}} e(R_{\overline{\rho}_{\underline{k}}}^{\mathrm{crys},\underline{\lambda},\tau}/\varpi) \cdot \overline{\mathcal{X}}^{\underline{k}} \\ &= \sum_{\underline{k}} \mu_{\underline{k}}(\overline{\mathcal{X}}_d^{\mathrm{crys},\underline{\lambda},\tau}) \cdot \overline{\mathcal{X}}^{\underline{k}} \\ &= Z_{\mathrm{crys},\underline{\lambda},\tau},\end{aligned}$$

where we used that $x_{\underline{k}}$ is a smooth point of $\overline{\mathcal{X}}_{d,\mathrm{red}}$ and is only contained in $\overline{\mathcal{X}}^{\underline{k}}$ to conclude that $\mu_{\underline{k}}(\overline{\mathcal{X}}_d^{\mathrm{crys},\underline{\lambda},\tau}) = e(R_{\overline{\rho}_{\underline{k}}}^{\mathrm{crys},\underline{\lambda},\tau}/\varpi)$. Similarly we have $Z_{\mathrm{ss},\underline{\lambda},\tau} = \sum_{\underline{k}} n_{\underline{k}}^{\mathrm{ss}}(\lambda,\tau) \cdot Z_{\underline{k}}$, and we conclude that the geometric Breuil–Mézard conjecture (Conjecture 8.2.2) is equivalent to the numerical conjecture.

8.3.7 Remark. The argument that we just made shows that the geometric conjecture follows from knowing the numerical conjecture for sufficiently generic $\overline{\rho}$ (indeed, it is enough to check it for a single sufficiently generic $\overline{\rho}$ for each irreducible component of $\mathcal{X}_{d,\mathrm{red}}$), while in turn the geometric conjecture implies the numerical conjecture for all $\overline{\rho}$.

8.4 THE WEIGHT PART OF SERRE'S CONJECTURE

We now briefly explain the relationship between Conjecture 8.2.2 and the weight part of Serre's conjecture. For more details, see [GHS18] (particularly Section 6).

We expect that the cycles $Z_{\underline{k}}$ will be effective, in the sense that they are combinations of the $\mathcal{X}_d^{\underline{k}}$ with non-negative coefficients. This expectation is borne out in all known examples (see the following sections), and in any case would be a consequence of standard conjectures about the Taylor–Wiles method. Indeed, the local cycles $Z_{\underline{k}}(\overline{\rho})$ of Section 8.3 are conjecturally the supports of certain "patched modules", and in particular are effective; see [EG14, Thm. 5.5.2]. The effectivity of the $Z_{\underline{k}}(\overline{\rho})$ would immediately imply the effectivity of the $Z_{\underline{k}}$.

The "weight part of Serre's conjecture" is perhaps more of a conjectural conjecture than an actual conjecture: it should assign to each $\overline{\rho}: G_K \to \mathrm{GL}_d(\mathbf{F})$ a set $W(\overline{\rho})$ of Serre weights, with the property that if $\overline{\rho}$ is the restriction to a decomposition group of a suitable global representation (for example, an irreducible representation coming from an automorphic form on a unitary group), then $\underline{k} \in W(\overline{\rho})$ if and only if $\underline{k}$ is a weight for the global representation (for example, in the sense that the global representation corresponds to some mod p cohomology class for a coefficient system corresponding to $F_{\underline{k}}$).

Many conjectural definitions of the sets $W(\overline{\rho})$ have been proposed. Following [GK14], one definition is to assume the Breuil–Mézard conjecture, for example in the form (8.3.2), and define $W(\overline{\rho})$ to be the set of $\underline{k}$ for which $\mu_{\underline{k}}(\overline{\rho}) > 0$. While this is less explicit than other definitions, it has the merit that it would follow from standard conjectures about modularity lifting theorems that it gives the correct set of weights; see [GHS18, §3, 4].

Assume Conjecture 8.2.2, and assume that the cycles $Z_{\underline{k}}$ are effective. As explained in Section 8.3, it follows that the numerical Breuil–Mézard holds for every $\overline{\rho}$, with $\mu_{\underline{k}}(\overline{\rho})$ being the Hilbert–Samuel multiplicity of the cycle $Z_{\underline{k}}(\overline{\rho})$, which (since $Z_{\underline{k}}(\overline{\rho})$ is effective) is positive if and only if $Z_{\underline{k}}(\overline{\rho})$ is nonzero, i.e., if and only if $Z_{\underline{k}}$ is supported at $\overline{\rho}$. Thus we can rephrase the Breuil–Mézard version of the weight part of Serre's conjecture as saying that $W(\overline{\rho})$ is the set of $\underline{k}$ such that $Z_{\underline{k}}$ is supported at $\overline{\rho}$.

Alternatively, we can rephrase this conjecture in the following way: to each irreducible component of $\mathcal{X}_{d,\mathrm{red}}$, we assign the set of weights $\underline{k}$ with the property that $Z_{\underline{k}}$ is supported on this component. Then $W(\overline{\rho})$ is simply the union of the sets of weights for the irreducible components of $\mathcal{X}_{d,\mathrm{red}}$ which contain $\overline{\rho}$. As we explain in Section 8.6, if $d=2$ then this description agrees with the other definitions of $W(\overline{\rho})$ in the literature, and therefore gives a geometrization of the weight part of Serre's conjecture.

8.5 THE CASE OF $\mathrm{GL}_2(\mathbf{Q}_p)$

The numerical Breuil–Mézard conjecture for $K = \mathbf{Q}_p$ and $d=2$ is completely known, thanks in large part to Kisin's paper [Kis09a] (which gave a proof in many cases by a mixture of local and global techniques), and Paškūnas' paper [Paš15] which reproved these results by purely local means (the p-adic local Langlands correspondence), relaxed the hypotheses on $\overline{\rho}$, and also proved the refined version of the correspondence. The remaining cases not handled by these papers are proved in the papers [HT15, San14, Tun18], culminating in the proof of the final cases for $p=2$ by Tung in [Tun21].

Consequently, by the discussion of Section 8.3, Conjecture 8.2.2 holds for $K = \mathbf{Q}_p$ and $d=2$. It follows easily from the explicit description of $\mu_{\underline{k}}(\overline{\rho})$ in the papers cited above (or alternatively from the description for $\mathrm{GL}_2(K)$ in Section 8.6 below) that $Z_{\underline{k}} = \mathcal{X}_2^{\underline{k}}$ unless $\underline{k} = (a+p-1, a)$ for some a, in which case $Z_{(a+p-1,a)} = \mathcal{X}_2^{(a+p-1,a)} + \mathcal{X}_2^{(a,a)}$.

8.6 $\mathrm{GL}_2(K)$: POTENTIALLY BARSOTTI–TATE TYPES

In this section we assume that p is odd, and explain some consequences of the results of [GK14] (which proved the numerical Breuil–Mézard conjecture for two-dimensional potentially Barsotti–Tate representations) and [CEGS19] (which studied moduli stacks of rank 2 Breuil–Kisin modules with tame descent data). We will take advantage of Remark 8.3.7.

We begin with the following slight extension of one of the main results of [GK14]. Let $\underline{\mathrm{BT}}$ denote the minimal regular Hodge type, i.e., we have $\underline{\mathrm{BT}}_{\sigma,1} = 1$, $\underline{\mathrm{BT}}_{\sigma,2} = 0$ for all $\sigma\colon K \hookrightarrow \overline{\mathbf{Q}}_p$.

8.6.1 Theorem. *Let $K/\mathbf{Q}_p$ be a finite extension with $p>2$, and let $\overline{\rho}\colon G_K \to \mathrm{GL}_2(\overline{\mathbf{Q}}_p)$ be arbitrary. Then the numerical Breuil–Mézard conjecture holds for potentially crystalline and potentially semistable lifts of $\overline{\rho}$ of Hodge type $\underline{\mathrm{BT}}$ and arbitrary inertial type τ.*

More precisely, there are unique non-negative integers $\mu_{\underline{k}}(\overline{\rho})$ such that (8.3.4) *and* (8.3.5) *both hold for $\underline{\lambda} = \underline{\mathrm{BT}}$ and τ arbitrary.*

Proof. If we remove the potentially semistable case and consider only potentially crystalline representations, the theorem is [GK14, Thm. A], which is proved as [GK14, Cor. 4.5.6]. We now briefly explain how to modify the proofs in [GK14] to prove the more general result; as writing out a full argument would be a lengthy exercise, and the arguments are completely unrelated to those of this book, we only explain the key points. Examining the proof of [GK14, Cor. 4.5.6], we see that we just need to verify that the assertion of the first sentence of the proof of [GK14, Thm. 4.5.5] holds in this setting, i.e., that the equivalent conditions of [GK14, Lem. 4.3.9] hold. Exactly as in the proof of [GK14, Cor. 4.4.3], it is enough to show that every irreducible component of a product of local deformation rings is witnessed by an automorphic representation.

By the usual Khare–Wintenberger argument, it suffices to prove this after making a solvable base change, and in particular we can suppose that all of the residual local Galois representations are trivial, the mod p cyclotomic character is trivial, the trivial mod p representation admits a non-ordinary crystalline lift, and the inertial types are all trivial. Now, any non-crystalline semistable representation of Hodge type $\underline{\mathrm{BT}}$ is necessarily ordinary (indeed, it follows from a direct computation of the possible weakly admissible modules that all such representations are unramified twists of an extension of the inverse of the cyclotomic character by the trivial character), and by results of Kisin and Gee (see [Sno18, Prop. 4.3.1], [Kis09b, Cor. 2.5.16], and [Gee06, Prop. 2.3]), as $p>2$ and $\overline{\rho}$ is trivial, the semistable ordinary deformation ring in question is a domain, as are the crystalline ordinary deformation ring and the crystalline non-ordinary deformation ring (all in Hodge type $\underline{\mathrm{BT}}$). It therefore suffices to show that in the situation of the proof of [GK14, Cor. 4.4.3], given any decomposition of the set S of places of F_1^+ lying over p as $S_{\mathrm{ss}} \coprod S_{\text{crys-ord}} \coprod S_{\text{non-ord}}$, we can arrange to have a congruence to an automorphic representation π'', having the properties that π'' is unramified at the places lying over a place in $S_{\text{crys-ord}} \coprod S_{\text{non-ord}}$, is an unramified twist of the Steinberg representation at the places lying over a place in S_{ss}, and is furthermore ordinary (resp. not ordinary) at the places lying over a place in $S_{\text{crys-ord}}$ (resp. $S_{\text{non-ord}}$).

The existence of such a representation if $S_{ss} = S$ follows from the construction of the global representation $\bar{r}$ used in the proof of the numerical Breuil–Mézard conjecture in [GK14]; more precisely, it follows from [Sno09, Prop. 8.2.1], which is applied in the proof of [GK14, Thm. A.2] (with the type function in the sense of [Sno09] being C at all places above p). Thus, to establish the general case, it suffices to establish the existence of suitable "level lowering" congruences. This is easily done by switching to a group which is ramified at the places lying over S_{ss}, and then making congruences to automorphic representations which are either unramified and ordinary (in the case of places lying over $S_{\text{crys-ord}}$) or have cuspidal type (in the case of places lying over $S_{\text{non-ord}}$), and then applying the Khare–Wintenberger argument again at the places lying over $S_{\text{non-ord}}$. (See, e.g., [Kis09b, Lem. 3.5.3] for a similar argument for places away from p.) □

We say that a Serre weight $\underline{k}$ for GL_2 is *Steinberg* if for each $\overline{\sigma}$ we have $k_{\overline{\sigma},1} - k_{\overline{\sigma},2} = p-1$. If $\underline{k}$ is Steinberg then we define $\underline{\tilde{k}}$ by $\tilde{k}_{\overline{\sigma},1} = \tilde{k}_{\overline{\sigma},2} = k_{\overline{\sigma},2}$.

8.6.2 Theorem. *Continue to assume that* $d=2$ *and* $p>2$. *Then Conjecture* 8.2.2 *holds for* $\underline{\lambda} = \underline{\mathrm{BT}}$ *and* τ *arbitrary, with the cycles* $Z_{\underline{k}}$ *being as follows: if* $\underline{k}$ *is not Steinberg, then* $Z_{\underline{k}} = \mathcal{X}_2^{\underline{k}}$, *while if* $\underline{k}$ *is Steinberg,* $Z_{\underline{k}} = \mathcal{X}_2^{\underline{k}} + \mathcal{X}_2^{\underline{\tilde{k}}}$.

Proof. We use the notation of Section 8.3. By Theorem 8.6.1 and the discussion of Section 8.3, we need only show that the cycles $Z_{\underline{k}}$ in the statement of the theorem are those determined by (8.3.6). Suppose firstly that $\underline{k}$ is not Steinberg. Then by [CEGS19, Thm. 5.2.2 (2)], $\mathcal{X}_2^{\underline{k}}$ has a dense set of finite type points with the property that their only non-Steinberg Serre weight is $\underline{k}$. It follows from this and [CEGS19, Thm. 5.2.2 (3)] that if neither $\underline{k}$ nor $\underline{k}'$ is Steinberg, then $\mu_{\underline{k}}(\overline{\rho}_{\underline{k}'}) = \delta_{\underline{k},\underline{k}'}$. By construction, if $\underline{k}'$ is Steinberg, then $\overline{\rho}_{\underline{k}'}$ is a twist of a très ramifiée extension of the trivial character by the mod p cyclotomic character. By [CEGS19, Lem. B.5], this implies that $\mu_{\underline{k}}(\overline{\rho}_{\underline{k}'}) = 0$. The cycles $Z_{\underline{k}}$ are therefore as claimed if $\underline{k}$ is non-Steinberg.

It remains to determine the values of $\mu_{\underline{k}}(\overline{\rho}_{\underline{k}'})$ in the case that $\underline{k}$ is Steinberg. By twisting, we can and do assume that $k_{\overline{\sigma},2} = 0$ for all $\overline{\sigma}$. If we apply Theorem 8.6.1 with τ being the trivial type, and recall that $L_{\underline{\mathrm{BT}}}$ is the trivial representation and that $\sigma^{\mathrm{ss}}(\tau)$ is the Steinberg type, the reduction of which is precisely the representation $F_{\underline{k}}$, we find that

$$Z_{\mathrm{ss},\underline{\mathrm{BT}},\tau} = Z_{\underline{k}}. \tag{8.6.3}$$

Thus $\mu_{\underline{k}}(\overline{\rho}_{\underline{k}'}) \neq 0$ if and only if $\overline{\rho}_{\underline{k}'}$ admits a semistable lift of Hodge type $\underline{\mathrm{BT}}$. If this lift is in fact crystalline, then $\overline{\rho}_{\underline{k}'}$ has $\underline{\tilde{k}}$ as a Serre weight, so by another application of [CEGS19, Thm. 5.2.2 (2)], we see that either $\underline{k}' = \underline{\tilde{k}}$ or else that $\underline{k}'$ is also Steinberg. In the former case, $\overline{\rho}_{\underline{k}'}$ is (by Remark 5.5.16) an unramified twist of an extension of the inverse of the mod p cyclotomic character by a non-trivial unramified character, so it does not admit a semistable non-crystalline lift of Hodge type $\underline{\mathrm{BT}}$ (as all such lifts are unramified twists of an extension of the inverse of the cyclotomic character by the trivial character); so we have $\mu_{\underline{k}}(\overline{\rho}_{\underline{\tilde{k}}}) = \mu_{\underline{\tilde{k}}}(\overline{\rho}_{\underline{\tilde{k}}}) = 1$.

Finally, we are left with the task of computing $\mu_{\underline{k}}(\overline{\rho}_{\underline{k}'})$ when both $\underline{k}$ and $\underline{k}'$ are Steinberg. Recall that by twisting, we are assuming that $k_{\overline{\sigma},2} = 0$ for all $\overline{\sigma}$. The weight $\underline{k}'$ is then a twist of $\underline{k}$. We have to show that in this case we again have $\mu_{\underline{k}}(\overline{\rho}_{\underline{k}'}) = \delta_{\underline{k},\underline{k}'}$. To see this, first note that since $\overline{\rho}_{\underline{k}'}$ is très ramifiée, it does not admit a crystalline lift of Hodge type $\underline{\mathrm{BT}}$, so all of its semistable lifts of Hodge type $\underline{\mathrm{BT}}$ are given by unramified twists of extensions of the inverse of the cyclotomic character by the trivial character. Furthermore, the reduction of any lattice in such an extension is an unramified twists of an extension of the inverse of the mod ϖ cyclotomic character by the trivial character, and hence cannot equal $\overline{\rho}_{\underline{k}'}$ unless $\underline{k}'$ equals $\underline{k}$ (rather than being a non-trivial twist of it). If we take into account (8.6.3), we find that indeed $\mu_{\underline{k}}(\overline{\rho}_{\underline{k}'}) = 0$ when $\underline{k}' \neq \underline{k}$.

Finally, any lift of $\rho_{\underline{k}}$ which is an unramified twist of an extension of the inverse of the cyclotomic character by the trivial character is automatically semistable of Hodge type $\underline{\mathrm{BT}}$, so that $R^{\mathrm{ss},\underline{\mathrm{BT}},\tau}_{\overline{\rho}_{\underline{k}}}$ is precisely the ordinary (framed) deformation ring parameterizing such lifts of $\overline{\rho}_{\underline{k}}$. A standard Galois cohomology calculation (that we leave to the reader) shows that this ring is formally smooth, and thus that $R^{\mathrm{ss},\underline{\mathrm{BT}},\tau}_{\overline{\rho}_{\underline{k}}}/\varpi$ is also formally smooth. It follows that $\mu_{\underline{k}}(\overline{\rho}_{\underline{k}}) = 1$, as claimed. ☐

The following lemma makes precise which Galois representations occur on Steinberg components. Note that the twist in the statement of the lemma is determined by the values of $k_{\underline{\sigma},2}$.

8.6.4 Lemma. *If $\underline{k}$ is Steinberg, then the $\overline{\mathbf{F}}_p$-points of $\mathcal{X}_2^{\underline{k}}$ are twists of an extension of the inverse of the mod p cyclotomic character by the trivial character.*

Proof. Twisting, we may assume that $k_{\underline{\sigma},2} = 0$ for all $\overline{\sigma}$. Let $\overline{\rho}\colon G_K \to \mathrm{GL}_2(\mathbf{F})$ correspond to a closed point of $\mathcal{X}_2^{\underline{k}}$. We claim that $\overline{\rho}$ is an unramified twist of an extension of $\overline{\epsilon}$ by 1. To see this, let $R^{\underline{\mathrm{BT}},\mathrm{St}}_{\overline{\rho}}$ denote the $\mathbf{Z}_p$-flat quotient of $R^{\underline{\mathrm{BT}},\mathrm{ss}}_{\overline{\rho}}$ determined by the irreducible components of the generic fiber which are not components of $R^{\underline{\mathrm{BT}},\mathrm{crys}}_{\overline{\rho}}$.

It follows from Theorem 8.6.2 that the cycle of $\operatorname{Spec} R^{\underline{\mathrm{BT}},\mathrm{St}}_{\overline{\rho}}/\varpi$ is nonzero, so that in particular $\overline{\rho}$ must admit a semistable non-crystalline lift of Hodge type BT. As in the proof of Theorem 8.6.1, any such representation is an unramified twist of an extension of ϵ^{-1} by the trivial character, so we are done. (We could presumably also phrase this argument on the level of the crystalline and semistable moduli stacks, but have chosen to present it in the more familiar setting of deformation rings.) ☐

8.6.5 Remark. The paper [CEGS19] constructs and studies moduli stacks of two-dimensional representations of $G_{K_{\pi^\flat,\infty}}$ (for a fixed choice of $\pi^\flat$) which are tamely potentially of height at most 1. Since restriction from Barsotti–Tate representations of G_K to representations of $G_{K_{\pi^\flat,\infty}}$ of height at most 1 is an equivalence (by the results of [Kis09b]), these stacks can be interpreted as stacks of G_K-representations (and indeed are interpreted as such in [CEGS19]). It is presumably straightforward (using that all the stacks under consideration are of finite type, and are $\mathbf{Z}_p$-flat and reduced, in order to reduce to a comparison of points over finite extensions of $\mathbf{Z}_p$) to identify them with the stacks considered in this book (for $\underline{\lambda}=\underline{\mathrm{BT}}$ and τ a tame inertial type). (We were able to apply the results of [CEGS19] in the proof of Theorem 8.6.2 without doing this because we only needed to use them on the level of versal rings, which are given by universal Galois lifting rings for both our stacks and those of [CEGS19].) It may well be the case that the stacks of Kisin modules considered in [CEGS19] can be identified with certain of the moduli stacks of Breuil–Kisin–Fargues modules that we defined in Section 4.5. Since we do not need to know this, we leave it as an exercise for the interested reader.

8.6.6 A lower bound

The patching arguments of [GK14] easily imply a general inequality, as we now record.

8.6.7 Proposition. *Assume that* $p>2$. *Let* $d=2$, *and let the* $Z_{\underline{k}}$ *be as in the statement of Theorem* 8.6.2. *Then for each regular Hodge type* $\underline{\lambda}$ *and each inertial type* τ, *we have* $Z_{\mathrm{crys},\underline{\lambda},\tau}\geq\sum_{\underline{k}} n_{\underline{k}}^{\mathrm{crys}}(\lambda,\tau)\cdot Z_{\underline{k}}$, *and* $Z_{\mathrm{ss},\underline{\lambda},\tau}\geq\sum_{\underline{k}} n_{\underline{k}}^{\mathrm{ss}}(\lambda,\tau)\cdot Z_{\underline{k}}$.

8.6.8 Remark. The meaning of the inequalities in the statement of Proposition 8.6.7 is the obvious one: each side is a linear combination of irreducible components, and the assertion is that for each irreducible component, the multiplicity on the left-hand side is at least the multiplicity on the right-hand side.

Proof of Proposition 8.6.7. As in Section 8.3, it is enough to show that for each $\overline{\rho}$, we have

$$e(\operatorname{Spec} R_{\overline{\rho}}^{\mathrm{crys},\underline{\lambda},\tau}/\varpi)\geq\sum_{\underline{k}} n_{\underline{k}}^{\mathrm{crys}}(\lambda,\tau)\mu_{\underline{k}}(\overline{\rho}),$$

$$e(\operatorname{Spec} R_{\overline{\rho}}^{\mathrm{ss},\underline{\lambda},\tau}/\varpi)\geq\sum_{\underline{k}} n_{\underline{k}}^{\mathrm{ss}}(\lambda,\tau)\mu_{\underline{k}}(\overline{\rho}),$$

where the $\mu_{\underline{k}}(\overline{\rho})$ are the uniquely determined integers from [GK14]. (In fact, it is enough to show these inequalities when $\overline{\rho}=\overline{\rho}_{\underline{k}}$ for some $\underline{k}$, but assuming this does not simplify our arguments.) Since [GK14] considers only potentially crystalline representations, we only give the proof for $R_{\overline{\rho}}^{\mathrm{crys},\underline{\lambda},\tau}$; the proof in the potentially semistable case is essentially identical, and we leave it to the reader.

We now examine the proof of [GK14, Thm. 4.5.5], setting $\overline{r}$ there to be our $\overline{\rho}$. As we've done throughout this book, we write $\underline{k}$ rather than σ for Serre weights. If we ignore the assumption that every potentially crystalline lift of Hodge type $\underline{\lambda}$ and inertial type τ is potentially diagonalizable, then we do not know that the equivalent conditions of [GK14, Lem. 4.3.9] hold, but examining the proof of [GK14, Lem. 4.3.9], we do know that the statement of part (4) of loc. cit. can be replaced with an inequality (with equality holding if and only if M_∞ is a faithful R_∞-module).

Returning to the proof of [GK14, Thm. 4.5.5], we note that the definition of $\mu_{\underline{k}}(\overline{\rho})$ is such that the quantity $\mu'_{\sigma_{\mathrm{gl}}}(\overline{r})$ is simply the product of the corresponding $\mu_{\underline{k}}(\overline{\rho})$. In particular, if we take $\lambda_v = \underline{\lambda}$ and $\tau_v = \tau$ for each v, then (bearing in mind the discussion of the previous paragraph), the main displayed equality becomes an inequality

$$e(\operatorname{Spec} R_{\overline{\rho}}^{\mathrm{crys},\underline{\lambda},\tau}/\varpi)^N \geq \left(\sum_{\underline{k}} n_{\underline{k}}^{\mathrm{crys}}(\lambda,\tau)\mu_{\underline{k}}(\overline{\rho})\right)^N,$$

where there are N places of F^+ lying over p. Since each side is non-negative, the result follows. □

8.6.9 Remark. The argument used to prove Proposition 8.6.7 is well known to the experts, and goes back to [Kis09a, Lem. 2.2.11]. Despite the relatively formal nature of the argument, it does not seem to be easy to prove an analogous statement for GL_d with $d > 2$; the difficulty is in the step where we used [GK14, Thm. 4.5.5] to replace a global multiplicity with a product of local multiplicities. Without this argument (which crucially relies on the modularity lifting theorems of [Kis09b]) one only obtains inequalities involving cycles in products of copies of $\mathcal{X}_d$.

8.7 BRIEF REMARKS ON GL_d, $d > 2$

We expect the situation for GL_d, $d > 2$, to be considerably more complicated than that for GL_2. Experience to date suggests that the weight part of Serre's conjecture in high dimension is consistently more complicated than is anticipated, and so it seems unwise to engage in much speculation. In particular, the results of [LHLM20] show that even for generic weights $\underline{k}$, we should not expect the cycles $Z_{\underline{k}}$ to have as simple a form as those for GL_2.

In the light of Lemma 5.5.4, it seems reasonable to expect $\mathcal{X}_d^{\underline{k}}$ to contribute to $Z_{\underline{k}}$, and it also seems reasonable to expect contributions from the "shifted" weights, as in Definition 5.5.2 (see also [GHS18, §7.4]). The comparative weakness of the automorphy lifting theorems available to us in dimension greater than 2 prevents us from saying much more than this, although we refer the reader to [GHS18, §§3,4,6] for a discussion of the conjectural general relationship between the numerical Breuil–Mézard conjecture, the weight part of Serre's conjecture, automorphy lifting theorems, and the stacks $\mathcal{X}_d$.

Appendix A

Formal algebraic stacks

The theory of formal algebraic stacks is developed in [Eme]. In this appendix we briefly summarize the parts of this theory that are used in the body of the book, and also introduce some additional terminology and establish some additional results which we will require.

Algebraic stacks

We follow the terminology of [Sta]; in particular, we write "algebraic stack" rather than "Artin stack". More precisely, an algebraic stack is a stack in groupoids in the *fppf* topology, whose diagonal is representable by algebraic spaces, which admits a smooth surjection from a scheme. See [Sta, Tag 026N] for a discussion of how this definition relates to others in the literature. If S is a scheme, then by "a stack over S" we mean a stack fibred in groupoids over the big *fppf* site of S.

We say that a morphism $\mathcal{X} \to \mathcal{Y}$ of stacks over S is *representable by algebraic stacks* if for any morphism of stacks $\mathcal{Z} \to \mathcal{Y}$ whose source is an algebraic stack, the fiber product $\mathcal{X} \times_{\mathcal{Y}} \mathcal{Z}$ is again an algebraic stack. (In the Stacks Project, the terminology *algebraic* is used instead [Sta, Tag 06CF].) Note that a morphism from a sheaf to a stack is representable by algebraic stacks if and only if it is representable by algebraic spaces (this is easily verified, or see, e.g., [Eme, Lem. 3.5]).

Following [Sta, Tag 03YK, Tag 04XB], we can define properties of morphisms representable by algebraic stacks in the following way.

A.1 Definition. If P is a property of morphisms of algebraic stacks which is *fppf* local on the target, and preserved by arbitrary base change, then we say that a morphism $f\colon \mathcal{X} \to \mathcal{Y}$ of stacks which is representable by algebraic stacks *has property* P if and only if for every algebraic stack $\mathcal{Z}$ and morphism $\mathcal{Z} \to \mathcal{Y}$, the base-changed morphism of algebraic stacks $\mathcal{Z} \times_{\mathcal{Y}} \mathcal{X} \to \mathcal{Z}$ has property P.

When applying Definition A.1, it suffices to consider the case when $\mathcal{Z}$ is actually a scheme, or even an affine scheme.

Some properties P to which we can apply Definition A.1 are being *locally of finite type*, *locally of finite presentation*, and *smooth.* The following lemma provides alternative descriptions of the latter two properties.

A.2 Lemma. *Let $f\colon \mathcal{X} \to \mathcal{Y}$ be a morphism of stacks which is representable by algebraic stacks.*

1. *The morphism f is locally of finite presentation if and only if it is* limit preserving on objects, *in the sense of [Sta, Tag 06CT].*
2. *The morphism f is smooth if and only if it is locally of finite presentation and* formally smooth, *in the sense that it satisfies the usual infinitesimal lifting property: for every affine $\mathcal{Y}$-scheme T, and every closed subscheme $T_0 \hookrightarrow T$ defined by a nilpotent ideal sheaf, the functor* $\mathrm{Hom}_{\mathcal{Y}}(T, \mathcal{X}) \to \mathrm{Hom}_{\mathcal{Y}}(T_0, \mathcal{X})$ *is essentially surjective.*

Proof. We first recall that a morphism of algebraic stacks is locally of finite presentation if and only if it is limit preserving on objects [EG21, Lem. 2.3.16].

Suppose now that f is locally of finite presentation, that T is an affine scheme, written as a projective limit of affine schemes $T \xrightarrow{\sim} \varprojlim T_i$, and suppose we are given a commutative diagram

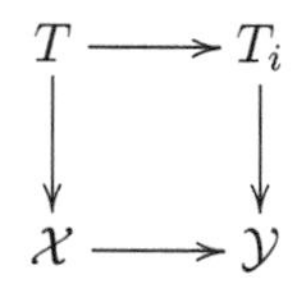

for some value of i. We must show that we can factor the left-hand vertical arrow through $T_{i'}$, for some $i' \geq i$, so that the evident resulting diagram again commutes. The original diagram induces a morphism $T \to T_i \times_{\mathcal{Y}} \mathcal{X}$, and obtaining a factorization of the desired type amounts to obtaining a factorization of this latter morphism through some $T_{i'}$. In other words, we must show that the morphism $T_i \times_{\mathcal{Y}} \mathcal{X} \to T_i$ is limit preserving on objects. But this is a morphism of algebraic stacks which is locally of finite presentation, and so it is indeed limit preserving on objects.

Conversely, suppose that f is limit preserving on objects. We must show that for any morphism $T \to \mathcal{Y}$, the base-changed morphism $T \times_{\mathcal{Y}} \mathcal{X} \to T$ is locally of finite presentation. However, this base change is again limit preserving on objects [Sta, Tag 06CV], and since it is a morphism of algebraic stacks, it is indeed locally of finite presentation.

We now turn to (2). By Definition A.1 and [Ryd11, Cor. B.9] (that is, by the result at hand in the case that $\mathcal{Y}$ is itself an algebraic stack), it is enough to show that f satisfies the infinitesimal lifting property if and only if the base-changed morphism $Z \times_{\mathcal{Y}} \mathcal{X} \to Z$ satisfies the infinitesimal lifting property for all morphisms $Z \to \mathcal{Y}$ whose source is a scheme.

Assume first that f satisfies the infinitesimal lifting property, and let $T_0 \to T$ be a nilpotent closed immersion whose target is an affine Z-scheme. Given a morphism $T_0 \to Z \times_{\mathcal{Y}} \mathcal{X}$, the infinitesimal lifting property lets us lift the composite $T_0 \to Z \times_{\mathcal{Y}} \mathcal{X} \to \mathcal{X}$ to a morphism $T \to \mathcal{X}$, and since we also have a morphism $T \to Z$, we obtain the required morphism $T \to Z \times_{\mathcal{Y}} \mathcal{X}$.

Conversely, assume that the base-changed morphism $Z \times_{\mathcal{Y}} \mathcal{X} \to Z$ satisfies the infinitesimal lifting property for all $Z \to \mathcal{Y}$ with Z a scheme, and let $T_0 \to T$ be as in the statement of the proposition. Taking $Z = T$, we can lift the induced morphism $T_0 \to T \times_{\mathcal{Y}} \mathcal{X}$ to a morphism $T \to T \times_{\mathcal{Y}} \mathcal{X}$, and the composite of this morphism with the second projection gives us the required lifting. $\square$

Formal algebraic spaces

Following [Sta, Tag 0AHW], an affine formal algebraic space over a base scheme S is a sheaf X on the *fppf* site of S which admits a description as an Ind-scheme $X \xrightarrow{\sim} \varinjlim_i X_i$, where the X_i are affine schemes and the transition morphisms are thickenings (in the sense of [Sta, Tag 04EX]). Here we allow the indexing set in the inductive limit to be arbitrary; if it can be chosen to be the natural numbers, then we say that X is *countably indexed.* A countably indexed affine formal algebraic space can be written in the form $X \cong \varinjlim_n \operatorname{Spec} A/I_n$, where A is a complete topological ring equipped with a decreasing sequence $\{I_n\}_n$ of open ideals which are weak ideals of definition, i.e., consist of topologically nilpotent elements ([Sta, Tag 0AMV]), and which form a fundamental basis of 0 in A (see the discussion of [Sta, Tag 0AIH]). We then write $X := \operatorname{Spf} A$ (following [Sta, Tag 0AIF]).

A particular example of a countably indexed affine formal algebraic space is an adic affine formal algebraic space. By definition, this is of the form $\operatorname{Spf} A$, where A is adic: that is, A is a topological ring which is complete (and separated, by our convention throughout this book), and which admits an ideal of definition (that is, an ideal I whose powers form a basis of open neighborhoods of zero). We say that $\operatorname{Spf} A$ is Noetherian if A is adic and Noetherian. We say that A (or $\operatorname{Spf} A$) is adic* if I can be taken to be finitely generated.

More generally, we say that a complete topological ring A is weakly admissible if A contains an open ideal I consisting of topologically nilpotent elements ([Sta, Tag 0AMV]).

A formal algebraic space over S is a sheaf X on the *fppf* site of S which receives a morphism $\coprod U_i \to X$ which is representable by schemes, étale, and surjective, and whose source is a disjoint union of affine formal algebraic spaces U_i. We say that X is *locally countably indexed* if the U_i can be chosen to be countably indexed. We say that X is *locally Noetherian* if the U_i can be taken to be Noetherian.

We will find the following lemmas useful.

A.3 Lemma. *Let X be an affine formal algebraic space over S, which is either countably indexed (e.g., an adic or adic* affine formal algebraic space), and hence of the form* $\operatorname{Spf} A$ *for some weakly admissible topological ring A, by [Sta, Tag 0AIK], or else is of the form* $\operatorname{Spf} A$, *for a pro-Artinian ring A. Then if $Y \to X$ is a closed immersion of formal algebraic spaces over S, we have that Y is of the form* $\operatorname{Spf} B$ *for some quotient B of A by a closed ideal, endowed with its quotient topology.*

Proof. In the countably indexed case, this is immediate from [Sta, Tag 0AIK,Tag 0ANQ,Tag 0APT]. If A is pro-Artinian, say $A = \varprojlim A_i$ with the A_i Artinian, then $\operatorname{Spec} A_i \times_{\operatorname{Spf} A} Y$ is a closed subscheme of $\operatorname{Spec} A_i$ for each index i, and hence is of the form $\operatorname{Spec} B_i$ for some quotient B_i of B. Thus

$$Y \xrightarrow{\sim} \varinjlim_i \operatorname{Spec} A_i \times_{\operatorname{Spf} A} Y \xrightarrow{\sim} \varinjlim \operatorname{Spec} B_i \xrightarrow{\sim} \operatorname{Spf} B,$$

where $B = \varprojlim_i B_i$. The general theory of pro-Artinian rings shows that B is a quotient of A, since each B_i is a quotient of the corresponding A_i [EG21, Rem. 2.2.7]. □

A.4 Lemma. ([Eme, Lem. 8.18]) *A morphism of Noetherian affine formal algebraic spaces* $\operatorname{Spf} B \to \operatorname{Spf} A$ *which is representable by algebraic spaces is (faithfully) flat if and only if the corresponding morphism* $A \to B$ *is (faithfully) flat.*

Ind-algebraic and formal algebraic stacks

A.5 Definition. ([Eme, Defn. 4.2]) An *Ind-algebraic stack* over a scheme S is a stack $\mathcal{X}$ over S which can be written as $\mathcal{X} \cong \varinjlim_{i \in I} \mathcal{X}_i$, where we are taking the 2-colimit in the 2-category of stacks of a 2-directed system $\{\mathcal{X}_i\}_{i \in I}$ of algebraic stacks over S.

By [Eme, Rem. 4.9], if $\mathcal{X} \to \mathcal{Y}$ is representable by algebraic spaces, and $\mathcal{Y}$ is an Ind-algebraic stack, then $\mathcal{X}$ is also an Ind-algebraic stack.

A.6 Definition. ([Eme, Defn. 5.3]) A *formal algebraic stack* over S is a stack $\mathcal{X}$ which admits a morphism $U \to \mathcal{X}$ which is representable by algebraic spaces, smooth, and surjective, and whose source is a formal algebraic space. If the source can be chosen to be locally countably indexed, then we say that $\mathcal{X}$ is locally countably indexed.

A.7 Definition. ([Eme, Defn. 7.6]) A *p-adic formal algebraic stack* is a formal algebraic stack $\mathcal{X}$ over $\operatorname{Spec} \mathbf{Z}_p$ which admits a morphism $\mathcal{X} \to \operatorname{Spf} \mathbf{Z}_p$ which is representable by algebraic stacks; so in particular, we may write $\mathcal{X}$ as an Ind-algebraic stack by writing $\mathcal{X} \cong \varinjlim_n \mathcal{X} \times_{\operatorname{Spf} \mathbf{Z}_p} \operatorname{Spec} \mathbf{Z}/p^n\mathbf{Z}$.

The relationship between formal algebraic stacks and Ind-algebraic stacks is discussed in detail in [Eme, §6]. Before explaining some of this material, we need to recall some preliminaries on finiteness properties and underlying reduced substacks.

Underlying reduced substacks

If $\mathcal{X}/S$ is any stack, then we let $(\mathcal{X}_{\mathrm{red}})'$ be the full subcategory of $\mathcal{X}$ whose set of objects consists of those $T \to \mathcal{X}$ for which T is a reduced S-scheme.

A.8 Definition. ([Eme, Lem. 3.27]) The *underlying reduced substack* $\mathcal{X}_{\mathrm{red}}$ of $\mathcal{X}$ is the intersection of all of the substacks of $\mathcal{X}$ which contain $(\mathcal{X}_{\mathrm{red}})'$.

A.9 Lemma. *([Eme, Lem. 4.16]) If $\mathcal{X}$ is an Ind-algebraic stack, and we write $\varinjlim_i \mathcal{X}_i \xrightarrow{\sim} \mathcal{X}$, for some 2-directed system $\{\mathcal{X}_i\}_{i\in\mathcal{I}}$ of algebraic stacks, then the induced morphism $\varinjlim_i (\mathcal{X}_i)_{\mathrm{red}} \to \mathcal{X}_{\mathrm{red}}$ is an isomorphism, and so in particular $\mathcal{X}_{\mathrm{red}}$ is again an Ind-algebraic stack.*

As the following lemma records, in the case that $\mathcal{X}$ is a formal algebraic stack, $\mathcal{X}_{\mathrm{red}}$ is algebraic, and (just as in the case of formal schemes, or formal algebraic spaces) $\mathcal{X}$ is a thickening of $\mathcal{X}_{\mathrm{red}}$.

A.10 Lemma. *([Eme, Lem. 5.26]) If $\mathcal{X}$ is is a formal algebraic stack over S, then $\mathcal{X}_{\mathrm{red}}$ is a closed and reduced algebraic substack of $\mathcal{X}$, and the inclusion $\mathcal{X}_{\mathrm{red}} \hookrightarrow \mathcal{X}$ is a thickening (in the sense that its base change over any algebraic space induces a thickening of algebraic spaces). Furthermore, any morphism $\mathcal{Y} \to \mathcal{X}$ with $\mathcal{Y}$ a reduced algebraic stack factors through $\mathcal{X}_{\mathrm{red}}$.*

Finiteness properties

Let $\mathcal{X}$ be a formal algebraic stack $\mathcal{X}$. We say that $\mathcal{X}$ is quasi-compact if the algebraic stack $\mathcal{X}_{\mathrm{red}}$ is quasi-compact. We say that $\mathcal{X}$ is quasi-separated if the diagonal morphism of $\mathcal{X}$ (which is automatically representable by algebraic spaces) is quasi-compact and quasi-separated.

We say that $\mathcal{X}$ is locally Noetherian if and only if it admits a morphism $U \to \mathcal{X}$ which is representable by algebraic spaces, smooth, and surjective, and whose source is a locally Noetherian formal algebraic space. Finally, we say that $\mathcal{X}$ is Noetherian if it is locally Noetherian, and it is quasi-compact and quasi-separated.

The relationship between Ind-algebraic and formal algebraic stacks

In one direction, we have the following lemma.

A.11 Lemma. *([Eme, Lem. 6.2]) If $\mathcal{X}$ is a quasi-compact and quasi-separated formal algebraic stack, then $\mathcal{X} \cong \varinjlim_i \mathcal{X}_i$ for a 2-directed system $\{\mathcal{X}_i\}_{i\in I}$ of quasi-compact and quasi-separated algebraic stacks in which the transition morphisms are thickenings.*

A partial converse to Lemma A.11 is proved as [Eme, Lem. 6.3]. In particular, we have the following useful criteria for being a formal algebraic stack.

A.12 Proposition. *([Eme, Cor. 6.6]) Suppose that $\mathcal{X}$ is an Ind-algebraic stack that can be written as the 2-colimit $\mathcal{X} \xrightarrow{\sim} \varinjlim \mathcal{X}_n$ of a directed sequence $(\mathcal{X}_n)_{n\geq 1}$ in which the $\mathcal{X}_n$ are algebraic stacks, and the transition morphisms are closed* *immersions. If $\mathcal{X}_{\mathrm{red}}$ is a quasi-compact algebraic stack, then $\mathcal{X}$ is a locally countably indexed formal algebraic stack.*

A.13 Proposition. *Suppose that $\{\mathcal{X}_n\}_{n\geq 1}$ is an inductive system of algebraic stacks over* $\operatorname{Spec} \mathbf{Z}_p$*, such that each $\mathcal{X}_n$ in fact lies over* $\operatorname{Spec} \mathbf{Z}/p^n\mathbf{Z}$*, and each of the induced morphisms $\mathcal{X}_n \to \mathcal{X}_{n+1} \times_{\operatorname{Spec} \mathbf{Z}/p^{n+1}\mathbf{Z}} \mathbf{Z}/p^n\mathbf{Z}$ is an isomorphism. Then the Ind-algebraic stack $\mathcal{X} := \varinjlim_n \mathcal{X}_n$ is a p-adic formal algebraic stack.*

Proof. This is a special case of [Eme, Lem. 6.3, Ex. 7.8]. □

Scheme-theoretic images

The notion of the scheme-theoretic image of a (quasi-compact) morphism of algebraic stacks is developed in [Sta, Tag 0CMH]; equivalent alternative presentations are given in [EG21, §3.1] and in [Eme, Ex. 9.9]. In [Eme, §6] this is extended to a definition of the scheme-theoretic image between formal algebraic stacks that are quasi-compact and quasi-separated. In fact, the definition there uses Lemma A.11 to write the formal stacks in question as Ind-algebraic stacks with transition morphisms being closed immersions, and this is the level of generality that is appropriate to us here.

A robust theory of scheme-theoretic images requires a quasi-compactness assumption on the morphism whose scheme-theoretic image is being formed, and we begin by introducing the notion of quasi-compactness which seems appropriate to our context.

A.14 Definition. Let $\mathcal{X}$ be an Ind-algebraic stack which can be written as a 2-colimit of quasi-compact algebraic stacks for which the transition morphisms are closed immersions, say $\mathcal{X} \cong \varinjlim \mathcal{X}_\lambda$. We say that a morphism of stacks $\mathcal{X} \to \mathcal{Y}$ is *Ind-representable by algebraic stacks* if each of the induced morphisms

$$\mathcal{X}_\lambda \to \mathcal{Y} \tag{A.15}$$

is representable by algebraic stacks. We say that such a morphism is furthermore *Ind-quasi-compact* if each of the morphisms $\mathcal{X}_\lambda \to \mathcal{Y}$ (representable by algebraic stacks by assumption) is quasi-compact.

Using the fact that any two descriptions of $\mathcal{X}$ as an Ind-algebraic stack as above (i.e., as the 2-colimit of quasi-compact algebraic stacks with respect to transition morphisms that are closed immersions) are mutually cofinal, one sees that the property of a morphism being Ind-representable by algebraic stacks (resp. Ind-representable by algebraic stacks and Ind-quasi-compact) is independent of the choice of such a description of $\mathcal{X}$.

Suppose now, in the context of the Definition A.14, that $\mathcal{Y}$ is also an Ind-algebraic stack, which admits a description as the 2-colimit of algebraic stacks with respect to transition morphisms that are closed immersions, say $\mathcal{Y} \cong \varinjlim \mathcal{Y}_\mu$. Then the induced morphism (A.15) factors through $\mathcal{Y}_\mu$ for some μ (since the $\mathcal{X}_\lambda$

are quasi-compact and the transition morphisms between the $\mathcal{Y}_\mu$ are monomorphisms). Since any morphism of algebraic stacks is representable by algebraic stacks, and since closed immersions are representable by algebraic stacks, we find that (A.15) is necessarily representable by algebraic stacks, and thus the morphism $\mathcal{X} \to \mathcal{Y}$ is necessarily Ind-representable by algebraic stacks. Furthermore, the morphism (A.15) is quasi-compact if and only if the induced morphism $\mathcal{X}_\lambda \to \mathcal{Y}_\mu$ is quasi-compact for some (or equivalently any) allowable choice of μ. For example, if the $\mathcal{Y}_\mu$ are all quasi-separated, then these induced morphisms are necessarily quasi-compact (via the usual graph argument, since each X_λ is quasi-compact), and so in this case any morphism $\mathcal{X} \to \mathcal{Y}$ is necessarily Ind-quasi-compact.

Suppose now that the morphism $\mathcal{X} \to \mathcal{Y}$ *is* Ind-quasi-compact, or, equivalently (as we have just explained), that the various induced morphisms $\mathcal{X}_\lambda \to \mathcal{Y}_\mu$ are quasi-compact. Then each of these induced morphisms has a scheme-theoretic image $\mathcal{Z}_\lambda$. One easily checks that $\mathcal{Z}_\lambda$, thought of as a closed substack of $\mathcal{Y}$, is independent of the particular choice of the index μ used in its definition. Evidently $\mathcal{Z}_\lambda$ is a closed substack of $\mathcal{Z}_{\lambda'}$ if $\lambda \leq \lambda'$. In particular, we may form the 2-colimit $\varinjlim \mathcal{Z}_\lambda$, which is an Ind-algebraic stack. There is a natural morphism $\varinjlim \mathcal{Z}_\lambda \to \mathcal{Y}$.

A.16 Definition. In the preceding context, we define the scheme-theoretic image of the Ind-quasi-compact morphism $\mathcal{X} \to \mathcal{Y}$ to be $\mathcal{Z} := \varinjlim \mathcal{Z}_\lambda$.

It is easily verified that the scheme-theoretic image, so defined, is independent of the chosen descriptions of $\mathcal{X}$ and $\mathcal{Y}$. There is a canonical monomorphism $\mathcal{Z} \hookrightarrow \mathcal{Y}$. (We don't claim that this monomorphism is necessarily a closed immersion in this level of generality.)

A very special case of this definition is the case that $\mathcal{X} \to \mathcal{Y}$ is representable by algebraic spaces and quasi-compact, and $\mathcal{Y} = \operatorname{Spf} A$ for a pro-Artinian ring A. In this case the definition of the scheme-theoretic image coincides with [EG21, Defn. 3.2.15] (i.e., the scheme-theoretic image can be computed via the scheme-theoretic images of the pull-backs of $\mathcal{X}$ to the discrete Artinian quotients of A). In this particular case the scheme-theoretic image *is* a closed formal subspace of $\mathcal{Y}$, i.e., of the form $\operatorname{Spf} B$ for some topological quotient B of A.

A.17 Definition. Let $\mathcal{X}$ and $\mathcal{Y}$ be Ind-algebraic stacks which satisfy the hypotheses introduced above, i.e., which may each be written as the 2-colimit with respect to closed immersions of algebraic stacks, which are furthermore quasi-compact in the case of $\mathcal{X}$. Then we say that an Ind-quasi-compact morphism $\mathcal{X} \to \mathcal{Y}$ is *scheme-theoretically dominant* if the induced map $\mathcal{Z} \to \mathcal{Y}$ is an isomorphism, where $\mathcal{Z}$ is the scheme-theoretic image of $\mathcal{X} \to \mathcal{Y}$.

A.18 Remark. If $\mathcal{X}$ is a quasi-compact and quasi-separated formal algebraic stack, then Lemma A.11 shows that $\mathcal{X} \cong \varinjlim \mathcal{X}_\lambda$ with the $\mathcal{X}_\lambda$ being quasi-compact and quasi-separated algebraic stacks, and the morphisms being thickenings, and so in particular closed immersions. The preceding discussion thus applies to morphisms $\mathcal{X} \to \mathcal{Y}$ of quasi-compact and quasi-separated formal algebraic stacks, and shows that such morphisms are necessarily Ind-representable by algebraic stacks and Ind-quasi-compact. In particular, Definition A.17 applies to such morphisms, and in this case recovers the definition of scheme-theoretic dominance given in [Eme, Def. 6.13].

A closely related context in which the preceding definition applies if the following: if A is an adic* topological ring, with finitely generated ideal of definition I, and $\mathcal{X}, \mathcal{Y} \to \operatorname{Spf} A$ are morphisms of quasi-compact formal algebraic stacks that are representable by algebraic stacks, then we may write $\mathcal{X} \cong \varinjlim_n \mathcal{X}_n$ and $\mathcal{Y} \cong \varinjlim_n \mathcal{Y}_n$, where $\mathcal{X}_n := \mathcal{X} \times_{\operatorname{Spf} A} \operatorname{Spec} A/I^n$ and $\mathcal{Y}_n := \mathcal{Y} \times_{\operatorname{Spf} A} \operatorname{Spec} A/I^n$ are quasi-compact algebraic stacks. A morphism $\mathcal{X} \to \mathcal{Y}$ of stacks over $\operatorname{Spf} A$ is then necessarily representable by algebraic stacks [Eme, Lem. 7.10], and so is Ind-quasi-compact if and only if it is quasi-compact in the usual sense. We may then define its scheme-theoretic image, following Definition A.16, or speak of such a morphism being scheme-theoretically dominant.

We now show (in Proposition A.21) that under certain hypotheses the scheme-theoretic image of a p-adic formal algebraic stack is also a p-adic formal algebraic stack. The deduction of this result from those of [Eme] will involve the following definition.

A.19 Definition. ([Eme, Defn. 8.26, Rem. 8.39]) We say that a formal algebraic stack $\mathcal{X}$ over a scheme S is *Ind-locally of finite type over* S if there exists an isomorphism $\mathcal{X} \xrightarrow{\sim} \varinjlim \mathcal{X}_i$, where each $\mathcal{X}_i$ is an algebraic stack locally of finite type over S, and the transition morphisms are thickenings.

A.20 Remark. If $\mathcal{X}$ is a quasi-compact and quasi-separated formal algebraic stack, then in order to verify that $\mathcal{X}$ is Ind-locally of finite type, it suffices to exhibit an isomorphism $\mathcal{X} \xrightarrow{\sim} \varinjlim \mathcal{X}_i$, where each $\mathcal{X}_i$ is an algebraic stack locally of finite type over S, and the transition morphisms are closed immersions (not necessarily thickenings) [Eme, Lem. 8.29].

A.21 Proposition. *Suppose that we have a commutative diagram*

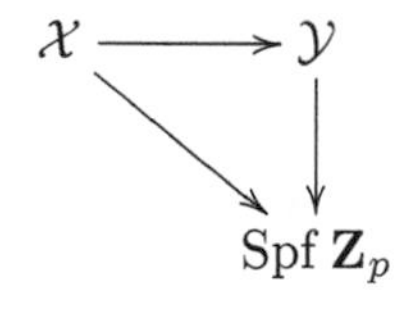

in which the diagonal arrow makes $\mathcal{X}$ into a p-adic formal algebraic stack of finite presentation, and where $\mathcal{Y}$ is an Ind-algebraic stack which can be written as the inductive limit of algebraic stacks, each of finite presentation over $\operatorname{Spec} \mathbf{Z}/p^a$ *for some $a \geq 1$, with the transition maps being closed immersions.*

Suppose also that the horizontal morphism $\mathcal{X} \to \mathcal{Y}$ (which, by the usual graph argument, is seen to be representable by algebraic stacks) is proper. Then the *scheme-theoretic image $\mathcal{Z}$ of this morphism is a p-adic formal algebraic stack of finite type. If $\mathcal{X}$ is flat over* $\operatorname{Spf} \mathbf{Z}_p$, *then so is $\mathcal{Z}$.*

Proof. We deduce this from [Eme, Prop. 10.5]. Note firstly that $\mathcal{X}$ is quasi-compact and quasi-separated, since it is of finite presentation over $\operatorname{Spf} \mathbf{Z}_p$. Examining the hypotheses of [Eme, Prop. 10.5], we need to show that $\mathcal{Z}$ is formal algebraic, that it is quasi-compact and quasi-separated, that it is Ind-locally of finite type over $\operatorname{Spec} \mathbf{Z}_p$, and that the induced morphism $\mathcal{X} \to \mathcal{Z}$ (which will be representable by algebraic stacks, by the usual graph argument) is proper.

By hypothesis, we can write $\mathcal{Y} \cong \varinjlim_\mu \mathcal{Y}_\mu$, where each $\mathcal{Y}_\mu$ is an algebraic stack of finite presentation over some $\mathbf{Z}/p^a$ (with a depending on μ), and the transition maps are closed immersions. If we write $\mathcal{X} \cong \varinjlim_a \mathcal{X}_a$, where $\mathcal{X}_a := \operatorname{Spec} \mathbf{Z}/p^a \otimes_{\operatorname{Spf} \mathbf{Z}_p} \mathcal{X}$, then by assumption each $\mathcal{X}_a$ is an algebraic stack. By definition, then, we have that

$$\mathcal{Z} := \varinjlim_a \mathcal{Z}_a, \tag{A.22}$$

where $\mathcal{Z}_a$ is the scheme-theoretic image of $\mathcal{X}_a \to \mathcal{Y}_\mu$, for μ sufficiently large. It follows from Lemma A.23 below that each of the closed immersions $\mathcal{Z}_a \hookrightarrow \mathcal{Z}_{a+1}$ is a finite order thickening. We conclude from [Eme, Lem. 6.3] that $\mathcal{Z}$ is in fact a formal algebraic stack.

Since $\mathcal{Y}_\mu$ is of finite presentation over $\operatorname{Spec} \mathbf{Z}_p$, each closed substack $\mathcal{Z}_a$ is of finite presentation over $\mathbf{Z}/p^a$, and in particular quasi-compact and quasi-separated; it follows that $\mathcal{Z}$ is quasi-compact and quasi-separated. The description (A.22) of $\mathcal{Z}$ furthermore exhibits $\mathcal{Z}$ as being Ind-locally of finite type over $\operatorname{Spec} \mathbf{Z}_p$.

It remains to show that $\mathcal{X} \to \mathcal{Z}$ is proper. If $T \to \mathcal{Z}$ is any morphism whose source is a scheme, then since $\mathcal{Z} \to \mathcal{Y}$ is a monomorphism, we find that there is an isomorphism $T \times_{\mathcal{Z}} \mathcal{X} \to T \times_{\mathcal{Y}} \mathcal{X}$. Since by assumption the target of this isomorphism is an algebraic stack, proper over T, the same is true of the source.

The claim regarding flatness follows from Lemma A.24 below. ◻

We used the following lemmas in the proof of the Proposition A.21.

A.23 Lemma. *Consider a commutative diagram of morphisms of algebraic stacks*

$$\begin{array}{ccc} \mathcal{X} & \longrightarrow & \mathcal{Z} \\ \downarrow & & \downarrow \\ \mathcal{X}' & \longrightarrow & \mathcal{Z}' \end{array}$$

with $\mathcal{Z}'$ being quasi-compact, in which the left-hand side vertical arrow is an nth order thickening for some $n \geq 1$, the right-hand side vertical arrow is a closed immersion, and the lower horizontal arrow is quasi-compact, surjective, and *scheme-theoretically dominant. Then the right-hand side vertical arrow is then also an nth order thickening.*

Proof. Since $\mathcal{X} \to \mathcal{X}'$ and $\mathcal{X}' \to \mathcal{Z}'$ are surjective, by assumption, the same is true of their composite $\mathcal{X} \to \mathcal{Z}'$, and hence of the closed immersion $\mathcal{Z} \to \mathcal{Z}'$; thus this closed immersion is a thickening. To see that it is of order n, we first note that since $\mathcal{Z}'$ is quasi-compact, we find a smooth surjection $U' \to \mathcal{Z}'$ whose source is an affine scheme; since the property of a thickening being of finite order may be checked *fppf* locally, and since the property of being quasi-compact and scheme-theoretically dominant is also preserved by flat base change, after pulling back our diagram over U', we may assume that $\mathcal{Z}' = U'$ is an affine scheme, and that $\mathcal{Z} = U$ is a closed subscheme. Let $\mathcal{I}$ be the ideal sheaf on U' cutting out U, and let a be a section of $\mathcal{I}$. If a' denotes the pull-back of a to $\mathcal{X}'$, then $a'_{|\mathcal{X}} = 0$, and so $(a')^n = 0$, by assumption. Since $\mathcal{X}' \to U'$ is scheme-theoretically dominant, we find that $a^n = 0$. Thus $U \hookrightarrow U'$ is indeed an nth order thickening. □

A.24 Lemma. *If $\mathcal{X} \to \mathcal{Y}$ is a quasi-compact scheme-theoretically dominant morphism of p-adic formal algebraic stacks which are locally of finite type, and if $\mathcal{X}$ is flat over $\mathbf{Z}_p$, then the same is true of $\mathcal{Y}$.*

Proof. Let $V \to \mathcal{Y}$ be a morphism which is representable by algebraic spaces and smooth, whose source is an affine formal algebraic space; so $V = \operatorname{Spf} B$ for some p-adically complete $\mathbf{Z}_p$-algebra B that is topologically of finite type. It suffices to show that B is flat over $\mathbf{Z}_p$. Since $V \to \mathcal{Y}$ is in particular flat (being smooth), the base-changed morphism $\mathcal{X} \times_{\mathcal{Y}} V \to V$ is again scheme-theoretically dominant. Since it is also quasi-compact, and since V is formally affine (and so quasi-compact), we may find a formal algebraic space $U = \operatorname{Spf} A$ endowed with a morphism $U \to \mathcal{X} \times_{\mathcal{Y}} V$ which is representable by algebraic spaces, smooth, and surjective; and thus also scheme-theoretically dominant. Thus we may replace our original situation with the composite morphism $\operatorname{Spf} A \to \operatorname{Spf} B$. But in this context, scheme-theoretic dominance amounts to the morphism $B \to A$ being injective; thus B is $\mathbf{Z}_p$-flat if A is. □

Versality and versal rings

We will sometimes find it useful to study scheme-theoretic images in terms of versal rings, and so we will recall some notation and results from [EG21, §2.2] related to this topic.

If Λ is a Noetherian ring, equipped with a finite ring map $\Lambda \to k$ whose target is a field, then we let $\mathcal{C}_\Lambda$ denote the category whose objects are Artinian local Λ-algebras A equipped with an isomorphism of Λ-algebras $A/\mathfrak{m}_A \xrightarrow{\sim} k$. We let pro-$\mathcal{C}_\Lambda$ be the corresponding category of formal pro-objects, which (via passage to projective limits) we identify with the category of topological pro-(discrete Artinian) local Λ-algebras A equipped with a Λ-algebra isomorphism

$A/\mathfrak{m}_A \xrightarrow{\sim} k$. By [EG21, Rem. 2.27], any morphism $A \to B$ in pro-$\mathcal{C}_\Lambda$ has closed image, and induces a topological quotient map from its source to its image, so that in particular $A \to B$ is surjective if and only if it is induced by a compatible system of surjective morphisms in $\mathcal{C}_\Lambda$.

Fix a locally Noetherian base scheme S, and let k be a finite type $\mathcal{O}_S$-field, i.e., k is a field equipped with a morphism $\operatorname{Spec} k \to S$ of finite type. Choose, as we may, an affine open subscheme $\operatorname{Spec} \Lambda \subseteq S$ for which $\operatorname{Spec} k \to S$ factors as $\operatorname{Spec} k \to \operatorname{Spec} \Lambda \to S$, with $\Lambda \to k$ being finite. (In what follows we fix such a choice of Λ, although the notions that we define in terms of it are independent of this choice.) We now define a category fibred in groupoids $\widehat{\mathcal{F}}_x$ in the following way; this category is an example of a deformation category in the sense of [Sta, Tag 06J9]. In the notation of [Sta, Tag 07T2], our category $\widehat{\mathcal{F}}_x$ is denoted $\mathcal{F}_{\mathcal{F},k,x}$ (with a slightly unfortunate clash of notation in the two instances of $\mathcal{F}$).

A.25 Definition. If $\mathcal{F}$ is a category fibred in groupoids over S, and if $x\colon \operatorname{Spec} k \to \mathcal{F}$ is a morphism, then we define a category $\widehat{\mathcal{F}}_x$, cofibred in groupoids over $\mathcal{C}_\Lambda$, as follows: For any object A of $\mathcal{C}_\Lambda$, the objects of $\widehat{\mathcal{F}}_x(A)$ consist of morphisms $y\colon \operatorname{Spec} A \to \mathcal{F}$, together with an isomorphism $\alpha\colon x \xrightarrow{\sim} \overline{y}$ compatible with the given identification of $A/\mathfrak{m}$ with k; here $\overline{y}$ denotes the induced morphism $\operatorname{Spec} A/\mathfrak{m} \to \mathcal{F}$. The set of morphisms between two objects (y,α) and (y',α') of $\widehat{\mathcal{F}}_x(A)$ consists of the subset of morphisms $\beta\colon y \to y'$ in $\mathcal{F}(A)$ for which $\alpha' \circ \overline{\beta} = \alpha$; here $\overline{\beta}$ denotes the morphism $\overline{y} \to \overline{y}'$ induced by β. If $A \to B$ is a morphism in $\mathcal{C}_\Lambda$, then the corresponding pushforward $\widehat{\mathcal{F}}_x(A) \to \widehat{\mathcal{F}}_x(B)$ is defined by pulling back morphisms to $\mathcal{F}$ along the corresponding morphism of schemes $\operatorname{Spec} B \to \operatorname{Spec} A$.

If $\mathcal{F}$ is a category fibred in groupoids over the locally Noetherian scheme S, and if $x\colon \operatorname{Spec} k \to \mathcal{F}$ is a k-valued point of $\mathcal{F}$, for some finite type $\mathcal{O}_S$-field, then the notion of a versal ring to $\mathcal{F}$ at x is defined in [EG21, Def. 2.2.9]. Rather than recalling that definition here, we will give a definition of versality with a greater level of generality that is convenient for us. We will then explain how the notion of versal ring is obtained as a particular case.

A.26 Definition. Let $f\colon \mathcal{F} \to \mathcal{G}$ be a morphism of categories fibred in groupoids over the locally Noetherian scheme S, let k be a finite type $\mathcal{O}_S$-field, let $x\colon \operatorname{Spec} k \to \mathcal{F}$ be a k-valued point of $\mathcal{F}$, and let $y\colon \operatorname{Spec} k \to \mathcal{G}$ be a k-valued point of $\mathcal{G}$ equipped with an isomorphism of k-valued points $\alpha\colon f \circ x \xrightarrow{\sim} y$. The morphism f induces in an evident way a morphism $\widehat{f}_x\colon \widehat{\mathcal{F}}_x \to \widehat{\mathcal{G}}_y$ of categories cofibred in groupoids. We say that f is *versal* at x if the morphism $\widehat{f}_x$ is *smooth* in the sense of [Sta, Tag 06HG]; that is, given a commutative diagram

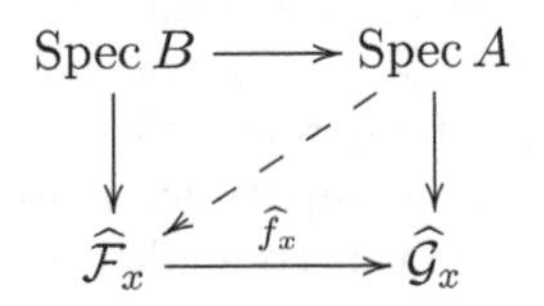

in which the upper arrow is the closed immersion corresponding to a surjection $A \to B$ in $\mathcal{C}_\Lambda$, we can fill in the dotted arrow so that the diagram remains commutative. (Clearly this notion is independent of the particular choice of y and α; more precisely it holds for any such choice if it holds for one such choice, such as $y = f \circ x$ and $\alpha = \mathrm{id}$.)

A.27 Example. Let S be a locally Noetherian scheme, let k be a finite type $\mathcal{O}_S$-field k, and let A be an object of pro-$\mathcal{C}_\Lambda$; recall in particular then that A comes equipped with a chosen isomorphism $\operatorname{Spec} A/\mathfrak{m} \xrightarrow{\sim} k$. We let x': $\operatorname{Spec} k \xrightarrow{\sim} \operatorname{Spec} A/\mathfrak{m} \hookrightarrow \operatorname{Spf} A$ denote the induced k-valued point of $\operatorname{Spf} A$.

Now let $\mathcal{F}$ be a category fibred in groupoids over S, let $x\colon \operatorname{Spec} k \to \mathcal{F}$ be a k-valued point of $\mathcal{F}$, and suppose that $f\colon \operatorname{Spf} A \to \mathcal{F}$ is a morphism, for which we can find an isomorphism $f \circ x' \xrightarrow{\sim} x$ of k-valued points of $\mathcal{F}$. Then the morphism f is versal at the point x' if and only if A is a versal ring to the k-valued point x of $\mathcal{F}$ in the sense of [EG21, Def. 2.2.9]

A.28 Example. Let S be a locally Noetherian scheme, let U be a locally finite type S-scheme, and let $f : U \to \mathcal{F}$ be a morphism to a category fibred in groupoids over S. If $u \in U$ is a finite type point, giving rise to a morphism $\operatorname{Spec} \kappa(u) \to U$, then f is versal at this $\kappa(u)$-valued point of U if and only if f is versal at the point u in the sense of [EG21, Def. 2.4.4 (1)]; cf. [EG21, Rem. 2.4.5].

A.29 Remark. If $\mathcal{X}$ is an algebraic stack which is locally of finite presentation over a locally Noetherian scheme S, then it admits (effective, Noetherian) versal rings at all finite type points [Sta, Tag 0DR1]. If X is an Ind-locally finite type algebraic space over S, then it admits (canonical) versal rings at all finite type points, by [EG21, Lem. 4.2.14]. We don't prove a general statement about the existence of versal rings for Ind-algebraic stacks, since in all the cases we consider in the body of the book we are able to construct them "explicitly" (for example, in terms of Galois lifting rings, or lifting rings for étale φ-modules).

The following lemma and its proof are essentially [EG21, Lem. 3.2.16], but since the setup there is different, we give the details here. Before stating the lemma, we introduce the setup. Let S be a locally Noetherian scheme. We suppose we are given a morphism $\mathcal{X} \to \mathcal{Y}$ of stacks over S which is representable by algebraic stacks and proper, and that $\mathcal{Y}$ is an Ind-algebraic stack which may be written as the 2-colimit of algebraic stacks which are of finite presentation over S (and so in particular quasi-compact and quasi-separated) with respect to transition morphisms that are closed immersions, say $\mathcal{Y} \cong \varinjlim \mathcal{Y}_\lambda$. Then $\mathcal{X}_\lambda := \mathcal{X} \times_{\mathcal{Y}} \mathcal{Y}_\lambda$ is an algebraic stack, and the projection $\mathcal{X}_\lambda \to \mathcal{Y}_\lambda$ is proper, so that $\mathcal{X}_\lambda$ is of finite type over S (and in particular quasi-compact, and also quasi-separated, although we won't use this latter fact). Furthermore, we have an induced isomorphism $\mathcal{X} \cong \varinjlim_\lambda \mathcal{X}_\lambda$, and so $\mathcal{X} \to \mathcal{Y}$ is a morphism of Ind-algebraic stacks whose scheme-theoretic image $\mathcal{Z}$ may be defined. Of course, in this context, since the Ind-structures on $\mathcal{X}$ and $\mathcal{Y}$ are compatible, if we let $\mathcal{Z}_\lambda$ denote the scheme-theoretic image of $\mathcal{X}_\lambda$ in $\mathcal{Y}_\lambda$, then this coincides with the scheme-theoretic image of $\mathcal{X}_\lambda$ in $\mathcal{Y}_{\lambda'}$ for any $\lambda' \geq \lambda$, and we may write $\mathcal{Z} \cong \varinjlim_\lambda \mathcal{Z}_\lambda$. Note that $\mathcal{Z}_\lambda$ is also of finite presentation over S, by Lemma 4.5.10.

A.30 Lemma. *Suppose that we are in the preceding situation, so that $\mathcal{X} \to \mathcal{Y}$ is a morphism of stacks over a locally Noetherian base S which is representable by algebraic stacks and proper, where $\mathcal{Y}$ is an Ind-algebraic stack which may be written as the 2-colimit of algebraic stacks which are of finite presentation over S, with respect to transition morphisms that are closed immersions, and we write $\mathcal{Z}$ for the scheme-theoretic image of $\mathcal{X} \to \mathcal{Y}$.*

Suppose that $x\colon \operatorname{Spec} k \to \mathcal{Z}$ is a finite type point, and that $\operatorname{Spf} A_x \to \mathcal{Y}$ is a versal morphism for the composite $x\colon \operatorname{Spec} k \to \mathcal{Z} \hookrightarrow \mathcal{Y}$. Let $\operatorname{Spf} B_x$ be the scheme-theoretic image of $\mathcal{X}_{\operatorname{Spf} A_x} \to \operatorname{Spf} A_x$. Then the morphism $\operatorname{Spf} B_x \to \mathcal{Y}$ factors through a versal morphism $\operatorname{Spf} B_x \to \mathcal{Z}$.

Proof. By definition, we may write $A_x = \varprojlim A_i$, $B_x = \varprojlim B_i$, where the A_i are objects of $\mathcal{C}_\Lambda$, and $\operatorname{Spec} B_i$ is the scheme-theoretic image of $\mathcal{X}_{A_i} \to \operatorname{Spec} A_i$. As explained immediately above, we write $\mathcal{X} \cong \varinjlim \mathcal{X}_\lambda$, $\mathcal{Y} \cong \varinjlim \mathcal{Y}_\lambda$, and $\mathcal{Z} \cong \varinjlim \mathcal{Z}_\lambda$, where the transition morphisms are closed immersions of algebraic stacks, and $\mathcal{Z}_\lambda$ is the scheme-theoretic image of the morphism $\mathcal{X}_\lambda \to \mathcal{Y}_\lambda$; in particular, the morphism $\mathcal{X}_\lambda \to \mathcal{Z}_\lambda$ is proper and scheme-theoretically dominant.

If follows that for each i, the composite $\operatorname{Spec} B_i \to \operatorname{Spec} A_i \to \mathcal{Y}$ factors through $\mathcal{Z}$. Indeed, for λ sufficiently large the morphism $(\mathcal{X}_\lambda)_{B_i} \to \operatorname{Spec} B_i$ is scheme-theoretically surjective, and the morphism $(\mathcal{X}_\lambda)_{B_i} \to \mathcal{Y}$ factors through $\mathcal{Z}_\lambda$, so the morphism $\operatorname{Spec} B_i \to \mathcal{Y}$ also factors through $\mathcal{Z}_\lambda$.

We now show that a morphism $\operatorname{Spec} A \to \operatorname{Spf} A_x$, with A an object of $\mathcal{C}_\Lambda$, factors through $\operatorname{Spf} B_x$ if and only if the composite $\operatorname{Spec} A \to \operatorname{Spf} A_x \to \mathcal{Y}$ factors through $\mathcal{Z}$. In one direction, if $\operatorname{Spec} A \to \mathcal{Y}$ factors through $\operatorname{Spf} B_x$, then it factors through $\operatorname{Spec} B_i$ for some i, and hence through $\mathcal{Z}$, as we saw above.

Conversely, if the composite $\operatorname{Spec} A \to \operatorname{Spf} A_x \to \mathcal{Y}$ factors through $\mathcal{Z}$, then we claim that there exists factorization $\operatorname{Spec} A \to \operatorname{Spec} B \to \mathcal{Y}$, where B is an object of $\mathcal{C}_\Lambda$, the morphism $\operatorname{Spec} A \to \operatorname{Spec} B$ is a closed immersion, and $\mathcal{X}_B \to \operatorname{Spec} B$ is scheme-theoretically dominant. To see this, note firstly that the composite $\operatorname{Spec} A \to \mathcal{Y}$ factors through $\mathcal{Z}_\lambda$ for some λ. Since $\mathcal{Z}_\lambda$ is an algebraic stack which is locally of finite presentation over a locally Noetherian base, it admits effective Noetherian versal rings by Remark A.29, so that $\operatorname{Spec} A \to \mathcal{Z}_\lambda$ factors through a versal morphism $\operatorname{Spec} C_x \to \mathcal{Z}_\lambda$ at the finite type point of $\mathcal{Z}_\lambda$ induced by x. By [EG21, Lem. 1.6.3], we can find a factorization $\operatorname{Spec} A \to \operatorname{Spec} R_x \to \operatorname{Spec} C_x$, where R_x is complete local Noetherian, $\operatorname{Spec} R_x \to \operatorname{Spec} C_x$ is faithfully flat, and $\operatorname{Spec} A \to \operatorname{Spec} R_x$ is a closed immersion. Since $\operatorname{Spec} R_x \to \operatorname{Spec} C_x$ is faithfully flat, and $\operatorname{Spec} C_x \to \mathcal{Z}_\lambda$ is flat by [Sta, Tag 0DR2], we see that the base-changed morphism $(\mathcal{X}_\lambda)_{R_x} \to \operatorname{Spec} R_x$ is scheme-theoretically dominant. By [EG21, Lem. 3.2.4] (applied to the proper morphism of algebraic stacks $\mathcal{X}_\lambda \to \mathcal{Z}_\lambda$), R_x admits a cofinal collection of Artinian quotients R_i for which $(\mathcal{X}_\lambda)_{R_i} \to \operatorname{Spec} R_i$ is scheme-theoretically dominant. The closed immersion $\operatorname{Spec} A \to \operatorname{Spec} R_x$ factors through $\operatorname{Spec} R_i$ for some R_i, so we may take $B = R_i$.

Now, by the versality of $\mathrm{Spf}\, A_x \to \mathcal{Y}$, we may lift the morphism $\mathrm{Spec}\, B \to \mathcal{Y}$ to a morphism $\mathrm{Spec}\, B \to \mathrm{Spf}\, A_x$, which furthermore we may factor as $\mathrm{Spec}\, B \to \mathrm{Spec}\, A_i \to \mathrm{Spf}\, A_x$, for some value of i. Since $\mathcal{X}_B \to \mathrm{Spec}\, B$ is scheme-theoretically dominant, the morphism $\mathrm{Spec}\, B \to \mathrm{Spec}\, A_i$ then factors through $\mathrm{Spec}\, B_i$, and thus through $\mathrm{Spf}\, B_x$, as required.

It follows in particular that the composite $\mathrm{Spf}\, B_x \to \mathrm{Spf}\, A_x \to \mathcal{Y}$ factors through a morphism $\mathrm{Spf}\, B_x \to \mathcal{Z}$. It remains to check that this morphism is versal. This is formal. Suppose we are given a commutative diagram

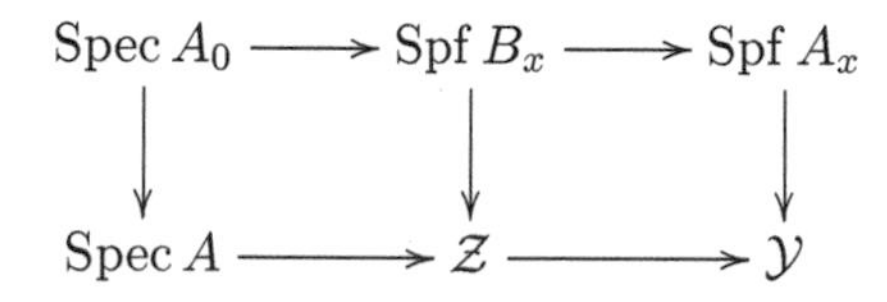

where the left-hand side vertical arrow is a closed immersion, and A_0, A are objects of $\mathcal{C}_\Lambda$. By the versality of $\mathrm{Spf}\, A_x \to \mathcal{Y}$, we may lift the composite $\mathrm{Spec}\, A \to \mathcal{Y}$ to a morphism $\mathrm{Spec}\, A \to \mathrm{Spf}\, A_x$. Since the composite $\mathrm{Spec}\, A \to \mathrm{Spf}\, A_x \to \mathcal{Y}$ factors through $\mathcal{Z}$, the morphism $\mathrm{Spec}\, A \to \mathrm{Spf}\, A_x$ then factors through $\mathrm{Spf}\, B_x$, as required. ☐

The following lemma records a useful property of versal rings in the Noetherian context.

A.31 Lemma. *Let $\mathcal{X}$ is a locally Noetherian formal algebraic stack, let R be a complete Noetherian local ring with residue field k, and let $f\colon \mathrm{Spf}\, R \to \mathcal{X}$ be a morphism for which the induced morphism $x\colon \mathrm{Spec}\, k \to \mathcal{X}_{\mathrm{red}}$ is a finite type point, and which is versal to $\mathcal{X}$ at x. Then the morphism f is flat, in the sense of* [Eme, Def. 8.42].

Proof. Since $\mathcal{X}$ is locally Noetherian, by assumption, we may find a morphism $\mathrm{Spf}\, A \to \mathcal{X}$ whose source is a Noetherian affine formal algebraic space, which is representable by algebraic spaces and smooth, and whose image contains the point x. By definition, we have to show that the base-changed morphism $\mathrm{Spf}\, A \times_{\mathcal{X}} \mathrm{Spf}\, R \to \mathrm{Spf}\, A$ is flat.

The fiber product $\mathrm{Spf}\, A \times_{\mathcal{X}} \mathrm{Spf}\, R$ is a formal algebraic space (which is locally Noetherian, since $\mathrm{Spf}\, R$ is so [Sta, Tag 0AQ7]), and so admits a morphism $\coprod \mathrm{Spf}\, B_i \to \mathrm{Spf}\, A \times_{\mathcal{X}} \mathrm{Spf}\, R$ whose source is the disjoint union of Noetherian affine formal algebraic spaces, and which is representable by algebraic spaces, étale, and surjective. Again, it suffices to show that each of the morphisms $\mathrm{Spf}\, B_i \to \mathrm{Spf}\, A$ is flat, which by definition is equivalent to each of the induced morphisms $A \to B_i$ being flat. If we let I denote an ideal of definition of the topology on A, then it suffices to show that each morphism $A/I^n \to B_i/I^n$ is flat (see, e.g., the proof of [Eme, Lem. 8.18]).

For this, it suffices in turn to show, for each maximal ideal $\mathfrak{n}$ of B_i, that the induced morphism $A/I^n \to (B_i/I^n)\hat{}$ is flat (where $(-)\hat{}$ denotes $\mathfrak{n}$-adic completion; note that if $\mathfrak{m}$ denotes the maximal ideal of R, then the topology on B_i is the $\mathfrak{m}$-adic topology, so that any maximal ideal $\mathfrak{n}$ of B_i contains $\mathfrak{m}$, and also contains IB). Now the induced morphism $\operatorname{Spf} \widehat{B}_i \to \operatorname{Spf} A$ is versal to the induced morphism $\operatorname{Spec} B_i/\mathfrak{n} \to \operatorname{Spf} A$ (as one sees by chasing through the constructions, beginning from the versality of $\operatorname{Spf} R \to \mathcal{X}$, and taking into account the infinitesimal lifting property for étale morphisms), and thus the induced morphism $\operatorname{Spf}(B_i/I^n)\hat{} \to \operatorname{Spec} A/I^n$ is also versal (to the induced morphism $\operatorname{Spec} B_i/\mathfrak{n} \to \operatorname{Spec} A/I^n$). Thus this morphism *is* flat, e.g., by [Sta, Tag 0DR2], and the lemma is proved. ☐

We will frequently find the following lemma useful.

A.32 Lemma. *Let $R \to S$ be a continuous surjection of objects in* pro-$\mathcal{C}_\Lambda$*, and let $X \to \operatorname{Spf} R$ be a finite type morphism of formal algebraic spaces.*

Make the following assumption: if A is any finite type Artinian local R-algebra for which the canonical morphism $R \to A$ factors through a discrete quotient of R, and for which the canonical morphism $X_A \to \operatorname{Spec} A$ admits a section, then the canonical morphism $R \to A$ furthermore factors through S.

Then the scheme-theoretic image of $X \to \operatorname{Spf} R$ is a closed formal subscheme of $\operatorname{Spf} S$.

Proof. Writing $\operatorname{Spf} T$ for the scheme-theoretic image of $X \to \operatorname{Spf} R$, we need to show that the surjection $R \to T$ factors through S. By definition, we can write T as an inverse limit of discrete Artinian quotients B for which the canonical morphism $X_B \to \operatorname{Spec} B$ is scheme-theoretically dominant. It follows from [EG21, Lem. 5.4.15] that for any such B, the surjection $R \to B$ factors through S; so the surjection $R \to T$ factors through S, as required. ☐

Immersed substacks

The following lemma is often useful for studying morphisms from a stack in terms of morphisms from a cover of the stack.

A.33 Lemma. *[Eme, Lem. 3.18, 3.19] Suppose that $\mathcal{X}$ is a stack over a base scheme S, and that $U \to \mathcal{X}$ is a morphism over S which is representable by algebraic spaces and whose source is a sheaf. Write $R := U \times_{\mathcal{X}} U$.*

1. *Suppose that the projections $R := U \times_{\mathcal{X}} U \rightrightarrows U$ (which are again representable by algebraic spaces, being the base change of morphisms which are so representable) are flat and locally of finite presentation. Then the morphism $U \to \mathcal{X}$ induces a monomorphism $[U/R] \to \mathcal{X}$.*
2. *Suppose that $U \to \mathcal{X}$ is flat, surjective, and locally of finite presentation. Then the morphism $U \to \mathcal{X}$ induces an isomorphism $[U/R] \xrightarrow{\sim} \mathcal{X}$.*

In particular, if $\mathcal{X}$ is a formal algebraic stack (over some base scheme S), then by definition there exists a morphism $U \to \mathcal{X}$ which is representable by algebraic spaces, smooth, and surjective, and whose source U is a formal algebraic space. In this case $R := U \times_{\mathcal{X}} U$ is a formal algebraic space which is endowed in a natural way with the structure of a groupoid in formal algebraic spaces over U, and (since smooth morphisms are in particular flat), by Lemma A.33, there is an isomorphism of stacks $[U/R] \xrightarrow{\sim} \mathcal{X}$.

Suppose now that $\mathcal{Z} \hookrightarrow \mathcal{X}$ is an immersion (in the sense that it is representable by algebraic spaces, and pulls back to an immersion over any test morphism $T \to \mathcal{X}$ whose source is a scheme). The induced morphism $W := U \times_{\mathcal{X}} \mathcal{Z} \hookrightarrow U$ is then an immersion of formal algebraic spaces, and W is R-invariant, in (an evident generalization of) the sense of [Sta, Tag 044F]. Conversely, if W is an R-invariant locally closed formal algebraic subspace of U, and if we write $R_W := R \times_U W = W \times_U R$ for the restriction of R to W, then the induced morphism $[W/R_W] \to [U/R] \xrightarrow{\sim} \mathcal{X}$ is an immersion. (In the context of algebraic stacks, this is [Sta, Tag 04YN].)

Flat parts

Let $\mathcal{O}$ be the ring of integers in a finite extension of $\mathbf{Q}_p$, and let $\mathcal{X} \to \operatorname{Spf} \mathcal{O}$ be a p-adic formal algebraic stack which is locally of finite type over $\operatorname{Spf} \mathcal{O}$. Then by [Eme, Ex. 9.11], there is a closed substack $\mathcal{X}_{\mathrm{fl}}$ of $\mathcal{X}$, the *flat part* of $\mathcal{X}$, which is the maximal substack of $\mathcal{X}$ which is flat over $\operatorname{Spf} \mathcal{O}$.

More precisely, $\mathcal{X}_{\mathrm{fl}} \to \operatorname{Spf} \mathcal{O}$ is flat, and if $\mathcal{Y} \to \mathcal{X}$ is a morphism of locally Noetherian formal algebraic stacks for which the composite $\mathcal{Y} \to \operatorname{Spf} \mathcal{O}$ is flat, then $\mathcal{Y} \to \mathcal{X}$ factors through $\mathcal{X}_{\mathrm{fl}}$.

Obstruction theory

Let $\mathcal{X}$ be a limit preserving Ind-algebraic stack over a locally Noetherian scheme S. If $x\colon \operatorname{Spec} A \to \mathcal{X}$ is a morphism for which the composite morphism $\operatorname{Spec} A \to S$ factors through an affine open subscheme of S, then in [Sta, Tag 07Y9] there is defined a functor T_x from the category of A-modules to itself, whose formation is also functorial in the pair (x, A) [Sta, Tag 07YA], and such that for any finitely generated A-module M, there is a natural identification of $T_x(M)$ with the set of lifts of x to morphisms $x'\colon \operatorname{Spec} A[M] \to \mathcal{X}$ (where $A[M]$ denotes the square zero extension of A by M). (These definitions apply to $\mathcal{X}$ by [Eme, Lem. 4.22].)

We make the following definition, which is a special case of [Eme, Defn. 11.6]; that definition incorporates an auxiliary module in its definition for technical reasons, but in our applications this module is zero, so we have suppressed it here. We have also incorporated [Eme, Rem. 11.9], which allows us to restrict to the case of finitely generated A-modules M.

A.34 Definition. We say that $\mathcal{X}$ admits a *nice obstruction theory* if, for each finite type S-algebra A which lies over an affine open subscheme of S, equipped with a morphism $x\colon \operatorname{Spec} A \to \mathcal{X}$, there exists a complex of A-modules $K^{\bullet}_{(x,A)}$ such that the following conditions are satisfied:

1. The complex $K^{\bullet}_{(x,A)}$ is bounded above and has finitely generated cohomology modules (or, equivalently, is isomorphic in the derived category to a bounded above complex of finitely generated A-modules).
2. The formation of $K^{\bullet}_{(x,A)}$ is compatible (in the derived category) with pull-back. More precisely, if $f\colon \operatorname{Spec} B \to \operatorname{Spec} A$, inducing the morphism $y\colon \operatorname{Spec} B \to \operatorname{Spec} A \to \mathcal{X}$, then there is a natural isomorphism in the derived category $f^* K^{\bullet}_{(x,A)} \xrightarrow{\sim} K^{\bullet}_{(y,B)}$.
3. For any pair (x, A) as above, and for any finitely generated A-module M, we have an isomorphism $T_x(M) \xrightarrow{\sim} H^1(K^{\bullet}_{(x,A)} \otimes^{\mathbf{L}} M)$ whose formation is functorial in M and in the pair (x, A).
4. For any pair (x, A) as above, and for any finitely generated A-module M, the cohomology module $H^2(K^{\bullet}_{(x,A)} \otimes^{\mathbf{L}} M)$ serves as an obstruction module. In particular, for any square zero extension

$$0 \to I \to A' \to A \to 0$$

for which A' is of finite type over S (or, equivalently, for which I is a finite A-module), we have a functorial obstruction element $o_x(A') \in H^2(K^{\bullet}_{(x,A)} \otimes^{\mathbf{L}} I)$, which vanishes if and only if x can be lifted to $\operatorname{Spec} A'$.

The following result generalizes the familiar fact that a pro-representing object for a formal deformation ring whose tangent space is finite dimensional is necessarily a complete Noetherian local ring.

A.35 Theorem. *([Eme, Thm. 11.13]) If $\mathcal{X}$ is a locally countably indexed and Ind-locally of finite type formal algebraic stack over a locally Noetherian scheme S, and if $\mathcal{X}$ admits a nice obstruction theory, then $\mathcal{X}$ is locally Noetherian.*

(When comparing this result with the statement of [Eme, Thm. 11.13], the reader should bear in mind that an Ind-locally of finite type formal algebraic stack over S is in particular the 2-colimit of algebraic stacks that are locally of finite type (or equivalently, locally of finite presentation) over S, and hence is limit preserving. Also, [Eme, Lem. 8.38] ensures that an Ind-locally of finite type formal algebraic stack over S is also locally of Ind-locally finite type—which is the condition that appears in [Eme, Thm. 11.13].)

Appendix B

Graded modules and rigid analysis

In this appendix we establish some (mostly straightforward) results that combine graded ring techniques with various results related to completions that are of a rigid analytic flavor.

B.1 ASSOCIATED GRADED ALGEBRAS AND MODULES

Let R be an Artinian local ring, with maximal ideal I and residue field k.

B.2 Definition. If M is an R-module, then we let $\mathrm{Gr}^\bullet M$ denote the graded k-vector space associated to the I-adic filtration on M (so $\mathrm{Gr}^i M := I^i M / I^{i+1} M$).

The formation of $\mathrm{Gr}^\bullet M$ is functorial in M. If M and N are two R-modules, then there is a natural surjection (of graded k-vector spaces)

$$\mathrm{Gr}^\bullet M \otimes_k \mathrm{Gr}^\bullet N \to \mathrm{Gr}^\bullet (M \otimes_R N). \tag{B.3}$$

In particular, it follows that if A is an R-algebra, then $\mathrm{Gr}^\bullet A$ is naturally a graded k-algebra, and if M is an A-module, then $\mathrm{Gr}^\bullet M$ is naturally a $\mathrm{Gr}^\bullet A$-module.

We recall the following basic lemmas.

B.4 Lemma. *If A is an R-algebra, then A is Noetherian if and only if $\mathrm{Gr}^\bullet A$ is Noetherian.*

Proof. This is easy and standard; see, e.g., [ST03, Prop. 1.1] for a statement of the "if" direction of this result (which is the less obvious of the two directions) in a significantly more general setting. □

B.5 Lemma. *If A is an R-algebra and if M is an A-module, then the natural morphism $\mathrm{Gr}^\bullet A \otimes_{\mathrm{Gr}^0 A} \mathrm{Gr}^0 M \to \mathrm{Gr}^\bullet M$ is surjective. Furthermore, M is nonzero if and only if $\mathrm{Gr}^0 M$ is nonzero, if and only if $\mathrm{Gr}^\bullet M$ is nonzero.*

Proof. The first claim is immediate. This implies in turn that if $\mathrm{Gr}^0 M$ is zero then $\mathrm{Gr}^\bullet M$ is zero (the converse being evident), and thus that M is zero (since

$\mathrm{Gr}^\bullet M$ is the associated graded of M with respect to a finite length filtration beginning at M and ending at 0). (Essentially equivalently, one can observe directly that since I is nilpotent, if $M/IM=0$ then $M=0$.) □

B.6 Lemma. *If A is an R-algebra and M is an A-module, then M is finitely generated over A if and only if $\mathrm{Gr}^0 M$ is finitely generated over $\mathrm{Gr}^0 A$, if and only if $\mathrm{Gr}^\bullet M$ is finitely generated over $\mathrm{Gr}^\bullet A$.*

Proof. Clearly if M is finitely generated over A, then $\mathrm{Gr}^0 M := M/IM$ is finitely generated over A/I; the converse assertion follows from Nakayama's lemma with respect to the nilpotent ideal I. If $\mathrm{Gr}^0 M$ is finitely generated over $\mathrm{Gr}^0 A$, then Lemma B.5 shows that $\mathrm{Gr}^\bullet M$ is finitely generated over $\mathrm{Gr}^\bullet A$; the converse assertion follows from the fact that (again by Lemma B.5) $\mathrm{Gr}^0 M$ may be obtained as the quotient of $\mathrm{Gr}^\bullet M$ by the ideal $\mathrm{Gr}^{\bullet>0} A$ of $\mathrm{Gr}^\bullet A$. □

B.7 Lemma. *Let $f\colon M \to N$ be a morphism of A-modules. Then f is an isomorphism if and only if the induced morphism $\mathrm{Gr}^\bullet f\colon \mathrm{Gr}^\bullet M \to \mathrm{Gr}^\bullet N$ is an isomorphism.*

Proof. If f is an isomorphism then certainly $\mathrm{Gr}^\bullet f$ is an isomorphism. Conversely, if $\mathrm{Gr}^\bullet f$ is an isomorphism, then in particular $M/IM \to N/IN$ is surjective, so f is surjective by Nakayama's lemma with respect to the nilpotent ideal I. In addition, for each i the morphism $I^i M/I^{i+1}M \to I^i N/I^{i+1}N$ is injective, so by an easy induction on i we see that any element of $\ker f$ is contained in $I^i M$ for all i, and is thus zero, as required. □

B.8 Lemma. *If A is an R-algebra, then an A-module M is (faithfully) flat over A if and only if $\mathrm{Gr}^\bullet M$ is (faithfully) flat over $\mathrm{Gr}^\bullet A$. Furthermore, if any of these conditions holds, then the natural morphism $\mathrm{Gr}^\bullet A \otimes_{\mathrm{Gr}^0 A} \mathrm{Gr}^0 M \to \mathrm{Gr}^\bullet M$ is an isomorphism.*

Proof. This is [EG21, Lem. 5.5.37]. □

B.9 Lemma. *If A is an R-algebra, and if M and N are A-modules, then the natural surjection* (B.3) *induces a natural surjection*

$$\mathrm{Gr}^\bullet M \otimes_{\mathrm{Gr}^\bullet A} \mathrm{Gr}^\bullet N \to \mathrm{Gr}^\bullet(M \otimes_A N).$$

Proof. The natural surjection $M \otimes_R N \to M \otimes_A N$ induces a surjection $\mathrm{Gr}^\bullet (M \otimes_R N) \to \mathrm{Gr}^\bullet(M \otimes_A N)$. It is immediate that its composite with the surjection (B.3) factors through $\mathrm{Gr}^\bullet M \otimes_{\mathrm{Gr}^\bullet A} \mathrm{Gr}^\bullet N$. □

B.10 Lemma. *If A is an R-algebra, and if M and N are A-modules with either M or N flat over A, then the surjection of Lemma* B.9 *is a natural isomorphism*

$$\mathrm{Gr}^\bullet M \otimes_{\mathrm{Gr}^\bullet A} \mathrm{Gr}^\bullet N \xrightarrow{\sim} \mathrm{Gr}^\bullet(M \otimes_A N).$$

Proof. Suppose (without loss of generality) that N is A-flat. Taking into account Lemma B.8, we see that we have to show that natural morphism $\mathrm{Gr}^\bullet M \otimes_{\mathrm{Gr}^0 A} \mathrm{Gr}^0 N \to \mathrm{Gr}^\bullet(M \otimes_A N)$ is an isomorphism, i.e., that for each $i \geq 0$, the natural morphism

$$I^i M/I^{i+1}M \otimes_{A/I} N/I \to I^i(M \otimes N)/I^{i+1}(M \otimes N)$$

is an isomorphism. This follows easily by tensoring N over A with the various short exact sequences

$$0 \to I^n M \to I^m M \to I^m M/I^n M \to 0,$$

taking into account the flatness of N over A. □

B.11 A GENERAL SETTING

We fix an Artinian local ring R with maximal ideal I and residue field k, as well as an R-algebra C^+, and an element $u \in C^+$, satisfying the following properties:

(A) u is a regular element (i.e., a nonzero divisor) of C^+.
(B) C^+/u is a flat R-algebra.
(C) C^+/I is a rank 1 complete valuation ring, and the image of u in C^+/I (which is necessarily nonzero, by (A)) is of positive valuation (i.e., lies in the maximal ideal of C^+/I).

B.12 Remark. In the preceding context, if $v \in C^+$ has nonzero image in C^+/I, then v^n divides u^m in C^+, for some $m, n > 0$. (Indeed, since C^+/I is a rank 1 valuation ring in which the image of u has positive valuation, we may write $u^a = vw + x$, for some $a > 0$, some $w \in C^+$, and some $x \in IC^+$. Raising both sides of this equation to a sufficiently large power, remembering that I is nilpotent, gives the claim.)

In particular, if $v \in C^+$ is another element satisfying conditions (A), (B), and (C) above, then the powers of u and of v are mutually cofinal with respect to divisibility (i.e., the u-adic and v-adic topologies on C^+ coincide), and consequently $C^+[1/u] = C^+[1/v]$.

B.13 Remark. Recall that since R is an Artinian local ring, an R-module is flat if and only if it is free [Sta, Tag 051G]. (We will use this fact in the proof of Lemma B.25 below. For a proof, note first that if M is any R-module, then any basis of M/IM lifts to a generating set of M. If M is furthermore flat, then lifting a basis of M/IM, we obtain a surjection $F \to M$ whose source is free, with the additional property that, if N denotes its kernel, then $N/IN = 0$. Thus $N = 0$, and so $F \xrightarrow{\sim} M$, as required.)

We begin by establishing some simple consequences of our assumptions on C^+ and u.

B.14 Lemma

(1) *For each $n \geq 1$, the quotient C^+/u^n is flat over R.*
(2) *C^+ itself is flat over R.*
(3) *The ring C^+ is u-adically complete.*

Proof. Claim (1) follows by an evident induction from assumptions (A) and (B) together with a consideration of the short exact sequences

$$0 \to C^+/u^n \xrightarrow{u\cdot} C^+/u^{n+1} \to C^+/u \to 0.$$

Now choose an increasing filtration (I_m) of R by ideals so that $I_0 = 0$, and such that each quotient I_{m+1}/I_m is of length 1 (i.e., is isomorphic to k). Taking into account (1), the short exact sequences

$$0 \to I_m \to I_{m+1} \to I_{m+1}/I_m (\cong k = R/I) \to 0$$

give rise to short exact sequences

$$0 \to I_m \otimes_R (C^+/u^n) \to I_{m+1} \otimes_R (C^+/u^n) \to C^+/(I, u^n) \to 0.$$

Suppose that for some m we know that the natural morphism

$$I_m \otimes_R C^+ \to \varprojlim_n I_m \otimes_R (C^+/u^n)$$

is an isomorphism. Then passing to the inverse limit over n, and taking into account both assumption (C) and the fact that the transition maps $I_m \otimes_R (C^+/u^{n+1}) \to I_m \otimes_R (C^+/u^n)$ are surjective, so that the relevant $\varprojlim^1$ vanishes, we obtain a short exact sequence

$$0 \to I_m \otimes_R C^+ \to \varprojlim_n I_{m+1} \otimes_R (C^+/u^n) \to C^+/I \to 0.$$

The five lemma shows that the natural morphism to this sequence from the exact sequence

$$I_m \otimes_R C^+ \to I_{m+1} \otimes_R C^+ \to C^+/I \to 0$$

is then an isomorphism. Proceeding by induction (the case $m=0$ being trivial), we find (once we reach the top of our filtration, so that $I_m = R$) that C^+ is u-adically complete, and we also find that each of the morphisms

$$I_m \otimes_R C^+ \to C^+$$

is injective. Since any ideal J of R can be placed in such a filtration (I_m), we find that C^+ is flat over R. Thus (2) and (3) are proved. □

We write $C := C^+[1/u]$. Since C is a localization of C^+, which is R-flat by Lemma B.14 (2), it is flat over R. Note also that $\mathrm{Gr}^0 C \cong (C^+/I)[1/u]$ is a field that is complete with respect to a non-Archimedean absolute value. In particular, we can do rigid geometry over $\mathrm{Gr}^0 C$; this is a key point, which we will exploit below.

B.15 Definition. If M is an R-module, then we write $M \widehat{\otimes}_R C^+$ to denote the u-adic completion of the tensor product $M \otimes_R C^+$. We also write $M \widehat{\otimes}_R C := (M \widehat{\otimes}_R C^+)[1/u]$.

B.16 Remark. Remark B.12 shows that the formation of $M \widehat{\otimes}_R C^+$ and $M \widehat{\otimes}_R C$ is independent of the choice of u satisfying (A), (B), and (C) above.

B.17 Remark. In applications, we will apply this construction primarily in the case when $M = A$ is an R-algebra, in which case $A \widehat{\otimes}_R C^+$ and $A \widehat{\otimes}_R C$ are again R-algebras.

B.18 Example. If we take $C^+ := R[[u]]$ (for an indeterminate u), then $A \widehat{\otimes}_R C^+ = A[[u]]$, and $A \widehat{\otimes}_R C = A((u))$.

We next state and prove some additional properties of u-adic completions that we will need.

B.19 Lemma. *If M is an R-module, then multiplication by u is injective on $M \widehat{\otimes}_A C^+$. Consequently (indeed, equivalently), the natural map $M \widehat{\otimes}_R C^+ \to M \widehat{\otimes}_R C$ is injective.*

Furthermore, for each $m \geq 0$, the natural map $M \widehat{\otimes}_R C^+ \to M \otimes_R (C^+/u^m)$ is surjective, and induces an isomorphism

$$(M \widehat{\otimes}_R C^+)/u^m(M \widehat{\otimes}_R C^+) \xrightarrow{\sim} M \otimes_R (C^+/u^m).$$

Proof. Note that the claim of the second paragraph is a general property of completion with respect to finitely generated ideals (see, e.g., the statement and proof of [Sta, Tag 05GG]). We will also deduce it as a by-product of the proof of the claims in the first paragraph, to which we now turn.

Given the definition of $M \widehat{\otimes}_R C$ as the localization $M \widehat{\otimes}_R C^+[1/u]$, the second claim of the first paragraph is clearly equivalent to the first. As for this first claim, we note that for each $m, n \geq 1$, Lemma B.14 (1) shows that the terms in the short exact sequence

$$0 \to C^+/u^n \xrightarrow{u^m \cdot} C^+/u^{n+m} \to C^+/u^m \to 0$$

are flat R-modules. Thus, tensoring this short exact sequence with M over R yields a short exact sequence

$$0 \to M \otimes_R (C^+/u^n) \to M \otimes_R (C^+/u^{n+1}) \to M \otimes_R (C^+/u) \to 0.$$

Passing to the inverse limit over n, we obtain the exact sequence

$$0 \to M \widehat{\otimes}_R C^+ \xrightarrow{u^m \cdot} M \widehat{\otimes}_R C^+ \to M \otimes_R C^+/u^m C^+$$

from which the claims of the first paragraph follow. If we use the surjectivity of the transition maps in the inverse limit to infer that the relevant $\varinjlim^1$ vanishes, then we see that this sequence is even exact on the right, so that we also obtain a confirmation of the claim of the second paragraph. □

B.20 Lemma

(1) The functors $M \mapsto M \widehat{\otimes}_R C^+$ and $M \mapsto M \widehat{\otimes}_R C$ are exact.

(2) If A is an R-algebra, if J is a finitely generated ideal in A, and if M is an A-module, then the natural map

$$J(M \widehat{\otimes}_R C^+) \to JM \widehat{\otimes}_R C^+$$

is an isomorphism, and consequently there is a short exact sequence

$$0 \to J(M \widehat{\otimes}_R C^+) \to M \widehat{\otimes}_R C^+ \to (M/JM) \widehat{\otimes}_R C^+ \to 0.$$

(3) If J_1 and J_2 are ideals in an R-algebra A, with $J_1 \supseteq J_2$, and if M is an A-module, then there is a natural isomorphism

$$J_1(M \widehat{\otimes}_R C^+)/J_2(M \widehat{\otimes}_R C^+) \xrightarrow{\sim} (J_1M/J_2M) \widehat{\otimes}_R C^+.$$

Proof. If $0 \to M_1 \to M_2 \to M_3 \to 0$ is a short exact sequence of R-modules, then Lemma B.14 (1) implies that

$$0 \to M_1 \otimes_R (C^+/u^n) \to M_2 \otimes_R (C^+/u^n) \to M_3 \otimes_R (C^+/u^n) \to 0$$

is exact for each n. If we pass to the inverse limit over n, and note that the transition morphisms are evidently surjective, so that the relevant $\varprojlim^1$ vanishes, we obtain a short exact sequence

$$0 \to M_1 \widehat{\otimes}_R C^+ \to M_2 \otimes_R C^+ \to M_3 \otimes_R C^+ \to 0,$$

proving the first exactness claim of (1). The second exactness claim follows from the first, together with the fact that the localization $C := C^+[1/u]$ is flat over C, and that there is an isomorphism $M \widehat{\otimes}_R C \xrightarrow{\sim} (M \widehat{\otimes}_R C^+) \otimes_{C^+} C$.

In order to prove (2), we apply (1) to the short exact sequence $0 \to JM \to M \to M/JM \to 0$, obtaining a short exact sequence

$$0 \to (JM) \widehat{\otimes}_R C^+ \to M \widehat{\otimes}_R C^+ \to (M/JM) \widehat{\otimes}_R C^+ \to 0. \tag{B.21}$$

Thus we may (and do) regard $(JM) \widehat{\otimes}_R C^+$ as a submodule of $M \widehat{\otimes}_R C^+$. There is a consequent inclusion

$$J(M \widehat{\otimes}_R C^+) \subseteq (JM) \widehat{\otimes}_R C^+, \tag{B.22}$$

whose image is u-adically dense in the target. Since J is a finitely generated ideal, we see that $J(M \widehat{\otimes}_R C^+)$ is furthermore u-adically complete, and thus that (B.22) is actually an equality, establishing the first claim of (2).

Replacing $(JM) \widehat{\otimes}_R C^+$ by $J(M \widehat{\otimes}_R C^+)$ in the short exact sequence (B.21) yields the short exact sequence whose existence is asserted in the second claim of (2).

Claim (3) follows directly from (2), and the exactness result of (1). □

The following lemma will allow us to use grading techniques to study u-adic completions.

B.23 Lemma. *If M is an R-module, then there are natural isomorphisms*

(1) $\mathrm{Gr}^\bullet M \otimes_k \mathrm{Gr}^0(C^+/u^n) \xrightarrow{\sim} \mathrm{Gr}^\bullet(M \otimes_R (C^+/u^n))$ *(for any $n \geq 1$),*
(2) $\mathrm{Gr}^\bullet(M \widehat{\otimes}_R C^+) \xrightarrow{\sim} \mathrm{Gr}^\bullet M \widehat{\otimes}_k \mathrm{Gr}^0 C^+$, *and*
(3) $\mathrm{Gr}^\bullet(M \widehat{\otimes}_R C) \xrightarrow{\sim} \mathrm{Gr}^\bullet M \widehat{\otimes}_k \mathrm{Gr}^0 C$.

Proof. Lemma B.14 (1) shows that C^+/u^n is flat over R, and (1) follows from this, together with Lemmas B.8 and B.10.

Lemma B.20 (3) (taking the inclusion $J_2 \subseteq J_1$ to be the various inclusions $I^{i+1} \subseteq I^i$ in turn) shows that each $\mathrm{Gr}^i(M \widehat{\otimes}_R C^+) \xrightarrow{\sim} \mathrm{Gr}^i M \widehat{\otimes}_R C^+$. Since I annihilates $\mathrm{Gr}^i(M)$, the target of this isomorphism can be rewritten as $\mathrm{Gr}^i M \widehat{\otimes}_k \mathrm{Gr}^0 \mathbf{C}^+$. This proves (2).

The R-algebra C may be described as the direct limit $C \xrightarrow{\sim} \varinjlim_n u^{-n} C^+$ (the transition maps being the obvious inclusions). Since the formation of direct limits is compatible with tensor products, we find that

$$\mathrm{Gr}^0 C \xrightarrow{\sim} \varinjlim_n \mathrm{Gr}^0 u^{-n} C^+.$$

The isomorphism of (3) is thus obtained from the isomorphisms

$$\mathrm{Gr}^\bullet(M \mathbin{\widehat{\otimes}}_R u^{-n} C^+) \xrightarrow{\sim} \mathrm{Gr}^\bullet M \mathbin{\widehat{\otimes}}_k \mathrm{Gr}^0 u^{-n} C^+$$

of (2) by passing to the direct limit over n. □

B.24 LOCI OF VANISHING AND OF EQUIVARIANCE

We maintain the notation of Section B.11, and prove some slightly technical results about the "locus of vanishing" and "locus of equivariance" of morphisms between finitely generated projective $A \mathbin{\widehat{\otimes}}_R C$-modules.

B.25 Lemma. *Let A be an R-algebra, and let M be a finitely generated projective module over $A \mathbin{\widehat{\otimes}}_R C$. If $m \in M$, then the "locus of vanishing" of m is a Zariski closed subset of* Spec A. *More precisely, there is an ideal J of A such that for any morphism $\phi\colon A \to B$ of R-algebras, the image of m in $M_B := (B \mathbin{\widehat{\otimes}}_R C) \otimes_{A \widehat{\otimes}_R C} M$ vanishes if and only if J is contained in the kernel of ϕ.*

Proof. Since M is a finitely generated projective over $A \mathbin{\widehat{\otimes}}_R C$, we may find another finitely generated projective $A \mathbin{\widehat{\otimes}}_R C$-module N such that $M \oplus N$ is free of finite rank over $A \mathbin{\widehat{\otimes}}_R C$. Replacing $m \in M$ by $m \oplus 0 \in M \oplus N$, we reduce to the case when M is free of finite rank, say $M = A \mathbin{\widehat{\otimes}}_R C^{\oplus r}$. Writing $m = (x_1, \ldots, x_r)$ with $x_i \in A \mathbin{\widehat{\otimes}}_R C$, we see that it suffices to construct a corresponding ideal J_i for each x_i; the ideal $J := \sum_i J_i$ will then satisfy the claim of the lemma. Thus we reduce to proving the lemma in the case of an element $x \in A \mathbin{\widehat{\otimes}}_R C = A \mathbin{\widehat{\otimes}}_R C^+[1/u]$. Recall from Lemma B.19 that $A \mathbin{\widehat{\otimes}}_R C^+$ is a subring of $A \mathbin{\widehat{\otimes}}_R C$. Thus, since u is a unit in $A \otimes_R C$, we may replace x by $u^n x$ for some sufficiently large value of n, and assume that $x \in A \mathbin{\widehat{\otimes}}_R C^+$.

Note that tensoring the inclusion $A \mathbin{\widehat{\otimes}}_R C^+ \hookrightarrow A \mathbin{\widehat{\otimes}}_R C$ with $B \mathbin{\widehat{\otimes}}_R C^+$ over $A \mathbin{\widehat{\otimes}}_R C^+$ induces the inclusion

$$B \mathbin{\widehat{\otimes}}_R C^+ \hookrightarrow B \mathbin{\widehat{\otimes}}_R C = B \mathbin{\widehat{\otimes}}_R C^+[1/u] = (B \mathbin{\widehat{\otimes}}_R C^+) \otimes_{A \widehat{\otimes}_R C^+} A \mathbin{\widehat{\otimes}}_R C.$$

Thus it suffices to construct an ideal J in A such that the image of x in $B \mathbin{\widehat{\otimes}}_R C^+$ vanishes if and only if the morphism $\phi\colon A \to B$ factors through A/J. Let x_n denote the image of x in $A \otimes_R (C^+/u^n)$, so that

$$x = (x_n) \in A \mathbin{\widehat{\otimes}}_R C^+ = \varprojlim_n A \otimes_R (C^+/u^n).$$

If we let y_n denote the image of x_n in $B \otimes_R (C^+/u^n)$, then the image y of x in $B \widehat{\otimes}_R C$ is equal to

$$(y_n) \in B \widehat{\otimes}_R C^+ = \varprojlim_n B \otimes_R (C^+/u^n).$$

Thus it suffices to construct an ideal J_n in A such that y_n vanishes if and only if the morphism $A \to B$ factors through A/J_n; indeed, we may then take $J := \sum_n J_n$.

Lemma B.14 shows that C^+/u^n is flat over R, and thus free over R. (See Remark B.13.) If we choose an isomorphism $C^+/u^n \cong R^{\oplus S}$, then we may write $x_n = (x_{n,s}) \in A^{\oplus S}$. If we let $J_{n,s}$ denote the ideal in A generated by $x_{n,s}$, then $J_n := \sum_{s \in S} J_{n,s}$ is the required ideal. □

B.26 Corollary. *Let A be an R-algebra, and let M and N be finitely generated projective modules over $A \widehat{\otimes}_R C$. If $f\colon M \to N$ is an $A \widehat{\otimes}_R C$-module homomorphism, then the "locus of vanishing" of f is a Zariski closed subset of* Spec A. *More precisely, there is an ideal J of A such that for any morphism $\phi\colon A \to B$ of R-algebras, the base change*

$$f_B\colon M_B := (B \widehat{\otimes}_R C) \otimes_{A \widehat{\otimes}_R C} M \to N_B := (B \widehat{\otimes}_R C) \otimes_{A \widehat{\otimes}_R C} N$$

vanishes if and only if J is contained in the kernel of ϕ.

Proof. We may regard $f \in \mathrm{Hom}_{A \widehat{\otimes}_R C}(M, N)$ as an element of the module

$$M^\vee \otimes_{A \widehat{\otimes}_R C} N.$$

The corollary then follows from Lemma B.25 applied to f so regarded. □

B.27 Corollary. *Let σ be a ring automorphism of C, let A be an R-algebra, and let M and N be finitely generated projective modules over $A \widehat{\otimes}_R C$ each endowed with a σ-semi-linear automorphism, denoted σ_M and σ_N respectively. If $f\colon M \to N$ is an $A \widehat{\otimes}_R C$-module homomorphism, then the "locus of σ-equivariance" of f is a Zariski closed subset of* Spec A. *More precisely, there is an ideal J of A such that for any morphism $\phi\colon A \to B$ of R-algebras, the base change*

$$f_B\colon M_B := (B \widehat{\otimes}_R C) \otimes_{A \widehat{\otimes}_R C} M \to N_B := (B \widehat{\otimes}_R C) \otimes_{A \widehat{\otimes}_R C} N$$

satisfies $f_B \circ \sigma_M = \sigma_N \circ f_B$ if and only if J is contained in the kernel of ϕ.

Proof. This follows from Corollary B.26, applied to the morphism $f - \sigma_N \circ f \circ \sigma_M^{-1}$. □

We also have the following variants on these results for $A\widehat{\otimes}_R C^+$-modules.

B.28 Lemma. *Let A be an R-algebra, and let M and N be finitely generated projective modules over $A\widehat{\otimes}_R C^+$. Let $f\colon M \to N[1/u]$ be an $A\widehat{\otimes}_R C^+$-module homomorphism. Then the locus where $f(M)\subseteq N$ is a Zariski closed subset of* Spec A. *More precisely, there is an ideal J of A such that for any morphism $\phi\colon A\to B$ of R-algebras, the base change*

$$f_B\colon M_B := (B\widehat{\otimes}_R C)\otimes_{A\widehat{\otimes}_R C} M \to N_B[1/u] := (B\widehat{\otimes}_R C)\otimes_{A\widehat{\otimes}_R C} N[1/u]$$

satisfies $f_B(M_B)\subseteq N_B$ if and only if J is contained in the kernel of ϕ.

Proof. We argue as in the proof of Lemma B.25. Since M, N are finitely generated projective over $A\widehat{\otimes}_R C^+$, we may find finitely generated projective modules P, Q over $A\widehat{\otimes}_R C^+$ such that $M\oplus P$ and $N\oplus Q$ are free. Replacing M by $M\oplus P$, N by $N\oplus Q$, and f by $(f,0)$, we reduce to the case that M, N are both finite free $A\widehat{\otimes}_R C^+$-modules. After choosing bases we may suppose that $M=(A\widehat{\otimes}_R C^+)^{\oplus r}$, $N=(A\widehat{\otimes}_R C^+)^{\oplus s}$, and considering the matrix representing f, we reduce to showing that for any element $x\in A\widehat{\otimes}_R C$, there is an ideal J of A such that $(\phi\otimes 1)(x)\in B\widehat{\otimes}_R C^+$ if and only if J is contained in the kernel of ϕ. This is a special case of Lemma B.29 below. □

B.29 Lemma. *Let A be an R-algebra, and let M be a finitely generated projective $A\widehat{\otimes}_R C^+$-module. Let x be an element of $M[1/u]$. Then the locus where $x\in M$ is a Zariski closed subset of* Spec A. *More precisely, there is an ideal J of A such that for any morphism $\phi\colon A\to B$ of R-algebras, the image of x in $M_B[1/u]$ lies in M_B if and only if J is contained in the kernel of ϕ.*

Proof. We begin by arguing as in the proof of Lemma B.28. Since M is finitely generated projective over $A\widehat{\otimes}_R C^+$, we may find a finitely generated projective module P, Q over $A\widehat{\otimes}_R C^+$ such that $M\oplus P$ is free. Replacing M by $M\oplus P$ and x by $(x,0)$, we reduce to the case that M is a finite free $A\widehat{\otimes}_R C^+$-module. After choosing a basis we may suppose that $M=(A\widehat{\otimes}_R C^+)^{\oplus r}$, so that we can reduce to the case that $M=A\widehat{\otimes}_R C^+$.

Choose n sufficiently large so that $u^n x\in A\widehat{\otimes}_R C^+$. Then $(\phi\otimes 1)(x)\in B\widehat{\otimes}_R C^+$ if and only if the image of $(\phi\otimes 1)(x)$ in

$$u^{-n}(B\widehat{\otimes}_R C^+)/(B\widehat{\otimes}_R C^+) = B\otimes_R (u^{-n}C^+)/C^+)$$

vanishes. As in the proof of Lemma B.25, $u^{-n}C^+/C^+$ is a free R-module, so we can and do choose an isomorphism $u^{-n}C^+/C^+\cong R^{\oplus S}$. Then if we write $x=(x_s)\in A^{\oplus S}\cong A\otimes_R(u^{-n}C^+/C^+)$, we can take J to be the ideal generated by the x_s. □

B.30 EXTENDING ACTIONS

We continue to remain in the context of Section B.11. We will show that various sorts of actions on C^+ or C can be extended to the completed tensor products $A\,\widehat{\otimes}_R\,C^+$ and $A\,\widehat{\otimes}_R\,C$.

Throughout our discussion, we endow C^+ with the u-adic topology, while we endow C with the unique topology which makes it a topological group, and in which C^+ is an open subgroup. More generally, if A is any R-algebra, we again endow $A\,\widehat{\otimes}_R\,C^+$ with the u-adic topology, while we endow $A\,\widehat{\otimes}_R\,C$ with the unique topology which makes it a topological group, and in which $A\,\widehat{\otimes}_R\,C^+$ (endowed with its u-adic topology) is an open subgroup.

B.31 Lemma. *Suppose that C^+ (resp. C) is equipped with a continuous R-linear endomorphism φ. Then there is a unique extension of φ to a continuous A-linear endomorphism of $A\,\widehat{\otimes}_R\,C^+$ (resp. $A\,\widehat{\otimes}_R\,C$).*

Proof. Lemma B.19 shows that u-adic topology on $A\,\widehat{\otimes}_R\,C^+$ coincides with its inverse limit topology, if we recall that $A\,\widehat{\otimes}_R\,C^+ := \varprojlim_n A\otimes_R (C^+/u^n)$, and we endow each of the objects appearing in the inverse limit with its discrete topology. Given this, it is immediate that a u-adically continuous R-linear endomorphism φ of C^+ extends uniquely to a u-adically continuous A-linear endomorphism of $A\,\widehat{\otimes}_R\,C^+$.

The case when φ is a continuous R-linear endomorphism of C is only slightly more involved. Namely, to say that $\varphi\colon C\to C$ is continuous is to say that, for some $m\geq 0$, we have an inclusion $\varphi(u^mC^+)\subseteq C^+$, and that the induced morphism $\varphi\colon u^mC^+\to C^+$ is continuous when each of the source and the target are endowed with their u-adic topologies. Again taking into account Lemma B.19, we then obtain a unique A-linear and continuous morphism

$$A\,\widehat{\otimes}_R\,C = \big(\varprojlim_n A\otimes_R (u^mC^+/u^{m+n})\big)[1/u] \longrightarrow \big(\varprojlim_n A\otimes_R (C^+/u^n)\big)[1/u] = A\,\widehat{\otimes}_R\,C,$$

as required. □

If G is a topological group acting continuously on C^+ (i.e., acting in such a way that the action map $G\times C^+\to C^+$ is continuous), then for any $g\in G$ and $m\geq 0$, we may find $n\geq 0$, and an open neighborhood U of g, such that $U\times u^nC^+\subseteq u^mC^+$. (This just expresses the continuity of the action map at the point $(g,0)\in G\times C^+$.) If G is furthermore compact, then G is covered by finitely many such open sets U, and thus for any given choice of m, we may in fact find n such that $G\cdot u^nC^+\subseteq u^mC^+$. An identical remark applies in the case of a compact topological group acting on C.

We will need to introduce an additional condition on G-actions on C, which is related to, but doesn't seem to follow from, the previous considerations.

B.32 Definition. If G is a group, we say that a G-action on C is *bounded* if, for any $M \geq 0$, there exists $N \geq 0$ such that $G \cdot u^{-M}C^+ \subseteq u^{-N}C^+$.

B.33 Remark. It follows from Remark B.12 that the notion of a group action being bounded is independent of the choice of $u \in C^+$ satisfying conditions (A), (B), and (C) of Section B.11.

We then have the following lemma.

B.34 Lemma. *Suppose that G is a compact topological group, and that C^+ (resp. C) is equipped with a continuous (resp. continuous and bounded) R-linear G-action. Then there is a unique extension of this action to a continuous A-linear action of G on $A \widehat{\otimes}_R C^+$ (resp. $A \widehat{\otimes}_R C$).*

Proof. The proof of this lemma is very similar to that of Lemma B.31. Indeed, applying that lemma (and taking into account its uniqueness statement) we obtain the desired action of G. It remains to confirm that this action is again continuous.

Consider first the case of a G-action on C^+. For each $m \geq 0$, we saw above that $G \cdot u^n C^+ \subseteq u^m C^+$ if n is sufficiently large, so that we obtain a well-defined map $G \times A \widehat{\otimes}_R C^+ = G \times \varprojlim_n A \otimes_R C^+/u^n \to A \otimes_R C^+/u^m$ which is continuous (since it is obtained from the continuous G-action on C^+ by tensoring with A). Passing to the projective limit in m then yields the desired continuity.

In the case of a G-action on C, we argue similarly: namely, the assumptions of continuity and boundedness of the G-action yield continuous morphisms

$$G \times u^{-M}C^+/u^n C^+ \to u^{-N}C^+/u^m C^+.$$

Passing to the projective limit in n, then in m, and then to the inductive limit in N, and then in M, we deduce the desired continuity of the G-action on $A \widehat{\otimes}_R C$. □

B.35 A RIGID ANALYTIC PERSPECTIVE

We keep ourselves in the setting of Section B.11. We are interested in studying finitely generated projective modules over $A \widehat{\otimes}_R C$, especially their descent properties. In the present general context, we don't know whether an analogue of Drinfeld's descent result for Tate modules (see [Dri06, Thm. 3.3, 3.11] and [EG21, Thm. 5.1.18]) holds for arbitrary R-algebras A. However, we will see that such a descent result is valid in the more restricted setting of finite type

R-algebras. In the case of finite type k-algebras, we will prove this using results from rigid analysis. For finite type R-algebras that are not necessarily annihilated by I, we will use grading techniques to reduce to the case of k-algebras.

If A is a finite type k-algebra, then $\operatorname{Spf}(A \widehat{\otimes}_k \operatorname{Gr}^0 C^+)$ (where the Spf is taken with respect to the u-adic topology on $A \widehat{\otimes}_k \operatorname{Gr}^0 C^+$) is a formal scheme of finite type over $\operatorname{Spf} \operatorname{Gr}^0 C^+$, whose rigid analytic generic fiber $\operatorname{Max} \operatorname{Spec}(A \widehat{\otimes}_k \operatorname{Gr}^0 C)$ is an affinoid rigid analytic space over the complete non-Archimedean field $\operatorname{Gr}^0 C$.

As an application of this rigid analytic point of view (together with graded techniques), we first establish the following proposition.

B.36 Proposition

(1) If A is a finite type R-algebra, then $A \widehat{\otimes}_R C$ is Noetherian.
(2) If $A \to B$ is a *(*faithfully*)* flat morphism of finite type R-algebras, then the induced morphisms $A \widehat{\otimes}_R C^+ \to B \widehat{\otimes}_R C^+$ and $A \widehat{\otimes}_R C \to B \widehat{\otimes}_R C$ are again *(*faithfully*)* flat.

Proof. In the case when A is in fact a k-algebra, it follows from the preceding discussion that $A \widehat{\otimes}_R C$ is an affinoid algebra over $\operatorname{Gr}^0 C$, and is therefore Noetherian by [BGR84, Thm. 1, §5.2.6]. In the general case, Lemma B.23 (3) gives an isomorphism $\operatorname{Gr}^\bullet(A \widehat{\otimes}_R C) \xrightarrow{\sim} \operatorname{Gr}^\bullet A \widehat{\otimes}_k \operatorname{Gr}^0 C$. Since $\operatorname{Gr}^\bullet A$ is a finite type k-algebra, it follows from the case already proved that the target of this isomorphism is Noetherian, and thus so is its source. Thus $A \widehat{\otimes}_R C$ is itself Noetherian, by Lemma B.4.

We explain how our rigid analytic perspective proves a part of (2). If we suppose first that $A \to B$ is a flat morphism of k-algebras, then so is the morphism $A \otimes_R C \to B \otimes_R C$. Now the preceding discussion shows that the morphisms $\operatorname{Max} \operatorname{Spec}(A \widehat{\otimes}_R C) \to \operatorname{Max} \operatorname{Spec}(A \otimes_R C)$ and $\operatorname{Max} \operatorname{Spec}(B \widehat{\otimes}_R C) \to \operatorname{Max} \operatorname{Spec} (B \otimes_R C)$ are open immersions of rigid analytic spaces over $\operatorname{Gr}^0 C$, so that the induced morphism $\operatorname{Max} \operatorname{Spec}(A \widehat{\otimes}_R C) \to \operatorname{Max} \operatorname{Spec}(B \widehat{\otimes}_R C)$ is also flat. This proves (the k-algebra case of) the second flatness claim of (2). If $A \to B$ is a flat morphism of R-algebras that are not necessarily k-algebras, then we deduce from Lemma B.8 that $\operatorname{Gr}^\bullet A \to \operatorname{Gr}^\bullet B$ is flat, and thus, from what we've already shown, that

$$\operatorname{Gr}^\bullet A \widehat{\otimes}_k \operatorname{Gr}^0 C \to \operatorname{Gr}^\bullet B \widehat{\otimes}_k \operatorname{Gr}^0 C$$

is flat. Lemma B.23 (3) then shows that the morphism

$$\operatorname{Gr}^\bullet(A \widehat{\otimes}_R C) \to \operatorname{Gr}^\bullet(B \widehat{\otimes}_R C)$$

is flat, and one more application of Lemma B.8 completes the proof of the second flatness claim of (2).

It is not clear to us whether one can deduce the second faithful flatness claim of (2) by this style of argument. In order to prove this result, as well as to obtain the first set of claims of (2), we appeal to the results of [FGK11]. Indeed, it is clear that the second set of claims in (2) follows immediately from the first, and it is the first set of claims that we will now prove.

We first note that if the morphism $A \to B$ is flat (resp. faithfully flat), then so are each of the morphisms $A \otimes_R C^+/u^n A \to B \otimes_R^+ C/u^n$. In the terminology of [FGK11, §5.2], the morphism $A \widehat{\otimes}_R C^+ \to B \widehat{\otimes}_R C^+$ is adically flat (resp. adically faithfully flat); here we regard the source and target as being endowed with their u-adic topologies. The discussion at the beginning of [FGK11, §5.2] (see also our Proposition 2.2.2) then shows that $A \widehat{\otimes}_R C^+ \to B \widehat{\otimes}_R C^+$ is flat. In the adically faithfully flat case, we deduce in addition from [FGK11, Prop. 5.2.1 (2)] (again, see also Proposition 2.2.2) that $A \widehat{\otimes}_R C^+ \to B \widehat{\otimes}_R C^+$ is faithfully flat; note that $A \widehat{\otimes}_R C^+$ and $B \widehat{\otimes}_R C^+$ are Noetherian outside u by (1) of the present proposition. □

B.37 DESCENT FOR $A \widehat{\otimes}_R C$-MODULES

We now establish the descent result alluded to above; it is analogous to Drinfeld's [EG21, Thm. 5.1.18 (1)] but is restricted to the context of finite type R-algebras. We use grading techniques to reduce to the case of finite type k-algebras, where we can then apply known results from rigid analysis.

B.38 Theorem. *Let $A \to B$ be a faithfully flat morphism of finite type R-algebras. Then the functor $M \mapsto M \otimes_{A \widehat{\otimes}_R C} (B \widehat{\otimes}_R C)$ induces an equivalence of categories between the category of finite type $A \widehat{\otimes}_R C$-modules and the category of finite type $B \widehat{\otimes}_R C$-modules equipped with descent data.*

B.39 Remark. Recall (for example from [BLR90, §6.1]) that if $S \to S'$ is a ring homomorphism, then a descent datum for an S'-module M' is an isomorphism of $S' \otimes_S S'$-modules $S' \otimes_S M' \xrightarrow{\sim} M' \otimes_S S'$, satisfying a certain cocycle condition. Given a descent datum, we have a pair of morphisms of S-modules $M' \to M' \otimes_S S'$, namely the obvious morphism and the composite of the obvious morphism $M' \to S' \otimes_S M'$ with the descent datum isomorphism. We let K denote the S-module given by the kernel of the difference of these two maps. We refer to the functor $M' \mapsto K$, from the category of S'-modules with descent data to the category of S-modules, as the *kernel functor*. (If $S \to S'$ is faithfully flat, then the theory of faithfully flat descent shows that M' is the extension of scalars from S to S' of K.)

Proof. We first treat the case when A and B are k-algebras. In this case, the category of finite type $A \widehat{\otimes}_R C$-modules (resp. finite type $B \widehat{\otimes}_R C$-modules) is equivalent to the category of coherent sheaves on the affinoid rigid analytic space $\operatorname{Max Spec}(A \widehat{\otimes}_R C)$ (resp. $\operatorname{Max Spec}(B \widehat{\otimes}_R C)$) over the field $\operatorname{Gr}^0 C$, and the statement follows from faithfully flat descent for rigid analytic coherent sheaves [BG98, Thm. 3.1].

We reduce the general case of R-algebras to the case of k-algebras via the usual graded arguments. To ease notation, we let $F\colon \mathcal{C}\to\mathcal{D}$ denote the base-change functor $M\mapsto M\otimes_{A\widehat{\otimes}_R C}(B\widehat{\otimes}_R C)$ from the category $\mathcal{C}$ of finite type $A\widehat{\otimes}_R C$-modules to the category $\mathcal{D}$ of finite type $B\widehat{\otimes}_R C$-modules equipped with descent data, and let $G\colon \mathcal{D}\to\mathcal{C}$ denote the kernel functor. There are evident natural transformations

$$G\circ F\to \mathrm{id}_{\mathcal{C}} \quad \text{and} \quad F\circ G\to \mathrm{id}_{\mathcal{D}}, \tag{B.40}$$

which we claim are natural isomorphisms (so that G provides a quasi-inverse to F, proving in particular that F induces an equivalence of categories).

We let $\overline{\mathcal{C}}$ denote the category of finite type $\mathrm{Gr}^0(A\widehat{\otimes}_R C)$-modules, and let $\overline{\mathcal{D}}$ denote the category of finite type $\mathrm{Gr}^0(B\widehat{\otimes}_R C)$-modules equipped with descent data to $\mathrm{Gr}^0(A\widehat{\otimes}_R C)$. We let $\overline{F}\colon \overline{\mathcal{C}}\to\overline{\mathcal{D}}$ denote the base change functor $-\otimes_{\mathrm{Gr}^0(A\widehat{\otimes}_R C)}\mathrm{Gr}^0(B\widehat{\otimes}_R C)$, and let $\overline{G}\colon \overline{\mathcal{D}}\to\overline{\mathcal{C}}$ denote the kernel functor. The case of k-algebras that we have already treated shows that there are natural isomorphisms

$$\overline{G}\circ\overline{F}\xrightarrow{\sim}\mathrm{id}_{\overline{\mathcal{C}}} \quad \text{and} \quad \overline{F}\circ\overline{G}\xrightarrow{\sim}\mathrm{id}_{\overline{\mathcal{D}}}. \tag{B.41}$$

We think of passage to the associated graded $\mathrm{Gr}^\bullet$ as inducing functors $\mathcal{C}\to\overline{\mathcal{C}}$ and also $\mathcal{D}\to\overline{\mathcal{D}}$. It follows from Lemmas B.8 and B.10, together with Proposition B.36, that there is a natural isomorphism

$$\mathrm{Gr}^\bullet\circ F\xrightarrow{\sim}\overline{F}\circ\mathrm{Gr}^\bullet;$$

there is also an evident natural isomorphism

$$\mathrm{Gr}^\bullet\circ G\xrightarrow{\sim}\overline{G}\circ\mathrm{Gr}^\bullet;$$

furthermore, these natural isomorphisms are compatible with the natural transformations (B.40) and (B.41). Since the natural transformations (B.41) are isomorphisms, the same is true of the natural transformations (B.40), by Lemma B.7. This completes the proof that F is an equivalence. □

Appendix C

Topological groups and modules

In this appendix we recall some more or less well-known facts regarding topological groups and modules for which we haven't located a convenient reference. We found the note [Kay96] to be a useful reference for the basic facts regarding Polish topological groups.

Recall the following definition.

C.1 Definition. A topological space is called *Polish* if it is separable (i.e., contains a countable dense subset) and completely metrizable. A topological group is called *Polish* if its underlying topological space is Polish. Similarly, a topological ring is called *Polish* if its underlying topological space is Polish.

Our interest in Polish groups is due to the following lemma and corollary; see also [Sta, Tag 0CQW] for a more algebraic proof of a closely related result.

C.2 Lemma. *If $\phi\colon G \to H$ is a continuous homomorphism between two Polish topological groups, then the following are equivalent:*

1. *ϕ is open.*
2. *$\phi(G)$ is not meagre in H.*

Proof. If (1) holds, then $\phi(G)$ is a non-empty open subset of the Polish, and hence Baire, space H, and so is not meagre. Thus (1) implies (2). For the converse, see, e.g., [Kay96, Thm. 18]. □

C.3 Corollary. *A continuous surjective homomorphism of Polish topological groups is necessarily open.*

Proof. A completely metrizable space is Baire, and hence is not a meagre subset of itself. The corollary thus follows from the implication "(2) $\Longrightarrow$ (1)" of Lemma C.2. □

Any metrizable space is first countable (i.e., each point contains a countable neighborhood basis). Recall that, conversely, if G is a topological group, then G is first countable if and only if the identity element 1 admits a countable neighborhood basis, and in this case, if (the underlying topological space of) G is Hausdorff then it is in fact metrizable (this is the Birkhoff–Kakutani theorem), and even admits a (left or right) translation invariant metric. In particular, (the underlying topological space of) a Hausdorff topological group is separable and metrizable if and only if it is *second countable* (i.e., admits a countable basis for its topology).

Recall also that a topological group G admits a canonical uniform structure, so that it makes sense to speak of G being complete. For the sake of completeness, we note that the underlying topological space of G being Polish forces G to be complete.

C.4 Lemma. *If G is a Polish topological group, then G is complete.*

Proof. Since G is metrizable, it is Hausdorff, and so we may consider the canonical embedding of topological groups $G \hookrightarrow \widehat{G}$ of G into its completion. Since G is separable and dense in $\widehat{G}$, we see that $\widehat{G}$ is separable. Also, $\widehat{G}$ is a complete metric space (it may be identified with the metric space completion of G with respect to any invariant metric inducing the topology on G). Thus $\widehat{G}$ is Polish. Any Polish subset of a Polish space is G_δ (i.e., a countable intersection of open sets), and so in particular G is G_δ inside $\widehat{G}$. Since $\widehat{G}$ is Polish, and thus Baire, we find that G is not meagre in $\widehat{G}$. It follows from Lemma C.2 that the inclusion $G \hookrightarrow \widehat{G}$ is open, and thus that G is an open subgroup of its completion $\widehat{G}$. Since an open subgroup of a topological group is also closed, we find that G is closed, as well as dense, in $\widehat{G}$, and thus that $G = \widehat{G}$, which is to say, G is complete, as claimed. □

C.5 Remark. It follows from Lemma C.4 and the discussion preceding it that a topological group is Polish if and only if it is complete and second countable.

We next recall a result about Noetherian and Polish topological rings, which is inspired by a result of Grauert and Remmert in the theory of classical Banach algebras [GR71, App. to §I.5]. (See [BGR84, Prop. 3, §3.7.3] for the analogous result in the context of non-Archimedean Banach algebras.)

C.6 Proposition. *Let A be a Polish (or, equivalently, a (Hausdorff) complete and second countable) topological ring which is Noetherian (as an abstract ring), and which contains an open additive subgroup that is closed under multiplication, and consists of topologically nilpotent elements. Then any finitely generated A-module M has a unique completely metrizable topology with respect to which it becomes a topological A-module. Furthermore, M is complete with respect to this topology, any submodule of M is closed, and any morphism of finitely generated A-modules is automatically continuous, has closed image, and induces an open mapping from its domain onto its image.*

Proof. To begin with, suppose that M is a finitely generated A-module which is furthermore endowed with a completely metrizable topology which makes it a topological A-module. We claim first that M is in fact complete, and that any surjection of A-modules $A^n \to M$ (for some $n \geq 1$) is continuous and open. It then follows that M is endowed with the quotient topology via this map, and thus the completely metrizable topology on M is uniquely determined (if it exists).

To see the claim, note that since M is a finitely generated A-module, we may choose a surjective homomorphism of A-modules $A^n \to M$ for some $n \geq 1$. Furthermore, since M is a topological A-module, any such surjection is continuous. Since A is separable as a topological space by assumption, we see that M is as well. (The image of any countable dense subset of A^n is a countable dense subset of M.) Thus M is Polish, as is A^n, and it follows from Lemma C.4 that M is complete (as a topological module), while it follows from Corollary C.3 that the given surjection $A^n \to M$ is open, as claimed.

We next claim that any A-submodule of M is closed. To see this, let N be a submodule of M, and let $\overline{N}$ denote its closure. Since A is Noetherian by assumption, so is its finitely generated module M, and thus $\overline{N}$ is finitely generated, say by the elements $x_1, \ldots, x_n$. Applying the results proved above for M to its closed (and hence completely metrizable topological) submodule $\overline{N}$, we find that the surjection $A^n \to \overline{N}$ given by $(a_1, \ldots, a_n) \mapsto \sum_{i=1}^n a_i x_i$ is an open mapping. Consequently, if we let I be an open additive subgroup of A which satisfies $I^2 \subseteq I$, and which consists of topologically nilpotent elements (such an I exists by assumption), then $Ix_1 + \cdots + Ix_n$ is an open subset of $\overline{N}$. Since N is dense in $\overline{N}$, we conclude that $N + Ix_1 + \cdots + Ix_n = \overline{N}$, and consequently we may write

$$x_i = \sum_{i=1}^{n} a_{ij} x_j + y_i$$

for each $1 \leq i \leq n$, for some $y_i \in N$ and $a_{ij} \in I$. Rearranging, we find that

$$\begin{pmatrix} 1-a_{11} & -a_{12} & \cdots & -a_{1n} \\ -a_{21} & 1-a_{22} & \cdots & -a_{2n} \\ \vdots & \vdots & \ddots & \vdots \\ -a_{n1} & -a_{2n} & \cdots & 1-a_{nn} \end{pmatrix} \begin{pmatrix} x_1 \\ x_2 \\ \vdots \\ x_n \end{pmatrix} = \begin{pmatrix} y_1 \\ y_2 \\ \vdots \\ y_n \end{pmatrix}.$$

The determinant of the matrix $\begin{pmatrix} 1-a_{11} & -a_{12} & \cdots & -a_{1n} \\ -a_{21} & 1-a_{22} & \cdots & -a_{2n} \\ \vdots & \vdots & \ddots & \vdots \\ -a_{n1} & -a_{2n} & \cdots & 1-a_{nn} \end{pmatrix}$ lies in $1+I$, and thus is a unit (since A is complete and I consists of topologically nilpotent elements). Thus we find that $x_1, \ldots, x_n$ lies in the A-span of $y_1, \ldots, y_n$, so that in fact $\overline{N} \subseteq N$. Thus N is indeed closed, as claimed.

Applying the preceding result to A^n, we find that any A-submodule of A^n is closed, and hence that any finitely generated A-module M is isomorphic to a quotient A^n/N, where N is a closed submodule of A^n. The quotient topology on A^n/N makes it into a complete and metrizable topological A-module, and thus M does indeed admit a complete and metrizable topological A-module structure. We have already seen that this topology is unique, and that any A-submodule of M is closed. Finally, we see that homomorphisms between finitely generated A-modules are necessarily continuous and that their images (being A-submodules of their targets) are necessarily closed, and a final application of Corollary C.3 shows that they induce open mappings onto their images. □

Appendix D

Tate modules and continuity

In this appendix we study continuity conditions for group actions on modules over Laurent series rings. We begin with some results on the topology of such modules, before introducing group actions and related notions.

D.1 TOPOLOGIES AND LATTICES

We fix a finite extension $E/\mathbf{Q}_p$ with ring of integers $\mathcal{O}$, uniformizer ϖ, and residue field $\mathbf{F}$, and we also fix a finite extension $k/\mathbf{F}_p$. If A is a p-adically complete $\mathcal{O}$-algebra, we write $\mathbf{A}_A^+ := (W(k) \otimes_{\mathbf{Z}_p} A)[[T]]$, and we let $\mathbf{A}_A$ be the p-adic completion of $\mathbf{A}_A^+[1/T]$.

D.2 Remark. Any finitely generated projective $\mathbf{A}_A$-module M has a natural topology. Indeed, we may write M as a direct summand of $\mathbf{A}_A^n$ for some $n \geq 1$. We then endow this latter module with its product topology, and endow M with the subspace topology. More intrinsically, since A is p-adically complete (by assumption), we may write $M = \varprojlim_a M/p^a M$. Each of the quotients $M/p^a M$ is then a projective $(A/p^a A)((T))$-module, and so has natural topology, making it a Tate $A/p^a A$-module in the sense of [Dri06]. The topology on M is then the projective limit of these Tate module topologies. We note that multiplication by T is topologically nilpotent on M.

Similarly, any Zariski locally finite free $\mathbf{A}_A^+$-module $\mathfrak{M}$ has a natural topology. We may describe this topology in an analogous manner to the case considered in the preceding paragraph. Namely, such a module is a finitely generated and projective $\mathbf{A}_A^+$-module, and thus is a direct summand of $(\mathbf{A}_A^+)^n$ for some $n \geq 0$. If we endow this latter module with its product topology then the natural topology on $\mathfrak{M}$ is its corresponding subspace topology. In this case, though, this topology admits a more intrinsic and succinct description: it is the (p, T)-adic topology on $\mathfrak{M}$.

We remind the reader that a topological group G is said to be *Polish* if its underlying topological space is Polish, i.e., is separable and completely metrizable. As explained in Remark C.5, this is equivalent to G being complete (as a topological group) and second countable (as a topological space).

D.3 Lemma. *If A is a p-adically complete $\mathcal{O}$-algebra for which A/p is countable, then $\mathbf{A}_A$ is Polish, and consequently any finitely generated projective $\mathbf{A}_A$-module is Polish when endowed with its canonical topology.*

Proof. We will use the fact that a countable product of Polish spaces is Polish, as is a closed subspace of a Polish space. It follows from these facts, and from the description of the natural topology on a finitely generated projective $\mathbf{A}_A$-module given in Remark D.2, that the canonical topology on any such module is Polish, provided that that $\mathbf{A}_A$ itself is Polish.

We may write $\mathbf{A}_A = \varprojlim_a \mathbf{A}_{A/p^aA}$, so that $\mathbf{A}_A$ is a closed subset of the (countable) product of the various spaces $\mathbf{A}_{A/p^aA}$. It suffices, then, to show that each of these spaces is Polish; in other words, we reduce to the case when A is a countable $\mathbf{Z}/p^a$-algebra for some $a \geq 1$.

Since A is countable, so is each of the quotients $\mathbf{A}_A^+/T^n$. Choose a subset $X_n \subseteq \mathbf{A}_A^+$ which maps bijectively onto $\mathbf{A}_A^+/T^n$, and write $X := \bigcup_{m,n} T^{-m}X_n$. Then X is a countable dense subset of $\mathbf{A}_A$, and thus $\mathbf{A}_A$ is separable.

The topology on $\mathbf{A}_A^+$ is the T-adic topology, and thus is metrizable (as is any I-adic topology on a ring). Since $\mathbf{A}_A^+$ is T-adically complete, it is in fact completely metrizable. The same is then evidently true for

$$\mathbf{A}_A := \mathbf{A}_A^+[1/T] = \bigcup_m T^{-m}\mathbf{A}_A^+. \qquad \square$$

D.4 Remark. Clearly $T\mathbf{A}_A^+$ is an open subgroup of $\mathbf{A}_A$ that is closed under multiplication and consists of topologically nilpotent elements. Thus if A/p is countable, then $\mathbf{A}_A$ satisfies the conditions of Proposition C.6, and thus the canonical topology on finitely generated projective $\mathbf{A}_A$-modules constructed in Remark D.2 is a particular case of the canonical topology constructed on any finitely generated $\mathbf{A}_A$-module in Proposition C.6.

We now present some additional facts related to the preceding concepts which will be needed in the sequel.

D.5 Lemma. *A finitely generated $\mathbf{A}_A^+$-module is projective of rank d if and only if for each $a \geq 1$, the quotient $\mathfrak{M}/p^a\mathfrak{M}$ is projective of rank d as an $\mathbf{A}_{A/p^aA}^+$-module. Similarly, a finitely generated $\mathbf{A}_A$-module M is projective of rank d if and only if for each $a \geq 1$, the quotient M/p^aM is projective of rank d as an $\mathbf{A}_{A/p^aA}$-module.*

Proof. This is immediate from [GD71, Prop. 0.7.2.10(ii)], applied to the p-adically complete rings $\mathbf{A}_A^+$ and $\mathbf{A}_A$. $\square$

D.6 Definition. If M is a finitely generated $\mathbf{A}_A$-module, then a *lattice* in M is a finitely generated $\mathbf{A}_A^+$-submodule $\mathfrak{M} \subseteq M$ whose $\mathbf{A}_A$-span is M.

Note that any finitely generated $\mathbf{A}_A$-module contains a lattice. (If $f\colon \mathbf{A}_A^r \to M$ is a surjection from a finitely generated free $\mathbf{A}_A$-module onto the finitely generated $\mathbf{A}_A$-module M, then the image of the restriction of f to $(\mathbf{A}_A^+)^r$ is a lattice in M.)

We record some additional lemmas and a remark which apply in the case when A is an $\mathcal{O}/\varpi^a$-algebra for some $a \geq 1$.

D.7 Lemma. *Let A be an $\mathcal{O}/\varpi^a$-module for some $a \geq 1$, and let M be a finitely generated $\mathbf{A}_A$-module.*

1. *If $\mathfrak{M}$ and $\mathfrak{N}$ are two lattices contained in a finitely generated $\mathbf{A}_A$-module M, then there exists $n \geq 0$ such that $T^n\mathfrak{M} \subseteq \mathfrak{N} \subseteq T^{-n}\mathfrak{M}$.*
2. *If in addition either A is Noetherian or M is projective, then any lattice in M is T-adically complete.*
3. *If A is Noetherian, if $\mathfrak{M}$ is a lattice in M, and if $\mathfrak{N}$ is an $\mathbf{A}_A^+$-submodule of M such that $T^n\mathfrak{M} \subseteq \mathfrak{N} \subseteq T^{-n}\mathfrak{M}$ for some $n \geq 0$, then $\mathfrak{N}$ is a lattice in M.*

Proof. To prove (1), it suffices to prove one of the inclusions; the reverse inclusion may then be obtained (possibly after increasing n) by switching the roles of $\mathfrak{M}$ and $\mathfrak{N}$. Since $\mathbf{A}_A = \mathbf{A}_A^+[1/T]$, we find that $\mathfrak{N} \subseteq \mathfrak{M}[1/T] = \bigcup_{n=0}^{\infty} T^{-n}\mathfrak{M}$. Since $\mathfrak{N}$ is finitely generated as an $\mathbf{A}_A^+$-module, we obtain that $\mathfrak{N} \subseteq T^{-n}\mathfrak{M}$ for some sufficiently large value of n, as required.

To prove (2), we note that if A is Noetherian, than $\mathbf{A}_A^+$ is also Noetherian, and thus any finitely generated $\mathbf{A}_A^+$-module is T-adically complete, since $\mathbf{A}_A^+$ itself is. If M is projective, then we write M as a direct summand of a finitely generated free $\mathbf{A}_A$-module, so that $\mathfrak{M}$ may then be embedded into a finitely generated free $\mathbf{A}_A^+$-module. Thus $\mathfrak{M}$ is T-adically separated, and hence the kernel of any surjection $(\mathbf{A}_A^+)^r \to \mathfrak{M}$ is T-adically closed. Since $(\mathbf{A}_A^+)^r$ is T-adically complete, we conclude that the same is true of $\mathfrak{M}$.

Suppose now that A is Noetherian, and that we are in the situation of (3). The inclusion $T^n\mathfrak{M} \subseteq \mathfrak{N}$ shows that $M = \mathfrak{M}[1/T] \subseteq \mathfrak{N}[1/T]$, so that $\mathfrak{N}$ generates M as an $\mathbf{A}_A$-module. We must show that $\mathfrak{N}$ is furthermore finitely generated over $\mathbf{A}_A^+$. For this, we first note that, since $\mathfrak{M}$ is T-adically complete (by (2)), the inclusion $T^n\mathfrak{M} \subseteq \mathfrak{N} \subseteq T^{-n}\mathfrak{M}$ shows that $\mathfrak{N}$ is also T-adically complete. It also shows that $\mathfrak{N}/T\mathfrak{N}$ is a subquotient of the finitely generated A-module $T^{-n}\mathfrak{M}/T^{n+1}\mathfrak{M}$, and thus is a finitely generated A-module (as A is Noetherian). Since $\mathfrak{N}$ is T-adically complete, an application of the topological Nakayama lemma shows that $\mathfrak{N}$ is finitely generated over $\mathbf{A}_A^+$, as required. □

D.8 Remark. By [EG21, Thm. 5.1.14] (a theorem of Drinfeld), if A is an $\mathcal{O}/\varpi^a$-algebra for some $a \geq 1$, then we may think of a finitely generated projective $\mathbf{A}_A$-module as a Tate A-module M together with a topologically nilpotent automorphism T. In this optic a lattice in our sense is precisely a lattice in the Tate module M which is also an $\mathbf{A}_A^+$-submodule (by definition, a lattice $L \subseteq M$ is an open submodule with the property that for every open submodule $U \subseteq L$, the A-module L/U is finitely generated), as the following lemma shows.

D.9 Lemma. *If A is an $\mathcal{O}/\varpi^a$-algebra for some $a \geq 1$, and if $\mathfrak{M}$ is an $\mathbf{A}_A^+$-submodule of a finitely generated projective $\mathbf{A}_A$-module M, then the following conditions on $\mathfrak{M}$ are equivalent:*

1. *$\mathfrak{M}$ is a lattice in M (in the sense of Definition* D.6)
2. *$\mathfrak{M}$ is open in M, and for every $\mathbf{A}_A^+$-submodule U of $\mathfrak{M}$ which is open in M, the quotient $\mathfrak{M}/U$ is finitely generated over A.*
3. *$\mathfrak{M}$ is open in M, and for every A-submodule U of $\mathfrak{M}$ which is open in M, the quotient $\mathfrak{M}/U$ is finitely generated over A (i.e., $\mathfrak{M}$ is a lattice in M, when M is thought of as a Tate module over A in the sense of Remark* D.8).

Proof. Suppose that $\mathfrak{M}$ is a lattice in M, in the sense of Definition D.6. If we choose a surjection $(\mathbf{A}_A^+)^r \to \mathfrak{M}$ for some $n \geq 0$, then the induced surjection $\mathbf{A}_A^r \to M$ is a continuous and open map, and so $\mathfrak{M}$ is open in M, since $(\mathbf{A}_A^+)^r$ is open in $\mathbf{A}_A^r$. Furthermore, since the submodules $T^n(\mathbf{A}_A^+)^r$ $(n \geq 0)$ form a neighborhood basis of 0 in $(\mathbf{A}_A^+)^r$, we see that their images $T^n\mathfrak{M}$ form a neighborhood basis of 0 in $\mathfrak{M}$. Thus if U is any open A-submodule of $\mathfrak{M}$, then $\mathfrak{M}/U$ is a quotient of $\mathfrak{M}/T^n\mathfrak{M}$ for some $n \geq 0$, and thus is finitely generated over A. Thus (1) implies (3).

Clearly (3) implies (2), and so we suppose that (2) holds. Let $\mathfrak{N}$ be an $\mathbf{A}_A^+$-submodule of M that is a lattice in the sense of Definition D.6, so that $M = \mathfrak{N}[1/T] = \bigcup_{n \geq 0} T^{-n}\mathfrak{N}$. Then $\mathfrak{M} = \bigcup_{n \geq 0} \mathfrak{M} \cap T^{-n}\mathfrak{N}$, and thus

$$\mathfrak{M}/(\mathfrak{M} \cap \mathfrak{N}) = \bigcup_{n \geq 0} (\mathfrak{M} \cap T^{-n}\mathfrak{N})/(\mathfrak{M} \cap \mathfrak{N}).$$

But $\mathfrak{M}/(\mathfrak{M} \cap \mathfrak{N})$ is finitely generated over A by assumption, and thus

$$\mathfrak{M}/\mathfrak{M} \cap \mathfrak{N} = (\mathfrak{M} \cap T^{-n}\mathfrak{N})/(\mathfrak{M} \cap \mathfrak{N}),$$

for some sufficiently large value of n, implying that $\mathfrak{M} \subseteq T^{-n}\mathfrak{N}$. In particular $\mathfrak{M}$ is T-adically complete, being open (and hence closed) in the lattice $T^{-n}\mathfrak{N}$, which is T-adically complete by Lemma D.7 (2). Since T is an automorphism of M, we find that $T\mathfrak{M}$ is an open submodule of $\mathfrak{M}$, so that $\mathfrak{M}/T\mathfrak{M}$ is finitely generated over A. As $\mathfrak{M}$ is T-adically complete, the topological Nakayama lemma implies that $\mathfrak{M}$ is finitely generated over $\mathbf{A}_A^+$.

Since T is topologically nilpotent and $\mathfrak{N}$ is finitely generated over $\mathbf{A}_A^+$, we also find that $T^n\mathfrak{N} \subseteq \mathfrak{M}$ for some sufficiently large value of n. Thus $\mathfrak{M}[1/T] = \mathfrak{N}[1/T] = M$. This completes the proof that $\mathfrak{M}$ is a lattice in M in the sense of Definition D.6, showing that (2) implies (1). $\square$

D.10 Remark. It follows from Lemmas D.7 and D.9 that if A is a $\mathcal{O}/\varpi^a$-algebra, and M is a finitely generated projective $\mathbf{A}_A$-module, then the lattices in M form a neighborhood basis of the origin in M.

We also note the following technical lemma.

D.11 Lemma. *If $A \subseteq B$ is an inclusion of $\mathcal{O}/\varpi^a$-algebras for some $a \geq 1$, with A Noetherian, if M is a projective $\mathbf{A}_A$-module M with extension of scalars $M_B := \mathbf{A}_B \otimes_{\mathbf{A}_A} M$ to B, and if $\mathfrak{M}_B$ is a lattice in M_B, then $\mathfrak{M} := M \cap \mathfrak{M}_B$ (the intersection takes place in M_B) is a lattice in M, having the additional property that $M \cap T^n \mathfrak{M}_B = T^n \mathfrak{M}$ for any integer n.*

Proof. Choose a projective complement to M, i.e., a finitely generated projective $\mathbf{A}_A$-module N such that $M \oplus N$ is free over $\mathbf{A}_A$, and let $\mathfrak{N}_B$ be a lattice in N_B. Then it suffices to prove the lemma with M replaced by $M \oplus N$ and with $\mathfrak{M}_B$ replaced by $\mathfrak{M}_B \oplus \mathfrak{N}_B$; thus we may and do suppose that M is free over $\mathbf{A}_A$. Choose a lattice $\mathfrak{M}'$ in M which is free over $\mathbf{A}_A^+$, and note that $\mathfrak{M}'_B := \mathbf{A}_B^+ \mathfrak{M}'$ is a lattice in M_B, and that $M \cap T^n \mathfrak{M}'_B = T^n \mathfrak{M}'$ for any integer n.

Lemma D.7 (1) shows that $T^n \mathfrak{M}'_B \subseteq \mathfrak{M}_B \subseteq T^{-n} \mathfrak{M}'_B$ for some sufficiently large value of n. Intersecting with M, we find that $T^n \mathfrak{M}' \subseteq \mathfrak{M} \subseteq T^{-n} \mathfrak{M}'$, so that $\mathfrak{M}$ is a lattice in M, by Lemma D.7 (3). Finally, by the definition of $\mathfrak{M}$, we find that $M \cap T^n \mathfrak{M}_B = T^n \mathfrak{M}$ for any integer n. □

D.12 GROUP ACTIONS

We now prove some lemmas which allow us to check whether the action of a topological group on a Tate module is continuous, and to extend such an action from a group to its completion.

D.13 Lemma. *Let G be a topological group acting on a Tate module M, and assume that G admits a neighborhood basis of the identity consisting of open subgroups. Then the following are equivalent:*

1. *The action $G \times M \to M$ is continuous.*
2. *a) For each $m \in M$, the map $G \to M$, $g \mapsto gm$ is continuous at the identity of G,*
   b) for each $g \in G$, the map $M \to M$, $m \mapsto gm$ is continuous, and
   c) for any lattice L in M, there is an open subgroup H of G that preserves L.

Furthermore, if these equivalent conditions hold, then the following stronger form of condition (2)(a) holds:

(2)(a') For each $m \in M$, the map $G \to M$, $g \mapsto gm$ is equicontinuous at the identity of G.

D.14 Remark. The assumption in Lemma D.13 that G admits a neighborhood basis of open subgroups holds in particular if G is a profinite group, or if G is a subgroup of a profinite group with the subspace topology. In particular, it holds for the groups $\mathbf{Z}_p$ and for $\mathbf{Z} \subset \mathbf{Z}_p$.

Proof of Lemma D.13. Suppose first that (1) holds; then clearly conditions (2)(a) and (2)(b) hold. Let L be a lattice in M. Since the action morphism $G \times M \to M$ is continuous, we may find an open subgroup H of G and an open submodule U of M, such that $HU \subseteq L$. Replacing G by H, we may thus suppose that $GU \subseteq L$. Now consider the morphism $G \times L/U \to M/L$ induced by the action morphism. Since L is a lattice, the A-module L/U is finitely generated. Let $\{m_i\}$ be a finite generating set. Since M/L is discrete, for each generator m_i, there is an open subgroup H_i of G such that $H_i m_i$ is constant, and thus equal to zero, modulo L. If we write $H := \bigcap_i H_i$, then H is an open subgroup of G such that $Hm_i \subseteq L$ for every i. Consequently, $HL \subseteq L$, verifying that (2)(c) holds as well.

Suppose conversely that (2) holds. We first note that we may strengthen (2)(a) as follows: if $m \in M$, then the orbit map $g \mapsto gm$ is a continuous map $G \to M$. Indeed, if $g_0 \in G$, then we may write this map as the composite of the continuous automorphism $g \mapsto gg_0^{-1}$ of G and the orbit map $g \mapsto gg_0m$. The latter orbit map is continuous at the identity of G, by (2)(a), and so the orbit map of m is continuous at g_0.

We now wish to prove that the action map $G \times M \to M$ is continuous. Let $(g, m) \in G \times M$. Any neighborhood of the image $gm \in M$ contains a neighborhood of the form $gm + L$, where L is a lattice in M. Given a lattice L, then, we must find a neighborhood of (g, m) whose image lies in $gm + L$.

Since g is a continuous automorphism of M, by (2)(a), we may find a lattice L' such that $gL' \subseteq L$. By (2)(c), we may find an open subgroup $H' \subseteq G$ that preserves L'. Then H' acts on M/L', and since this quotient is discrete, and since the orbit maps for the action are continuous (by what we proved above), it follows that the action of H' on M/L' is smooth, and so we may find an open subgroup H of H' that fixes the image of m in M/L'. Then we find that $(gH)(m + L') \subseteq gm + L$, and since $gH \times (m + L')$ is an open neighborhood of $(g, m) \in G \times M$, we have proved the required continuity.

Finally, suppose that conditions (1) and (2) hold; we must show that the continuity condition of (2)(a) can be upgraded to the equicontinuity condition of (2)(a'). To do this, we need to show that for any two lattices $L, L' \subseteq M$, there is an open subgroup H of G such that for each $m \in L$, we have $Hm \subseteq m + L'$.

Replacing L' by $L \cap L'$, we can assume that $L' \subseteq L$. By (2)(c), we can choose H such that $HL' \subseteq L'$ and $HL \subseteq L$, so that we have an induced continuous map $H \times (L/L') \to L/L'$. Since L/L' is a finitely generated A-module and is discrete, after replacing H by an open subgroup we can assume that H acts trivially on L/L', as required. □

D.15 Lemma. *Let G be a Hausdorff topological group, which admits a neighborhood basis of the identity consisting of open subgroups, and suppose furthermore* *that for any open subgroup H of G, the quotient G/H is finite; equivalently, suppose that the completion $\widehat{G}$ of G is profinite.*

If M is a Tate module, endowed with a continuous action $G\times M\to M$, then the action $G\times M\to M$ extends to a continuous action $\widehat{G}\times M\to M$.

Proof. We have to show that the continuous action $G\times M\to M$ extends (necessarily uniquely, since G is dense in $\widehat{G}$) to a continuous map $\widehat{G}\times M\to M$. (The fact that this map will induce a $\widehat{G}$-action on M follows from the corresponding fact for G, and the density of G in $\widehat{G}$.) Since M is the union of its lattices, it suffices to show that the induced map $G\times L\to M$ extends to a continuous map $\widehat{G}\times L\to M$, for each lattice $L\subseteq M$. For this, it suffices to show that the map $G\times L\to M$ is uniformly continuous, for each lattice L. That is, for any lattice $L'\subseteq M$, we have to find an open subgroup $H\subseteq G$ and a sublattice $L''\subseteq L$ such that $gHm+HL''\subseteq gm+L'$ for all $g\in G$ and all $m\in L$.

By Lemma D.13, the conditions (2)(a)–(c) of that lemma hold. We begin by showing that we may find a sublattice L'' of L such $GL''\subseteq L'$. For this, we first note that, by (2)(c), we may find an open subgroup $H\subseteq G$ such that $HL'=L'$. If we let $\{g_i\}$ denote a (finite!) set of coset representatives for $H\backslash G$, then since each g_i induces a continuous automorphism of M, we may find a lattice L_i such that $g_iL_i\subseteq L'$. Taking into account Remark D.10, we may then find a lattice $L''\subseteq L'\cap\bigcap_i L_i$, and by construction $GL''\subseteq L'$.

Condition (2)(a') allows us to choose an open subgroup $H\subseteq G$ such that $Hm\subseteq m+L''$ for all $m\in L$. We then find that

$$gHm+HL''\subseteq gm+GL''\subseteq gm+L'$$

for all $g\in G$, as required. □

D.16 T-QUASI-LINEAR ENDOMORPHISMS

In order to apply the preceding results to the case of interest to us (the semilinear action of Γ on (φ,Γ)-modules), we now introduce and study the notion of a T-quasi-linear endomorphism of a finite projective $\mathbf{A}_A$-module. The relevance of this notion to (φ,Γ)-modules is explained in Lemma D.27 below.

D.17 Definition. If M is a finite projective $\mathbf{A}_A$-module, then a *T-quasi-linear endomorphism* of M is a morphism $f\colon M\to M$ which is $W(k)\otimes_{\mathbf{Z}_p}A$-linear, and which furthermore satisfies the following (T-quasi-linearity) condition: there exist power series $a(T)\in(\mathbf{A}_A^+)^\times$ and $b(T)\in(p,T)\mathbf{A}_A^+$ such that

$$f(Tm)=a(T)Tf(m)+b(T)Tm$$

for every $m\in M$.

D.18 Lemma. *If f is T-quasi-linear, then for all $n \in \mathbf{Z}$, we may write*

$$f(T^n m) = a(T)^n T^n f(m) + b_n(T) T^n m$$

for all $m \in M$, where $a(T) \in (\mathbf{A}_A^+)^\times$ and $b_n(T) \in (p, T)\mathbf{A}_A^+$.

Proof. Note that the case $n = 0$ is trivial (taking $b_0(T) = 0$), while if the claim holds for some $n \geq 1$, then we may write

$$f(T^{-n} m) = a(T)^{-n} T^{-n} f(m) - a(T)^{-n} b_n(T) T^{-n} m.$$

It therefore suffices to prove the result for $n \geq 1$. This may be proved by induction on n, the case $n = 1$ being the definition of T-quasi-linearity. Indeed, if the claim holds for n, then we have

$$\begin{aligned} f(T^{n+1} m) = f(T^n(Tm)) &= a(T)^n T^n f(Tm) + b_n(T) T^n (Tm) \\ &= a(T)^n T^n (a(T) T f(m) + b(T) T m) + b_n(T) T^{n+1} m \\ &= a(T)^{n+1} T^{n+1} f(m) + (b_n(T) + b(T) a(T)^n) T^{n+1} m, \end{aligned}$$

as required. □

D.19 Lemma. *Let A be an $\mathcal{O}/\varpi^a$-algebra for some $a \geq 1$, and let M be a finite projective $\mathbf{A}_A$-module. Let f be a T-quasi-linear endomorphism of M, and let $\mathfrak{M}$ be a lattice in M. Then there is an integer $m \geq 0$ such that for each $s \in \mathbf{Z}$ and $n \geq 0$ we have $f^n(T^s \mathfrak{M}) \subseteq T^{s-mn} \mathfrak{M}$.*

Proof. Choose a finite set $\{m_i\}$ of generators for $\mathfrak{M}$ as an $\mathbf{A}_A^+$-module. Choose m sufficiently large that we have $T^m f(m_i) \in \mathfrak{M}$ for each i. Then by Lemma D.18 and the $W(k) \otimes_{\mathbf{Z}_p} A$-linearity of f, we see that for each $s \in \mathbf{Z}$ we have $f(T^{s+m}\mathfrak{M}) \subseteq T^s \mathfrak{M}$, which gives the result in the case $n = 1$. The general case follows by induction on n. □

As noted in Remark D.2, any finite projective $\mathbf{A}_A$-module M has a natural topology, so it makes sense to speak of an endomorphism of M being continuous, or topologically nilpotent.

D.20 Lemma. *If A is an $\mathcal{O}/\varpi^a$-algebra for some $a \geq 1$, and M is a finite projective $\mathbf{A}_A$-module, then any T-quasi-linear endomorphism of M is necessarily continuous.*

Proof. This is immediate from Lemma D.19. □

D.21 Lemma. *If A is an $\mathcal{O}/\varpi^a$-algebra for some $a \geq 1$, and M is a finite projective $\mathbf{A}_A$-module, and if f is a T-quasi-linear endomorphism of M, then the following are equivalent:*

1. *f is topologically nilpotent.*
2. *There exists a lattice $\mathfrak{M}$ in M and some $n \geq 1$ such that $f^n(\mathfrak{M}) \subseteq T\mathfrak{M}$.*
3. *There exists a lattice $\mathfrak{M}$ in M and some $n \geq 1$ such that $f^n(\mathfrak{M}) \subseteq (p,T)\mathfrak{M}$.*
4. *There exists a lattice $\mathfrak{M}$ in M such that for any $m \geq 1$, there exists n_0 such that for any $s \in \mathbf{Z}$ and any $n \geq n_0$, we have $f^n(T^s\mathfrak{M}) \subseteq T^{s+m}\mathfrak{M}$.*
5. *For any lattice $\mathfrak{M}$ in M and any $m \geq 1$, there exists n_0 such that for any $s \in \mathbf{Z}$ and any $n \geq n_0$, we have $f^n(T^s\mathfrak{M}) \subseteq T^{s+m}\mathfrak{M}$.*

Proof. By Lemma D.7 (1), we see that (4) $\implies$ (5) $\implies$ (1) $\implies$ (2) $\implies$ (3), so we only need to show that (3) $\implies$ (4). To this end, note firstly that it follows from Lemma D.18 and the $\mathbf{Z}_p$-linearity of f that for each $i, j \geq 0$ and $s \in \mathbf{Z}$ we have

$$f((p,T)^j T^s f^i(\mathfrak{M})) \subseteq (p,T)^j T^s f^{i+1}(\mathfrak{M}) + (p,T)^{j+1} T^s f^i(\mathfrak{M}).$$

It follows (by induction on m) that for each $m \geq 0$ and $s \in \mathbf{Z}$ we have

$$f^m(T^s\mathfrak{M}) \subseteq \sum_{i=0}^{m} (p,T)^{m-i} T^s f^i(\mathfrak{M}).$$

In particular, if we take $m = n$ and recall that $f^n(\mathfrak{M}) \subseteq (p,T)\mathfrak{M}$ by hypothesis, we find that

$$f^n(T^s\mathfrak{M}) \subseteq (p,T)T^s\mathfrak{M} + \sum_{i=1}^{n-1} (p,T)^{n-i} T^s f^i(\mathfrak{M}). \tag{D.22}$$

The same argument shows that if $\mathfrak{N}$ is an $\mathbf{A}_A^+$-submodule of M with the property that

$$\mathfrak{N} \subseteq \sum_{i=0}^{n-1} (p,T)^{a_i} T^s f^i(\mathfrak{M})$$

for non-negative integers $a_0, \ldots, a_{n-1}$, then

$$f(\mathfrak{N}) \subseteq \sum_{i=0}^{n-1} (p,T)^{b_i} T^s f^i(\mathfrak{M})$$

where $b_0 = \min(a_0 + 1, a_{n-1} + 1)$, and $b_i = \min(a_i + 1, a_{i-1})$ if $i > 0$. It follows by an easy induction on N (with the base case being given by (D.22)) that for all $N \geq n$, if we write $N + 1 = (q+1)n + r$ with $0 \leq r < n$, then we have

$$f^N(T^s\mathfrak{M}) \subseteq \sum_{i=0}^{n-1} (p,T)^{c_i} T^s f^i(\mathfrak{M})$$

where

$$(c_{n-1},\dots,c_0) = (q,q,\dots,q) + (r,r+1,\dots,n-1,1,2,\dots,r).$$

In particular we see that for all $N \geq n$ we have

$$f^N(T^s\mathfrak{M}) \subseteq (p,T)^{\lfloor (N+1)/n \rfloor -1} T^s \sum_{i=0}^{n-1} f^i(\mathfrak{M}).$$

Now, by Lemma D.19, for any sufficiently large t we have $f^i(\mathfrak{M}) \subseteq T^{-t}\mathfrak{M}$ for $0 \leq i \leq n-1$, so it follows that for $N \geq n$ we have

$$f^N(T^s\mathfrak{M}) \subseteq (p,T)^{\lfloor (N+1)/n \rfloor -1} T^{s-t}\mathfrak{M}.$$

Since $p^a = 0$ in A, we also have $(p,T)^{n+a-1} \subseteq (T^n)$ for all $n \geq 0$, so that if $N \geq an-1$ then we have

$$f^N(T^s\mathfrak{M}) \subseteq T^{s+\lfloor (N+1)/n \rfloor - a - t}\mathfrak{M}.$$

Since $\lfloor (N+1)/n \rfloor - a - t \to \infty$ as $N \to \infty$, we have (4), as required. □

D.23 Corollary. *Suppose that A is an $\mathcal{O}/\varpi^a$-algebra for some $a \geq 1$. If M is a finite projective $\mathbf{A}_A$-module, and if f is a T-quasi-linear endomorphism of M, then the following are equivalent:*

1. *f is topologically nilpotent.*
2. *The action of f on $M \otimes_{\mathcal{O}/\varpi^a} \mathbf{F}$ is topologically nilpotent.*

Proof. Obviously (1) $\implies$ (2). Conversely, if (2) holds, then by the equivalence of conditions (1) and (2) of Lemma D.21 for the action of f on $M \otimes_{\mathcal{O}/\varpi^a} \mathbf{F}$, we see that condition (3) of Lemma D.21 holds (for the action of f on M), and therefore condition (1) of Lemma D.21 holds, as required. □

The following lemmas provide the key examples of T-quasi-linear endomorphisms, and explain our interest in the concept.

We suppose that $\mathbf{A}_A$ is endowed with a continuous action of $\mathbf{Z}_p$ by A-algebra automorphisms which preserve $\mathbf{A}_A^+$, and suppose further that, for some topological generator γ of $\mathbf{Z}_p$, that

$$\gamma(T) - T \in (p,T)T\mathbf{A}_A^+. \qquad \text{(D.24)}$$

D.25 Lemma. *If* (D.24) *holds, then* γ *preserves the ideals* (T) *and* (p,T) *of* $\mathbf{A}_A^+$. *Furthermore for each integer* $n \geq 1$ *we have*

$$\gamma^n(T) - T \in (p,T)T\mathbf{A}_A^+. \tag{D.26}$$

Proof. The first claim follows immediately from (D.24), which shows that $\gamma(T)$ is a unit multiple of T, together with the fact that γ preserves $\mathbf{A}_A^+$. Then (D.26) follows by induction on n (the case $n=1$ being (D.24)). □

D.27 Lemma. *If* M *is a finite projective* $\mathbf{A}_A$*-module which is endowed with an action of the subgroup* $\langle\gamma\rangle$ *of* $\mathbf{Z}_p$ *which is semi-linear with respect to the given action of this group on* $\mathbf{A}_A$ *(obtained by restricting the* $\mathbf{Z}_p$*-action), then for any integer* $n \geq 1$, $f := \gamma^n - 1$ *is a* T*-quasi-linear endomorphism of* M.

Proof. We have $f(Tm) = \gamma^n(T)f(m) + (\gamma^n(T) - T)m$, so it follows from D.26 that f is T-quasi-linear. □

D.28 Lemma. *Suppose that* A *is an* $\mathcal{O}/\varpi^a$*-algebra for some* $a \geq 1$*, and that* $\mathbf{A}_A$ *is endowed with an action of* $\mathbf{Z}_p$ *satisfying* (D.24)*. Let* M *be a finite projective* $\mathbf{A}_A$*-module, equipped with a semi-linear action of* $\langle\gamma\rangle \subset \mathbf{Z}_p$*. Then the following are equivalent:*

1. *The action of* $\langle\gamma\rangle$ *extends to a continuous action of* $\mathbf{Z}_p$.
2. *The action of* $\langle\gamma\rangle$ *on* $M \otimes_{\mathcal{O}/\varpi^a} \mathbf{F}$ *extends to a continuous action of* $\mathbf{Z}_p$.
3. *For any lattice* $\mathfrak{M} \subseteq M$*, and any* $n \geq 1$*, there exists* $s \geq 0$ *such that* $(\gamma^{p^s} - 1)^i (\mathfrak{M}) \subseteq (p,T)^n \mathfrak{M}$ *for all* $i \geq 1$.
4. *For any lattice* $\mathfrak{M} \subseteq M$*, there exists* $s \geq 0$ *such that* $(\gamma^{p^s} - 1)(\mathfrak{M}) \subseteq T\mathfrak{M}$.
5. *For some lattice* $\mathfrak{M} \subseteq M$ *and some* $s \geq 0$*, we have* $(\gamma^{p^s} - 1)(\mathfrak{M}) \subseteq (p,T)\mathfrak{M}$.
6. *The action of* $\gamma - 1$ *on* $M \otimes_{\mathcal{O}/\varpi^a} \mathbf{F}$ *is topologically nilpotent.*
7. *The action of* $\gamma - 1$ *on* M *is topologically nilpotent.*

Proof. Noting that if $n \geq a$ then $(p,T)^n \subseteq (T)$, we see that $(3) \Longrightarrow (4)$, and by Lemma D.21 and Corollary D.23 we have $(4) \Longrightarrow (5) \Longrightarrow (6) \Longrightarrow (7)$. (For $(5) \Longrightarrow (6)$, we also use that $(\gamma - 1)^{p^s} \equiv (\gamma^{p^s} - 1) \pmod{\varpi}$.)

We next show that $(7) \Longrightarrow (3)$. Suppose that (7) holds. Recalling again that for each $s \geq 0$ we have $(\gamma - 1)^{p^s} \equiv (\gamma^{p^s} - 1) \pmod{\varpi}$, it follows from Corollary D.23 and Lemma D.27 that the action of $(\gamma^{p^s} - 1)$ on M is topologically nilpotent for each $s \geq 0$. We now argue by induction on a, noting that for $a = 1$, the implication $(7) \Longrightarrow (3)$ is immediate from Lemma D.21. We may therefore assume that

$$(\gamma^{p^s} - 1)^i (\mathfrak{M}) \subseteq (p,T)^n \mathfrak{M} + \varpi^{a-1} M$$

for all $i \geq 1$. It follows in particular that $p(\gamma^{p^s} - 1)^i(\mathfrak{M}) \subseteq (p,T)^n \mathfrak{M}$ for all $i \geq 1$, so that for any $t \geq 0$ and any $i \geq 1$, we have

$$((\gamma^{p^{s+t}}-1)^i-(\gamma^{p^s}-1)^{ip^t})(\mathfrak{M})\subseteq (p,T)^n\mathfrak{M}.$$

(To see this, write $(\gamma^{p^{s+t}}-1)=((\gamma^{p^s}-1)+1)^{p^t}-1$ and use the binomial theorem.) It therefore suffices to show that there is some $t\geq 1$ for which $(\gamma^{p^s}-1)^{ip^t}(\mathfrak{M})\subseteq (p,T)^n\mathfrak{M}$ for all $i\geq 1$; but we have already seen that $(\gamma^{p^s}-1)$ acts topologically nilpotently on M, so by Lemma D.21 we can even arrange that $(\gamma^{p^s}-1)^{ip^t}(\mathfrak{M})\subseteq T^n\mathfrak{M}$ for all $i\geq 1$.

We have shown the equivalence of conditions (3)–(7). Suppose now that (1) holds. Then Lemma D.13 shows that, for each lattice $\mathfrak{M}\subseteq M$, we have $\gamma^{p^s}(\mathfrak{M})\subseteq \mathfrak{M}$ for all sufficiently large $s\geq 0$, and that furthermore, for each $m\in\mathfrak{M}$, there is some $t(m)$ such that if $t\geq t(m)$, then $\gamma^{p^t}(m)\in m+(p,T)\mathfrak{M}$. Letting $m_1,\dots,m_n$ be generators for $\mathfrak{M}$ as an $\mathbf{A}_A^+$-module, we see that if s is sufficiently large, then $(\gamma^{p^s}-1)(m_i)\in(p,T)\mathfrak{M}$ for each i. Then if $\lambda_i\in\mathbf{A}_A^+$, we have

$$(\gamma^{p^s}-1)(\sum_i\lambda_i m_i)=\sum_i\gamma^{p^s}(\lambda_i)(\gamma^{p^s}-1)(m_i)+\sum_i(\gamma^{p^s}-1)(\lambda_i)m_i.$$

It follows from (D.26) that $(\gamma^{p^s}-1)(\lambda_i)\in T\mathbf{A}_A^+$; hence $(\gamma^{p^s}-1)(\mathfrak{M})\subseteq(p,T)\mathfrak{M}$, and so (5) holds.

Suppose now that the equivalent conditions (3)–(7) hold. We will show that (1) holds. By Lemma D.15 (taking the group G there to be $\langle\gamma\rangle\cong\mathbf{Z}$ endowed with its p-adic topology, so that $\widehat{G}\cong\mathbf{Z}_p$), it is enough to show that the conditions of Lemma D.13 (2) hold.

We begin with Lemma D.13 (2)(b), the condition that any $g\in\langle\gamma\rangle$ acts continuously on M. It is enough to show that if $\mathfrak{M}\subseteq M$ is a lattice, then there is a lattice $\mathfrak{N}$ with $g(\mathfrak{N})\subseteq\mathfrak{M}$. This is in fact a general property of semi-linear automorphisms of $\mathbf{A}_A$ which preserve $\mathbf{A}_A^+$. Indeed, let $m_1,\dots,m_n$ be generators of $\mathfrak{M}$ as an $\mathbf{A}_A^+$-module, and let $\mathfrak{N}$ be the $\mathbf{A}_A^+$-module generated by $g^{-1}(m_1),\dots,g^{-1}(m_n)$. This is a lattice, because g is an automorphism of M, and it follows easily from the semi-linearity of the action of g on M that $g(\mathfrak{N})\subseteq\mathfrak{M}$, as required.

We now check Lemma D.13 (2)(c). Let $\mathfrak{M}\subseteq M$ be some lattice, fix a choice of $n\geq 1$, and then choose s as in (3). It suffices to show that the subgroup $H=\langle\gamma^{p^s}\rangle$ of $\langle\gamma\rangle$ preserves $\mathfrak{M}$. Since $(\gamma^{p^s}-1)(\mathfrak{M})\subseteq(p,T)^n\mathfrak{M}\subseteq\mathfrak{M}$, we certainly have $\gamma^{p^s}(\mathfrak{M})\subseteq\mathfrak{M}$, so it suffices to show that $\gamma^{-p^s}(\mathfrak{M})\subseteq\mathfrak{M}$. For this, note that since (as recalled above) it follows from the congruence $(\gamma-1)^{p^s}\equiv(\gamma^{p^s}-1) \pmod{\varpi}$ that $\gamma^{p^s}-1$ acts topologically nilpotently on M, and preserves $\mathfrak{M}$, for any $m\in\mathfrak{M}$ we have

$$\gamma^{-p^s}(m)=(1-(1-\gamma^{p^s}))^{-1}(m)=m+(1-\gamma^{p^s})(m)+(1-\gamma^{p^s})^2(m)+\cdots\in\mathfrak{M},$$

as required.

To complete the verification of the conditions of Lemma D.13 (2), we need to show that for any lattice $\mathfrak{M}\subseteq M$, the orbit maps $\langle\gamma\rangle\to M$, for the various $m\in\mathfrak{M}$, are (equi)continuous at the identity of $\langle\gamma\rangle$. It is enough to show that for each $n \geq 1$, we can find s sufficiently large such that $H = \langle \gamma^{p^s} \rangle$ satisfies $Hm \subseteq m + (p,T)^n \mathfrak{M}$, for each $m \in \mathfrak{M}$, or equivalently, that $(h-1)(\mathfrak{M}) \subseteq (p,T)^n \mathfrak{M}$, for all $h \in H$. We accomplish this by choosing s as in (3).

It then suffices to prove the stronger claim that $(\gamma^{rp^s} - 1)(\mathfrak{M}) \subset (p,T)^n \mathfrak{M}$ for all $r \in \mathbf{Z}$. We have already seen that $\gamma^{rp^s}(\mathfrak{M}) \subseteq \mathfrak{M}$. If $r \geq 1$ we may write $(\gamma^{rp^s} - 1) = (\gamma^{p^s} - 1)(1 + \gamma^{p^s} + \cdots + \gamma^{(r-1)p^s})$, and since $(1 + \gamma^{p^s} + \cdots + \gamma^{(r-1)p^s})(\mathfrak{M}) \subset \mathfrak{M}$, we have $(\gamma^{rp^s} - 1)(\mathfrak{M}) \subseteq (\gamma^{p^s} - 1)(\mathfrak{M}) \subset (p,T)^n \mathfrak{M}$, as required. If $r \leq 0$, then the result follows by writing $(\gamma^{-rp^s} - 1) = -(\gamma^{rp^s} - 1)(\gamma^{-p^s})^r$.

Finally, applying the equivalence of (1) and (6) with M replaced by $M \otimes_{\mathcal{O}/\varpi^a} \mathbf{F}$, we see that (2) and (6) are equivalent, as required. □

The following lemmas will allow us to reduce the problem of investigating the topological nilpotency of a T-quasi-linear endomorphism from the projective case to the free case.

D.29 Lemma. *If M_1 and M_2 are Tate modules over a ring A, endowed with continuous endomorphisms f_1 and f_2 respectively, then f_1 and f_2 are both topologically nilpotent if and only if the direct sum $f := f_1 \oplus f_2$ is a topologically nilpotent endomorphism.*

Proof. This is immediate from the definitions. □

D.30 Lemma. *If M is a finite projective $\mathbf{A}_A$-module endowed with a T-quasi-linear endomorphism f, then we may find a finite projective $\mathbf{A}_A$-module N, endowed with a T-quasi-linear endomorphism g which is furthermore topologically nilpotent, such that $M \oplus N$ is a free $\mathbf{A}_A$-module.*

Proof. By definition, there is an $\mathbf{A}_A$-module N such that $M \oplus N$ is free. We may then take g to be the zero endomorphism of N; this is evidently both T-quasi-linear and topologically nilpotent. □

If A is an $\mathcal{O}/\varpi^a$-algebra for some $a \geq 1$, and M is a finite projective $\mathbf{A}_A$-module, equipped with a T-quasi-linear endomorphism f, and if $A \to B$ is a morphism of $\mathcal{O}/\varpi^a$-algebras, then we have a base-changed $W(k) \otimes_{\mathbf{Z}_p} B$-linear endomorphism of $B \otimes_A M$. This base-changed endomorphism is T-adically continuous (by Lemma D.19) and it therefore induces an endomorphism f_B of the base-changed projective $\mathbf{A}_B$-module M_B (which by definition is the completion of $B \otimes_A M$ for the T-adic topology). The endomorphism f_B is evidently also T-quasi-linear. (Note that since f is not necessarily $A[[T]]$-linear, we do not define f_B by viewing M_B as $B[[T]] \otimes_{A[[T]]} M$.) If f is furthermore topologically nilpotent, then so is f_B (for example, by Lemma D.21).

D.31 Lemma. *Let $\{A_i\}_{i \in I}$ be a directed system of $\mathcal{O}/\varpi^a$-algebras, with $A = \varinjlim_{i \in I} A_i$, and suppose that I admits a least element i_0. Let M be a finite projective $\mathbf{A}_{A_{i_0}}$-module, let f be a T-quasi-linear endomorphism of M, and suppose* *that the base-changed endomorphism f_A is topologically nilpotent. Then for some $i \in I$, the base-changed endomorphism f_{A_i} is topologically nilpotent.*

Proof. By Lemma D.30, we may choose a finite projective $\mathbf{A}_{A_{i_0}}$-module N, endowed with a topologically nilpotent T-quasi-linear endomorphism g, such that $M \oplus N$ is free. By Lemma D.29, the base-changed endomorphism $f_A \oplus g_A = (f \oplus g)_A$ of $M_A \oplus N_A$ is topologically nilpotent, and it suffices to show that $f_{A_i} \oplus g_{A_i} = (f \oplus g)_{A_i}$ is topologically nilpotent for some $i \in I$. Thus we may reduce to the case when M is free, in which case we may also choose a free lattice $\mathfrak{M}$ contained in M. Taking into account Lemma D.21, for some sufficiently large n we have $f_A^n(\mathfrak{M}_A) \subseteq T\mathfrak{M}_A$, and it suffices to show that

$$f_{A_i}^n(\mathfrak{M}_{A_i}) \subseteq T\mathfrak{M}_{A_i} \tag{D.32}$$

for some $i \in I$. Lemma D.19 shows that $f^n(T^r\mathfrak{M}) \subseteq T\mathfrak{M}$ for some $r \geq 0$, and that $f^n(\mathfrak{M}) \subseteq T^{-s}\mathfrak{M}$ for some $s \geq 0$. Thus f^n induces a morphism $\mathfrak{M}/T^r\mathfrak{M} \to T^{-s}\mathfrak{M}/T\mathfrak{M}$ of finite rank free A_{i_0}-modules, which, by assumption, vanishes after base change to A. Thus this morphism in fact vanishes after base change to some A_i, and consequently (D.32) does indeed hold for this choice of A_i. □

Appendix E

Points, residual gerbes, and isotrivial families

Recall that if $\mathcal{X}$ is an algebraic stack, then the underlying set $|\mathcal{X}|$ of points of $\mathcal{X}$ is defined as the set of equivalence classes of morphisms $\operatorname{Spec} K \to \mathcal{X}$, with K being a field; two such morphisms are regarded as equivalent if they may be dominated by a common morphism $\operatorname{Spec} L \to \mathcal{X}$.

If X is a scheme (regarded as an algebraic stack), then the set of underlying points $|X|$ is naturally identified with the underlying set of points of X in the usual sense, and the equivalence class of morphisms representing a given point $x \in |X|$ has a canonical representative, namely the morphism $\operatorname{Spec} k(x) \to X$, where $k(x)$ is the residue field of x. More abstractly, this morphism is a monomorphism, and this property characterizes it uniquely, up to unique isomorphism, among the morphisms in the equivalence class corresponding to x.

If X is an algebraic space, then it is not the case that every equivalence class in $|X|$ admits a representative which is a monomorphism (see, e.g., [Sta, Tag 02Z7]). However, under mild assumptions on X, the points of $|X|$ do admit such representatives (which are then unique up to unique isomorphism); this in particular is the case if X is quasi-separated. See, e.g., the discussion at the beginning of [Sta, Tag 03I7].

If $\mathcal{X}$ is genuinely an algebraic stack, then it is not reasonable to expect the equivalence classes in $|\mathcal{X}|$ to admit monomorphism representatives in general (even if $\mathcal{X}$ is quasi-separated), since points of $\mathcal{X}$ typically admit non-trivial stabilizers. The aim of the theory of residual gerbes is to provide a replacement for such representatives.

We recall the relevant definition [Sta, Tag 06MU]):

E.1 Definition. Let $\mathcal{X}$ be an algebraic stack. We say that the *residual gerbe* at a point $x \in |\mathcal{X}|$ exists if we may find a monomorphism $\mathcal{Z}_x \hookrightarrow \mathcal{X}$ such that $\mathcal{Z}_x$ is reduced and locally Noetherian, $|\mathcal{Z}_x|$ is a singleton, and the image of $|\mathcal{Z}_x|$ in $|\mathcal{X}|$ is equal to x. If such a monomorphism exists, then $\mathcal{Z}_x$ is unique up to unique isomorphism, and we refer to it as the *residual gerbe* at x.

In the case that X is an algebraic space, the residual gerbe $\mathcal{Z}_x$ exists at every point $x \in |X|$ [Sta, Tag 06QZ], and is itself an algebraic space, called the *residual space* of X at x [Sta, Tag 06R0]. However, as already noted, in this case, if X

is furthermore quasi-separated, then the residual space at a point is simply the spectrum of a field.

If $\mathcal{X}$ is an algebraic stack with quasi-compact diagonal (e.g., if $\mathcal{X}$ is quasi-separated), then the residual gerbe exists at every point of $|\mathcal{X}|$ [Sta, Tag 06RD]).

If X is a scheme, then a *finite type point* of X is a point $x \in X$ that is locally closed. Equivalently, these are the points for which the morphism $\operatorname{Spec} k(x) \to X$ is a finite type morphism, and may be characterized more abstractly as those equivalence classes of morphisms $\operatorname{Spec} K \to X$ which admit a representative which is locally of finite type. This latter notion makes sense for an arbitrary algebraic stack, and allows us to define the notion of a *finite type point* of an algebraic stack (see, e.g., [EG21, 1.5.3]).

We then have the following results, which provide analogues of the topological characterization of the finite type points of a scheme as being those points that are locally closed.

E.2 Lemma. *Suppose that X is an algebraic space, and that $x \in |X|$ is a finite type point with the property that for any étale morphism $U \to X$ whose source is an affine scheme, the fiber over x is finite. Then, if Z_x denotes the residual space at x in X, the canonical monomorphism $Z_x \hookrightarrow X$ is an immersion.*

Proof. Consider the construction of Z_x given in [Sta, Tag 06QZ]: we choose a surjective étale morphism $U \to X$, form the union

$$U' = \coprod_{u \in U \text{ lying over } x} \operatorname{Spec} k(u),$$

and then realize $Z_x \hookrightarrow X$ as a descent of the monomorphism $U' \hookrightarrow U$. In making this construction, we may replace U by any open subscheme containing x in its image, and thus we may assume that U is affine. By assumption, there are then only finitely many points u in U lying over x, these points are all finite type points of U, and furthermore, none of these points are specializations of any of the others [Sta, Tag 03IM]. Thus $U' \hookrightarrow U$ is in fact an immersion, and thus the same is true of $Z_x \hookrightarrow X$. □

E.3 Lemma. *If $\mathcal{X}$ is an algebraic stack whose diagonal is quasi-compact, if $x \in |\mathcal{X}|$ is a finite type point, and if $\mathcal{Z}_x$ is the residual gerbe of $\mathcal{X}$ at x, then the canonical monomorphism $\mathcal{Z}_x \hookrightarrow \mathcal{X}$ is in fact an immersion.*

Proof. We prove this by examining the construction of $\mathcal{Z}_x$ carried out in the proof of [Sta, Tag 06RD]. The first step of the proof is to replace $\mathcal{X}$ by the closure of x in $|\mathcal{X}|$, regarded as a closed substack of $\mathcal{X}$ with its induced reduced structure. The assumption that $\mathcal{X}$ has quasi-compact diagonal implies that the morphism $\mathcal{I}_{\mathcal{X}} \to \mathcal{X}$ is quasi-compact, and thus that we may find a dense open substack $\mathcal{U}$ of $\mathcal{X}$ over which this morphism is flat and locally of finite presentation. Since x

is dense in $|\mathcal{X}|$, we find that $x \in |\mathcal{U}|$, and so replacing $\mathcal{X}$ by $\mathcal{U}$, we may assume that $\mathcal{I}_{\mathcal{X}} \to \mathcal{X}$ is flat and locally of finite presentation. By [Sta, Tag 06QJ], this implies that $\mathcal{X}$ is a gerbe over some algebraic space X. If we let Z_x denote the residual space at (the image in $|X|$ of) x in $|X|$, then $\mathcal{Z}_x$ is obtained as the base change of Z_x over the morphism $\mathcal{X} \to X$.

Note that, in the various reduction steps undertaken in the preceding argument, we replaced $\mathcal{X}$ by an open substack of a closed substack; it thus suffices to verify that $\mathcal{Z}_x \to \mathcal{X}$ is an immersion after making these reductions. Note that since immersions are monomorphisms, the diagonal of the stack obtained after these reductions are made is the base change of the diagonal of the original stack $\mathcal{X}$, and thus continues to be quasi-compact. Consequently, we may assume that we are in the case where $\mathcal{X}$ is a gerbe over X. Since $\mathcal{Z}_x \hookrightarrow \mathcal{X}$ is then obtained as the base change of $Z_x \hookrightarrow X$, it suffices to show that this latter morphism is an immersion. For this, if we take into account Lemma E.2, it suffices to show that X can be chosen to be quasi-separated (as every étale morphism from an affine scheme to a quasi-separated algebraic space has finite fibres over every point of $|\mathcal{X}|$; see the discussion at the beginning of [Sta, Tag 03I7]).

If we examine the proof of [Sta, Tag 06QJ], we see that X is constructed as follows: We choose a smooth surjective morphism $U \to \mathcal{X}$ whose source is a scheme, and write $R = U \times_{\mathcal{X}} U$, so that $\mathcal{X} = [U/R]$. We then factor the morphism $R \to U \times_{\operatorname{Spec} \mathbf{Z}} U$ through a morphism $R' \to U \times_{\operatorname{Spec} \mathbf{Z}} U$, where $R' \to U \times_{\operatorname{Spec} \mathbf{Z}} U$ is a flat and locally of finite presentation equivalence relation, and set $X = U/R'$. We note one additional aspect of the situation, namely that the morphism $R \to R'$ is surjective. The assumption that $\mathcal{X}$ has quasi-compact diagonal then implies that $R \to U \times_{\operatorname{Spec} \mathbf{Z}} U$ is quasi-compact, and since R surjects onto R', we find that $R' \to U \times_{\operatorname{Spec} \mathbf{Z}} U$ is again quasi-compact. Thus $X = U/R'$ is quasi-separated, as required. □

E.4 Example. We note that quasi-separatedness (or some such hypothesis) is necessary for the truth of Lemma E.3. To illustrate this, let G be an algebraic group of positive dimension over an algebraically closed field k, and let $X := G/G(k)$. Then X contains a unique finite type point x—namely, the equivalence class of the monomorphism $\operatorname{Spec} k = G(k)/G(k) \hookrightarrow G/G(k)$—and this monomorphism realizes $\operatorname{Spec} k$ as the residual space Z_x. This monomorphism is *not* an immersion, since its pull-back to G induces the monomorphism $G(k) \hookrightarrow G$, which is not an immersion.

Traditionally, in the theory of moduli problems, a family of some objects parameterized by a base scheme T is called *isotrivial* if the isomorphism class of the members of the family is constant over T. From the viewpoint of morphisms to a moduli stack $\mathcal{X}$, this corresponds to the image of T in $|\mathcal{X}|$ being a singleton, say x. We would like to conclude that the morphism $T \to \mathcal{X}$ factors through the residual gerbe $\mathcal{Z}_x$. In practice, we often verify the "constancy" of the morphism $T \to \mathcal{X}$ only at finite type points. The following result gives sufficient conditions, under such a constancy hypothesis on the finite type points, for a morphism to factor through the residual gerbe.

E.5 Lemma. *Let $\mathcal{X}$ be an algebraic stack, and let $x \in |\mathcal{X}|$ be a point for which the residual gerbe $\mathcal{Z}_x$ exists, and for which the canonical monomorphism $\mathcal{Z}_x \hookrightarrow \mathcal{X}$ is an immersion. If $f: T \to \mathcal{X}$ is a morphism whose domain is a reduced scheme, and for which all the finite type points of T map to the given point x, then f factors through $\mathcal{Z}_x$.*

Proof. Consider the fiber product $T \times_{\mathcal{X}} \mathcal{Z}_x$. By assumption, the projection from this fiber product to T is an immersion whose image contains every finite type point of T. A locally closed subset of T that contains every finite type point is necessarily equal to T, and thus, since T is reduced, this immersion is in fact an isomorphism. Consequently the morphism f factors through $\mathcal{Z}_x$, as claimed. □

We end this discussion by explaining how the preceding discussion generalizes to certain Ind-algebraic stacks. Let $\{\mathcal{X}_i\}_{i \in I}$ be a 2-directed system of algebraic stacks, and assume that the transition morphisms are monomorphisms. Let $\mathcal{X} := \varinjlim_i \mathcal{X}_i$ be the Ind-algebraic stack obtained as the 2-direct limit of the $\mathcal{X}_i$. We may define the underlying set of points $|\mathcal{X}|$ as equivalence classes of morphisms from spectra of fields in the usual way, and since any such morphism factors through some $\mathcal{X}_i$, we find that $|\mathcal{X}| = \varinjlim_i |\mathcal{X}_i|$.

Now suppose that $\mathcal{X}$ has quasi-compact diagonal, or, equivalently (since the transition morphisms are monomorphisms) that each $\mathcal{X}_i$ has quasi-compact diagonal. If $x \in |\mathcal{X}|$, then $x \in |\mathcal{X}_i|$ for some i, and the residual gerbe $\mathcal{Z}_x$ at x in $\mathcal{X}_i$ exists. The composite monomorphism $\mathcal{Z}_x \hookrightarrow \mathcal{X}_i \hookrightarrow \mathcal{X}_{i'}$ realizes $\mathcal{Z}_x$ as the residual gerbe at x in $\mathcal{X}_{i'}$, for any $i' \geq i$, and so we may regard $\mathcal{Z}_x$ as being the residual gerbe at x in $\mathcal{X}$. Lemmas E.3 and E.5 immediately extend to the context of such Ind-algebraic stacks $\mathcal{X}$.

Appendix F

Breuil–Kisin–Fargues modules and potentially semistable representations (by Toby Gee and Tong Liu)

In this appendix we briefly discuss the relationship between Breuil–Kisin–Fargues modules with semi-linear Galois actions and potentially semistable Galois representations. As explained in Remark F.12 below, we do not expect our results to be optimal, but they suffice for our applications in the body of the book. The results of this appendix were originally inspired by [Car13]; the recent paper [Gao19] corrects a mistake in [Car13] and independently proves related (and in some cases stronger) versions of some of our results, by different methods. We do not make any use of the arguments of either [Car13, Gao19], but instead combine [Liu18] with a result of Fargues ([BMS18, Thm. 4.28]).

We use the notation introduced in the body of the book, in particular in Section 2.1. The kernel of the usual ring homomorphism $\theta\colon \mathbf{A}_{\mathrm{inf}} \to \mathcal{O}_{\mathbf{C}}$ is a principal ideal (ξ); one possible choice of ξ is $\mu/\varphi^{-1}(\mu)$, where $\mu = [\varepsilon] - 1$ (for some compatible choice of roots of unity $\varepsilon = (1, \zeta_p, \zeta_{p^2}, \dots) \in \mathcal{O}_{\mathbf{C}}^\flat$). Recall that $\mathrm{B}_{\mathrm{dR}}^+$ is the $\ker(\theta)$-adic completion of $\mathbf{A}_{\mathrm{inf}}[\frac{1}{p}]$ and that $t := \log[\varepsilon] \in \mathrm{B}_{\mathrm{dR}}^+$ is a generator of $\ker(\theta)$ in $\mathrm{B}_{\mathrm{dR}}^+$.

As usual, we let K be a finite extension of $\mathbf{Q}_p$ with residue field k. Recall that for each choice of uniformizer π of K, and each choice $\pi^\flat \in \mathcal{O}_{\mathbf{C}}^\flat$ of p-power roots of π, we write $\mathfrak{S}_{\pi^\flat}$ for $\mathfrak{S} = W(k)[[u]]$, regarded as a subring of $\mathbf{A}_{\mathrm{inf}}$ via $u \mapsto [\pi^\flat]$. Write $E_\pi(u)$ for the Eisenstein polynomial for π, and $E_{\pi^\flat}$ for its image in $\mathbf{A}_{\mathrm{inf}}$; then $E_{\pi^\flat} \in (\xi)$, because $\theta(E_{\pi^\flat}) = E_\pi(\pi) = 0$. Indeed, $(E_{\pi^\flat}) = (\xi)$ (this follows for example from the criterion given in [BMS18, Rem. 3.11] and the definition of an Eisenstein polynomial).

In contrast to the body of the book, we do not use coefficients in most of this appendix (we briefly consider $\mathcal{O}$-coefficients at the end). Accordingly, we have the following definitions.

F.1 Definition. An étale (φ, G_K)-module is a finite free $W(\mathbf{C}^\flat)$-module M equipped with a φ-semi-linear map $\varphi\colon M \to M$ which induces an isomorphism $\varphi^* M \xrightarrow{\sim} M$, together with a continuous semi-linear action of G_K which commutes with φ.

There is an equivalence of categories between the category of étale (φ, G_K)-modules M of rank d and the category of free $\mathbf{Z}_p$-modules T of rank d which are equipped with a continuous action of G_K. The Galois representation corresponding to M is given by $T(M) = M^{\varphi=1}$. (See Section 3.6.4.)

F.2 Definition. Fix a choice of $\pi^\flat$. We define a Breuil–Kisin module of height at most h to be a finite free $\mathfrak{S}_{\pi^\flat}$-module $\mathfrak{M}$ equipped with a φ-semi-linear morphism $\varphi\colon \mathfrak{M} \to \mathfrak{M}$, with the property that the corresponding morphism $\Phi_{\mathfrak{M}}: \varphi^*\mathfrak{M} \to \mathfrak{M}$ is injective, with cokernel killed by $E^h_{\pi^\flat}$.

F.3 Definition. A Breuil–Kisin–Fargues module of height at most h is a finite free $\mathbf{A}_{\mathrm{inf}}$-module $\mathfrak{M}^{\mathrm{inf}}$ equipped with a φ-semi-linear morphism $\varphi\colon \mathfrak{M}^{\mathrm{inf}} \to \mathfrak{M}^{\mathrm{inf}}$, with the property that the corresponding morphism $\Phi_{\mathfrak{M}^{\mathrm{inf}}}: \varphi^*\mathfrak{M}^{\mathrm{inf}} \to \mathfrak{M}^{\mathrm{inf}}$ is injective, with cokernel killed by ξ^h.

F.4 Remark. If $\mathfrak{M}$ is a Breuil–Kisin module we write $\mathfrak{M}^{\mathrm{inf}}$ for the Breuil–Kisin–Fargues module $\mathbf{A}_{\mathrm{inf}} \otimes_{\mathfrak{S}_{\pi^\flat}} \mathfrak{M}$ (this is indeed a Breuil–Kisin–Fargues module, because $\mathfrak{S}_{\pi^\flat} \to \mathbf{A}_{\mathrm{inf}}$ is faithfully flat, and $(E_{\pi^\flat}) = (\xi)$). As noted in Remark 4.1.1, we are not twisting the embedding $\mathfrak{S} \to \mathbf{A}_{\mathrm{inf}}$ by φ, and accordingly, in Definition F.3, we demand that the cokernel of φ be killed by a power of ξ, rather than a power of $\varphi(\xi)$ as in [BMS18].

Leaving this difference aside, our definition of a Breuil–Kisin–Fargues module is less general than that of [BMS18], in that we require φ to take $\mathfrak{M}$ to itself; this corresponds to only considering Galois representations with non-negative Hodge–Tate weights. This definition is convenient for us, as it allows us to make direct reference to the literature on Breuil–Kisin modules. The restriction to non-negative Hodge–Tate weights is harmless in our main results, as we can reduce to this case by twisting by a large enough power of the cyclotomic character (the interpretation of which on Breuil–Kisin–Fargues modules is explained in [BMS18, Ex. 4.24]). (We are also only considering free Breuil–Kisin–Fargues modules, rather than the more general possibilities considered in [BMS18].)

F.5 Definition. A Breuil–Kisin–Fargues G_K-module of height at most h is a Breuil–Kisin–Fargues module of height at most h which is equipped with a semi-linear G_K-action which commutes with φ.

F.6 Remark. Note that if $\mathfrak{M}^{\mathrm{inf}}$ is a Breuil–Kisin–Fargues G_K-module, then $W(\mathbf{C}^\flat) \otimes_{\mathbf{A}_{\mathrm{inf}}} \mathfrak{M}^{\mathrm{inf}}$ is naturally an étale (φ, G_K)-module in the sense of Definition F.1.

Recall that for each choice of $\pi^\flat$ and each $s \geq 0$ we write $K_{\pi^\flat,s}$ for $K(\pi^{1/p^s})$, and $K_{\pi^\flat,\infty}$ for $\cup_s K_{\pi^\flat,s}$.

F.7 Definition. Let $\mathfrak{M}^{\mathrm{inf}}$ be a Breuil–Kisin–Fargues G_K-module of height at most h with a semi-linear G_K-action. Then we say that $\mathfrak{M}^{\mathrm{inf}}$ *admits all descents* if the following conditions hold.

1. For every choice of π and $\pi^\flat$, there is a Breuil–Kisin module $\mathfrak{M}_{\pi^\flat}$ of height at most h with $\mathfrak{M}_{\pi^\flat} \subset (\mathfrak{M}^{\mathrm{inf}})^{G_{K_{\pi^\flat,\infty}}}$ for which the induced morphism $\mathbf{A}_{\mathrm{inf}} \otimes_{\mathfrak{S}_{\pi^\flat}} \mathfrak{M}_{\pi^\flat} \to \mathfrak{M}^{\mathrm{inf}}$ is an isomorphism.

2. The $W(k)$-submodule $\mathfrak{M}_{\pi^\flat}/[\pi^\flat]\mathfrak{M}_{\pi^\flat}$ of $W(\overline{k})\otimes_{\mathbf{A}_{\mathrm{inf}}}\mathfrak{M}^{\mathrm{inf}}$ is independent of the choice of π and $\pi^\flat$.
3. The $\mathcal{O}_K$-submodule $\varphi^*\mathfrak{M}_{\pi^\flat}/E_{\pi^\flat}\varphi^*\mathfrak{M}_{\pi^\flat}$ of $\mathcal{O}_{\mathbf{C}}\otimes_{\theta,\mathbf{A}_{\mathrm{inf}}}\varphi^*\mathfrak{M}^{\mathrm{inf}}$ is independent of the choice of π and $\pi^\flat$.

F.8 Remark. In fact condition (2) in Definition F.7 is redundant; see Remark F.20 below. However, we include the condition as it is useful when considering versions of the theory with coefficients and descent data.

F.9 Definition. Let $\mathfrak{M}^{\mathrm{inf}}$ be a Breuil–Kisin–Fargues G_K-module which admits all descents. We say that $\mathfrak{M}^{\mathrm{inf}}$ is furthermore *crystalline* if for each choice of π and $\pi^\flat$, and each $g\in G_K$, we have

$$(g-1)(\mathfrak{M}_{\pi^\flat})\subset\varphi^{-1}(\mu)[\pi^\flat]\mathfrak{M}^{\mathrm{inf}}. \tag{F.10}$$

Definitions F.7 and F.9 are motivated by the following result, whose proof occupies most of the rest of this appendix.

F.11 Theorem. *Let M be an étale (φ,G_K)-module. Then $V(M)$ is semistable with Hodge–Tate weights in $[0,h]$ if and only if there is a (necessarily unique) Breuil–Kisin–Fargues G_K-module $\mathfrak{M}^{\mathrm{inf}}$ which is of height at most h, which admits all descents, and which satisfies $M=W(\mathbf{C}^\flat)\otimes_{\mathbf{A}_{\mathrm{inf}}}\mathfrak{M}^{\mathrm{inf}}$.*

Furthermore, $V(M)$ is crystalline if and only if $\mathfrak{M}^{\mathrm{inf}}$ is crystalline.

F.12 Remark. As already noted in Remark F.8, condition (2) of Definition F.7 is redundant, and it is plausible that condition (3) is redundant as well (i.e., that both condition (2) and condition (3) in Definition F.7 are consequences of condition (1)); but this does not seem to be obvious. Note though that it is not sufficient to demand the existence of a descent for a single choice of π, as there are representations of finite height which are potentially semistable but not semistable (see for example [Liu10, Ex. 4.2.1]).

We begin with some preliminary results. The following proposition and its proof are due to Heng Du, and we thank him for allowing us to include them here.

F.13 Proposition. *Let $\mathfrak{M}^{\mathrm{inf}}$ be a Breuil–Kisin–Fargues G_K-module with the property that $\mathbf{C}\otimes_{\theta,\mathbf{A}_{\mathrm{inf}}}\varphi^*\mathfrak{M}^{\mathrm{inf}}$ has a basis consisting of G_K-fixed vectors. Let $M=W(\mathbf{C}^\flat)\otimes_{\mathbf{A}_{\mathrm{inf}}}\mathfrak{M}^{\mathrm{inf}}$; then the G_K-representation $V(M)$ is de Rham.*

F.14 Remark. Using the equivalence of categories of [BMS18, Thm. 4.28] (a theorem of Fargues), one can easily check that Proposition F.13 admits a converse: namely that if M is an étale (φ,G_K)-module with the property that $T(M)$ is a $\mathbf{Z}_p$-lattice in a de Rham representation of G_K with non-negative Hodge–Tate weights, then there is a Breuil–Kisin–Fargues module $\mathfrak{M}^{\inf}$ with the properties in the statement of Proposition F.13.

Proof of Proposition F.13. By [BMS18, Thm. 4.28] (and our assumption that $\varphi(\mathfrak{M}^{\inf}) \subseteq \mathfrak{M}^{\inf}$), we have injections

$$\mathrm{B}^+_{\mathrm{dR}} \otimes_{\mathbf{A}_{\inf}} \varphi^*\mathfrak{M}^{\inf} \hookrightarrow \mathrm{B}^+_{\mathrm{dR}} \otimes_{\mathbf{A}_{\inf}} \mathfrak{M}^{\inf} \hookrightarrow \mathrm{B}_{\mathrm{dR}} \otimes_{\mathbf{Z}_p} T(M).$$

To show that $V(M)$ is de Rham, we need to show that the B_{dR}-vector space $\mathrm{B}_{\mathrm{dR}} \otimes_{\mathbf{Z}_p} T(M)$ has a basis consisting of G_K-fixed vectors, so it suffices to show that the $\mathrm{B}^+_{\mathrm{dR}}$-lattice $\mathrm{B}^+_{\mathrm{dR}} \otimes_{\mathbf{A}_{\inf}} \varphi^*\mathfrak{M}^{\inf}$ has a basis consisting of G_K-fixed vectors. Since $\mathrm{B}^+_{\mathrm{dR}}$ is a complete discrete valuation ring with maximal ideal (ξ), it is enough to show that there are compatible bases of $\mathrm{B}^+_{\mathrm{dR}}/(\xi^n) \otimes_{\mathbf{A}_{\inf}} \varphi^*\mathfrak{M}^{\inf}$ for all $n \geq 1$, consisting of G_K-fixed vectors.

In the case $n=1$, since ξ generates the kernel of θ, we have such a basis by hypothesis. Suppose that $\{e_i^{(n)}\}_{i=1,\dots,d}$ is a basis of G_K-fixed vectors for $\mathrm{B}^+_{\mathrm{dR}}/(\xi^n) \otimes_{\mathbf{A}_{\inf}} \varphi^*\mathfrak{M}^{\inf}$, and let $\{\widetilde{e}_i^{(n+1)}\}_{i=1,\dots,d}$ be any basis of $\mathrm{B}^+_{\mathrm{dR}}/(\xi^{n+1}) \otimes_{\mathbf{A}_{\inf}} \varphi^*\mathfrak{M}^{\inf}$ lifting $\{e_i^{(n)}\}$. Recall that $t = \log[\varepsilon] \in \mathrm{B}^+_{\mathrm{dR}}$ is a generator of (ξ), and for each $g \in G_K$, write

$$g \cdot (\widetilde{e}_1^{(n+1)}, \dots, \widetilde{e}_d^{(n+1)}) = (\widetilde{e}_1^{(n+1)}, \dots, \widetilde{e}_d^{(n+1)})(1_d + t^n B_g^{(n+1)})$$

where we can view $B_g^{(n+1)}$ as an element of $M_d(B^+_{\mathrm{dR}}/(\xi)) = M_d(\mathbf{C})$.

A simple calculation shows that $g \mapsto B_g^{(n+1)}$ is a continuous 1-cocycle valued in $M_d(\mathbf{C}(n))$. Since $n \geq 1$, the corresponding cohomology group $H^1(G_K, M_d(\mathbf{C}(n)))$ vanishes, because $H^1(G_K, \mathbf{C}(n)) = 0$ by a theorem of Tate–Sen. There is therefore some $A \in M_d(\mathbf{C})$ such that for all g we have

$$B_g^{(n+1)} = \chi^n(g)g(A) - A,$$

where χ is the p-adic cyclotomic character. Then

$$(e_1^{(n+1)}, \dots, e_d^{(n+1)}) = (\widetilde{e}_1^{(n+1)}, \dots, \widetilde{e}_d^{(n+1)})(1_d - t^n A)$$

is the required basis of $\mathrm{B}^+_{\mathrm{dR}}/(\xi^{n+1}) \otimes_{\mathbf{A}_{\inf}} \varphi^*\mathfrak{M}^{\inf}$ consisting of G_K-fixed vectors lifting $\{e_i^{(n)}\}$. □

F.15 Lemma. *There is no proper closed subgroup of G_K containing all of the subgroups $G_{K_{\pi^\flat,\infty}}$.*

Proof. Let s be the greatest integer with the property that K contains a primitive p^sth root of unity; equivalently, it is the greatest integer with the property that $K_{\pi^\flat,s}/K$ is Galois over K (for one, or equivalently every, choice of π

and $\pi^\flat$). Then $K_{\pi^\flat,s'}$ depends only on π if $s' \leq s$, and so we write $K_{\pi,s'} = K_{\pi^\flat,s'}$ for such s'.

The subgroups $G_{K_{\pi^\flat,\infty}}$ for any fixed choice of π topologically generate $G_{K_{\pi,s}}$. It therefore suffices to show that as π varies, the various normal subgroups $G_{K_{\pi,s}}$ of G_K collectively generate G_K. Let H be the normal subgroup that they generate; then for every uniformizer π, G_K/H is a quotient of $G_K/G_{K_{\pi,s}}$, a cyclic group of order p^s. If H were a proper subgroup of G_K, then necessarily $s \geq 1$, and the subgroups $G_{K_{\pi,1}}$ would coincide for every π (since they would coincide with the unique index p subgroup of H). Thus the extensions $K_{\pi,1}/K$ would have to all coincide. By Kummer theory, this would imply that the ratio of any two uniformizers of K is a pth power in K, which is nonsense (for example, consider the uniformizers π and $\pi + \pi^2$). □

Proof of Theorem F.11. We begin with the semistable case. If $V(M)$ is semistable then the existence of $\mathfrak{M}^{\text{inf}}$ is a straightforward consequence of the results of [Liu18]. In particular, the existence of a unique $\mathfrak{M}^{\text{inf}}$ satisfying condition (1) of Definition F.7 follows from [Liu18, Thm. 2.2.1].

The proofs that conditions (2) and (3) hold are implicit in the proof of [Liu18, Prop. 4.2.1], as we now explain. Write $\overline{\mathfrak{M}}_{\pi^\flat} = \mathfrak{M}_{\pi^\flat}/[\pi^\flat]\mathfrak{M}_{\pi^\flat}$. Since $W(\overline{k}) \otimes_{W(k)} \overline{\mathfrak{M}}_{\pi^\flat} = W(\overline{k}) \otimes_{\mathbf{A}_{\text{inf}}} \mathfrak{M}^{\text{inf}}$ is independent of $\pi^\flat$, it suffices to show that the K_0-vector space $D_{\pi^\flat} := \overline{\mathfrak{M}}_{\pi^\flat}[\frac{1}{p}]$ is independent of $\pi^\flat$. Let $S_{\pi^\flat}$ be the p-adic completion of $\mathfrak{S}_{\pi^\flat}[\frac{E^i_{\pi^\flat}}{i!}, i \geq 1]$. By [Bre97, Prop. 6.2.1.1] and [Liu12, §2], $D_{\pi^\flat}$ admits a section $s_{\pi^\flat} : D_{\pi^\flat} \to S_{\pi^\flat}[\frac{1}{p}] \otimes_{\varphi,\mathfrak{S}_{\pi^\flat}} \mathfrak{M}_{\pi^\flat}$ so that

$$S_{\pi^\flat}[\frac{1}{p}] \otimes_{K_0} s_{\pi^\flat}(D_{\pi^\flat}) = S_{\pi^\flat}[\frac{1}{p}] \otimes_{\varphi,\mathfrak{S}_{\pi^\flat}} \mathfrak{M}_{\pi^\flat}. \tag{F.16}$$

In particular, we may regard $s_{\pi^\flat}(D_{\pi^\flat})$ as a submodule of $\mathrm{B}^+_{\text{cris}} \otimes_{\varphi,\mathfrak{S}_{\pi^\flat}} \mathfrak{M}_{\pi^\flat} \subseteq \mathrm{B}^+_{\text{cris}} \otimes_{\mathbf{Z}_p} T(M)$.

Write $\mathfrak{u} = \log([\pi^\flat]) \in \mathrm{B}^+_{\text{st}} = \mathrm{B}^+_{\text{cris}}[\mathfrak{u}]$. Note that the *set* B^+_{st} does not depend on the choice of $\pi^\flat$, because, if $\varpi^\flat$ is another choice and we write $\mathfrak{u}' = \log([\varpi^\flat])$, then $\mathrm{B}^+_{\text{cris}}[\mathfrak{u}'] = \mathrm{B}^+_{\text{cris}}[\mathfrak{u}]$. Indeed, we have $\log([\varpi^\flat]) = \log([\pi^\flat]) + \lambda$ with $\lambda = \log([\frac{\varpi^\flat}{\pi^\flat}]) \in \mathrm{B}^+_{\text{cris}}$. In particular, $D_{\text{st}}(T(M)) := (\mathrm{B}^+_{\text{st}} \otimes_{\mathbf{Z}_p} T(M))^{G_K}$ is independent of $\pi^\flat$. Furthermore, [Liu12, Prop. 2.6] shows that we have a commutative diagram (where $i_{\pi^\flat}$ is a K_0-linear isomorphism)

$$\begin{array}{ccccc} D_{\text{st}}(T(M)) & \hookrightarrow & \mathrm{B}^+_{\text{st}} \otimes_{K_0} D_{\text{st}}(T(M)) & \hookrightarrow & \mathrm{B}^+_{\text{st}} \otimes_{\mathbf{Z}_p} T(M) \\ \wr\downarrow i_{\pi^\flat} & & \downarrow \text{mod } \mathfrak{u} & & \downarrow \text{mod } \mathfrak{u} \\ s_{\pi^\flat}(D_{\pi^\flat}) & \hookrightarrow & \mathrm{B}^+_{\text{cris}} \otimes_{\varphi,\mathfrak{S}_{\pi^\flat}} \mathfrak{M}_{\pi^\flat} & \hookrightarrow & \mathrm{B}^+_{\text{cris}} \otimes_{\mathbf{Z}_p} T(M) \end{array}$$

To show (2), it therefore suffices to show that the image of the composite

$$D_{\mathrm{st}}(T(M)) \to \mathrm{B}^+_{\mathrm{st}} \otimes_{K_0} D_{\mathrm{st}}(T(M)) \xrightarrow{\mathrm{mod}\ u} \mathrm{B}^+_{\mathrm{cris}} \otimes_{\varphi, \mathfrak{S}_{\pi^\flat}} \mathfrak{M}_{\pi^\flat} \to W(\bar{k})[1/p] \otimes_{\varphi, \mathbf{A}_{\inf}} \mathfrak{M}^{\inf}$$

is independent of $\pi^\flat$. Here the last map is induced by $\nu\colon \mathrm{B}^+_{\mathrm{cris}} \to W(\overline{k})[\frac{1}{p}]$, which extends the natural projection $\mathbf{A}_{\inf} \to W(\overline{k})$. It suffices in turn to show that the composite $\mathrm{B}^+_{\mathrm{st}} \xrightarrow{\mathrm{mod}\ u} \mathrm{B}^+_{\mathrm{cris}} \xrightarrow{\nu} W(\overline{k})[\frac{1}{p}]$ is independent of $\pi^\flat$. This follows from the fact that $\lambda = \log([\frac{\varpi^\flat}{\pi^\flat}])$ is in $\ker(\nu)$ (see the proof of [Liu12, Lem. 2.10]).

To prove (3), it again suffices to prove the statement after inverting p. For any subring $A \subset \mathrm{B}^+_{\mathrm{dR}}$, set $F^1 A = A \cap \xi \mathrm{B}^+_{\mathrm{dR}}$. It is easy to see that $F^1 \mathfrak{S}_{\pi^\flat} = E_{\pi^\flat} \mathfrak{S}_{\pi^\flat}$ and also that $S_{\pi^\flat}/F^1 S_{\pi^\flat} = \mathcal{O}_K$. (Note that the inclusion $E_{\pi^\flat} S_{\pi^\flat} \subsetneq F^1 S_{\pi^\flat}$ is strict.) Now we again use the isomorphism (F.16). By reducing modulo $F^1 S_{\pi^\flat}$ on the both sides of this identification, we conclude that

$$K \otimes_{K_0} D_{\pi^\flat} = K \otimes_{K_0} (s_{\pi^\flat}(D_{\pi^\flat}) \mod [\pi^\flat]) = \varphi^* \mathfrak{M}_{\pi^\flat} / E_{\pi^\flat} \varphi^* \mathfrak{M}_{\pi^\flat} [\frac{1}{p}].$$

On the other hand, by tensoring $\mathbf{A}_{\inf}$ via $\mathfrak{S}_{\pi^\flat}$ to (F.16), we obtain the isomorphism $\mathrm{B}^+_{\mathrm{cris}} \otimes_{K_0} s_{\pi^\flat}(D_{\pi^\flat}) = \mathrm{B}^+_{\mathrm{cris}} \otimes_{\mathbf{A}_{\inf}} \varphi^* \mathfrak{M}^{\inf}$. Modulo $F^1 \mathrm{B}^+_{\mathrm{cris}}$ on both sides, a similar argument to the above shows that $\mathbf{C} \otimes_{K_0} D_{\pi^\flat} = (\varphi^* \mathfrak{M}^{\inf} / \xi \varphi^* \mathfrak{M}^{\inf})[\frac{1}{p}] = \mathbf{C} \otimes_{W(\overline{k})} \overline{\varphi^* \mathfrak{M}^{\inf}}$, where we write $\overline{\varphi^* \mathfrak{M}^{\inf}} = W(\overline{k}) \otimes_{\mathbf{A}_{\inf}} \varphi^* \mathfrak{M}^{\inf}$. In summary we see that

$$(\varphi^* \mathfrak{M}_{\pi^\flat} / E_{\pi^\flat} \varphi^* \mathfrak{M}_{\pi^\flat})[\tfrac{1}{p}] = K \otimes_{K_0} D_{\pi^\flat} \subseteq \mathbf{C} \otimes_{W(\bar{k})} \overline{\varphi^* \mathfrak{M}^{\inf}} = (\varphi^* \mathfrak{M}^{\inf} / \xi \varphi^* \mathfrak{M}^{\inf})[\tfrac{1}{p}].$$

Since we have shown that $D_{\pi^\flat} \subseteq \overline{\varphi^* \mathfrak{M}^{\inf}}$ is independent of $\pi^\flat$, it follows that $\varphi^* \mathfrak{M}_{\pi^\flat} / E_{\pi^\flat} \varphi^* \mathfrak{M}_{\pi^\flat}$ is independent of $\pi^\flat$, as required.

Suppose conversely that we are given $\mathfrak{M}^{\inf}$ as in Definition F.7. Write $\overline{\mathfrak{M}}$ for the $W(k)$-module of Definition F.7 (2), and $\overline{\mathfrak{M}}'$ for the $\mathcal{O}_K$-module of Definition F.7 (3). Since for each $\pi^\flat$ the action of $G_{K_{\pi^\flat, \infty}}$ on $\mathfrak{M}_{\pi^\flat}$ is trivial by assumption, it follows from Lemma F.15 that G_K acts trivially on $\overline{\mathfrak{M}}$ and $\overline{\mathfrak{M}}'$.

Since $\mathbf{C} \otimes_{\mathbf{A}_{\inf}, \theta} \varphi^* \mathfrak{M}^{\inf} = \mathbf{C} \otimes_{\mathcal{O}_K} \varphi^* \overline{\mathfrak{M}}'$, we see that the hypotheses of Proposition F.13 are satisfied, so that $V(M)$ is de Rham, and consequently potentially semistable. To show that $V(M)$ is semistable, we may replace K with an (infinite) unramified extension and assume that K is a complete discretely valued field with residue field $k = \overline{k}$. In particular, we now have $\overline{\mathfrak{M}} = W(\overline{k}) \otimes_{\mathbf{A}_{\inf}} \mathfrak{M}^{\inf}$, and we claim that $\overline{\mathfrak{M}}[\frac{1}{p}]$ with its G_K-action is isomorphic to $D_{\mathrm{pst}}(V(M))$. If the claim holds, then since G_K acts trivially on $\overline{\mathfrak{M}}$, G_K acts trivially on $D_{\mathrm{pst}}(V(M))$, so that $V(M)$ is semistable, as required.

We now prove the claim. Let L/K be a finite Galois extension so that $V(M)|_{G_L}$ is semistable, and let $\mathfrak{M}_L$ be the Breuil–Kisin module attached to $T(M)|_{G_L}$ for some choice of $\pi^\flat_L$. By [Liu12, (2.12)], $\mathbf{A}_{\inf} \otimes_{\mathfrak{S}_{\pi^\flat_L}} \varphi^* \mathfrak{M}_L$ injects into $\mathbf{A}_{\inf} \otimes_{\mathbf{Z}_p} T(M)$. Furthermore, by [Liu12, Lem. 2.9], $\varphi^* \mathfrak{M}^{\inf}_L := \mathbf{A}_{\inf} \otimes_{\mathfrak{S}_{\pi^\flat_L}}$

$\varphi^*\mathfrak{M}_L$ is stable under the G_K-action on $\mathbf{A}_{\inf}\otimes_{\mathbf{Z}_p}T(M)$, and by [Liu12, Cor. 2.12], $W(k)\otimes_{\mathfrak{S}_{\pi_L^\flat}}\varphi^*\mathfrak{M}_L=W(k)\otimes_{\mathbf{A}_{\inf}}\varphi^*\mathfrak{M}_L^{\inf}$ (recall that $k=\overline{k}$) together with its G_K-action is isomorphic to a $W(k)$-lattice inside $D_{\mathrm{st},L}(T(M))=(\mathrm{B}_{\mathrm{st}}^+\otimes_{\mathbf{Z}_p}T(M))^{G_L}$. In summary, $W(k)\otimes_{\mathbf{A}_{\inf}}\varphi^*\mathfrak{M}_L^{\inf}[1/p]$ is isomorphic to $D_{\mathrm{st},L}(T(M))$ as G_K-modules.

It remains to show that $\varphi^*\mathfrak{M}_L^{\inf}=\varphi^*\mathfrak{M}^{\inf}$. By [BMS18, Thm. 4.28], it suffices to show that $\mathrm{B}_{\mathrm{dR}}^+\otimes_{\mathbf{A}_{\inf}}\varphi^*\mathfrak{M}_L^{\inf}=\mathrm{B}_{\mathrm{dR}}^+\otimes_{\mathbf{A}_{\inf}}\varphi^*\mathfrak{M}^{\inf}$. By the proof of Proposition F.13, we see that

$$\mathrm{B}_{\mathrm{dR}}^+\otimes_{\mathbf{A}_{\inf}}\varphi^*\mathfrak{M}^{\inf}=\mathrm{B}_{\mathrm{dR}}^+\otimes_K D_{\mathrm{dR}}(T(M)). \tag{F.17}$$

Since $T(M)$ is semistable over L, it is well known that

$$\mathrm{B}_{\mathrm{st}}^+\otimes_{\mathfrak{S}_{\pi_L^\flat}}\varphi^*\mathfrak{M}_L=\mathrm{B}_{\mathrm{st}}^+\otimes_{W(k_L)}D_{\mathrm{st}}(T(M)|_{G_L}). \tag{F.18}$$

(This can be easily seen from the proof of [Kis06, Cor.1.3.15], or see [Liu12, §2] for a more detailed discussion; the key point, in terms of the diagram above, is that [Liu12, §2] shows that $K_0[\mathfrak{u}]\otimes_{K_0}D_{\mathrm{st}}(T(M))=K_0[\mathfrak{u}]\otimes_{K_0}s_{\pi^\flat}(D_{\pi^\flat})$.)

From (F.18) we obtain

$$\mathrm{B}_{\mathrm{dR}}^+\otimes_{\mathfrak{S}_{\pi_L^\flat}}\varphi^*\mathfrak{M}_L=\mathrm{B}_{\mathrm{dR}}^+\otimes_L D_{\mathrm{dR}}(T(M)|_{G_L}). \tag{F.19}$$

Comparing (F.17) and (F.19) we have $\mathrm{B}_{\mathrm{dR}}^+\otimes_{\mathbf{A}_{\inf}}\varphi^*\mathfrak{M}_L^{\inf}=\mathrm{B}_{\mathrm{dR}}^+\otimes_{\mathbf{A}_{\inf}}\varphi^*\mathfrak{M}^{\inf}$, as required.

Finally, we turn to the crystalline case. Given the above, the result is a consequence of [Oze18, Thm. 3.8], as we now explain. In our case, $f(u)$ in [Oze18] is equal to u^p, so the assumptions of [Oze18, Thm. 3.8] are automatically satisfied (as noted at the beginning of [Oze18, §3.2]). More precisely, fix some $\pi^\flat$, and let $\widehat{K}$ be the Galois closure of $K_{\pi^\flat}$. There is a subring $\widehat{\mathcal{R}}\subseteq\mathbf{A}_{\inf}$ (constructed in [Oze18]) such that $\mathfrak{S}_{\pi^\flat}\subset\widehat{\mathcal{R}}$ and G_K acts on $\widehat{\mathcal{R}}$ through $\widehat{G}:=\mathrm{Gal}(\cup_{n\geq 1}K(\zeta_{p^n},\pi^{1/p^n})/K)$. When $T(M)$ is crystalline, [Oze18, Thm. 3.8] shows that $T(M)$ admits a $(\varphi,\widehat{G})$-module, which by definition consists of the following data:

- The Breuil-Kisin module $\mathfrak{M}_{\pi^\flat}$ attached to $T(M)|_{G_{\pi^\flat,\infty}}$.
- A $\widehat{G}$-action on $\widehat{\mathfrak{M}}:=\widehat{\mathcal{R}}\otimes_{\mathfrak{S}_{\pi^\flat}}\varphi^*\mathfrak{M}_{\pi^\flat}$ which commutes with the action of φ and satisfies (F.10).
- $(W(\mathbf{C}^\flat)\otimes_{\widehat{\mathcal{R}}}\widehat{\mathfrak{M}})^{\varphi=1}=T(M)$ as G_K-modules.

(In fact [Oze18] uses contravariant functors, which can be easily translated to the covariant functors used here.) This proves that if $T(M)$ is crystalline then $\mathfrak{M}^{\inf}$ satisfies (F.10). Conversely, if $\mathfrak{M}^{\inf}$ satisfies (F.10) for *one* fixed $\pi^\flat$, then [Oze18, Lem. 3.15] shows that $s_\pi(D_{\pi^\flat})\subset(\mathrm{B}_{\mathrm{cris}}^+\otimes_{\mathfrak{S}_{\pi^\flat}}\varphi^*\mathfrak{M}_{\pi^\flat})^{G_K}$. Hence $T(M)$ is crystalline, as required. □

F.20 Remark. Note that condition (2) in Definition F.7 is redundant for proving Theorem F.11. In fact, if we remove (2) from Definition F.7 then the necessity part of Theorem F.11 of course still holds. For the sufficiency, note that in the above proof, we extended K so that the residue field is $\bar{k}$; so we only use that G_K (the inertia subgroup in this situation) acts trivially on $W(\bar{k}) \otimes_{\mathbf{A}_{\mathrm{inf}}} \mathfrak{M}^{\mathrm{inf}}$, and we do not need that $\mathfrak{M}_{\pi^\flat}/[\pi^\flat]\mathfrak{M}_{\pi^\flat} \subseteq W(\bar{k}) \otimes_{\mathbf{A}_{\mathrm{inf}}} \mathfrak{M}^{\mathrm{inf}}$ is independent of $\pi^\flat$.

However, it is not immediately clear that the analogous condition is redundant in the version of the theory with coefficients that we consider in the body of the book (see Definition 4.2.4), and this condition is used in defining our moduli stacks of potentially semistable representations of given inertial type, so we include it here.

F.21 POTENTIALLY SEMISTABLE REPRESENTATIONS

It will be convenient for us to have a slight refinement of these results, allowing us to discuss potentially semistable (and potentially crystalline) representations. To this end, fix a finite Galois extension L/K.

F.22 Definition. Let $\mathfrak{M}^{\mathrm{inf}}$ be a Breuil–Kisin–Fargues G_K-module of height at most h. Then we say that $\mathfrak{M}^{\mathrm{inf}}$ *admits all descents over* L if the corresponding Breuil–Kisin–Fargues G_L-module (obtained by restricting the G_K-action on $\mathfrak{M}^{\mathrm{inf}}$ to G_L) admits all descents (in the sense of Definition F.7).

F.23 Corollary. *Let M be an étale (φ, G_K)-module. Then $V(M)|_{G_L}$ is semistable with Hodge–Tate weights in $[0,h]$ if and only if there is a (necessarily unique) Breuil–Kisin–Fargues G_K-module $\mathfrak{M}^{\mathrm{inf}}$ which is of height at most h, which admits all descents over L, and which satisfies $M = W(\mathbf{C}^\flat) \otimes_{\mathbf{A}_{\mathrm{inf}}} \mathfrak{M}$. Furthermore $T(M)|_{G_L}$ is crystalline if and only if $\mathfrak{M}^{\mathrm{inf}}$ is crystalline as a Breuil–Kisin–Fargues G_L-module.*

Proof. The sufficiency of the condition is immediate from Theorem F.11. For the necessity, by Theorem F.11 there is a Breuil–Kisin–Fargues G_L-module $\mathfrak{M}_L^{\mathrm{inf}}$ which admits all descents and satisfies $M = W(\mathbf{C}^\flat) \otimes_{\mathbf{A}_{\mathrm{inf}}} \mathfrak{M}_L$, so we need only show that $\mathfrak{M}_L^{\mathrm{inf}}$ is G_K-stable. This follows from [Liu12, Lem. 2.9], or one can argue as follows: note that the proof of Theorem F.11 shows that the Breuil–Kisin–Fargues module $\mathfrak{M}_L^{\mathrm{inf}} := \mathbf{A}_{\mathrm{inf}} \otimes_{\mathfrak{S}_{\pi_L^\flat}} \mathfrak{M}_L$ corresponds via [BMS18, Thm. 4.28] to the pair $(T(M), \mathrm{B}_{\mathrm{dR}}^+ \otimes_L D_{\mathrm{dR}}(T(M)|_{G_L}))$. This pair has a G_K-action, because $V(M)$ is de Rham, hence $\mathfrak{M}_L^{\mathrm{inf}}$ is G_K-stable by [BMS18, Thm. 4.28]. □

F.24 HODGE AND INERTIAL TYPES

We finally recall how to interpret Hodge–Tate weights and inertial types in terms of Breuil–Kisin modules. Fix a finite extension $E/\mathbf{Q}_p$ with ring of integers $\mathcal{O}$, which is sufficiently large that E contains the image of every embedding $\sigma\colon K\hookrightarrow \overline{\mathbf{Q}}_p$. The definitions above admit obvious extensions to the case of Breuil–Kisin–Fargues modules with $\mathcal{O}$-coefficients (see Definition 4.2.4), and since $\mathcal{O}$ is a finite free $\mathbf{Z}_p$-module, the proofs of Theorem F.11 and Corollary F.23 go over unchanged in this setting.

Suppose that $\mathfrak{M}^{\mathrm{inf}}$ is as in the statement of Corollary F.23; so it is a Breuil–Kisin–Fargues G_K-module of height at most h, which admits all descents over L. Write l for the residue field of L, write $\overline{\mathfrak{M}}$ for the $W(l)$-module $\mathfrak{M}_{\pi^\flat}/[\pi^\flat]\mathfrak{M}_{\pi^\flat}$ of Definition F.7 (2), and $\overline{\mathfrak{M}}'$ for the $\mathcal{O}_L$-module $\varphi^*\mathfrak{M}_{\pi^\flat}/E_{\pi^\flat}\varphi^*\mathfrak{M}_{\pi^\flat}$ of Definition F.7 (2). These modules are independent of the choice of π (which now denotes a uniformizer of L) and of $\pi^\flat$, by definition.

The semi-linear G_K-action on $\mathfrak{M}^{\mathrm{inf}}$ induces semi-linear actions on $\overline{\mathfrak{M}}$ and on $\overline{\mathfrak{M}}'$, and as noted in the proof of Theorem F.11, the action of G_L on both modules is trivial. Furthermore, by [Liu12, Cor. 2.12], the inertial type $D_{\mathrm{pst}}(V(M))|_{I_K}$ is given by $\overline{\mathfrak{M}}[1/p]$ with its action of $I_{L/K}$. More precisely, [Liu12, §2.3] shows that $W(\overline{k})[1/p]\otimes_{\mathbf{A}_{\mathrm{inf}}}\varphi^*\mathfrak{M}^{\mathrm{inf}}$ is isomorphic to $D_{\mathrm{pst}}(T(M))\simeq W(\overline{k})\otimes_{W(l)} D_{\mathrm{st}}(T(M)|_{G_L})$ as $W(\overline{k})[1/p][G_K]$-modules. Furthermore, this isomorphism is compatible with the isomorphism (from the proof of Theorem F.11)

$$\iota_{\pi^\flat}:\ \overline{\mathfrak{M}}[\tfrac{1}{p}]=D_{\pi^\flat}\simeq s_{\pi^\flat}(D_{\pi^\flat})\xleftarrow[\sim]{i_{\pi^\flat}} D_{\mathrm{st}}(T(M)|_{G_L})\ .$$

Hence $\overline{\mathfrak{M}}[\frac{1}{p}]\subset W(\overline{k})[1/p]\otimes_{\mathbf{A}_{\mathrm{inf}}}\varphi^*\mathfrak{M}^{\mathrm{inf}}$ is endowed with an action of $I_{L/K}$ and is isomorphic to $D_{\mathrm{pst}}(V(M))$.

We now turn to the Hodge–Tate weights. We have a filtration on the $L\otimes_{\mathbf{Q}_p} E$-module $D_L=\overline{\mathfrak{M}}'[1/p]$, which is defined as follows. For each $\pi^\flat$ we write $\Phi_{\mathfrak{M}_{\pi^\flat}}:\varphi^*\mathfrak{M}_{\pi^\flat}\to\mathfrak{M}_{\pi^\flat}$ and $f_\pi\colon \varphi^*\mathfrak{M}_{\pi^\flat}[\frac{1}{p}]\twoheadrightarrow D_L$. For each $i\geq 0$ we define $\mathrm{Fil}^i\varphi^*\mathfrak{M}_{\pi^\flat}=\Phi_{\mathfrak{M}_{\pi^\flat}}^{-1}(E^i_{\pi^\flat}\mathfrak{M}_{\pi^\flat})$ and $\mathrm{Fil}^i D_L=f_\pi(\mathrm{Fil}^i\varphi^*\mathfrak{M}_{\pi^\flat})$. Considering the isomorphism $\iota_{\pi^\flat}:D_{\mathrm{st}}(V(M)|_{G_L})\simeq D_{\pi^\flat}$, we obtain an isomorphism $D_{\mathrm{dR}}(V(M)|_{G_L})\simeq L\otimes_{W(l)[\frac{1}{p}]}D_{\mathrm{st}}(V(M)|_{G_L})\simeq D_L$. By [Liu08b, Cor. 3.2.3, Thm. 3.4.1], this isomorphism respects the filtrations on each side. In particular, $\mathrm{Fil}^i D_L$ is independent of the choice of $\pi^\flat$.

By Hilbert 90, D_L and its filtration descend to a filtration on a $K\otimes_{\mathbf{Q}_p}E$-module $D_K\simeq D_{\mathrm{dR}}(V(M))$. Then for each $\sigma\colon K\hookrightarrow E$, and each $i\in[0,h]$, the multiplicity of i in the multiset $\mathrm{HT}_\sigma(V(M))$ is the dimension of the ith graded piece of $e_\sigma D_K$, where $e_\sigma\in(K\otimes_{\mathbf{Q}_p}E)$ is the idempotent corresponding to σ.

We summarize the preceding discussion in the following corollary.

F.25 Corollary. *In the setting of Corollary F.23, the inertial type and Hodge type of $V(M)$ are determined as follows: the inertial type $D_{\mathrm{pst}}(V(M))|_{I_K}$ is given by $\overline{\mathfrak{M}}[1/p]$ with its action of $I_{L/K}$, while the Hodge type of $V(M)$ is given by the jumps in the filtration on D_K described above.*

Bibliography

[AHR20] Jarod Alper, Jack Hall, and David Rydh, *A Luna étale slice theorem for algebraic stacks*, Ann. of Math. (2) **191** (2020), no. 3, 675–738.

[All19] Patrick B. Allen, *On automorphic points in polarized deformation rings*, American Journal of Mathematics **141** (2019), no. 1, 119–167.

[AM69] M. F. Atiyah and I. G. Macdonald, *Introduction to commutative algebra*, Addison-Wesley Publishing Co., Reading, Mass.-London-Don Mills, Ont., 1969.

[Ax70] James Ax, *Zeros of polynomials over local fields—The Galois action*, J. Algebra **15** (1970), 417–428.

[BDJ10] Kevin Buzzard, Fred Diamond, and Frazer Jarvis, *On Serre's conjecture for mod ℓ Galois representations over totally real fields*, Duke Math. J. **155** (2010), no. 1, 105–161.

[Ber11] Laurent Berger, *La correspondance de Langlands locale p-adique pour* $\mathrm{GL}_2(\mathbf{Q}_p)$, Astérisque (2011), no. 339, Exp. No. 1017, viii, 157–180, Séminaire Bourbaki. Vol. 2009/2010. Exposés 1012–1026.

[Ber14] ———, *Lifting the field of norms*, J. Éc. polytech. Math. **1** (2014), 29–38.

[BG98] Siegfried Bosch and Ulrich Görtz, *Coherent modules and their descent on relative rigid spaces*, J. Reine Angew. Math. **495** (1998), 119–134.

[BGR84] S. Bosch, U. Güntzer, and R. Remmert, *Non-Archimedean analysis*, Grundlehren der Mathematischen Wissenschaften [Fundamental Principles of Mathematical Sciences], vol. 261, Springer-Verlag, Berlin, 1984, A systematic approach to rigid analytic geometry.

[BIP21] Gebhard Böckle, Ashwin Iyengar, and Vytautas Paškūnas, *On local Galois deformation rings*, arXiv e-prints (2021), arXiv:2110.01638.

[BL95] Arnaud Beauville and Yves Laszlo, *Un lemme de descente*, C. R. Acad. Sci. Paris Sér. I Math. **320** (1995), no. 3, 335–340.

[BLGGT14] Thomas Barnet-Lamb, Toby Gee, David Geraghty, and Richard Taylor, *Potential automorphy and change of weight*, Ann. of Math. (2) **179** (2014), no. 2, 501–609.

[BLR90] Siegfried Bosch, Werner Lütkebohmert, and Michel Raynaud, *Néron models*, Ergebnisse der Mathematik und ihrer Grenzgebiete (3) [Results in Mathematics and Related Areas (3)], vol. 21, Springer-Verlag, Berlin, 1990.

[BM02] Christophe Breuil and Ariane Mézard, *Multiplicités modulaires et représentations de* $\mathrm{GL}_2(\mathbf{Z}_p)$ *et de* $\mathrm{Gal}(\overline{\mathbf{Q}}_{\mathrm{p}}/\mathbf{Q}_{\mathrm{p}})$ *en* $l=p$, Duke Math. J. **115** (2002), no. 2, 205–310, with an appendix by Guy Henniart.

[BM14] Christophe Breuil and Ariane Mézard, *Multiplicités modulaires raffinées*, Bull. Soc. Math. France **142** (2014), no. 1, 127–175.

[BMS18] Bhargav Bhatt, Matthew Morrow, and Peter Scholze, *Integral p-adic Hodge theory*, Publ. Math. Inst. Hautes Études Sci. **128** (2018), 219–397.

[Bou98] Nicolas Bourbaki, *Algebra I. Chapters 1–3*, Elements of Mathematics (Berlin), Springer-Verlag, Berlin, 1998, Translated from the French, Reprint of the 1989 English translation [MR0979982 (90d:00002)].

[Bre97] Christophe Breuil, *Représentations p-adiques semi-stables et transversalité de Griffiths*, Math. Ann. **307** (1997), no. 2, 191–224.

[Bre98] ———, *Schémas en groupes et corps des normes*, available at https://www.math.u-psud.fr/~breuil/PUBLICATIONS/groupes normes.pdf.

[BS21] Bhargav Bhatt and Peter Scholze, *Prismatic f-crystals and crystalline Galois representations*, 2021.

[Car13] Xavier Caruso, *Représentations galoisiennes p-adiques et* (φ,τ)*-modules*, Duke Math. J. **162** (2013), no. 13, 2525–2607.

[CC98] F. Cherbonnier and P. Colmez, *Représentations p-adiques surconvergentes*, Invent. Math. **133** (1998), no. 3, 581–611.

[CDT99] Brian Conrad, Fred Diamond, and Richard Taylor, *Modularity of certain potentially Barsotti-Tate Galois representations*, J.A.M.S. **12** (1999), 521–567.

[CEG+16] Ana Caraiani, Matthew Emerton, Toby Gee, David Geraghty, Vytautas Paškūnas, and Sug Woo Shin, *Patching and the p-adic local Langlands correspondence*, Camb. J. Math. **4** (2016), no. 2, 197–287.

[CEGS19] Ana Caraiani, Matthew Emerton, Toby Gee, and David Savitt, *Moduli stacks of two-dimensional Galois representations*, arXiv e-prints (2019), arXiv:1908.07019.

[CL11] Xavier Caruso and Tong Liu, *Some bounds for ramification of p^n-torsion semi-stable representations*, J. Algebra **325** (2011), 70–96.

[Col10] Pierre Colmez, *Représentations de* $\mathrm{GL}_2(\mathbf{Q}_p)$ *et* (ϕ,Γ)*-modules*, Astérisque (2010), no. 330, 281–509.

[DDR16] Lassina Dembélé, Fred Diamond, and David P. Roberts, *Serre weights and wild ramification in two-dimensional Galois representations*, Forum Math. Sigma **4** (2016), e33, 49.

[Dee01] Jonathan Dee, Φ-Γ *modules for families of Galois representations*, J. Algebra **235** (2001), no. 2, 636–664.

[Dri06] Vladimir Drinfeld, *Infinite-dimensional vector bundles in algebraic geometry: An introduction*, The unity of mathematics, Progr. Math., vol. 244, Birkhäuser Boston, Boston, MA, 2006, pp. 263–304.

[EG14] Matthew Emerton and Toby Gee, *A geometric perspective on the Breuil-Mézard conjecture*, J. Inst. Math. Jussieu **13** (2014), no. 1, 183–223.

[EG19] Matthew Emerton and Toby Gee, *Dimension theory and components of algebraic stacks*, arXiv e-prints (2019), arXiv:1704.07654.

[EG20] ———, *Moduli stacks of* (φ,Γ)*-modules: A survey*, arXiv e-prints (2020), arXiv:2012.12719.

[EG21] Matthew Emerton and Toby Gee, *"Scheme-theoretic images" of certain morphisms of stacks*, Algebraic Geometry **8** (2021), no. 1, 1–132.

[Eme] Matthew Emerton, *Formal algebraic stacks*, Available at http://www.math.uchicago.edu/~emerton/pdffiles/formal-stacks.pdf.

[FGK11] Kazuhiro Fujiwara, Ofer Gabber, and Fumiharu Kato, *On Hausdorff completions of commutative rings in rigid geometry*, J. Algebra **332** (2011), 293–321.

[FK18] Kazuhiro Fujiwara and Fumiharu Kato, *Foundations of rigid geometry. I*, EMS Monographs in Mathematics, European Mathematical Society (EMS), Zürich, 2018.

[Fon85] Jean-Marc Fontaine, *Il n'y a pas de variété abélienne sur* **Z**, Invent. Math. **81** (1985), no. 3, 515–538.

[Fon90] ———, *Représentations p-adiques des corps locaux. I*, The Grothendieck Festschrift, Vol. II, Progr. Math., vol. 87, Birkhäuser Boston, Boston, MA, 1990, pp. 249–309.

[FS21] Laurent Fargues and Peter Scholze, *Geometrization of the local Langlands correspondence*, arXiv e-prints, arXiv:2012.13459.

[FW79] Jean-Marc Fontaine and Jean-Pierre Wintenberger, *Le "corps des normes" de certaines extensions algébriques de corps locaux*, C. R. Acad. Sci. Paris Sér. A-B **288** (1979), no. 6, A367–A370.

[Gao19] Hui Gao, *Breuil–Kisin modules and integral p-adic Hodge theory*, arXiv e-prints (2019), arXiv:1905.08555.

[GD71] A. Grothendieck and J. A. Dieudonné, *Eléments de géométrie algébrique. I*, Grundlehren der Mathematischen Wissenschaften [Fundamental Principles of Mathematical Sciences], vol. 166, Springer-Verlag, Berlin, 1971.

[Gee06] Toby Gee, *A modularity lifting theorem for weight two Hilbert modular forms*, Math. Res. Lett. **13** (2006), no. 5-6, 805–811.

[Ger19] David Geraghty, *Modularity lifting theorems for ordinary Galois representations*, Math. Ann. **373** (2019), no. 3-4, 1341–1427.

[GHLS17] Toby Gee, Florian Herzig, Tong Liu, and David Savitt, *Potentially crystalline lifts of certain prescribed types*, Doc. Math. **22** (2017), 397–422.

[GHS18] Toby Gee, Florian Herzig, and David Savitt, *General Serre weight conjectures*, J. Eur. Math. Soc. (JEMS) **20** (2018), no. 12, 2859–2949.

[GK14] Toby Gee and Mark Kisin, *The Breuil–Mézard conjecture for potentially Barsotti–Tate representations*, Forum Math. Pi **2** (2014), e1 (56 pages).

[GLS15] Toby Gee, Tong Liu, and David Savitt, *The weight part of Serre's conjecture for* GL(2), Forum Math. Pi **3** (2015), e2, 52.

[GR71] H. Grauert and R. Remmert, *Analytische Stellenalgebren*, Springer-Verlag, Berlin-New York, 1971, Unter Mitarbeit von O.

Riemenschneider, Die Grundlehren der mathematischen Wissenschaften, Band 176.

[GR03] Ofer Gabber and Lorenzo Ramero, *Almost ring theory*, Lecture Notes in Mathematics, vol. 1800, Springer-Verlag, Berlin, 2003.

[GR04] O. Gabber and L. Ramero, *Foundations for almost ring theory – Release 7.5*, ArXiv Mathematics e-prints (2004).

[Hel13] Eugen Hellmann, *On arithmetic families of filtered ϕ-modules and crystalline representations*, J. Inst. Math. Jussieu **12** (2013), no. 4, 677–726.

[Hel16] ______, *Families of p-adic Galois representations and (φ, Γ)-modules*, Comment. Math. Helv. **91** (2016), no. 4, 721–749.

[Her98] Laurent Herr, *Sur la cohomologie galoisienne des corps p-adiques*, Bull. Soc. Math. France **126** (1998), no. 4, 563–600.

[Her01] ______, *Une approche nouvelle de la dualité locale de Tate*, Math. Ann. **320** (2001), no. 2, 307–337.

[Her09] Florian Herzig, *The weight in a Serre-type conjecture for tame n-dimensional Galois representations*, Duke Math. J. **149** (2009), no. 1, 37–116.

[HH20] Urs Hartl and Eugen Hellmann, *The universal family of semistable p-adic Galois representations*, Algebra Number Theory **14** (2020), no. 5, 1055–1121.

[HT15] Yongquan Hu and Fucheng Tan, *The Breuil-Mézard conjecture for non-scalar split residual representations*, Ann. Sci. Éc. Norm. Supér. (4) **48** (2015), no. 6, 1383–1421.

[Kay96] Richard Kaye, *Polish groups*, Unpublished note.

[Kis06] Mark Kisin, *Crystalline representations and F-crystals*, Algebraic geometry and number theory, Progr. Math., vol. 253, Birkhäuser Boston, Boston, MA, 2006, pp. 459–496.

[Kis08] ______, *Potentially semi-stable deformation rings*, J. Amer. Math. Soc. **21** (2008), no. 2, 513–546.

[Kis09a] ______, *The Fontaine-Mazur conjecture for* GL_2, J. Amer. Math. Soc. **22** (2009), no. 3, 641–690.

[Kis09b] ______, *Moduli of finite flat group schemes, and modularity*, Ann. of Math. (2) **170** (2009), no. 3, 1085–1180.

[Kis10] ———, *The structure of potentially semi-stable deformation rings*, Proceedings of the International Congress of Mathematicians. Volume II, Hindustan Book Agency, New Delhi, 2010, pp. 294–311.

[KL15] Kiran S. Kedlaya and Ruochuan Liu, *Relative p-adic Hodge theory: foundations*, Astérisque (2015), no. 371, 239.

[Knu71] Donald Knutson, *Algebraic spaces*, Lecture Notes in Mathematics, Vol. 203, Springer-Verlag, Berlin, 1971.

[KPX14] Kiran S. Kedlaya, Jonathan Pottharst, and Liang Xiao, *Cohomology of arithmetic families of* (φ, Γ)*-modules*, J. Amer. Math. Soc. **27** (2014), no. 4, 1043–1115.

[LHLM20] Daniel Le, Bao V. Le Hung, Brandon Levin, and Stefano Morra, *Local models for Galois deformation rings and applications*, 2020.

[Liu08a] Ruochuan Liu, *Cohomology and duality for* (ϕ, Γ)*-modules over the Robba ring*, Int. Math. Res. Not. IMRN (2008), no. 3, Art. ID rnm150, 32.

[Liu08b] Tong Liu, *On lattices in semi-stable representations: A proof of a conjecture of Breuil*, Compos. Math. **144** (2008), no. 1, 61–88.

[Liu10] ———, *A note on lattices in semi-stable representations*, Math. Ann. **346** (2010), no. 1, 117–138.

[Liu12] ———, *Lattices in filtered* (ϕ, N)*-modules*, J. Inst. Math. Jussieu **11** (2012), no. 3, 659–693.

[Liu18] ———, *Compatibility of Kisin modules for different uniformizers*, J. Reine Angew. Math. **740** (2018), 1–24.

[Maz89] Barry Mazur, *Deforming Galois representations*, Galois groups over **Q** (Berkeley, CA, 1987), Math. Sci. Res. Inst. Publ., vol. 16, Springer, New York, 1989, pp. 385–437.

[Mul13] Alain Muller, *Relèvements cristallins de représentations galoisiennes*, Université de Strasbourg Ph.D. thesis, 2013.

[Nek93] Jan Nekovář, *On p-adic height pairings*, Séminaire de Théorie des Nombres, Paris, 1990–91, Progr. Math., vol. 108, Birkhäuser Boston, Boston, MA, 1993, pp. 127–202.

[Oze18] Yoshiyasu Ozeki, *Lattices in crystalline representations and Kisin modules associated with iterate extensions*, Doc. Math. **23** (2018), 497–541.

[Paš15] Vytautas Paškūnas, *On the Breuil–Mézard conjecture*, Duke Math. J. **164** (2015), no. 2, 297–359.

[Pil20] Vincent Pilloni, *Higher coherent cohomology and p-adic modular forms of singular weights*, Duke Math. J. **169** (2020), no. 9, 1647–1807.

[Pot13] Jonathan Pottharst, *Analytic families of finite-slope Selmer groups*, Algebra Number Theory **7** (2013), no. 7, 1571–1612.

[PR09] G. Pappas and M. Rapoport, *Φ-modules and coefficient spaces*, Mosc. Math. J. **9** (2009), no. 3, 625–663, back matter.

[Pyv20] Alexandre Pyvovarov, *Generic smooth representations*, Doc. Math. **25** (2020), 2473–2485.

[Rie86] Christine Riedtmann, *Degenerations for representations of quivers with relations*, Ann. Sci. École Norm. Sup. (4) **19** (1986), no. 2, 275–301.

[Ryd11] David Rydh, *The canonical embedding of an unramified morphism in an étale morphism*, Math. Z. **268** (2011), no. 3-4, 707–723.

[San14] Fabian Sander, *Hilbert-Samuel multiplicities of certain deformation rings*, Math. Res. Lett. **21** (2014), no. 3, 605–615.

[Sch12] Peter Scholze, *Perfectoid spaces*, Publ. Math. Inst. Hautes Études Sci. **116** (2012), 245–313.

[Sen72] Shankar Sen, *Ramification in p-adic Lie extensions*, Invent. Math. **17** (1972), 44–50.

[Ser79] Jean-Pierre Serre, *Local fields*, Graduate Texts in Mathematics, vol. 67, Springer-Verlag, New York, 1979, Translated from the French by Marvin Jay Greenberg.

[Sho18] Jack Shotton, *The Breuil-Mézard conjecture when $l \neq p$*, Duke Math. J. **167** (2018), no. 4, 603–678.

[Sno09] Andrew Snowden, *On two dimensional weight two odd representations of totally real fields*, arXiv e-prints (2009), arXiv:0905.4266.

[Sno18] Andrew Snowden, *Singularities of ordinary deformation rings*, Math. Z. **288** (2018), no. 3-4, 759–781.

[ST03] Peter Schneider and Jeremy Teitelbaum, *Algebras of p-adic distributions and admissible representations*, Invent. Math. **153** (2003), no. 1, 145–196.

[Sta] The Stacks project authors, *The Stacks project*, http://stacks.math.columbia.edu.

[SZ99] P. Schneider and E.-W. Zink, *K-types for the tempered components of a p-adic general linear group*, J. Reine Angew. Math. **517** (1999), 161–208, With an appendix by Schneider and U. Stuhler.

[Tun18] S.-N. Tung, *On the automorphy of 2-dimensional potentially semi-stable deformation rings of $G_{\mathbf{Q}_p}$*, arXiv e-prints (2018).

[Tun21] Shen-Ning Tung, *On the modularity of 2-adic potentially semi-stable deformation rings*, Math. Z. **298** (2021), no. 1-2, 107–159.

[Wac96] Nathalie Wach, *Représentations p-adiques potentiellement cristallines*, Bull. Soc. Math. France **124** (1996), no. 3, 375–400.

[Wan17] X. Wang, *Weight elimination in two dimensions when $p=2$*, arXiv e-prints (2017).

[WE18] Carl Wang-Erickson, *Algebraic families of Galois representations and potentially semi-stable pseudodeformation rings*, Math. Ann. **371** (2018), no. 3-4, 1615–1681.

[Win83] Jean-Pierre Wintenberger, *Le corps des normes de certaines extensions infinies de corps locaux; applications*, Ann. Sci. École Norm. Sup. (4) **16** (1983), no. 1, 59–89.

[Yos10] Manabu Yoshida, *Ramification of local fields and Fontaine's property* (P_m), J. Math. Sci. Univ. Tokyo **17** (2010), no. 3, 247–265 (2011).

[Zwa00] Grzegorz Zwara, *Degenerations of finite-dimensional modules are given by extensions*, Compositio Math. **121** (2000), no. 2, 205–218.

Index

Milton Keynes UK
Ingram Content Group UK Ltd.
UKHW020834010923
427885UK00009B/311